Lecture Notes in Mechanical Engineering

Lecture Notes in Mechanical Engineering (LNME) publishes the latest developments in Mechanical Engineering - quickly, informally and with high quality. Original research reported in proceedings and post-proceedings represents the core of LNME. Volumes published in LNME embrace all aspects, subfields and new challenges of mechanical engineering. Topics in the series include:

- Engineering Design
- Machinery and Machine Elements
- Mechanical Structures and Stress Analysis
- Automotive Engineering
- Engine Technology
- Aerospace Technology and Astronautics
- Nanotechnology and Microengineering
- Control, Robotics, Mechatronics
- MEMS
- Theoretical and Applied Mechanics
- Dynamical Systems, Control
- Fluid Mechanics
- Engineering Thermodynamics, Heat and Mass Transfer
- Manufacturing
- Precision Engineering, Instrumentation, Measurement
- Materials Engineering
- Tribology and Surface Technology

To submit a proposal or request further information, please contact the Springer Editor in your country:

China: Li Shen at li.shen@springer.com
India: Dr. Akash Chakraborty at akash.chakraborty@springernature.com
Rest of Asia, Australia, New Zealand: Swati Meherishi at swati.meherishi@springer.com
All other countries: Dr. Leontina Di Cecco at Leontina.dicecco@springer.com

To submit a proposal for a monograph, please check our Springer Tracts in Mechanical Engineering at http://www.springer.com/series/11693 or contact Leontina.dicecco@springer.com

Indexed by SCOPUS. The books of the series are submitted for indexing to Web of Science.

More information about this series at http://www.springer.com/series/11236

Muhammed Nafis Osman Zahid ·
Radhiyah Abd. Aziz · Ahmad Razlan Yusoff ·
Nafrizuan Mat Yahya · Fazilah Abdul Aziz ·
Mohd Yazid Abu
Editors

iMEC-APCOMS 2019

Proceedings of the 4th International Manufacturing Engineering Conference and The 5th Asia Pacific Conference on Manufacturing Systems

iMEC-APCOMS
A joint conference of
4th International Manufacturing Engineering Conference
5th Asia-Pacific Conference on Manufacturing Systems
2019

Springer

Editors
Muhammed Nafis Osman Zahid
Faculty of Manufacturing Engineering
Universiti Malaysia Pahang
Pekan, Pahang, Malaysia

Radhiyah Abd. Aziz
Faculty of Manufacturing Engineering
Universiti Malaysia Pahang
Pekan, Pahang, Malaysia

Ahmad Razlan Yusoff
Faculty of Manufacturing Engineering
Universiti Malaysia Pahang
Pekan, Pahang, Malaysia

Nafrizuan Mat Yahya
Faculty of Manufacturing Engineering
Universiti Malaysia Pahang
Pekan, Pahang, Malaysia

Fazilah Abdul Aziz
Faculty of Manufacturing Engineering
Universiti Malaysia Pahang
Pekan, Pahang, Malaysia

Mohd Yazid Abu
Faculty of Manufacturing Engineering
Universiti Malaysia Pahang
Pekan, Pahang, Malaysia

ISSN 2195-4356 ISSN 2195-4364 (electronic)
Lecture Notes in Mechanical Engineering
ISBN 978-981-15-0949-0 ISBN 978-981-15-0950-6 (eBook)
https://doi.org/10.1007/978-981-15-0950-6

This Springer imprint is published by the registered company Springer Nature Singapore Pte Ltd.
The registered company address is: 152 Beach Road, #21-01/04 Gateway East, Singapore 189721, Singapore

Foreword

For the third time, the 4th International Manufacturing Engineering Conference (iMEC) 2019 is co-organized with 5th Asia-Pacific Conference on Manufacturing System (APCOMS) 2019, owned by Fakulti Teknologi Industri, Institut Teknologi Bandung (ITB), Indonesia. Starting from 2019, the collaboration has been extended to the other institutions including Universiti Teknikal Malaysia Melaka (UTEM), Malaysia and Universitas Sebelas Maret, Indonesia. This extended collaboration aims to intensify knowledge sharing and experiences between higher learning institutions in Malaysia and Republic of Indonesia.

We are immensely pleased to welcome all delegates and distinguished guests to the iMEC-APCOMS 2019, held in the heart of Putrajaya, Malaysia. The conference aims to bring the researchers, academicians, scientists, students, engineers and practitioners around the world to present their latest findings, ideas, development and applications in manufacturing engineering and other related areas. With rapid advancements in manufacturing engineering that currently gearing towards Industry 4.0, iMEC provides an excellent avenue for the community to keep pace with the changes. In 2019, the conference theme is "Intelligent Engineering & Sustainable Development" which reflects to the acceleration of knowledge and technology in global manufacturing. In addition to three keynote speeches, there are 93 papers will be presented in 13 technical sessions. The papers published in these proceedings have underwent an intense peer review from the member of Technical Review Committee. The accepted submissions were categorized based on the conference topics which related to manufacturing systems, manufacturing processes, manufacturing automation and materials.

We are honoured to collaborate with respective institutions to make this conference a grand success. A sincere thanks to all members of the Organizing Committee for their infinite contribution. Not forgetting to all sponsors—Atomic Solutions, Crest, FESTO and others—for their kind gesture and continuous support.

Further, we would like to extend our appreciation to all authors for participation and high-quality contribution to the proceedings. Last but not least, we are grateful to publisher support especially to Dr. Christoph Baumann and Ms. Megana Dinesh. We hope this book will escalate the knowledge sharing and resources in the field of manufacturing engineering.

August 2019

Muhammed Nafis Osman Zahid
Radhiyah Abd. Aziz
Ahmad Razlan Yusoff
Nafrizuan Mat Yahya
Fazilah Abdul Aziz
Mohd Yazid Abu

Organizing Committee

Chairman of the Organizing Committee

Ahmad Razlan Yusoff	Universiti Malaysia Pahang, Malaysia

Vice Chairman of the Organizing Committee

Nafrizuan Mat Yahya	Universiti Malaysia Pahang, Malaysia
Wisnu Aribowo	Institut Teknologi Bandung, Indonesia

Chief Editor

Muhammed Nafis Osman Zahid	Universiti Malaysia Pahang, Malaysia

Secretaries

Nurrina Rosli	Universiti Malaysia Pahang, Malaysia
Nurul Fatma Nadirah Umar	Universiti Malaysia Pahang, Malaysia

Treasurers

Nurul Akmal Che Lah	Universiti Malaysia Pahang, Malaysia
Saidatul Ulfah Munamad Sahidin	Universiti Malaysia Pahang, Malaysia

Publication Committee

Radhiyah Abd Aziz	Universiti Malaysia Pahang, Malaysia
Mohd Yazid Abu	Universiti Malaysia Pahang, Malaysia
Fazilah Abdul Aziz	Universiti Malaysia Pahang, Malaysia

Moh. Mi'radj Isnaini	Institut Teknologi Bandung, Indonesia
Cucuk Nur Rosyidi	Universitas Sebelas Maret, Indonesia
Mohd Shahadan Mohd Suan	Universiti Teknikal Malaysia Melaka, Malaysia

Promotion and Sponsorship Committee

Zulhelmi Ismail	Universiti Malaysia Pahang, Malaysia
Ahmad Fakhri Ab. Nasir	Universiti Malaysia Pahang, Malaysia
Aidil Shafiza Safiee	Universiti Malaysia Pahang, Malaysia
Muhammad Nasrullah Alias	Universiti Malaysia Pahang, Malaysia
Wildan Trusaji	Institut Teknologi Bandung, Indonesia
Muh. Hisjam	Universitas Sebelas Maret, Indonesia
Zamberi Jamaludin	Universiti Teknikal Malaysia Melaka
Mohd Shahir Kasim	Universiti Teknikal Malaysia Melaka

Logistic Committee

Mohd Nizar Mhd Razali	Universiti Malaysia Pahang, Malaysia
Ahmad Najmuddin Ibrahim	Universiti Malaysia Pahang, Malaysia
Zubair Khalil	Universiti Malaysia Pahang, Malaysia

Technical Committee

Ahmad Rosli Abdul Manaf	Universiti Malaysia Pahang, Malaysia
Fariz Muharram Hasby	Institut Teknologi Bandung, Indonesia
Ismayuzri Ishak	Universiti Malaysia Pahang, Malaysia
Nursyazwan Md Talip	Universiti Malaysia Pahang, Malaysia
Mohd Faisal Mohd Saari	Universiti Malaysia Pahang, Malaysia
Shahandzir Baharom	Universiti Malaysia Pahang, Malaysia

Event Management

Ahmad Redza Ahmad Mokhtar	Universiti Malaysia Pahang, Malaysia
Mohd Khairulnazri Saidi	Universiti Malaysia Pahang, Malaysia

Contents

Manufacturing Systems

Formulation of Marketing Strategies in Expedition Services Company with SWOT and QSPM Methods

Nanang Alamsyah(✉), Arina Luthfini Lubis, and Dede Hamdi

Industrial Engineering, Sekolah Tinggi Teknik Ibnu Sina, Jl. Teuku Umar, Lubuk Baja, Batam, Kepulauan Riau, Indonesia
nanang@stt-ibnusina.ac.id

Abstract. This research was conducted to determine what strategies the company should do in increasing project load. The company engaged in export and import expedition services has problems with unstable project load, especially in export projects in the past two years. The stages of research are divided into 3 stages: (1) input stage using Internal Factor Evaluation (IFE) Matrix and External Factor Evaluation (EFE) Matrix, (2) matching stage using Internal-External (IE) Matrix, the Strategic Position and Action Evaluation (SPACE) Matrix & the Strength-Weakness-Opportunity-Threat (SWOT) Matrix, (3) decision stage using the Quantitative Strategic Planning Matrix (QSPM) Matrix. In addition, the Analytic Hierarchy Process (AHP) method is used for weighting variables at the input stage. Based on the input stage, there are 8 key questions external factors and 12 key questions internal factors. From IE & SPACE Matrix, it shows the current position of the company in a conservative profile where the strategy that must be carried out is hold and maintain. The results of the SWOT matrix analysis are alternative strategies that are raised based on the comparison between S-O, S-T, W-O, and W-T. Then the QSPM matrix analysis will compare the value of interests between existing strategies and alternative strategies produced by the SWOT matrix. It is show that the proposed weighted strategy value of 6.27 while the existing strategic weighting is 5.56. Based on the results of the analysis, the things that must be done by the company are improving the internal relations, expanding marketing network, and adjusting prices.

Keywords: Strategic management · Strategy formulation · Expedition services company

1 Introduction

One of the processes in manufacturing is the distribution of goods through the supply chain. An expedition company in 2020 the vision, mission and goals for the future were first announced in the 2015 strategy and recently focused and re-emphasized in "Strategy 2020: Focus, Connect, Grow." A strategy that is simple but not mediocre. Built on three pillars of Focus, Connect, and Develop, this strategy sets out a clear plan for the coming years with ambitious but achievable goals. It is a global logistics

M. N. Osman Zahid et al. (Eds.): iMEC-APCOMS 2019, LNME, pp. 3–9, 2020.
https://doi.org/10.1007/978-981-15-0950-6_1

company with a position that right in the developing world market. The company remains focused on the logistics sector as the core of our business while continuing to contribute to a better world, which we call "Undergoing Responsibility." But based on the data and information the authors receive has a lack of stability in marketing, especially in the export section that is not stuck, the following are Project-Load data for 2 years, namely 2016 and 2017 (Table 1 and Fig. 1).

Table 1. Project load

Year	Jan	Feb	Mar	Apr	May	Jun	Jul	Aug	Sep	Oct	Nov	Dec	Average
2016	36	28	36	33	28	23	13	24	19	30	22	12	25.33
2017	13	10	12	20	18	17	13	28	8	18	34	16	17.25

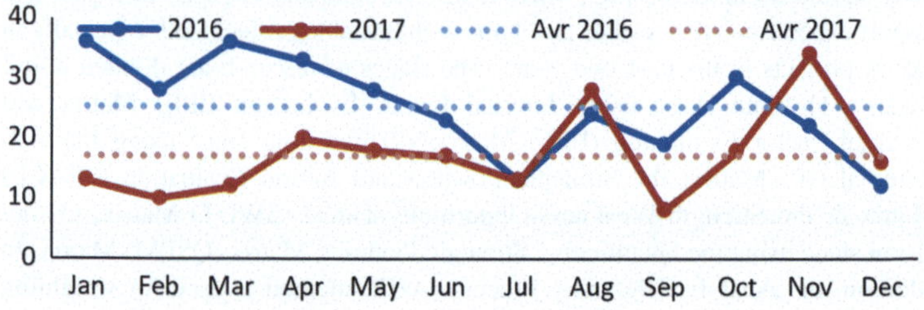

Fig. 1. Project-load fluctuation

Therefore, we need to know how to determine a good strategy for the company to stabilize and improve the company's project load, is to make observations on the internal and external parties of the company. Then do data processing with the SWOT and QSPM methods.

2 Research Methodology

The stages of research that have been carried out begin with the identification of problems through observation. Then proceed with identifying internal and external factors that affect the company through interviews with the head of the company's branch. Factors, both internal and external, that have been identified are given scores based on questionnaires that have been filled by customers and branch heads and given weights using the calculation of Analytical Hierarchy Process (AHP) [2]. With these scores and weights, we can create an IFE & EFE matrix, so that we can find out the scores of Strength (S), Weakness (W), Opportunity (O), and Threat (T). The IE and SPACE matrix can inform us of the position of the company in what quadrant. Each quadrant has a different strategy recommendation. The SWOT matrix helps us in choosing strategies that are in accordance with the comparison between SO, ST, WO, and WT. The QSPM matrix serves to find out whether the strategy we are proposing is better than the existing strategy [3] (Figs. 2 and 3).

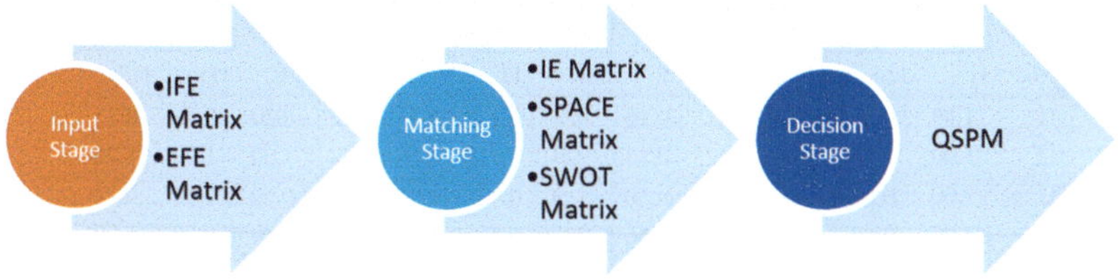

Fig. 2. Strategy-formulation analytical framework [1]

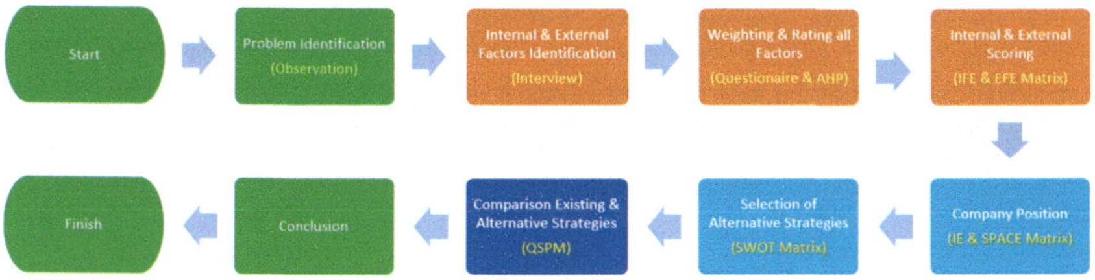

Fig. 3. Stages of research

3 Finding and Discussion

See Tables 2, 3, 4, 5, 6 and 7 and Figs. 4 and 5.

Table 2. Key questions internal & external factors

No.	Item description	Forces
I1	More than ever emphasize low price and value versus rivals	Marketing
I2	More than ever emphasize how the product/service will make your life better	Marketing
I3	Does the firm have good relations with its investors and stockholders?	Finance
I4	Are the firm's financial managers experienced and will trained?	Finance
I5	Is the firm's debt situation excellent?	Finance
I6	Are facilities, equipment, machinery, and offices in good condition?	Operation
I7	Are facilities, resources, and markets strategically located?	Operation
I8	Is communication between R&D and other organizational units effective?	R&D*
I9	Is there a chief information officer or director of information systems positions in the firm?	MIS**
I10	Do managers from all functional areas of the firm contribute input to the information system?	MIS
I11	Are strategists of the firm familiar with the information systems of rival firms?	MIS
I12	Is the firm's information system continually being improved in content and user-friendliness?	MIS
E1	Worker productivity levels	Economic
E2	Value of dollar in world markets	Economic
E3	Tax rates	Economic
E4	Attitudes toward work	Social
E5	Ethical concerns	Social
E6	Attitudes toward product quality	Social
E7	Attitudes toward customer service	Social
E8	Import – Export regulations	Political

Note: *R&D: Research and Development; **MIS: Management Information System

Table 3. Pairwise comparisons of IFE

No.	I1	I2	I3	I4	I5	I6	I7	I8	I9	I10	I11	I12
I1	**1.00**	0.20	0.33	0.25	0.16	0.50	0.14	0.33	0.12	0.20	0.20	0.25
I2	5.00	**1.00**	1.66	1.25	0.83	2.50	0.71	1.66	0.62	1.00	1.00	1.25
I3	3.00	0.60	**1.00**	0.75	0.50	1.50	0.42	1.00	0.37	0.60	0.60	0.75
I4	4.00	0.80	1.33	**1.00**	0.66	2.00	0.57	1.33	0.50	0.80	0.80	1.00
I5	6.00	1.20	2.00	1.50	**1.00**	3.00	0.85	2.00	0.75	1.20	1.20	1.50
I6	2.00	0.40	0.66	0.50	0.33	**1.00**	0.28	0.66	0.25	0.40	0.40	0.50
I7	7.00	1.40	2.33	1.75	1.16	3.50	**1.00**	2.33	0.87	1.40	1.40	1.75
I8	3.00	0.60	1.00	0.75	0.50	1.50	0.42	**1.00**	0.37	0.60	0.60	0.75
I9	8.00	1.60	2.66	2.00	1.33	4.00	1.14	2.66	**1.00**	1.60	1.60	2.00
I10	5.00	1.00	1.66	1.25	0.83	2.50	0.71	1.66	0.62	**1.00**	1.00	1.25
I11	5.00	1.00	1.66	1.25	0.83	2.50	0.71	1.66	0.62	1.00	**1.00**	1.25
I12	4.00	0.80	1.33	1.00	0.66	2.00	0.57	1.33	0.50	0.80	0.80	**1.00**
Sum	53.00	10.60	17.66	13.25	8.83	26.50	7.57	17.66	6.62	10.60	10.60	13.25

Table 4. Pairwise comparisons of EFE

No.	E1	E2	E3	E4	E5	E6	E7	E8
E1	**1.00**	0.20	0.20	0.20	3.00	3.00	0.25	5.00
E2	5.00	**1.00**	1.00	1.00	1.67	1.67	1.25	1.00
E3	5.00	1.00	**1.00**	1.00	1.67	1.67	1.25	1.00
E4	5.00	1.00	1.00	**1.00**	1.67	1.67	1.25	1.00
E5	0.33	0.60	0.60	0.60	**1.00**	1.00	0.75	0.60
E6	0.33	0.60	0.60	0.60	1.00	**1.00**	0.75	0.60
E7	4.00	0.80	0.80	0.80	1.33	1.33	**1.00**	0.80
E8	0.20	1.00	1.00	1.00	1.67	1.67	1.25	**1.00**
Sum	20.87	6.20	6.20	6.20	13.00	13.00	7.75	11.00

Table 5. IFE & EFE matrix

Key internal factors	Weight	Rating	Weighted score	Key external factors	Weight	Rating	Weighted score
Strengths				Opportunities			
I1	0.02	3	0.06	E1	0.13	3	0.41
I4	0.09	3	0.28	E4	0.15	3	0.46
I6	0.06	3	0.17	E6	0.15	3	0.46
I8	0.08	3	0.23	E7	0.15	3	0.46
I10	0.11	3	0.34				**1.8**
I11	0.04	3	0.11				
			1.19				
Weaknesses				Threats			
I2	0.13	2	0.26	E2	0.08	2	0.15
I3	0.06	2	0.11	E3	0.08	2	0.15
I5	0.15	2	0.3	E5	0.12	2	0.25
I7	0.09	2	0.19	E8	0.13	2	0.25
I9	0.09	2	0.19				**0.8**
I12	0.08	2	0.15				
			1.21				
Total	**1**		**2.4**	Total	**1**		**2.6**

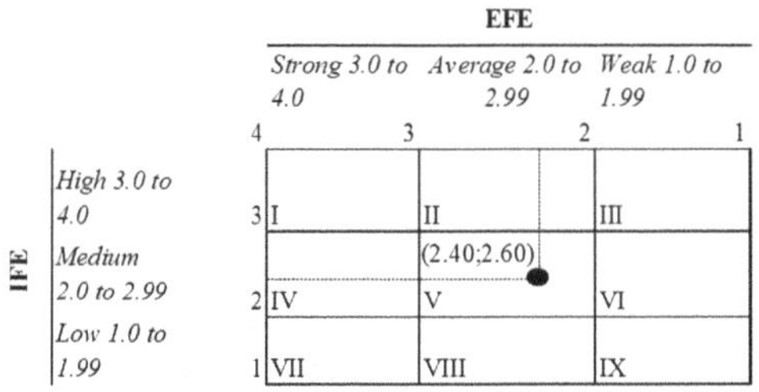

Fig. 4. IE matrix

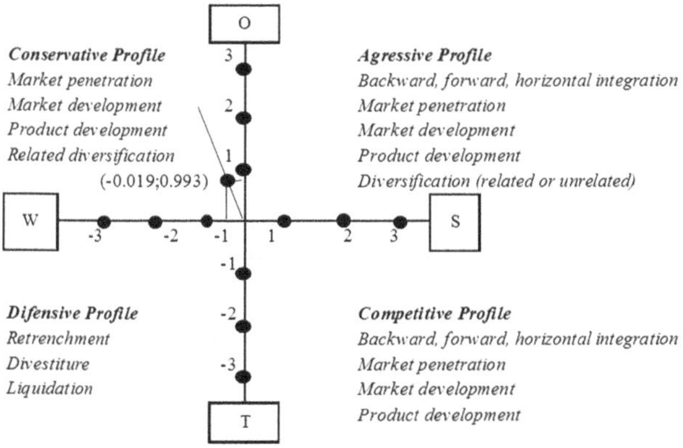

Fig. 5. SPACE matrix

Table 6. SWOT matrix

	Strength	Weakness
Opportunities	**SO strategies** All notifications let post and distribute to all department by system information of there (S5, O1) Expand market and coming to customer for do cooperations (S3, S4, O2, O3, O4)	**WO strategies** Increasing the company information systems (W5, W6, O2) Repair relationships with the cooperations (W2, O2) Do calculation price on the input, process, output for repair debt situation (W3, O3, O4)
Threats	**ST Strategies** Create the ethical rule for employee toward work (S4, T3) Do changed price for get more profit toward service import & export (S2, T4)	**WT Strategies** Determine BEP price of service/product to minimize the loss value of dollar in world markets (W5, T1, T4)

Table 7. QSPM matrix

Key factors	Strategic alternative				
	Weight	Focus		Company repair	
		Connected		Expand market	
		Evolved		Change price	
		AS	TAS	AS	TAS
Opportunities					
E1	0.14	3	0.41	4	0.55
E4	0.15	2	0.31	4	0.62
E6	0.15	3	0.46	3	0.46
E7	0.15	3	0.46	3	0.46
Threats					
E2	0.08	3	0.23	4	0.31
E3	0.08	2	0.15	3	0.23
E5	0.12	3	0.37	3	0.37
E8	0.13	3	0.38	3	0.38
	1.00				
Strength					
I1	0.02	2	0.04	2	0.04
I4	0.09	4	0.38	4	0.38
I6	0.06	3	0.17	3	0.17
I8	0.08	2	0.15	2	0.15
I10	0.11	2	0.23	2	0.23
I11	0.04	2	0.08	1	0.04
Weakness					
I2	0.13	3	0.40	3	0.40
I3	0.06	3	0.17	3	0.17
I5	0.15	2	0.30	3	0.45
I7	0.09	3	0.28	3	0.28
I9	0.09	4	0.38	4	0.38
I12	0.08	3	0.23	3	0.23
Total	**1.00**	**55**	**5.56**	**60**	**6.27**

4 Conclusion

Based on the input stage, there are 8 key questions external factors and 12 key questions internal factors. From IE & SPACE Matrix, it shows the current position of the company in a conservative profile where the strategy that must be carried out is hold and maintain. The results of the SWOT matrix analysis are alternative strategies that are raised based on the comparison between S-O, S-T, W-O, and W-T. Then the QSPM matrix analysis will compare the value of interests between existing strategies and alternative strategies produced by the SWOT matrix. It is show that the proposed weighted strategy value of 6.27 while the existing strategic weighting is 5.56. Based on the results of the analysis, the things that must be done by the company are improving the internal relations, expanding marketing network, and adjusting prices.

References

1. David, F.R.: Strategic Management: Concepts and Cases, 13th edn. Pearson, New Jersey (2011)
2. Alamsyah, N.: Pengurutan skala prioritas perspektif balance scorecard dan KPI perusahan startup dengan metode AHP (analytical hierarchy process). Jurnal Teknik Ibnu Sina JT-IBSI **1** (01), 51–61 (2016)
3. Alamsyah, N., Lubis, A.L.: Analisa strategi pemasaran menggunakan matriks IEF dan EFE (Studi Kasus PT. Kian Ho Indonesia). Jurnal Industri Kreatif (JIK) **2**(2), 51–58 (2018)

Optimization of CNG Multi-depot Distribution to Determine Model Routes and GTM Totals Using Tabu Search and Differential Evolution Methods

Afni Khadijah(✉) and Huswatun Hasanah

Universitas Banten Jaya, Serang, Banten, Indonesia
{afnikhadijah,huswatunhasanah}@unbaja.ac.id

Abstract. Fuel Gas Filling Station Sadikun Bekasi and Fuel Gas Filling Station Sadikun Sukabumi multi-depots have 30 consumers of distribution area such as Bogor, Tangerang, and Cilegon. They distribute Compressed Natural Gas (CNG) to industrial consumers by using a trailer-based Gas Transport Module (GTM). The CNG purchase agreement between the provider of CNG to consumers who will fulfill the request per day without time limitation (time windows) and the distribution frequency set by the company. To optimize the distribution of CNG in this research is conducted using Tabu Search (TS) and the Different Evolution (DE) Methods. Meanwhile, TS Method has reducing total distribution costs as Rp 45.294.844, the reducing of mileage distribution as 1532, 3 km, and the route of distribution is reduced. Starting from 17-trips to 6-trips as the offer route of distribution using 6 GTM. It affects increasing the utility value of GTM as 22.70%. DE Method has reduced total distribution costs as Rp 36.571.190, the reducing of mileage distribution as 1.441,7 km, and the route of distribution is reduced. Starting from 17-trips to 8-trips as the offer route of distribution using 8 GTM. It affects increasing the utility value of GTM as 29.95%. This study shows that Tabu Search (TS) Method is more reliable.

Keywords: Compressed Natural Gas · Tabu Search · Differential Evolution

1 Introduction

Today, the largest contribution of gas utilization is industrial sector reached 44% of totals and will increase in 2050 to 69%. In the industrial sector, natural gas is not only consumed to fuel, but also as raw material. By 2050, the power generation sector, commercial, and transportation of each section of gas utilization are 26%, 13% and 1%. At the same time, the part of household sector is below 1% [1]. To satisfy the demand of industrial gas needs which to increase annually, then the perpetrators of the natural gas business needs to expand the infrastructure of pipelines, increase the optimization value of the production process and supply chain from upstream to downstream in order to the availability of gas supply awake in various regions, price stability, and timeliness of the distribution of precision so that the cycle of industrial production in Indonesia continues to perform well.

M. N. Osman Zahid et al. (Eds.): iMEC-APCOMS 2019, LNME, pp. 10–16, 2020.
https://doi.org/10.1007/978-981-15-0950-6_2

The study focuses on multi-depot of Fuel Gas Filling Station Sadikun Bekasi and Sukabumi which are 30 consumer distribution areas such as Bogor, Tangerang, and Cilegon. The distribution of Compressed Natural Gas (CNG) to industrial consumers uses a trailer-based Gas Transport Module (GTM) that is specifically used for the transport of gas. The Gas Sales Purchase Agreement between the CNG provider to consumers who will perform the request of CNG per day without a time limit (time windows) and the distribution frequency set by the company depot. The Problem in research is related to quantity different in each point, charge, the limited capacity of fleet limited fleet, the distance, those who to traffic congestion. The distribution still uses the one-on-one. To minimize the total cost of distribution made the determination of the route and the number of the distribution of CNG optimal industry to reduce fixed cost and variable cost are conducted by using Tabu Search (TS) and the Different Evolution (DE) algorithm of The Vehicle Routing Problem with Times Window (VRPTW).

2 Literature Review

2.1 Tabu Search (TS)

Tabu Search is a method that is incorporated in a class called meta-heuristics. TS method has proven successful in solving problems related to the combinatorial optimization problems. The basis of TS meta-heuristic is to use aggressive strategies escort to cut local search procedure to carry out exploration in the solution set in order to avoid being stuck in local optima. When local optima encountered, an aggressive strategy to move to the best solution in every neighbor even if it resulted in a decrease in the value of interest.

2.2 Different Evolution (DE)

Differential Evolution algorithm is not much different from other Evolutionary algorithms. DE uses vectors that represent all of candidate solutions which its search technique performed simultaneously on a number of solutions called the population. Initial population (zero generation) is formed by generating a random number, while the next population is the result of the evolution of vectors that have been through the stages of reproduction, mutation, recombination and selection. Each individual is defined as a D-dimensional vector in which the vectors are denoted as $x_{i.g}$ which is a member of the population in g-generation. Population denoted as Px comprising the vectors are Np dimension where Np is the population size.

2.3 The Vehicle Routing Problem with Times Window (VRPTW)

VRP with time windows (VRPTW) is the development of CVRP which has a capacity constraint applied and each i-consumer associated with interval $[a_i, b_i]$ called time window. The instant time when the vehicle leaves the depot, travel time TIJ, for each notation (i, j) € A (or te notation for e € E) and the addition of the service time for each i-consumer have been determined [2].

Service to customers should be initiated which is associated in the time window and the vehicle must stop at customer locations are for the time instant. Sometimes, in the case of an earlier arrival at the location of the i-consumer, vehicles are generally allowed to wait until the instant time a_i, i.e. until the service begins.

3 Research Methodology

The data used in the research is the historical data in 2015 regarding the distribution of CNG from CNG provider company depot to various consumer depots which have a demand regardless of time of delivery (time windows). Here is a secondary data that is required in this study.

(a) Fleet data used
(b) Data of distribution costs (fixed cost and variable cost)
(c) The number of CNG by consumer demand
(d) Data of mileage of the vehicle from the depot to the consumer
(e) Data of matrix consumer depot locations

Overall the data processed by designing mathematical models and algorithms VRPTW translated into programming language Matlab R2015b (ver.7.9.0) by entering the objective function and constraints are predetermined.

Constraint

Q = The number of transported for customers j
j Point = 1, 2, 3, …., 32
d_{ij} = The distance from i to j
FC_l = Fixed costs vehicles
VC_l = Variable costs vehicles

Variable

$$X_{ijk} = \begin{cases} 1\ (i \text{ point connected to } j \text{ point on the day to } k) \\ 0\ \text{No} \end{cases}$$

$$Y_k = \begin{cases} \text{Vehicle used} \\ 0\ \text{No} \end{cases}$$

The Function

$$\min Z = \sum_{i}\sum_{j}\sum_{k} VC_k d_{ij} x_{ijk} + \sum_{k} FC_k x_{ijk} \; i \neq j$$

4 Result

4.1 Tabu Search (TS)

Tabu Search method before processing the data then must set control parameters to be used for coding Matlab program is determining the length tabu list, the number of neighboring solutions, and the number of iterations (stopping criterion) [3].

In the experimental results use the determination of the number of iterations parameter value that tabu list is 20, the value of the number of neighbor solution is 10, and also the number of Matlab program iterations is 800 iterations. The results are used to process the data distribution of CNG by using TS. Before processing the data to determine whether the preparation of programming language is conformity with the model algorithm has been formulated, then the verification and validation testing performed on coding VRPTW algorithm is using TS. From the results of running Matlab use DE as much as 5 times running the obtained all of running results (Table 1).

Table 1. Whole running result

Run to-	Computation time (s)	Routes number	GTM numbers		Range total (km)	Cost distribution totals (Rp)
			20″	10″		
1	**0.05**	**6**	**4**	**2**	**1452.6**	**Rp 38,028,074**
2	0.07	6	4	2	1484	Rp 38,545,274
3	0.05	6	4	2	1589.7	Rp 38,846,674
4	0.05	6	4	2	1410.8	Rp 38,322,874
5	0.04	6	4	2	1780.5	Rp 39,366,174

From results of data processing use Matlab R2015b program for 5 times, experimental results can be seen. Run to-1 has the lowest total cost of distribution of Rp 38,028,074. Then, the results of an experiment to-1 were determined as the best solution in this study using TS having a number of routes as many as 6 using 6 GTM fleet size of 20 “by 4 units and 10” as much as 2 pieces, so obtain a total distance of 1452.6 distribution km (Table 2).

Table 2. Best result

Trip	Route					Demand (m^3)	Truck capacity (m^3)	Range (km)
1	2	4	8	12	2	3754.59	4104	361
2	1	17	15	1		1603.84	1728	287.80
3	1	14	13	16	1	3002.28	4104	183.8
4	2	5	2			1205.5	1728	292
5	1	19	7	18	1	3609.61	4104	181.2
6	1	11	6	1		3043.82	4104	146.8
Range total								**1452.60**
Cost distribution total								**Rp 38,028,074**

4.2 Differential Evolution (DE)

Preparation of differential evolution algorithm, needed some control parameters must be set from the beginning. The control parameters consisting of the parameter size of the population (NP), the control parameters mutation (F), and the control parameters crossovers (Cr) [4]. In this experiment conducted random simulation process running as much as 15 times the experiment where the nominal range 100–1500 iteration using the control parameter F = 0.6 and Cr = 0.9. We get the results of the iteration in Table 3 that the optimal number of iterations 600 iterations.

Table 3. The result of the overall run test

Run to-	Computation time (s)	Route number	GTM number		Range total (km)	Cost distribution totals (Rp)
			20″	10″		
1	9.176309	8	4	4	1,840.90	Rp 47,289,128
2	**7.54527**	**8**	**4**	**4**	**1,543.20**	**Rp 46,751,728**
3	6.40959	8	4	4	1,734.00	Rp 47,158,228
4	8.052023	8	4	4	1,865.90	Rp 47,362,628
5	9.642917	8	4	4	1,611.60	Rp 46,825,928

The experimental results of the 2nd determined as the best solution in this study using DE. This solution has a number of CNG distribution of 8 times traveling with the use of GTM 20 "by 4 units and GTM 10" by 4 units and the distribution of a total distance of 1543.20 km (Table 4).

Table 4. Best result

2nd run result							
Computation time: 7.545270 s							
Route number	GTM type	Route					Range (km)
1	20″	D-l	SMI	API	D-l		147.8
2	20″	D-l	DA	BI	DI	D-l	288
3	20″	D-2	AHB	ASI	NF	D-2	308.4
4	20″	D-1	MPF	FAA	GKI	D-l	301.8
5	10″	D-2	TBT	AMB	D-2		72.9
6	10″	D-l	SMU	D-l			125.1
7	10″	D-2	NI	QA	D-2		160.6
8	10″	D-l	ACI	D-l			138.6
Range total							**1543.2**
GTM number 20″							**4**
GTM number 10″							**4**
Cost total							**Rp 46,751,728**

4.3 Analyze TS and DE

From all data analysis of distribution route, fleet distribution utility data, and cost distribution total, it can build comparison of the current route and proposed route in the Tables 5 and 6.

Table 5. Comparison of the current route and TS proposed route

	Route numbers	Mileage total (km)	GTM total	GTM utility (%)	Cost distribution total (Rp)
Current route	17	2984.9	17	58.86	83,322,918
Proposed route	6	1452.6	6	81.56	38,028,074

Table 6. Comparison of data analysis result of DE and TS methods

	Route numbers	Mileage total (km)	GTM totals	GTM utility (%)	Cost distribution total (Rp)
Proposed route of TS	6	1452.6	6	81.56	38,028,074
Proposed route of DE	8	1543.2	8	88.81	46,751,728

5 Conclusion

In the TS methods is obtained a new route which is more optimal as it can lower total cost of distribution of Rp 45,294,844, which at the moment total distribution costs to be incurred amounted to Rp 83,322,918 after the trial found the total cost of the proposed distribution of Rp 38,028,074 more efficient due to reduced mileage distribution. Mileage distribution today is 2984.9 km and the mileage proposal i.e. 1452.60 km. So we get the mileage difference in the comparison of current and proposed at 1532.3 km. This happens due to CNG distribution system does not use a system of one-to-one but apply the system of one-to-many. So that the distribution of travel is reduced from the current 17-trip, but the trip to 6 times the proposed distribution of travel by using 6 pieces GTM 20 "by 4 units and 10" as much as 2 pieces. It affects the increase in the utility value of 22.70% GTM fleets where the original 58.86% to 81.56%.

Meanwhile, the method of DE obtained new route is more optimal as it can lower total cost of distribution of Rp 36,571,190, which at the moment s total distribution costs to be incurred amounted to Rp 83,322,918 after the trial found the total cost of the proposed distribution of Rp 46 751.728 is more efficient due to reduced mileage distribution. Mileage distribution today is 2984.9 km and the mileage proposal is 1543.2 km. So we get the mileage difference in the comparison of current and proposed at 1441.7 km. This happens due to CNG distribution system does not use a system of one-to-one but apply the system of one-to-many. So that the distribution of travel is reduced from the current 17-trip, but on the way the proposed distribution to 8 times the journey by 8 fleet GTM. GTM is 4 pieces of 20 "and 4 pieces GTM 10". It affects the increase in utility value of GTM by 29.95%., Where the original 58.86% to 88.81%. In the data processing results can be inferred TS is more reliable because during the data processing time required is shorter, the number of trips is minimal, and lower total cost of distribution.

References

1. Teknologi, B. P. Outlook Energi Indonesia 2015. Jakarta (2015). ISBN 978–602-1328-04-0
2. Toth, P., Vigo, D.: The Vehicle Routing Problem. SIAM, Philadelphia (2001)
3. Lai, D.S., Demirag, O.C., Leung, J.M.: A tabu search heuristic for the heterogeneous vehicle routing. Transp. Res. Part E **86**, 32–52 (2016)
4. Ho, S., Haugland, D.: A tabu search heuristic for the vehicle routing problem with time windows and split deliveries. Comput. Oper. Res. **31**, 1947–1964 (2004)

Capacity Planning and Assembly Line Balancing for Long-Term Routine and Short-Term Intermittent Demand in Small Medium Enterprises

Disa Agatha Willim[1], Wildan Trusaji[2](✉), and Anas Ma'ruf[1]

[1] Industrial Engineering Department, Institut Teknologi Bandung, Bandung, Indonesia

[2] Engineering Management Department, Institut Teknologi Bandung, Bandung, Indonesia

wildan@mail.ti.itb.ac.id

Abstract. Small Medium Enterprise (SME) of Doll Producer in Bandung has two types of buyers. The first type of buyers is long-term & routine doll design buyer (the client) and the second one is short-term & intermittent buyers (customers). Although most of the SME's revenue comes from the client, the SME unable to fulfill the demand of the client at a time. However, demand from customers can be met, since the SME always prioritize the shortest due time contract. To solve the problem, balancing of the assembly line by using the U-Line mathematical model is done. The U-Line is appropriate for the problem since the SME has a small room for the assembly activities. After that, the assembly line capacity is calculated as the basis to allocate resources for both the client and customers project. The model suggests that the SME must allocate 2 different assembly line (4 work stations & 3 work stations) for the client projects and 3 parallel assembly line (3 work station each) for customers. Based on the simulation, the SME's capacity is improved up to 36.5% and the smoothness index are 22.14, 10.94, and 7.5 for the first client, the second client, and customers project assembly line respectively.

Keywords: Line balancing · U-Line model · Capacity planning · Small Medium Enterprise · Doll Producer

1 Introduction

The unmet demand is a common problem that a company usually faces. One of the main causes of the problem is located in the production line. Beigel [1] stated that there is two production line's main problem, which is (1) system obstacles and (2) unbalance workload. The scheduled and the unscheduled maintenance become sources of system obstacle, and the unbalance workload make the production line deficient to meet the production plan.

The SME serves two types of customer, the long-term contract customer (LCC) and the short-term contract customer (SCC). The LCC always order in fix amount to SME

M. N. Osman Zahid et al. (Eds.): iMEC-APCOMS 2019, LNME, pp. 17–23, 2020.
https://doi.org/10.1007/978-981-15-0950-6_3

periodically every month along the year. On the other hand, the SCC intermittently order to SME with unfix amount of the doll. The SME always prioritized and met the SCC's demand. Nevertheless, the LCC's demand is hardly met and always late. Based on field observation and further analysis, the unmet demand in Small Medium Enterprises (SME) of Doll Producer is caused by two causes, there is the unbalance workload between work stations and the absence of capacity planning for the intermittent demand. Thus, line balancing and capacity planning is necessary to solve the SME problem [2–4].

In manufacturing industry, the straight line and U-line model is usually adapted [5]. However, the SME is lack of investment and they have very limited shop floor space. Therefore, an U-line model is chosen to design the production line, since it is suitable for the SME limitation. In other hands, the U-Line model has some advantages such as: can find the minimum amount of work stations, can accommodate work element that not in sequence, increases visibility, increases flexibility, and can produce at least same line efficiency compare with straight-line model [6–9]. This paper is organized in three parts, introduction, U-line model development, result & analysis, and conclusion

2 Methodology and U-Line Model Development

Ten step method of Gasperz [10] from identified every work element and adaptation of chosen solution become this research methodology. The limitation of shop floor space and number of workstation require a model that can minimize the number of work stations. Therefore, Urban [11] model is chosen since the model has minimised the number of work station as an objective function. The explanation of Urban model is as follows: The model has three decision variables that are an assignment of work element i in the original diagram to work station j (1), assignment of work element i in phantom diagram to work station j (2), and assignment of work station j (3). The objective function is to minimize the number of the workstation (4). Constraints are each work element only assigned to work station (5), process time in each work station cannot exceed the takt time (6), the sequence of the work element (7), and make sure the decision variables are an integer (8). The mathematical notation used in the model is as the following (Table 1):

Table 1. The notation used in the Model

i	Work element	r	Predecessor work element
j	Work station	s	Successor work element
n	Number of work element	p	Precedence constraint
m	Number of work station	m_{Max}	Number of maximum work station
C	Takt time	t_i	Process time of work element i
ST_j	Total work element duration in work station j	ST_{max}	The duration in the longest work station
L	Line efficiency	K	Number of assembly line
P	Working hour/week (46,5 h)	CT	Cycle time

Decision Variable

$$x_{ij}\begin{cases} 1, \text{work element } i \text{ in original precedence diagram is assigned to work station } j \\ 0, \text{Other} \end{cases} \quad (1)$$

$$y\begin{cases} 1, \text{work element } i \text{ in phantom precedence diagram is assigned to work station } j \\ 0, \text{Other} \end{cases} \quad (2)$$

$$z_j\begin{cases} 1, \text{work station } j \text{ has assigned work element} \\ 0, \text{Other} \end{cases} \quad (3)$$

Objective Function

$$Z = \min \sum_{j=1}^{Mmax} z_j \quad (4)$$

Constrains

$$\sum_{j=1}^{Mmax} \left(x_{ij} + y_{ij}\right) = 1, \text{for } i = 1, \ldots, n \quad (5)$$

$$\sum_{i=1}^{n} t_i\left(x_{ij} + y_{ij}\right) \leq C, \text{for } j = 1, \ldots, m \quad (6)$$

$$\sum_{j=1}^{m_{max}} (m_{max} - j + 1)\left(x_{rj} - x_{sj}\right) \geq 0, \text{for } (r, s) \in P \quad (7)$$

$$\sum_{j=1}^{m_{max}} (m_{max} - j + 1)\left(y_{sj} - y_{rj}\right) \geq 0, \text{for } (r, s) \in P \quad (8)$$

$$x_{ij}, y_{ij}, z_j = \{0, 1\} \text{ for every } i \text{ and } j$$

Since to solve the problem we need not only to minimize the number of work station but to maximize the smoothness index too. Thus, the second model is developed to accommodate the second step optimization, which is to maximize the smoothness index. The developed model has the same decision variable with the Urban. The objective function of the developed model is to maximize the smoothness index and is has three additional constrains to Urban model. The mathematical formulation of the developed model is as the following.

Objective Function

$$Z = \min\sqrt{\sum_{i=1}^{m}(\mathrm{ST}_{max} - \mathrm{ST}_j)^2} \tag{9}$$

Additional Constraints

$$ST_j = \sum_{i=1}^{n} t_i(x_{ij} + y_{ij}) \text{ for } i = 1,\ldots,n \tag{10}$$

$$ST_{max} = \max\left(ST_j\right) \tag{11}$$

3 Result and Analysis

The SME manufactures 3 different doll model, the white bear (WB), the bear head (BH), and Souvenir Doll (SD). The WB and BH are the LCC demand, and the SD is the SCC demand. Therefore, the optimal assembly line for the WB and BH need to be decided first. After that, the remaining work station is calculated to get the optimal assembly for the SD. In addition to that, the capacity planning of the SD is calculated, so the SD order will not disrupt the LCC demand. There are 16 existing work station, and the other existing condition is explained in Table 2.

Table 2. The existing condition

Doll type	Work stations	Smoothness index (SI)	Capacity (unit/week)
WB	6	77.74	909
BH	6	59.36	925
SD	4	120.64	4300

The precedence diagram of each doll type is illustrated in Figs. 1, 2, and 3 respectively. The cycle time of each work element in WB, BH, and SD is listed in Tables 3, 4, and 5 respectively.

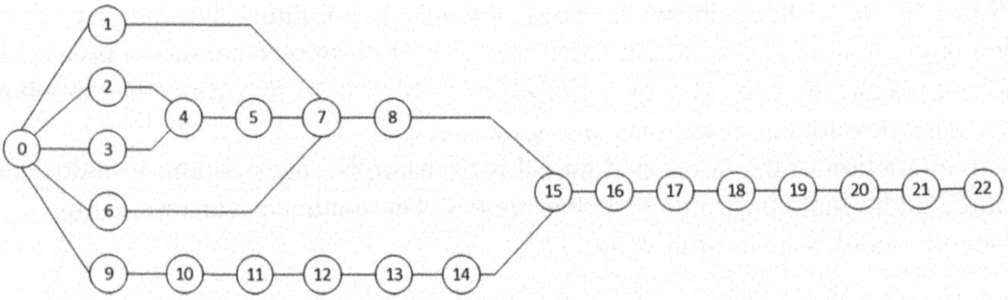

Fig. 1. The precedence diagram of white bear doll

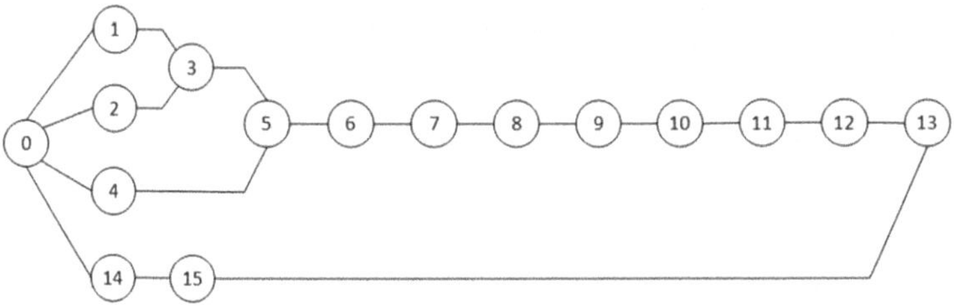

Fig. 2. The precedence diagram of bear head doll

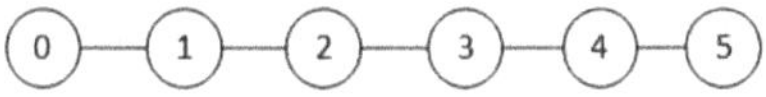

Fig. 3. The precedence diagram of white bear doll

Table 3. Cycle time of WB work element (Second)

Work element	1	2	3	4	5	6	7	8
Cycle time	6.58	1.03	1.03	11.30	5.86	4.16	12.83	6.01
Work element	9	10	11	12	13	14	15	16
Cycle time	8.32	5.94	6.07	7.94	7.94	5.28	11.89	10.78
Work element	17	18	19	20	21	22		
Cycle time	4.83	19.01	51.52	22.11	29.05	19.04		

Table 4. Cycle time of BH work element (Second)

Work element	1	2	3	4	5	6	7	8
Cycle time	6.14	6.14	5.43	5.02	8.85	16.85	12.76	23.10
Work element	9	10	11	12	13	14	15	
Cycle time	65.04	11.63	56.82	3.45	4.37	5.84	9.86	

Table 5. Cycle time of SD work element (Second)

Work element	1	2	3	4	5
Cycle time	6.77	6.77	56.83	85.89	115.08

The takt time for BH and WB is 111.6 and 133.92 s respectively. There is no takt time of SD since there is no fixed amount of SD demand. We process the data with the Urban model and developed model by using the LINGO Software. The optimal number of work stations for BH, WB, and SD are 4, 3, and 3 respectively. Since there are 6 works stations left, the SD assembly line is tripled.

Based on the determined number of work station, the second optimization is done. Every work element is grouped into work stations while maintaining the lowest score of the smoothness index. The result of work grouping can be seen in Table 6.

Table 6. Grouping of work element

Work stations	Work element	Work station duration
White bear doll type		
1	1, 2, 3, 4, 5, 6, 7, 8, 9, 10, 11	86.33 s
2	12, 13, 14, 15, 16, 17, 18	86.97 s
3	19	65.99 s
4	20, 21, 22	88.04 s
Bear head doll type		
1	1, 2, 3, 4, 5, 6, 7, 8	107.74 s
2	9, 14, 15	102.27 s
3	10, 11, 12, 13	98.25 s
Souvenir doll type		
1	1, 2, 3	81.62 s
2	4	85.89 s
3	5	88.61 s

The smoothness index for WB, BH, and SD are 22.14, 10.94, and 7.50 respectively. To make the capacity planning for the SD assembly line, the line efficiency must be calculated by Eq. 12, and the line efficiency is 96.35%. The capacity of one SD assembly line is calculated by Eq. 13, and, the capacity is 1820 unit/week. Since there are three SD assembly line, then the total capacity is 5460 unit/week. In another word, the SME should not accept the SCC order if the order more than 5460/week.

$$L = \frac{\sum_{i=1}^{K} ST_i}{K \times CT} \times 100\% \tag{12}$$

$$Capacity = \frac{P \times L}{ST_{i_{max}}} \tag{13}$$

4 Conclusion

This paper uses the U-line model from Urban to solve the SME's assembly line balancing and capacity planning. The model is used to find the minimum work stations. Furthermore, the second step optimization is done to minimize the smoothness index. The minimum work station for WB, BH, and SD are 4, 3, and 3 respectively, and the smoothness index are 22.14, 10.94, and 7.5 respectively. Based on this assembly line design, the capacity of the total SD assembly line is 5460 unit/week. Thus, the SME should not accept the SCC order more than 5460 unit/week if they want to keep the LCC undisrupted.

References

1. Beigel, J.E.: Production Control. Prentice-Hall Inc., New York (1999)
2. Yogesh, M., Prabagaran, S.: A rational approach of lean tool implementation to SME - a case study. Indian J. Sci. Technol. **10**(1), 1–8 (2017)
3. Yaman, R.: An assembly line design and construction for a small manufacturing company. Assem. Autom. **28**(2), 163–172 (2008)
4. Chan, S., Ramlan, R.: Line balancing scorecard for production line in small and medium enterprise. Int. J. Manag. Appl. Sci. **2**(1), 98–102 (2016)
5. Becker, C., Scholl, A.: A survey on problems and methods in generalized assembly line balancing. Eur. J. Oper. Res. **168**(3), 694–715 (2006)
6. Gokcen, H., Agpak, K.: A goal programming approach to simple U-line balancing problem. Eur. J. Oper. Res. **170**(2), 577–585 (2006)
7. Kriengkorakot, N., Pianthong, N.: The U-line assembly line balancing problem. KKU Eng. J. **34**(3), 267–274 (2007)
8. Widyana, G.A.: Multi objective model for balancing U-Type assembly line with permanent and temporary workers. Jurnal Teknik Industri **11**(1), 33–42 (2009)
9. Elsayed, E.A.: Analysis and Control of Production System. Prentice-Hall Inc., New Jersey (1994)
10. Gasperz, V.: Production Planning and Inventory Control. Gramedia Pustaka Umum, Jakarta (2004)
11. Urban, T.L.: Note: optimal balancing of U-Shaped assembly lines. Manag. Sci. **44**, 738–741 (1998)

Optimal Down-Time Target for Performance Based Remanufactured Lease Contract

Hennie Husniah[1(✉)], Andi Cakravastia[2], and Bermawi P. Iskandar[2]

[1] Department of Industrial Engineering, Langlangbuana University, Jl. Karapitan no 116, Bandung 40261, Indonesia
hennie.husniah@gmail.com

[2] Department of Industrial Engineering, Institut Teknologi Bandung, Jl. Ganesha 10, Bandung 40132, Indonesia
bermawi@mail.ti.itb.ac.id

Abstract. To reduce the high cost of purchasing equipment, many companies tend to lease a new or a remanufactured equipment rather than to purchase it. The consumer willing to lease a remanufactured equipment if the lessor ensures a maximum availability of the equipment during a lease contract period. In this paper, the equipment is leased for a period of time with maximum usage per period. If the total usage per period exceeds the maximum usage allowed in the contract, then the customer (as a lessee) will be charged an additional cost. To reach a high availability, the lessor needs a performance based preventive maintenance (pm) policy. Since the lessor objective function is to maximize their profit with keep the reliability as high as possible, then the decision variable for the lessor is to find the optimal down-time allowed and the optimal pm efforts for a given price of the lease. Numerical examples are given to illustrate the optimal solution.

Keywords: Remanufactured equipment · Preventive maintenance · Lease contract

1 Introduction

In the last few decades, the demand on remanufactured equipment has been growing in the last decade, and this due to the increase in environmental concern, which often called as green technology. In this regards, buying a refurbished or remanufactured equipment is significantly saving our natural environment (https://www.curvature.com/GreenIT) and led to a circular economy. Remanufactured products for heavy equipment are also available in the market and are commonly advertised [5, 6]. The price of a remanufactured equipment is a much cheaper compared to the new one (45%−65% lower of the price of a new equipment). To keep the remanufactured equipment in good condition, an effective maintenance strategy is needed.

Leasing an equipment becomes a favourable option for many companies rather than to purchase it [3]. A comprehensive review can be found in [4] and the research works on lease contract can be classified into two classes – i.e. study from (i) the viewpoint of

M. N. Osman Zahid et al. (Eds.): iMEC-APCOMS 2019, LNME, pp. 24–29, 2020.
https://doi.org/10.1007/978-981-15-0950-6_4

lessor or (ii) from the viewpoint of lessee. In this paper, we propose a remanufactured lease contract for a period of time with maximum usage per period (e.g. 1 year) as in [1], whilst in this paper the lessor controls the down-time target. The decision variable for the lessor is to find the optimal down-time allowed and the optimal pm efforts for a given price of the lease.

The paper is organized as follows. In Sect. 2 we give model formulation for the LC studied. Sections 3 and 4 deal with model analysis and the optimal decisions for the lessor through numerical examples. Finally, we conclude with topics for further research in Sect. 5.

2 Model Formulation

In this section, we first define a new LC, describe failure model, formulate a preventive maintenance policy and its effect on reliability, and then obtain the expected profit for a lessor and a lessee.

Notations:

$\Omega = [0, \Gamma) \times [0, \infty)$	Lease coverage region
δ_y	Preventive maintenance level
$\tilde{D}$	Down time target or pre-specified time limit for performing a minimal repair
U_y	Total usage in $[0, m\Gamma)$
$G(x)$	Distribution function of downtime X_i
$m\Gamma$	Lease contract periods, $\Gamma > 0$ and $m = 1, 2, \ldots$
y	Usage rate
N	Number of PM during lease contract
T_r	The time to the first failure of the remanufactured equipment
P	Lease contract price
C_b	Annual cost
C_p	Preventive maintenance cost
C_r	Average repair cost
$C_{\tilde{D}}$	Penalty cost if the minimal repair time exceeds the limit $\tilde{D}$
$F(t, \alpha_r), f(t, \alpha_r)$	Distribution function and density function for T_r
$\lambda(t, \alpha_r)$, $\Lambda(t, \alpha_r)$	Hazard function and cumulative hazard function
$\Pi_r(m, y)$	Expected profit for a lessor
$J^1(K, \delta_y)$	Expected minimal repair cost
$J^2(K, \delta_y)$	Expected preventive maintenance cost
$J^3(K, \delta_y)$	Expected penalty cost

2.1 Modelling Failure

In general, most products at the end of the first life have a reliability below the threshold value of reliability R^*. Remanufacturing improves the reliability of the equipment to at least the same level of the threshold reliability. Let T_r be the time of the first failure the remanufactured product. If the reliability of the remanufactured product is $R(t,\alpha_r) = 1 - F(t,\alpha_r) = e^{-(t/\alpha_r)^{\beta}}$. As $R(t,\alpha_r) \geq R^*$, then we have $\alpha_r \geq t/\{(\sqrt[\beta]{-\ln R^*}\}$.

We model the conditional hazard function, $\lambda_y(t)$ by using

$$\begin{aligned} \lambda_y(t) &= e^{\rho((y-y_0)/y_0)}\lambda_0(t,\alpha_0) \\ \alpha_0 &= \frac{t}{\sqrt[\beta]{-\ln R^*}}\left(\frac{u_0}{u}\right)^r \end{aligned} \tag{1}$$

$\rho \geq 1$ and $\lambda_0(t,\alpha_0)$ is the base line failure rate function when the usage rate, y is equal to y_0 (a nominal usage of the equipment). If the lessee uses the equipment with the usage rate exceeding the normal value, $y > y_0$, then $\alpha_y > \alpha_r$ or the equipment will deteriorate faster, otherwise when $y < y_0$ then it goes slower.

PM for a given customer with $Y = y$, is done periodically at $k\tau_y$, $k = 1, 2, \ldots$ where k is an integer value, and hence we have k disjoint intervals $-[0,\tau_y)$, $\ldots, [k\tau_y, (k+1)\tau_y = \Gamma)$. As in and Iskandar and Husniah [2], we model the impact of PM through the reduction in the intensity function – i.e. the reduction is δ_{yj} after PM at $t_j, j \geq 1$, and $\delta_{yj} = \delta_y$. As any failure occurring between PM is minimally repaired, then the expected total number of minimal repairs over $[0,\Gamma)$ or $([t_{j-1}, t_j), 1 \leq j \leq k)$ is given by $N = \sum_{j=1}^{k} \int_{t_{j-1}}^{t_j} r_{j-1}(t')dt'$.

2.2 Analysis

The lessor's expected total cost consists of preventive maintenance cost, corrective maintenance cost, and the penalty cost which incurs to the lessor due to the down time exceeding the threshold value. If $J^1(k,\delta_y)$, $J^2(k,\tau_y,\delta_y)$ and $J^3(k,\tau_y,\delta_y)$ are the expected preventive maintenance cost, the expected total repair cost, and the expected penalty cost over the LC period $(0, m\Gamma], m = 1, 2, \ldots$ for a given usage rate y, respectively, then the expected total cost, $\Psi[k,\tau_y,\delta_y]$ is given by

$$\Psi[k,\tau_y,\delta_y] = J^i(k,\delta_y) + J^2(k,\tau_y,\delta_y) + J^3(k,\tau_y,\delta_y) \tag{2}$$

where

$$J^1(k,\delta_y) = C_r R_y(\Gamma) \tag{3}$$

$$J^2(k,\tau_y,\delta_y) = kC_0 - \sum_{j=1}^{k+1} [C_r(\Gamma_0 - j\tau_y) - C_v][r(j\tau_y) - r((j-1)\tau_y)] \tag{4}$$

$$J^3(k, \tau_y, \delta_y) = C_{\tilde{D}} \int_{\tilde{D}}^{\infty} (z - \tilde{D})dF(z)\left(R_y(\Gamma) - \sum\nolimits_{j=1}^{k} (\Gamma - j\tau_y)\delta_y\right) \quad (5)$$

Expected Profit Total Revenue:
The expected profit is over LC is given by

$$E\left[\Pi_y\right] = P(m) + \Phi(m, y) - \Psi\left[k, \tau_y, \delta_y\right] - mC_b^r \quad (6)$$

$$P(m) = P\{e^{-\varphi m}/e^{-\varphi}\} = P\left\{e^{-\varphi(m-1)}\right\} \quad (7)$$

$$\Phi(m, y) = C_d \sum_{j=1}^{m} Max\{0, (y \times j\Gamma - jU_{\max})\} \quad (8)$$

where $0 \leq \varphi < 1$ represents a discount parameter when $m \geq 2$, C_b^r is the annual cost of the remanufactured equipment, and C_d is the additional cost charged (e.g. $/km or $/page copied) to the lessee.

3 Optimization

In this section, we consider that the lessor would like to maximize their profit. In other words, the profit earned for the lessor depends on the lessor's cost and revenue structure. Then, the strategy set of the lessor is given by $Q_y = \left\{(\delta_y, \tau_y, k_y, \tilde{D})\middle| 0 \leq \delta_y \leq 1, \tau_y \geq 0, k_y \geq 0\right\}$. Here the decision variables are the set of strategies $q_y^* \in Q_y$ consist $\delta_y, \tau_y, k_y, \tilde{D}$ that solves $\max_q E\left[\Pi_y\right]$ for the lessor.

Finding Optimal Solution:
The procedure to obtain the optimal decision involves two main stages as follows.

Stage 1: $\tilde{D}^*$, find $\tilde{D}^*$ that minimizes $\Psi(k, \tau_y, \delta_y)$ by taking the first derivative of $\Psi(k, \tau_y, \delta_y)$ with the respect to $\tilde{D}$ and set the resulting derivative to zero.

Stage 2: $[k^*, \tau_y^*, \delta_y^*]$, for a fixed $\tilde{D}^*$, we find $[k^*, \tau_y^*, \delta_y^*]$ that maximize $E\left[\Pi_y\right]$ with procedure as in Iskandar and Husniah [2].

Finally, we repeat the two stages to find $\tilde{D}^*$ and $(k^*, \tau_y^*, \delta_j^{y*})$ for other values of $y, 0 < y < \infty$.

4 Numerical Example

In this section, the optimal maintenance policy and the number of lease periods is evaluated through some numerical examples to illustrate the features of the profit model. Suppose that the lifetime distribution of equipment follows the Weibull distribution with α as in Eq. (1), β = 2.5. The other parameter values used are: R = 0.4,

$U_{max} = 12(1 \times 10^4)$ km, $(\gamma = U/\Gamma = 1)$, $y_0 = 1$, $\rho = 2$ and $C_v = C_d = C_0 = 0.5C_r$, $C_r = 100$, $C_{\tilde{D}} = 75$, $C_b^r = 16870 \times e^{-0.05(m-1)}$.

Table 1. The optimal policy with K = 2025 × 50 and m = 1, 2

LC region m = 1 (12 months)					LC region m = 2 (24 months)			
y	$k^*; \tilde{D}^*$	$\tau^*; \delta^*$	$E[\Pi_y] \times 10^5$	$P \times 10^5$	$k^*; \tilde{D}^*$	$\tau^*; \delta^*$	$E[\Pi_y] \times 10^5$	$P \times 10^5$
1.0	2; 60	4.36; 0.12	1.0122	1.34	2; 60	4.37; 0.12	0.8999	1.23
1.5	4; 59.99	*2.504; 1.82*	1.0422	3.35	6; 60	3.53; 0.99	0.8617	3.14
2.0	4; 9.34	*2.506; 1.82*	1.0722	5.83	12; 4.41	1.87; 7.51	(2.1471)	6.30

Table 1 shows the profit for the higher usage rate where additional charges to the lessee is very large reduces significantly the profit of the lessor. This is expected as for a large usage rate, it requires a large total cost for the lessor (for maintenance and penalty costs), and this in turn will cause a more reduction in the profit for a higher usage rate. The maintenance degree, and the optimal price of the LC increases as y increases while the down-time target and the length of preventive maintenance decreases when y increases.

5 Conclusion

In this paper we have studied a performance usage based multi-period lease contract for remanufactured equipment such as dump trucks. Under this lease contract, the equipment is leased for a period of $m\Gamma$ and a maximum usage, $U_{\max}$. We find the optimal down-time target and maintenance policy solution for the lessor. We can extend this multi-period LC by considering the lessee various usage pattern and the lessor bargaining position point of view (i.e. a Stackelberg game theory formulation). These two topics are currently under investigation.

Acknowledgements. This work is funded by the Ministry of Research, Technology, and Higher Education of the Republic of Indonesia through the scheme of "PUPT 2019".

References

1. Husniah, H., Pasaribu, U.S., Iskandar, B.P.: Lease contract in usage based remanufactured equipment service system. In: IFIP WG 5.7 Proceedings Part II International Conference, APMS, Seoul (2018)
2. Iskandar, B.P., Husniah, H.: Optimal preventive maintenance for a two dimensional lease contract. Comput. Ind. Eng. **133**, 693–703 (2017)
3. Jaturonnatee, J., Murthy, D.N.P., Boondiskulchok, R.: Optimal preventive maintenance of leased equipment with corrective minimal repairs. Eur. J. Oper. Res. **174**, 201–215 (2006)

4. Murthy, D.N.P., Jack, N.: Extended Warranties Maintenance Service and Lease Contracts. Springer Series in Reliability Engineering. Springer, London (2014)
5. Nnorom, I.C., Osibanjo, O.: Electronic waste (e-waste): material flows and management practices in Nigeria. Waste Manag. Res. **26**(4), 317–326 (2008)
6. Webster, S., Mitra, S.: Competitive strategy in remanufacturing and the impact of take-back laws. J. Oper. Manag. **25**, 1123–1140 (2007)

Fuzzy Initial Condition in a Technology Transfer Model with Competing Followers

Hennie Husniah[1](✉), Rachmawati Wangsaputra[2], and Asep K. Supriatna[3]

[1] Department of Industrial Engineering, Langlangbuana University, Jl. Karapitan no 116, Bandung 40261, Indonesia
hennie.husniah@gmail.com

[2] Department of Industrial Engineering, Institut Teknologi Bandung, Jl. Ganesha 10, Bandung 40132, Indonesia

[3] Department of Mathematics, Padjadjaran University, Jl Raya Bandung-Sumedang Km 21, Jatinangor 45363, Indonesia
a.k.supriatna@unpad.ac.id

Abstract. In this paper we present a mathematical model of technology transfer. The model takes form in a system of differential equations. We assume that there are two competing followers. We also assume a logistic growth function of technological development for all parties involved. Most papers of similar model assume a crisp initial condition for each of the followers. Here we use the assumption that the fuzziness of the input in the process propagates to the output. Some numerical examples are presented to illustrate properties of the model. We found that in some circumstances fuzzy initial condition might result in a crisp long-term solution.

Keywords: Technology transfer · Mathematical model · Dynamical system · Fuzzy initial condition

1 Introduction

Technology transfer is a process of the implementation of scientific/technological information developed in one area into another area. It is also defined as a process of migration and redeployment of technology from one area into different area [1, 6, 7, and 9]. In regards to technology transfer, new competition rules for technology transfer agreements have been created by the European Commission to regulate a good practice in the technology transfer implementation [5]. This fact is among the reason to develop a technological transfer model considering competition among related parties. Specifically, in this paper we develop a mathematical model of technology transfer from a leader to two competing followers. We assume that the competition takes place in the accessing process of the source of the technology being transferred.

In this paper we model the process of technology transfer in the form of a dynamical system. Different from many literatures, which use a standard analysis of its equilibriums (equilibrium points) and their stability properties, here we present a numerical computation to show the graphical solution of the model. The model is

M. N. Osman Zahid et al. (Eds.): iMEC-APCOMS 2019, LNME, pp. 30–35, 2020.
https://doi.org/10.1007/978-981-15-0950-6_5

generic, hence it can be applied either for "physical" technology transfer or for knowledge transfer. An example of knowledge transfer is the transfer of knowledge among researchers in universities as a result of a long-standing collaboration. The details of the model are described in Sect. 2.

2 Model Formulation

Following [2, 3, 10], we assume that in the absence of technology transfer both the leader and the followers have a logistic curve of technological development as a function of time, i.e. $X_L(t)$ for the leader and $X_{Fi}(t)$ with $i = 1, 2$ for the followers, respectively. The model adopts the idea of [8]. The full equations in the presence of competition are given by:

$$\frac{dX_L(t)}{dt} = k_L X_L(t)\left(1 - \frac{X_L(t)}{u_L}\right), \tag{1}$$

$$\frac{dX_{F1}(t)}{dt} = (1 + (1 - C_2)k_T(X_L(t) - X_{F1}(t)))Y_{F1}, \tag{2}$$

$$\frac{dX_{F2}(t)}{dt} = (1 + (1 - C_1)k_T(X_L(t) - X_{F2}(t)))Y_{F2}, \tag{3}$$

where $Y_{F1} = k_{F1}X_{F1}(t)\left(1 - \frac{X_{F1}(t)}{u_{F1}}\right)$ and $Y_{F2} = k_{F2}X_{F2}(t)\left(1 - \frac{X_{F2}(t)}{u_{F2}}\right)$.

The notations and their descriptions are the same as in [3] as explained below in Table 1. Furthermore, we see straight away from the model that if $C_i = 1$ then the follower i completely outcompetes the other follower from utilizing the technology transfer facilities, hence the other only grows naturally as if there was no technology transfer process. Likewise, if k_T is zero then exactly all followers grow according to their natural growth, because there is no transfer rate. The model describing the competing technology transfer above takes form as a system of nonlinear differential equations:

$$\begin{gathered} X_L'(t) = G(X_L, X_{F1}, X_{F2}), X_{F1}'(t) = H_1(X_L, X_{F1}, X_{F2}), \\ \text{and } X_{F2}'(t) = H_2(X_L, X_{F1}, X_{F2}). \end{gathered} \tag{4}$$

There are some methods to analyze the system, i.e. by solving the system explicitly to obtain the functions $X_L(t), X_{F1}(t),$ and $X_{F2}(t)$ satisfying the system. Upon solving the system, the solution is able to describe both the transient and the steady state behaviour of the system. However, this need a high sophisticated mathematics, and sometimes an explicit solution is unable to find. Other method is by analyzing the equilibrium solution of the system. It is a state in which opposing forces or influences are balanced resulting in a zero growth rate of the technological development. The sections that follow gives the equilibrium points (equilibria) as the main result of model and discuss some insight from the analysis of the equilibriums of the model using

Table 1. Notation and description.

Notation	Description
$X_L(T)$	The measure of technological development for the leader
$X_{Fi}(t)$	The measure of technological development for the follower
k_L	The indigenous ability of the leader to develop
u_L	The upper limit of the technological development of the leader
k_F	The indigenous ability of the follower to develop
u_F	The upper limit of the technological development of the follower
k_T	The technology transfer rate
C_i	Competition coefficient of *i-th party* with i = 1, 2

dynamical system approach. Specifically, first we search for all the equilibria, then we analyze the stability of each equilibrium to gain the long-term behaviour of the system. The transient solution can still be obtained using some numerical methods.

3 Result and Discussion

Our system has 18 equilibrium points as shown elsewhere in which 10 of them have been already found in [3]. The authors in [3] only give a graphical visualization of these equilibrium points for certain data set of parameters and we found a complete list of these 18 equilibrium points. Among all the equilibrium points one is always stable and the other one is stable for certain conditions. The analysis of these equilibriums will able to give some information regarding the long-term behavior of the system. Since the complete analysis of their stability requires a deep and rigour mathematical treatment, in this short paper we only restrict to show the effect of fuzzy initial condition to the dynamic property of the stable equilibrium point. Specifically, we will compare the crisp and the fuzzy solutions of this equilibrium.

3.1 Crisp Solution

Figure 1 shows technological development of three parties (one leader and two followers) in the presence of technology transfer with the follower 1 have a very higher competitiveness, so that it can exploits the resources of the leader to make its own development faster compare to its competitor. The parameters of the model and the initial conditions used in the figure are $k_L = 0.5, k_{F1} = k_{F2} = 0.2,\ \ U_L = 6$, $U_{F1} = 2, U_{F2} = 3,\ \ X_L(0) = 0.6, X_{F1}(0) = X_{F2}(0) = 0.02,$ with $k_T = 1, C_1 = 0.95$, $C_2 = 1 - C_1$ and $k_T = 1$. These parameters are given hypothetically just to illustrate the model. The figure shows that all the parties eventually reach their respective upper limits of the technological development, as suggested by the stable coexistence equilibrium $\left(X_L^* = u_L, X_{F1}^* = u_{F1}, X_{F2}^* = u_{F2}\right)$. The unit of technological development could be anything, e.g. the number of products, the level of product quality, etc. In this case, the technology transfer is taking place in the situation of moderate differences of technological development between the leader and the followers. Other case, in which the technology transfer is taking place in a situation of extreme differences of technological development between the leader and the followers, can also be investigated. Complete dynamic behaviour is worth to explore to obtain further understanding of the system.

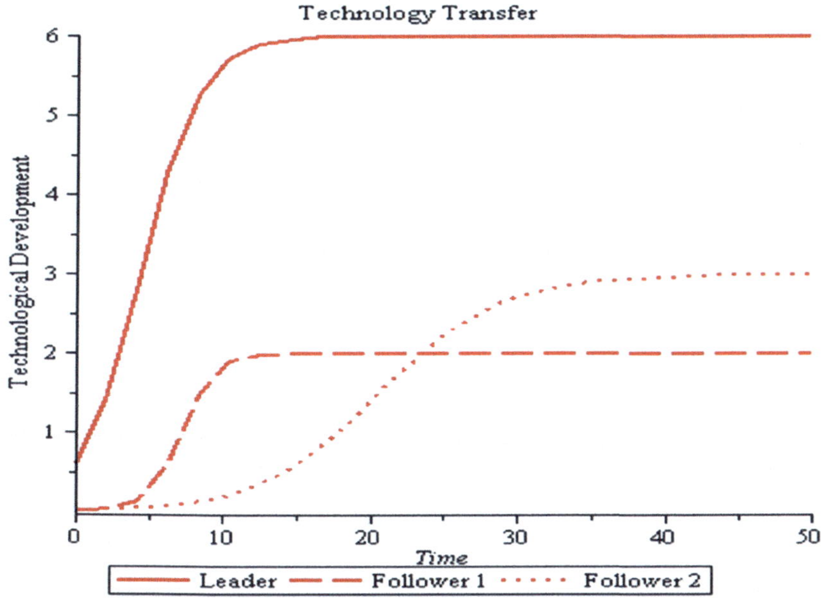

Fig. 1. The growth of technological development for three parties with one as a leader and the others as followers.

3.2 Fuzzy Initial Solution

Figure 1 shows the dynamic of the technological transfer by assuming that the parameters or the initial values are crisp numbers. Further, the following figure assumes that the initial value of the leader is somewhat vague represented by a triangular fuzzy number around its original value ($X_L(0) = 0.6$ in Fig. 1). We chose arbitrarily such a triangular fuzzy number given by (0.1; 0.6; 0.7). We discretize the dynamic equation using the fourth order Runge-Kutta method and use the same parameters as in the previous examples, except the initial value for the leader technological development given by the triangular fuzzy number (0.1; 0.6; 0.7). We assume that the crisp function propagates the fuzziness of the initial condition as the independent variable to the solution of the differential equation as the dependent variable as on of the method in handling fuzziness [4: p. 153]. We then plot the time series as the solution of the dynamical system as in Fig. 2.

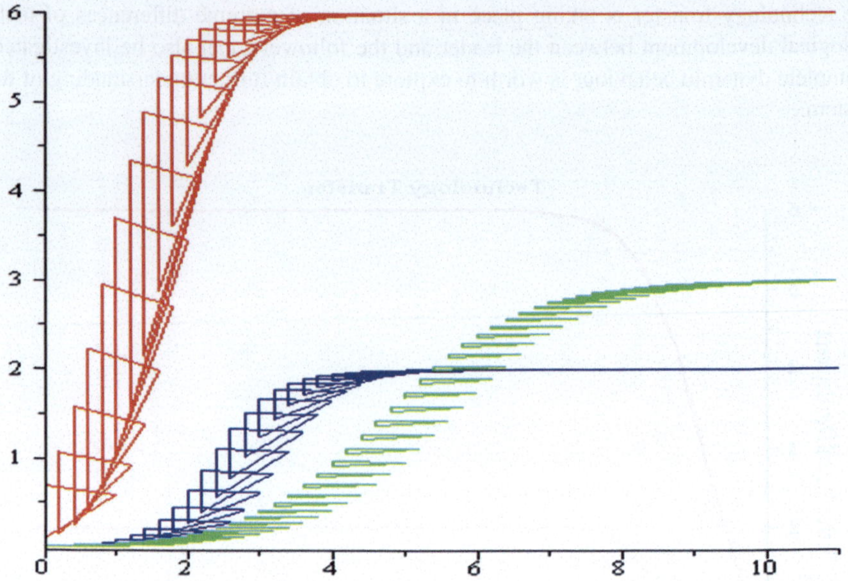

Fig. 2. The visual appearance of the technology transfer model with fuzzy initial value of the measure of technological development for the leader.

4 Conclusion

This paper is used to investigate a model of technology transfer with two competing followers. The equilibria and their stability are discussed by showing the solution numerically for certain data sets and the effect of fuzzy initial value of the leader technological development measure on the overall solution of the system is also presented. The resulting solution of technology transfer model with the leader fuzzy initial value for its technological development resembles those with crisp initial value in the long-term. The disappearance of the fuzziness in the system might due to the method used in solving the dynamical system. Other methods and choices of fuzzy numbers, such as Trapezoidal, Gauss, etc. are worth to investigate to reveal the robustness of the current method choice of the fuzzy number.

Acknowledgement. This work is funded by Kemenristekdikti - the Government of the Republic of Indonesia - through the scheme of PDUPT 2019 with contract number 2827/UN6.D/LT/2019 to AKS.

References

1. Bar-Zakay, S.N.: A technology transfer model. Technol. Forecast. Soc. Change **2**, 321–337 (1971)
2. Husniah, H., Sebrina, S., Supriatna, A.K.: A dynamical system approach in modeling technology transfer: a review and a new model. J. Indones. Math. Soc. **22**, 37–58 (2016)

3. Husniah, H., Supriatna, A.K.: A simple mathematical model of technological transfer with two competing followers (a preliminary result). In: Proceeding of 9-th International Seminar on Industrial Engineering and Management, Padang (2016). ISSN 1978-774X
4. Lee, K.H.: First Course on Fuzzy Theory and Applications, 1st edn. Springer, Berlin (2005)
5. Mariniello, M., Antonielli, M.: The new competition rules for technology transfer agreements (2013). http://bruegel.org/2013/05/the-new-competition-rules-for-technology-transfer-agreements/. Accessed 25 Apr 2016
6. Ramanathan, K.: The polytrophic components of manufacturing technology. Technol. Forecast. Soc. Change **46**, 221–258 (1994)
7. Ramanathan, K.: An overview of technology transfer and technology transfer models (2017). http://tto.boun.edu.tr/files/1383812118_An%20overview%20of%20TT%20and%20TT%20Models.pdf. Accessed 10 May 2017
8. Raz, B., Assa, I.: A model of "coupled" technology transfer: a logistic curve approach. Technol. Forecast. Soc. Change **33**, 251–265 (1988)
9. Souder, W.E., Nashar, A.S., Padmanathan, V.: A guide to the best technology transfer practices. J. Technol. Transf. **15**, 1–2 (1990)
10. Supriatna, A.K., Husniah, H., Frey, S.: Raz and Assa model of technology transfer: revisited. In: Prosiding Seminar Nasional Industrial Services (SNIS) III, Cilegon (2013)

Order Acceptance and Scheduling Model for Small-Sized Metal Manufacturing Company

Silfia Nurul Ariyani, Fariz Muharram Hasby(✉), and Anas Ma'ruf

Production Systems Lab, Institut Teknologi Bandung, Bandung, Indonesia
fariz@mail.ti.itb.ac.id

Abstract. One of the characteristics of small-sized metal manufacturing companies in Indonesia is that they tend to accept all orders without considering available production capacity. This research aims to propose an order acceptance and scheduling model based on available production capacity that can be used to improve the capacity planning of such companies. A suitable capacity planning model is chosen (Chen et al. 2009) and then used with real data gathered from Company K, a small-sized metal manufacturing company in Bandung, Indonesia. It can be shown that the solution from the model will increase profit as much as 2.52%. The increase results from a change in order acceptance as the model correctly identifies one of the historically received order will not generate any profit.

Keywords: Order acceptance · Capacity planning · Scheduling

1 Introduction

The manufacturing industry sector in Indonesia is starting to play a large role in the economic growth of Indonesia. In the third quarter of 2017 the growth of non-oil-and-gas industry has achieved 5.49% with the majority (10.6%) consisting of growth from the basic metal manufacturing industry [1]. The basic metal manufacturing industry not only consists of large companies, but also numerous small-medium enterprises which frequently act as component and parts supplier for the larger enterprises. As such, it is important to ensure that small-medium metal manufacturing company can operate efficiently to also ensure the efficient growth of the whole industry sector.

Small-medium metal manufacturing companies face several problems during their daily operation, most of which are based on the relative inexperience of such companies to efficiently manage their operation. Widespread problems such as a lack of adequate planning methods and operation control, coupled with a lack of standard operating procedures will frequently cost the company through unmet delivery schedules. Reliable delivery time is a very important parameter for a make-to-order company to be able to compete in the market [2].

Such problems also occurred at Company K, a small metal manufacturing company located in Bandung, Indonesia. This company used to accept all orders that came to them with the thinking that it will boost their overall profit. Unfortunately, without

M. N. Osman Zahid et al. (Eds.): iMEC-APCOMS 2019, LNME, pp. 36–41, 2020.
https://doi.org/10.1007/978-981-15-0950-6_6

proper capacity planning and scheduling, the job shop nature of metal manufacturing company will induce a higher complexity along with the number of received orders. Figure 1 shows that on average, in December 2017 to January 2018, all orders in Company K are late by 5.7 days. This will create losses for the company because every late order will incur a penalty fee.

Fig. 1. Lateness on orders of company K

This research aims to find a suitable simultaneous order acceptance, capacity planning and scheduling model such that it can be used in small metal manufacturing companies to help in maximizing the effectiveness of their operation. The model will be used with real order data from Company K to judge its suitability in maximizing the profit. It is hoped that the chosen model can be used as a base to find a general model capable to be used by the majority of such small-medium metal manufacturing companies in Indonesia.

2 Literature Research and Model Selection

As mentioned before, one of the problems that small-medium metal manufacturing companies are facing is in capacity planning, specifically regarding the production decisions such as order acceptance process, scheduling and inventory planning [3]. This topic has also been a research topic with lots of interest in the past 20 years, for example order acceptance model with profit maximizing objective function where the profit is defined as the difference between revenue and penalty costs or order acceptance model with an objective function to minimize lost revenue from order rejection and early order inventory costs by [4]. More recently, Ndiaye et al. [5] has also suggested that due to the heterogeneity of small firms, modeling the capacity utilization of small-medium enterprises generically is hard to do. Therefore, an approach that can be done is to develop a model that can be implemented in a small-medium enterprise and continue to iterate the model for a broader use by more and more firms.

2.1 Model Selection

Through a literature review process, this research has narrowed down several candidate models to be chosen as the most suitable model for the problems in small-medium metal manufacturing company in Indonesia [6–9]. The models are then compared by several different parameters, such as type of production system, possibility of different product routings, and the possibility of overtimes and extra shifts. The chosen model [7] is as follows:

Objective Function

$$Maximize\ i \in I_{si} Z_i - i \in I_j \in J_{ik} \in K_r \in R_t \in T_{ckr} X_{ijkrt} \tag{1}$$

Subject To

$$\sum_{i \in I} \sum_{j \in J_i} X_{ijkrt} \le b_{krt} \quad \forall\, k \in K,\ r \in R,\ t \in T \tag{2}$$

$$\sum_{r \in R \setminus \{|R|\}} \sum_{t \in T} X_{ijkrt} + \sum_{t \in T} X_{ijk|R|t} = p_{ijk} Z_i \quad \forall\, i \in I,\ j \in J_i,\ k \in K \tag{3}$$

$$\sum_{t \in T} X_{ijk|R|t} = p_{ijk} O_{ij} \quad \forall\, i \in I,\ j \in J_i,\ k \in K \tag{4}$$

$$\sum_{j \in J_i} \sum_{k \in K} X_{ijkrt} \le l_{rt} \quad \forall\, i \in I,\ t \in T,\ r \in R / \{|R|\} \tag{5}$$

$$X_{ijk|R|t} \le l_{|R|t} O_{ij} \quad \forall\, i \in I, j \in J_i, k \in K, t \in T \tag{6}$$

$$\sum_{j \in J_i} \sum_{r \in R} \sum_{k \in K} X_{ijkrt} \le 24 \quad \forall\, i \in I, t \in T \tag{7}$$

$$X_{ijkrt} \ge \tau Y_{ijkrt} \quad \forall\, i \in I,\ j \in J_i,\ k \in K,\ r \in R,\ t \in T \tag{8}$$

$$X_{ijkrt} \le p_{ijk} Y_{ijkrt} \quad \forall\, i \in I,\ j \in J_i,\ k \in K,\ r \in R,\ t \in T \tag{9}$$

$$\sum_{k \in K} t\, Y_{i|J_i|krt} \le d_i Z_i \quad \forall\, i \in I,\ r \in R,\ t \in T \tag{10}$$

$$\sum_{r' \in R} \sum_{k \in K} \sum_{t'=1}^{t=1} X_{i(j-1)kr't'} + \sum_{r'=1}^{r} \sum_{k \in K} X_{i(j-1)kr't} \ge \sum_{k \in K} p_{i(j-1)k} Y_{ijkrt}$$
$$\forall\, i \in I,\ j \in J_i \setminus \{1\},\ r \in R \setminus \{|R|\},\ t \in T \tag{11}$$

$$\sum_{r \in R} \sum_{k \in K} \sum_{t'=1}^{t} X_{i(j-1)krt'} \ge \sum_{k \in K} p_{i(j-1)k} Y_{ijk|R|t}$$
$$\forall\, i \in I,\ j \in J_i \setminus \{1\},\ t \in T \tag{12}$$

$$\sum_{j \in J_i} \sum_{k \in K} X_{ijk|R|t} \le 24 - \sum_{j \in J_i} \sum_{k \in K} \sum_{r \in R\{|R|\}} rl_{rt} Y_{i(j-1)krt}$$
$$\forall\, i \in I, t \in T \tag{13}$$

$$X_{ijkrt} \geq 0 \quad \forall\, i \in I,\ j \in J_i,\ k \in K,\ r \in R,\ t \in T \tag{14}$$

$$Y_{ijkrt} \in \{0, 1\} \quad \forall\, i \in I,\ j \in J_i,\ k \in K,\ r \in R,\ t \in T \tag{15}$$

$$Z_i \in \{0, 1\} \quad \forall\, i \in I \tag{16}$$

$$O_{ij} \in \{0, 1\} \quad \forall\, i \in I,\ j \in J_i \tag{17}$$

This model aims to maximize total profit by accepting certain order and calculating the profit and manufacturing costs from each order, subject to the routing and shifts (regular or overtime) in which the orders will be processed.

2.2 Model Adjustment

A small adjustment is made to the model, which is the definition of the parameter τ, the smallest time window that can be allocated to each machine, into 15 min. This value is obtained through observation of the system and interviews with the workers of the company.

3 Data Collection and Computations

Before the model is used with real data, we have done a small experiment by running the chosen model with several sets of hypothetical data. The chosen model was proven to be able to reject unprofitable orders, owing to the high production cost of the order. Besides that, the chosen model was also proven to be able to distinguish between infeasible orders that would create infeasible schedules and feasible orders that would generate lower profit. As such, we are confident that the model can be used to create a valid solution.

Real order data that came to the company are collected from 26/3/2018 to 30/3/2018. The order consists of different products and components to be supplied to another larger company, all with different routings, materials, product revenues, due dates and order arrival dates. Additional model parameters must also be gathered for the model to be run, such as length of each shifts, machine capacity, and machining costs. In total there are 27 different orders with differing quantities that will be considered.

Because the orders are collected in a period of 5 days, the model is also iterated at the start of each day, with the orders that have not been finished combined with the new orders for the next day to be scheduled again. Figure 2 shows a small sample of the schedule obtained from the model in the 3rd iteration.

It can be seen that compared to the previous method which accepts all incoming orders, the chosen model rejected an order of the product 'Flange-2 Clamping RPV'. This rejection combined with better scheduling resulted in an improvement in profit by 2.52% compared with the historical data as shown in Table 1.

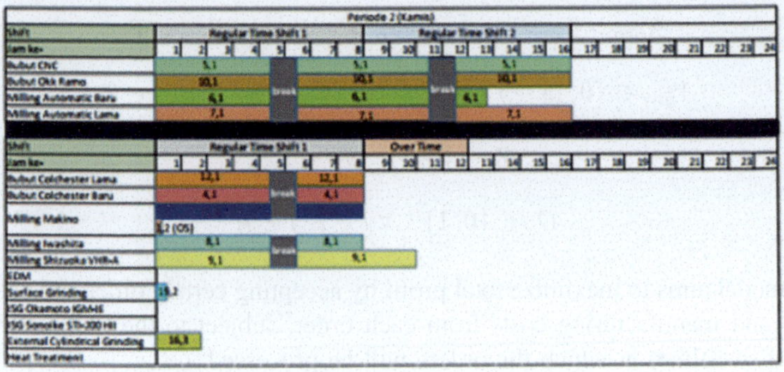

Fig. 2. Partial schedule from the 3rd iteration.

Table 1. Difference of historical and model profit

Historical profit	Model profit	Difference
Rp. 42,735,663	Rp. 43,811,200	Rp. 1,075,537

4 Discussion and Conclusions

The results above have shown that the model is able to maximize the profit of the company. Although the improvement is quite small (2.52%) it should be noted that the sample size for the model is also relatively small. During the observed 5 days period the company actually processes 127 different orders, but only the orders that are started and finished during that period are included in the final data set. It should be clear that hopefully the model will be able to give a greater increase in profit should the model be fed with the real larger order data.

Although the model is made with the objective function to maximize profit, there is a slight difficulty in the calculation of the total profit for 5 days. This is because the unfinished orders that carry over from the previous day will be counted once more if all the objective functions from all the iterations are summed directly. As such, during this research a separate calculation must be done independently from the model to arrive at the total profit. Although this approach is effective for the small problem set used in this research, this approach will not be efficient for larger data sets. As such, it is clear that an improvement to the model can be made by accounting for the iterative approach that must be used if the model is to be used in a real market environment.

Another improvement that can be made to the model concerns the resulting schedule of the model. The resulting schedule is not a non-delay schedule. This feature of the model is detrimental to its real-world adoption by small medium enterprises, mainly because spare capacity is needed to maximize profit in case of rush order arrivals and because idle workers are very costly for small enterprises with limited capital. The resulting schedule could be converted into a non-delay schedule by using the left shift heuristics as shown in Fig. 3, but once again improvements in the model can be made such that the resulting schedule will be a non-delay schedule.

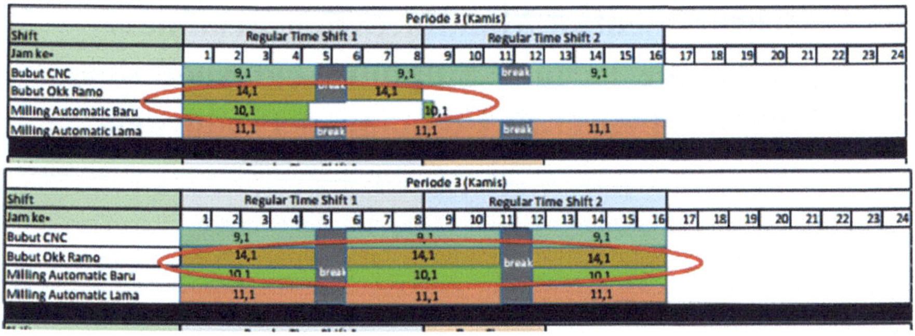

Fig. 3. (top) Original model output schedule, (bottom) Adjusted non-delay schedule through left-shift heuristics

As such, it can be concluded that with some adjustments, such as the few mentioned above, the chosen model will be suitable for use by small-medium metal manufacturing enterprises. Next step of this research will focus on the developments mentioned above and the testing of its validity and feasibility of use by further testing with real world cases.

References

1. Indonesian ministry of industry website. http://www.kemenperin.go.id/artikel/18379/Capai-5,49-Persen,-Pertumbuhan-Industri-Kembali-Meroket-di-atas-Perekonomian. Accessed 4 Mar 2018
2. Ebadian, M., Torabi, S.A., Rabbani, M., Jolai, F.: Hierarchical production planning and scheduling in make-to-order environments: reaching short and reliable delivery dates. Int. J. Prod. Res. **47**(20), 5761–5789 (2009)
3. Martinez-Costa, C., Mas-Machuca, M., Benedito, E., Corominas, A.: A review of mathematical programming models for strategic capacity planning in manufacturing. Int. J. Prod. Econ. **153**, 66–85 (2014)
4. Slotnick, S.A.: Order acceptance and scheduling: a taxonomy and review. Eur. J. Oper. Res. **212**, 1–11 (2011)
5. Ndiaye, N., Razak, L.A., Nagayev, R., Ng, A.: Demystifying small and medium enterprises' (SMEs) performance in emerging and developing economies. Borsa Istanb. Rev. **18**(4), 269–281 (2018)
6. Mestry, S., Damodaran, P., Chen, C.S.: A branch and price solution approach for order acceptance and capacity planning in make-to-order operations. Eur. J. Oper. Res. **211**, 480–495 (2011)
7. Chen, C.S., Mestry, S., Damodaran, P., Wang, C.: The capacity planning problem in make-to-order enterprises. Eur. J. Math. Comput. Model. **50**, 1461–1473 (2009)
8. Oguz, C., Salman, F.S., Yalcin, Z.B.: Order acceptance and scheduling decision in make-to-order systems. Int. J. Prod. Econ. **125**, 200–211 (2010)
9. Aouam, T., Geryl, K., Kumar, K., Brahimi, N.: Production planning with order acceptance and demand uncertainty. Comput. Oper. Res. **91**, 145–149 (2018)

Measurement of Manufacturing Readiness Level for the Tartaric Sulphur Acid Anodizing Method in Aircraft Component Production

Fariz Muharram Hasby(✉), Hanifa Laila Novianti, and Iwan Inrawan Wiratmadja

Production Systems Lab, Institut Teknologi Bandung, Bandung, Indonesia
fariz@mail.ti.itb.ac.id

Abstract. The trend of green industry is currently developing in every sector of industry, including in the aerospace industry. Airbus as a leading company in the industry has developed a program that aims to eliminate the use of chromate in any aircraft component manufacturing. As such, Indonesian Aerospace of Indonesia, a supplier of Airbus, must also reorganize their manufacturing capabilities to achieve that goal, especially in the anodizing process with the Tartaric Sulphur Acid Anodizing (TSAA) Program. This research aims to measure the level of manufacturing readiness of the TSAA technology using the adjusted Manufacturing Readiness Level (MRL) Model. The development of the model was done using the expert judgment from among practitioners and academicians, validated by using the Content Validity Ratio (CVR). The MRL developed for this study consists of 10 levels, 9 criteria and 22 sub-criteria. Measurement results show that the current level of the technology is at MRL2, while the goal of the project is at MRL8. This shows a huge gap that must be addressed before Indonesian Aerospace can reach the level that is permissible by Airbus.

Keywords: Tartaric Sulphur Acid Anodizing (TSAA) Technology · Manufacturing Readiness Level (MRL) · Content Validity Ratio (CVR) · MRL gap

1 Introduction

Nowadays, a trend for green industry is currently developing in various industry sectors located in many countries. The term 'green design and manufacturing' has become a common term in industrial activities with an emphasis on Design for Environment (DFE) [1]. The same trend is also growing in Indonesia and the Indonesian government has also decided to capture this issue in its National Industry Development Master Plan (RencanaInduk Pembangunan Industri Nasional/RIPIN) for the year 2015–2035. More specifically, the government has created several green programs for the Transportation Vehicle Sector in the period 2015–2019, namely the development of product, process and management standards (respectively ISO 9000, ISO 14000, and ISO 26000) and green industry, along with technical specifications and guidelines for the transportation industry [2].

M. N. Osman Zahid et al. (Eds.): iMEC-APCOMS 2019, LNME, pp. 42–47, 2020.
https://doi.org/10.1007/978-981-15-0950-6_7

These trends are also affecting Indonesian Aerospace, the only aerospace industry in Indonesia, such as in their joint program with Airbus of Europe. Airbus has decided to implement a program that aims to develop eco-efficient alternatives for their current production processes that are using chromate materials [3]. Even though local firms that act as suppliers for multinational companies usually have a strategic choice of adopting certain technological advancements of multinational companies [4], in this case they have to comply with the same specifications, procedures and guidelines to secure their contract with Airbus. To this end, Indonesian Aerospace, as a supplier of Airbus, must develop a new procedure for the surface treatment of aircraft components by using Tartaric Sulfuric Acid Anodizing (TSAA) to replace the Chromic Acid Anodizing process.

Preliminary interviews with Indonesian Aerospace have indicated that the transfer of technology process for the new manufacturing process has not been successful. The new process is still only at the pre-qualification level and further development of the process to reach its intended level will be complicated and costly. Indonesian Aerospace will need a tool to reliably assess the development of the new process so that it can minimize the risks associated with the development of the new process.

Although the Indonesian government through the Ministry of Research and Higher Education has developed a guideline to measure and assess the manufacturing readiness level of technology transfer projects in 2016, until this research is done Indonesian Aerospace has not started a program to measure the manufacturing readiness level of their technology transfer projects. This research then aims to develop such a tool to help Indonesian Aerospace in its assessment of their new procedure, specifically the TSAA technology.

2 Literature Research

Technology transfer is the process of transferring technical knowledge, data, designs, prototypes, material, inventions, software, and/or trade secrets from an organization to another of from a function to another [5]. The process of technology transfer is an interactive two-way communication between a 'sender' and a 'receiver' with messages flowing in those two directions [6]. The 'receiver' is not just a passive entity but it should possess the required capabilities for knowledge transfer, such that it can use that knowledge in its own production environment [7].

The transfer of technology project is inherently a very costly project [7], as such companies attempting such projects must be aware of all the risks that are included in such projects. Companies often need to measure the manufacturing readiness level of the technology in question to have a decent understanding of the project that will be done and all its risks. Transfer of technology is heavily linked with the concept of innovation [8]. It has been shown that without proper assessment of technology readiness, innovation projects can lead to time wasting, customer disappointment, and loss of income [9].

The US Deputy Under Secretary of Defense at the Joint Defense Manufacturing Panel in 2009 stated that manufacturing readiness is the ability to utilize manufacturing, production, quality assurance and industrial functions to achieve a certain operational capability that can fulfill a mission's needs. Manufacturing readiness is one of the key success factors in a technology transfer project [10], as success in technology transfer projects can be achieved through effective management of two main risk areas, namely risks from product technology and manufacturing capability immaturities [11].

The Department of Defense has developed a procedure to assess the manufacturing readiness parameter of certain industries through the Manufacturing Readiness Assessment (MRA) Procedure. This procedure uses the Manufacturing Readiness Level (MRL), which is a metric designed to illustrate manufacturing readiness level and used for assessing maturity and manufacturing risks of a technology [10]. MRL is comprised of 10 levels, denoted by MRL1 to MRL10. Each level is illustrated through a number of criteria which were designed as a prerequisite for the corresponding manufacturing readiness level. There exists a total of 9 criteria and 22 sub-criteria that can be used as general metrics for the assessment of manufacturing readiness level. The definition of each MRL levels can be seen in Table 1.

Table 1. Definition of MRL levels

MRL level	Definition
1	*Basic manufacturing implications identified*
2	*Manufacturing concept identified*
3	*Manufacturing proof of concept developed*
4	*Capability to produce the technology in a laboratory environment*
5	*Capability to produce prototype components in a production relevant environment*
6	*Capability to produce a prototype system or subsystem in a production relevant environment*
7	*Capability to produce a system, subsystem, or components in a production representative environment*
8	*Pilot line capability demonstrated. Ready to begin low rate production*
9	*Low rate production demonstrated. Capability in place to begin full rate production*
10	*Full rate production demonstrated and lean practices in place*

3 Research Methodology

This research is done in Indonesian Aerospace, the only aerospace manufacturing company in Indonesia, specifically by using the TSAA technology as the object to be assessed. Figure 1 illustrates the main steps of this research. The research is started by a preliminary survey and interview. The aim of this step is to gather the context for the research, namely the aspects of the TSAA Project. This includes interviews with the stakeholders of the program and gathering of relevant data such as the aims of the project, steps of the project, etc.

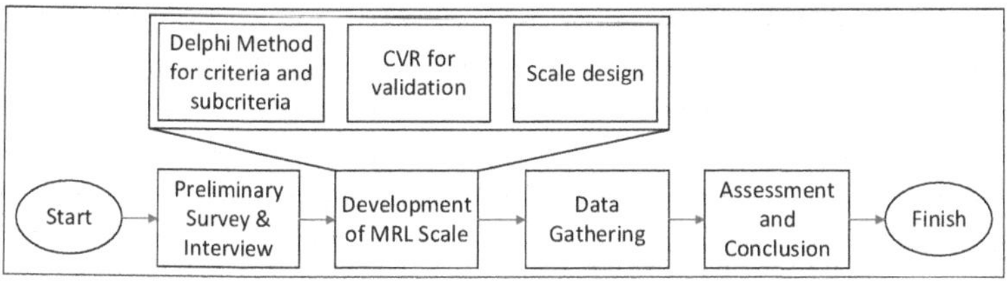

Fig. 1. Main steps of the research

3.1 Development of MRL Scale

The generic criteria and sub-criteria from the original MRL model are once again tested through the Delphi method. Experts chosen for the method consists of 6 people (4 experts from the company and 2 academic experts). These experts are given questionnaires indicating their approval of the criteria and sub-criteria to be included in the final measuring tool. The Delphi method is used mainly because it facilitates the formation of consensus without face-to-face meeting between the experts [12].

The questionnaires are validated using Content Validity Ratio Method (CVR), where each criterion is declared valid and thus not eliminated if its CVR value is greater than 0. The CVR value is computed through Eq. 1 (N = number of experts, n_e = number of experts agreeing that an item is 'important')

$$\mathrm{CVR} = \frac{n_e - N/2}{N/2} \tag{1}$$

This value is reached only when more than half of the experts have agreed that the item is important to be measured [13]. The final criteria to be included in the assessment tool are: Technology and Industry Base, Design, Production Cost and Financing, Materials, Process and Control Capability, Quality Management, Manufacturing Labor, Facilities, and Manufacturing Management. Each criterion is further broken down into up to 4 sub-criteria.

3.2 Data Collection and Results

The developed assessment tool consists of 10 manufacturing readiness level with corresponding indicators that illustrates each level. The indicators are formed as questions that were derived from the corresponding criteria and sub-criteria for each level. In total there are 524 indicator questions for all 10 levels of the tool. The questions were then collected into questionnaires which were sent to 10 other experts consisting of the stakeholders (top and middle management) for the TSAA program. Each respondent must then assess the level of each indicator question, with a scale from 0% to 100%, where 0% indicates that the program has never done the activity and 100% indicates that the activity has been fully done by the program. The scale is further divided into increments of 20% (k = 0%, 20%, 40%, …, 100%) so that the resulting data could be further collected into one value by using geometric mean [14]. Afterwards, the resultant value of each indicator will be used to calculate the mean of the corresponding level.

The assessment is done sequentially for each MRL level (starting from MRL1 to MRL10) and the assessment must be stopped if the requirements for the current level is judged as inadequate to continue to the next level (lower than 90%). This limit is obtained through the consensus of experts by way of focus group discussions involving senior managers of the TSAA Project. Table 2 shows the result for the MRL assessment, while Table 3 shows the truncated table from the assessment for the MRL3 level.

Table 2. Result of MRL assessment

MRL level	Mean value	Status
MRL1	90.5%	Achieved
MRL2	90.1%	Achieved
MRL3	88.6%	Not yet achieved

Table 3. Truncated table for MRL3 level assessment

Indicator code	Respondents										Geometric mean
	1	2	3	4	5	6	7	8	9	10	
3.K1.SK.1.1	80%	100%	100%	100%	80%	100%	100%	80%	100%	100%	94%
3.K1.SK.1.2	80%	100%	100%	100%	60%	80%	100%	80%	100%	100%	89%
Mean of criteria 1											**91.2%**
3.K2.SK.1.1	80%	100%	100%	100%	60%	100%	100%	80%	100%	100%	91%
...											
3.K2.SK.2.3	80%	100%	80%	100%	80%	80%	80%	80%	40%	80%	78%
3.K2.SK.2.4	80%	100%	80%	100%	80%	80%	100%	80%	100%	100%	89%
Mean of criteria 2											**87.0%**
3.K3.SK.1.1	100%	80%	100%	80%	80%	100%	100%	80%	100%	80%	89%
...											
3.K3.SK.2.2	80%	80%	100%	80%	80%	80%	100%	80%	100%	80%	86%
3.K3.SK.3.1	80%	100%	100%	100%	80%	80%	100%	80%	80%	80%	87%
Mean of criteria 3											**85.9%**
3.K4.SK.1.1	80%	100%	100%	100%	80%	80%	100%	80%	80%	80%	87%
...											
3.K4.SK.4.3	80%	100%	100%	100%	80%	100%	100%	80%	100%	100%	94%
3.K4.SK.4.4	100%	100%	80%	100%	80%	80%	80%	80%	100%	100%	89%
Mean of criteria 4											**93.6%**
3.K5.SK.1.1	60%	80%	100%	80%	80%	100%	100%	80%	100%	80%	85%
...											
3.K5.SK.2.2	100%	80%	100%	80%	80%	80%	100%	80%	100%	80%	87%
3.K5.SK.3.1	80%	80%	100%	80%	80%	80%	100%	80%	100%	80%	86%
Mean of criteria 5											**86.6%**
3.K6.SK.1.1	80%	100%	100%	100%	80%	100%	100%	80%	80%	100%	91%
3.K6.SK.1.2	80%	100%	100%	100%	80%	80%	100%	80%	80%	100%	89%
Mean of criteria 6											**87.3%**
3.K7.SK.1.1	80%	100%	100%	100%	100%	100%	100%	80%	100%	100%	95.6%
Mean value for MRL3											**89.6%**

It can be seen that the during the measurement of the MRL3 level the mean of its indicator value is only 89.6% (lower than 90%). As such, the requirements for the MRL3 level has not been met and assessment for the levels above MRL3 cannot be conducted. It can be concluded that the TSAA project has only achieved the MRL2 level.

4 Remarks and Conclusion

The preliminary survey has previously indicated that the TSAA technology would be used for low production rate (corresponding to MRL level 8), while the assessment concluded that the current MRL level of the technology is only at MRL level 2. As such, there is still a need for a vast improvement of the technology's capabilities.

In total this research has produced an MRL assessment tool with 10 levels, 9 criteria, 22 sub-criteria and a total of 524 indicator questions from the level MRL1 to MRL10. In this research the participants are only asked 49 indicator questions because the assessment is stopped at the MRL3 level. It is clear that the rest of the questions are still valid and can be used if in the next assessment the company has met the goals for achieving the MRL3 level. The assessment tool must be used periodically during the course of the transfer technology project, although the optimal period for the assessment should vary with the complexity and scope of the assessed project.

References

1. Kalpakjian, S., Schmid, S.R.: Manufacturing Engineering and Technology, 6th edn. Pearson, New Jersey (2009)
2. Pusat Komunikasi Publik: Rencana Induk Pembangunan Industri Nasional 2015–2035. Pusat Komunikasi Publik Kementerian Perindustrian. Jakarta (2015)
3. Museux, F., Theilmann, R.: Introducing more eco-efficient chemical treatment for aircraft structure towards a chromate free airbus. Flight Airworth. Support Technol. 2–9 (2009)
4. Kwon, C., Chun, B.G.: The effect of strategic technology adoptions by local firms on technology spillover. Econ. Model. **51**, 13–20 (2015)
5. TRsG Inc., Excellence in Technology Transfer & Intellectual Property. http://teamtrsg.com/tt%20def.html
6. Rogers, E.M.: The nature of technology transfer. Sci. Commun. **23**, 323–341 (2002)
7. Maskus, K.E.: Encouraging International Technology Transfer. ICTSD and UNCTAD, Geneva (2004)
8. Dubickis, M., Gaile-Sarkane, E.: Perspectives on innovation and technology transfer. Procedia Soc. Behav. Sci. **215**, 965–970 (2015)
9. Clausing, D., Holmes, M.: Technology readiness. Res. Technol. Manag. **53**, 52–59 (2015)
10. Joint Defense Manufacturing Technology Panel: Manufacturing Readiness Assessment (MRA) Deskbook. US Department of Defense (2009)
11. US Department of Defense: Manufacturing Readiness Level (MRL) Deskbook. US Department of Defense (2017)
12. Meyer, M.A., Booker, J.M.: Eliciting and Analyzing Expert Judgment. Academic Press Limited, London (2001)
13. Lawshe, C.: A quantitative approach to content validity. Pers. Psychol. **28**, 563–575 (1975)
14. Sekaran, U., Bougie, R.: Research Methods for Business, 7th edn. Wiley, Chichester (2016)

Measurement of Technoware and Humanware Readiness to Fulfill SNI 07-2052-2002 in a Steel Manufacturing Company

Violla Tania, Praditya Ajidarma(✉), Mohammad Mi'radj Isnaini, and Dradjad Irianto

Industrial Engineering Program Study, Bandung Institute of Technology, Jl. Ganesha10, Bandung, West Java 40132, Indonesia
ajidarmap@itb.ac.id

Abstract. In manufacturing industry, technoware and humanware play a key role in production floor. Both factors define the extent of technological advancement in a certain manufacturing company. This study took place in a producer of reinforcing steel bars (rebar) in Jakarta, Indonesia. This research aims to analyze the gap between the industry standard and actual capabilities of technoware and humanware in the company. Both elements are measured by Quality Function Deployment (QFD), which consists of three subsequent steps. First, the required level of technoware and humanware in the industry standard is studied according to the Indonesian National Standard (SNI). The output quality and production process are standardized based on the customer's needs and benefits for the former, and the technical response for the latter. Second, the actual technological advancement in the firm is measured. The data gathered from the first two steps is further analyzed, based on technoware and humanware readiness measurement with regard to the SNI standards using two matrices. The calculation on planning matrix shows the level of technological component advancement with regard to industry standard, while the calculation on technical matrix indicates the effectivity of production process with regard to the fulfillment of quality standards. The result shows that the company's readiness level of technoware and humanware has fulfilled the required standard, and it becomes a basis to further develop technological components and to improve the production process of reinforcing steel bars.

Keywords: Quality Function Deployment · Technoware · Humanware · Technological readiness · Steel industry

1 Introduction

In steel manufacturing industry, high percentage of billet defects consequently entails profit loss. This research is conducted at steel reinforcing bar (rebar) manufacturer in Jakarta. The steel manufacturing company on which this study took place has a maximum 0.25% billet-defect policy, a general standard in the industry [1]. However, the company has not been able to maintain such standard. The current high defect rate indicates a threatening issue in the fulfillment of product quality standards.

M. N. Osman Zahid et al. (Eds.): iMEC-APCOMS 2019, LNME, pp. 48–53, 2020.
https://doi.org/10.1007/978-981-15-0950-6_8

There are four aspects which potentially contribute such problem: material, machine and equipment, human labor, and procedure. Based on previous research, expert interviews, and direct observation on the company suggested that the two defect-inducing factors are machine and human labor. Therefore, a method of Quality Function Deployment (QFD) is proposed to measure the levels of technoware and humanware in the steel manufacturing company, and further, to compare the current state of facility and operator capability with the industry standards. The industry follows a nationwide Indonesian National Standard (SNI) 07-2052-2002, which is specifically designed for concrete-reinforcing bars (rebar) industry.

The production processes of rebar manufacturing in the company are as follows:

Table 1. Core processes in reinforcing bar (rebar) production [2].

Process	Description	Process	Description
1	Loading steel scrap into buckets	8	Billet cutting
2	Steel scrap melting	9	Billet cooling and inspection
3	Pouring molten EAF steel into ladle	10	Billet reheating process
4	Bubbling process	11	Rolling process
5	Alloying process	12	Rebar cutting
6	Temperature and sample inspection	13	Tempcore process
7	Molten steel casting	14	Rebar cooling and inspection

The problem of this research is formulated as follows: is there any gap, in either technoware and humanware, between the steel manufacturing company and industry standards? If any gap is identified, what are the necessary steps to reduce such discrepancy? Thus, the objective of this research is as follows:

1. To identify the gap between technoware and humanware level currently possessed by the company and the levels required by industry standards
2. To recommend solutions that increase the fulfillment of customer requirement by improving current technological components of the company.

2 Literature Review

Standard is a technical specification or a procedure which includes rules and method, arranged based on a consensus of every party involved which considers the requirement of safety, security, well-being, environment, advancement of science and technology, as well as experience, recent and future development, that is established to achieve an optimal output from a process. By definition, standardization aims to achieve overall economical condition that is optimal and beneficial for every segment in society [3].

United Nations Economic and Social Commission for Asia and the Pacific (UNESCAP) in 1989 divided technology into four rudimentary components: technoware, humanware, infoware, and orgaware. Identification of the extent of technological advancement can be conducted by observing dynamical interaction between technological components [4]. Technology is a crucial aspect for any industry as it becomes a basis to develop business competitiveness with respect to market competition [5].

Quality Function Deployment is a structured methodology used to plan and develop a product specification based on consumer needs and wants; it is also a tool to systematically evaluate if a product or a service offers a capability to fully satisfy the consumers' requirements. Models to measure the level of technoware and humanware technological components in industry with regard to Indonesian national standard has been studied: one model is developed to measure the technoware aspect [7], and another is made to measure the humanware aspect [8].

3 Methodology

The methodology of this study comprises of three major steps. First, this study measured technoware and humanware levels, both in its current industry and the required standard. Second, the House of Quality (HOQ) framework is applied and Technical Matrices of both components are attained. Lastly, analysis is drawn from the result.

3.1 Measurement of Technoware and Humanware Levels on Industry and the National Standards

The production of steel rebar must adhere to the clauses of SNI 07-2052-2002 which regulates quality and tagging requirement. This study summarizes every clause on the standard into 9 primary clauses, which is referred as customer needs and benefits. This research also defines 14 core production process of steel rebar (shown in Table 1) based on observation, secondary data, and expert interview, which is henceforth defined as technical response. Further, a mapping is conducted on the correlation between the technical response and the customer needs and benefit on Table 2. The level of correlation varies from strong (9), medium (3), and weak (1).

Table 2. Correlation table between customer needs-benefits and the technical response.

Standard of SNI 07-2052-2002 (customer needs and benefit)	Rebar production process (technical response)													
	Process number (according to Table 1)													
	1	2	3	4	5	6	7	8	9	10	11	12	13	14
Survey of steel bar			3	1	1	1			3		9		3	3
Surface of threaded and plain rebar			3	1	1	1			3		9			3
Diameter of threaded and plain rebar											9			3
Dimension of threaded and plain rebar											9			3
Weight of threaded and plain rebar											9			3
Length of steel bar								1				9		3
Mechanical trait based on pull-out test	3			1	9	9							9	
Mechanical trait based on bend test	3			1	9	9							9	
Markings on rebar											9			

Furthermore, this research calculates the level of technoware and humanware attainment based on experts' judgments and technological advancement criteria of the United Nations Economic and Social Commission for Asia-Pacific (UNESCAP). For the customer needs and benefit, this study filters any standard clauses for steel rebar which are related to the production process, and further defines them as standard requirement. For the technical responses, the technoware and humanware level in industry are measured based on UNESCAP criteria. The assessment is conducted on the aspects of machinery, equipment, and human resource, in relation with the production process of steel rebar.

3.2 Fulfillment of Industrial Technoware and Humanware with Regard to the National Standards

The fulfillment of technoware and humanware in a company with regard to the standard is calculated by the Technical Matrix, which is the final step of House of Quality (HOQ) framework. The matrix shows how much the production process of steel manufacturing industry has managed to attain the necessary standard. The technical matrices for technoware and humanware are shown on Tables 3 and 4, respectively.

Table 3. Technical matrix for technoware.

Production process (technoware)	Process number (according to Table 1)													
	1	2	3	4	5	6	7	8	9	10	11	12	13	14
Contribution value	0.31	0	0.06	0.12	0.96	0.96	0	0	0.05	0.10	0.33	0	0.97	0.11
Normalized contribution value	0.08	0	0.01	0.03	0.24	0.24	0	0	0.01	0.03	0.08	0	0.24	0.03
Raw score of the national standard	0.94	0	0.13	0.72	5.73	4.77	0	0	0.19	0.52	1.64	0	5.78	0.44
Raw score of the industry	0.63	0	0.05	0.49	1.92	0.96	0	0	0.11	0.31	1.31	0	3.87	0.44
Gap	0.31	0	0.08	0.23	3.81	3.81	0	0	0.08	0.21	0.32	0	1.90	0
Gap percentage	33%	–	60%	32%	67%	80%	–	0%	42%	40%	20%	0%	33%	−0.6%

Table 4. Technical matrix for humanware.

Production process (humanware)	Process number (according to Table 1)													
	1	2	3	4	5	6	7	8	9	10	11	12	13	14
Contribution value	0.31	0	0.06	0.12	0.96	0.96	0	0	0.05	0.10	0.33	0	0.97	0.11
Normalized contribution value	0.08	0	0.01	0.03	0.24	0.24	0	0	0.01	0.03	0.08	0	0.24	0.03
Raw score of the national standard	1.57	0	0.28	0.59	5.75	4.77	0	0	0.25	0.42	1.77	0.02	5.78	0.49
Raw score of the industry	1.57	0	0.39	0.61	6.70	1.92	0	0	0.28	0.63	1.97	0.05	5.80	0.66
Gap	0	0	−0.1	−0.02	−0.96	2.86	0	0	−0.03	−0.21	−0.2	−0.02	−0.03	−0.2
Gap percentage	0%	–	−40%	−3%	−17%	60%	–	33%	−11%	−50%	−11%	−1	0%	−35%

4 Analysis

In this section, recommendations are made to further improve the production activity. According to Table 4, not every production process fulfills the necessary standards of steel rebar manufacturing. This study focuses on improving technoware and humanware aspects related to the unfulfilled production processes.

4.1 Recommendations on Technoware Components

Out of 14 production processes, nine did not abide by the standard technological advancement level. Those nine processes account for 96.4% cumulatively in terms of technological contribution. Further, pareto analysis is conducted to select the most influential processes need to be improved. The analysis narrows down the 9 processes into 3: tempcore process, alloying process, and temperature and sample inspection.

The recommendation must account for the current state of the industry. On the tempcore and alloying process, the parameter tuning is currently conducted by experienced operators based on their expertise. However, this activity is never logged systematically. This study recommends the industry to use a comprehensive record-keeping procedure to log the parameter tuning under different conditions. Compared to one-time tuning based on best practices, the formal record-keeping procedure is better; it is more traceable and made troubleshooting easier.

The next improvement could be made on temperature and sample inspection. Currently, the industry used thermocouple and sampoline to manually conduct the process. A better alternative would be SSC liquid sampler which automatically does the job.

4.2 Recommendations on Humanware Components

In terms of humanware, out of 14 production process, two were unable to comply with the required standard: temperature and sample inspection and billet cutting process. The two process contribute to 23,03% of the humanware aspects cumulatively.

Potential improvements could be made by proposing technical training for human operator to match the level of current technological advancement of the machinery. For instance, if the company manage to accommodate a more sophisticated technoware element, it must ensure that human operators are prepared for the change. Otherwise, the expected increase in technological competitiveness by upgrading technoware is not attainable due to humanware being the bottleneck in the process.

5 Conclusion

This study measures, compares, and summarizes the extent of technological readiness in steel rebar manufacturer with regard to the required national standard. The technical matrices, as a final step of HOQ, shows the gap between the technoware and humanware level of the industry and those of the national standard SNI 07-2052-2002. A negative gap indicates a condition when industry exceeds the required standard of expectation, while a positive gap indicates a state when an industry is unable to comply with the required standard. Meanwhile, zero gap implies a condition when industry standard is sufficiently met.

On the technoware aspect, there are one production process with negative gap, four processes with zero gap, and nine processes with positive gap. The processes with positive gap are temperature and sample inspection, alloying process, pouring molten EAF steel into ladle, rebar cooling and inspection, billet reheating, loading steel scrap into buckets, tempcore process, bubbling process, and rolling process. On the humanware components, there are nine production processes with negative gap, three processes with zero gap, and two processes with positive gap. The two processes are temperature and sample inspection, and billet cutting process.

Recommendations are made to improve each technological component. In terms of technoware, improvements are proposed by conducting record-keeping procedure to keep track of varied conditions during tempcore and alloying process. Secondly, further investment in automation is also proposed, particularly in the process related to sampling and inspection. In terms of humanware, the recommendations are focused on the human resource needed in the temperature and sample inspection. A series of training programs must be provided to ensure every operator will be able to operate each technology and machinery optimally.

References

1. Rahadi, D., Uliana, A.: SNI 2052-2017 Baja Tulangan Beton. Badan Standarisasi Nasional, Jakarta (2017)
2. Metal processing and metal working industry. In: Encyclopaedia of Occupational Health and Safety http://www.ilocis.org/documents/chpt82e.htm. Accessed 26 July 2019
3. Purwanggono, B., Abduh, S., Nurjanah, Habibie, F.H., Trilaksani, W., Bakhtiar, A., Haryono, T.: Pengantar Standarisasi. Badan Standarisasi Nasional, Jakarta (2009)
4. Purwasasmita, M.: Konsep Teknologi. Penerbit ITB, Bandung (2010)
5. Wiratmadja, I.I., Govindaraju, R.: Analysis of the influence of technology on the business performance of Rattan processing SME's in South Kalimantan In: Proceedings on the 11th Asia Pacific Industrial Engineering and Management Systems, Melaka (2010)
6. Cohen, L.: Quality Function Deployment: How to Make QFD Work for You. Addison Wesley, Boston (1995)
7. Putra, V.C.: Perancangan Model Kesesuaian Komponen Teknologi (Technoware) Industri terhadap Pemenuhan Standar Produk. Institut Teknologi Bandung, Bandung (2012)
8. Sinuhaji, L.H.: Perancangan Model Pengukuran Kompetensi Sumber Daya Manusia (SDM) Industri dalam Memenuhi Standar Produk. Institut Teknologi Bandung, Bandung (2012)
9. Saaty, T.L.: Decision making with the analytic hierarchy process. Int. J. Serv. Sci. **1**(1), 83 (2008)
10. Indrawati, S.W.: Analisis Pengaruh Komponen Teknologi Technoware, Humanware, Infoware, dan Orgaware Terhadap Faktor Utama Daya Saing Industri Kecil. Institut Teknologi Bandung, Bandung (2003)
11. Koch, R.: The 80/20 Principle. Nicholas Brealey Publishing, London (1998)
12. Fauzi, I., Muharram, H., Irianto, D.: Design of measurement for evaluating readiness of technoware components to meet the required standard of products. In: APCOMS-IMEC 2017, p. 319. IOP Publishing, Bandung (2017)

Fuzzy Analytical Hierarchy Process with Unsymmetrical Triangular Fuzzy Number for Supplier Selection Process

Irene Septin Maharani, Ririn Diar Astanti, and The Jin Ai(✉)

Department of Industrial Engineering, Universitas Atma Jaya Yogyakarta, Sleman 55281, Indonesia
{ririn.astanti, the.jinai}@uajy.ac.id

Abstract. In the area of Multi Criteria Decision Making, it is well known that Analytical Hierarchy Process (AHP) is widely used. In the AHP, the weight of criteria and alternatives are obtained from the expert judgment doing pairwise comparison. Fuzzy AHP (FAHP) is developed in order to overcome situation in which expert has difficulty to give clear judgment on the comparison. FAHP is usually developed using a set of Triangular Fuzzy Number (TFN). This research is proposed a FAHP with unconventional TFN, which allow the expert give his/her own fuzzy number for each comparison. We name the number as the Unsymmetrical Triangular Fuzzy Number, in which the membership function is still in the shape of triangular but the shape is not symmetric. The complete methodology of the proposed FAHP is developed with reference to the fuzzy logarithmic least square method (LLSM). A case study for supplier selection process is provided in order to show the applicability of the proposed method.

Keywords: Analytic Hierarchy Process · Fuzzy AHP · Triangular Fuzzy Number · Unsymmetrical TFN · Multi Criteria Decision Making

1 Introduction

This research is motivated by a case study of supplier selection process, which is very common problem in the area of Multi Criteria Decision Making (MCDM), especially with Analytical Hierarchy Process (AHP) methodology. It is known that the simplest structure of decision making with AHP consists of three level, which are objective, criteria, and alternatives. The AHP methodology helps the decision maker to set the weight of each criterion and alternative based on the pairwise comparisons among criteria and alternatives. The pairwise comparisons are obtained from the expert judgment, and comparison on two items are expressed [1].

Sometimes expert has difficulty to give clear judgment on the pairwise comparison. In order to overcome this situation, the AHP was extended into Fuzzy AHP (FAHP) [2] in which the comparison is expressed in the form of Triangular Fuzzy Number (TFN), i.e. represented by three values (l, m, u) [3, 4]. The TFN are usually selected from these values: (1, 1, 1), (1, 2, 3), (2, 3, 4), …, (7, 8, 9), (9, 9, 9). In the perspective of the expert, l, m, and u are the lower, middle, and upper value of judgment, respectively.

M. N. Osman Zahid et al. (Eds.): iMEC-APCOMS 2019, LNME, pp. 54–59, 2020.
https://doi.org/10.1007/978-981-15-0950-6_9

They represent the smallest, most-likely, and the biggest possible judgment of the expert.

Our case study presented in Sect. 4 show that experts may use different value of TFN whenever they are questioned in detail about their smallest, most-likely, and biggest possible judgment, i.e. in the form of (1, 1, 2), (1/3, 2, 3). Also, among all comparisons, expert confident to give crisp judgment instead of fuzzy judgment for some comparisons. For example, the judgment is in the form of (3, 3, 3) or (7, 7, 7).

In order to accommodate this situation, this research explore the possibility to use Unsymmetrical Triangular Fuzzy Number (UTFN) in the FAHP. The UTFN has triangular shape membership function, but the shape is not symmetric. The UTFN is applied in the established FAHP methodology, which is the fuzzy logarithmic least square method (LLSM) [5], in order to be utilized to make decision.

This paper is organized as follow. Section 2 introduces the FAHP. Section 3 presents the definition of UTFN and how it is integrated into FAHP. Section 4 demonstrates the case study using FAHP and UTFN. Finally, Sect. 5 concludes the result obtained in this paper.

2 Fuzzy Analytical Hierarchy Process

In term of methodology, various fuzzy AHP methods have been proposed in the literature, such as: extent analysis method [6–8], fuzzy logarithmic least square method (LLSM) [5, 9, 10], and grey relational analysis [11–13]. In term of applications, FAHP has been applied in various fields such as: personnel selection [14], part-machine grouping [15], weapon selection [16], selection of energy alternatives [17], supply chain risk assessment [18], and passenger aircraft type selection [19], and supplier selection [20–22].

This research is using fuzzy LLSM method [5]. This method is started with triangular fuzzy comparison matrices. For each matrix, the fuzzy comparison is then converted into normalized local fuzzy weight by solving a constrained nonlinear optimization model, in which the parameter of this model are the value of the matrix. After all the local fuzzy weight is obtained, another constrained nonlinear optimization model is solved in order to obtain the global weight. In order to avoid duplication of the models, reader may refer to the two formulations of this LLSM model [5].

3 Unsymmetrical Triangular Fuzzy Number

As defined in introduction, UTFN is used for representing judgment of preference. UTFN is defined by three values (l, m, u) where $l \leq m \leq u$. It is noted that the value of $(m - l)$ is not necessary the same as the value of $(u - m)$ as TFN. The membership function of UTFN is similar with general triangular fuzzy number.

$$\mu_M(x) = \begin{cases} 0, & x \leq l \\ 1 - \frac{m-x}{m-l}, & l < x < m \\ 1, & x = m \\ 1 - \frac{x-m}{u-m}, & m < x < u \\ 0, & x \geq u \end{cases} \quad (1)$$

In order to obtain these three values, expert is being asked the smallest possible, the most-likely, and the biggest possible judgment of preference. By this definition, UTFN is able to provide various possibilities for expert judgment, i.e. whenever expert confident with the judgment, i.e. (3, 3, 3), expert face situation similar with TFN, i.e. (2, 3, 4), expert give interchange preference on l and m, i.e. (1/2, 2, 3), or expert give the same value on l and m (5, 5, 7).

It is noted that this UTFN can be directly applied as the parameter of the constrained nonlinear optimization model for obtaining the local fuzzy weight in the fuzzy LLSM method.

4 The Application of FAHP-UTFN for Supplier Selection

In order to show the applicability of the proposed UTFN, it is applied to a case of selection of resin supplier in a fiberglass-related-product company. The decision maker of this case is the purchasing manager of the company. The decision hierarchy structure of this case only consists of criteria and alternatives. There are four criteria involved, which are price, delivery lead time, payment method, and quality. While there are three alternatives of supplier involved, namely supplier A, supplier B, and supplier C. The decision hierarchy is shown in Fig. 1.

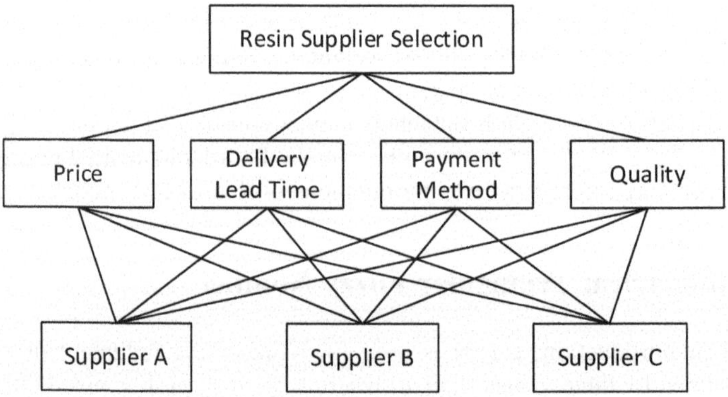

Fig. 1. Decision hierarchy of the resin supplier selection.

It is noted that for each pairwise comparison, the expert is being asked for three different judgments, which are the smallest possible, the most likely, and the biggest possible judgment of preference. After comprehensively interviewing the expert for this case, which is the purchasing manager of the company, the pairwise comparison among criteria and alternatives within each criterion are presented in Tables 1, 2, 3, 4 and 5. It is implied that for this case actually expert have difficulties to give judgment only on few comparisons. Therefore, the UTFN only appear in Tables 1, 3, and 5, in which only 6 UTFN are appearing in the whole comparisons in various form. This example is proving our criticisms mentioned in the Sect. 1.

Table 1. Pairwise comparison of criteria using UTFN.

	Price	Delivery lead time	Payment method	Quality
Price	(1, 1, 1)	**(1, 1, 2)**	(1, 1, 1)	(1/4, 1/4, 1/4)
Delivery lead time	**(1/2, 1, 1)**	(1, 1, 1)	(1, 1, 1)	(1/4, 1/4, 1/4)
Payment method	(1, 1, 1)	(1, 1, 1)	(1, 1, 1)	(1/4, 1/4, 1/4)
Quality	(4, 4, 4)	(4, 4, 4)	(4, 4, 4)	(1, 1, 1)

Table 2. Pairwise comparison of suppliers based on price criteria using UTFN.

	Supplier A	Supplier B	Supplier C
Supplier A	(1, 1, 1)	(6, 6, 6)	(7, 7, 7)
Supplier B	(1/6, 1/6, 1/6)	(1, 1, 1)	(1, 1, 1)
Supplier C	(1/7, 1/7, 1/7)	(1, 1, 1)	(1, 1, 1)

Table 3. Pairwise comparison of suppliers based on delivery lead time criteria using UTFN.

	Supplier A	Supplier B	Supplier C
Supplier A	(1, 1, 1)	**(1/3, 1/2, 3)**	(1/9, 1/9, 1/9)
Supplier B	**(1/3, 2, 3)**	(1, 1, 1)	(1/9, 1/9, 1/9)
Supplier C	(9, 9, 9)	(9, 9, 9)	(1, 1, 1)

Table 4. Pairwise comparison of suppliers based on payment method criteria using UTFN.

	Supplier A	Supplier B	Supplier C
Supplier A	(1, 1, 1)	(9, 9, 9)	(9, 9, 9)
Supplier B	(1/9, 1/9, 1/9)	(1, 1, 1)	(9, 9, 9)
Supplier C	(1/9, 1/9, 1/9)	(1/9, 1/9, 1/9)	(1, 1, 1)

Table 5. Pairwise comparison of suppliers based on quality criteria using UTFN.

	Supplier A	Supplier B	Supplier C
Supplier A	(1, 1, 1)	**(1, 1, 2)**	(9, 9, 9)
Supplier B	**(1/2, 1, 1)**	(1, 1, 1)	(9, 9, 9)
Supplier C	(1/9, 1/9, 1/9)	(1/9, 1/9, 1/9)	(1, 1, 1)

Using the mathematical programming of Fuzzy LLSM, i.e. formulation in [5], and using mathematical programming software, i.e. LINGO version 18, the fuzzy global weight of Supplier A, B, C are 0.5267, 0.3057, and 0.1676, respectively.

5 Conclusion

This paper proposed an unsymmetrical triangular fuzzy number which allow the experts give their own fuzzy number while doing pairwise comparison. From the application example, it is known that this kind of number is rarely being used even it is provided by the methodology. Current existing FAHP methodology is able to handle this unsymmetrical triangular fuzzy number.

References

1. Saaty, T.L.: The Analytic Hierarchy Process. McGraw-Hill, New York (1980)
2. van Laarhoven, P.J.M., Pedrycz, W.: A fuzzy extension of Saaty's priority theory. Fuzzy Sets Syst. **11**(1–3), 229–241 (1983)
3. Mardani, A., Jusoh, A., Zavadskas, E.K.: Fuzzy multiple criteria decision-making techniques and applications - two decades review from 1994 to 2014. Expert Syst. Appl. **42**(8), 4126–4148 (2015)
4. Singh, B.: Analytical hierarchical process (AHP) and fuzzy AHP applications - a review paper. Int. J. Pharm. Technol. **8**(4), 4925–4946 (2016)
5. Wang, Y.M., Elhag, T.M., Hua, Z.: A modified fuzzy logarithmic least squares method for fuzzy analytic hierarchy process. Fuzzy Sets Syst. **157**(23), 3055–3071 (2006)
6. Chang, D.-Y.: Applications of the extent analysis method on fuzzy AHP. Eur. J. Oper. Res. **95**(3), 649–655 (1996)
7. Zhu, K.-J., Jing, Y., Chang, D.-Y.: Discussion on extent analysis method and applications of fuzzy AHP. Eur. J. Oper. Res. **116**(2), 450–456 (1999)
8. Wang, Y.-M., Luo, Y., Hua, Z.: On the extent analysis method for fuzzy AHP and its applications. Eur. J. Oper. Res. **186**(2), 735–747 (2008)
9. Mikhailovv, L.: A fuzzy programming method for deriving priorities in the analytic hierarchy process. J. Oper. Res. Soc. **51**(3), 341–349 (2000)
10. Wang, Y.-M., Yang, J.-B., Xu, D.-L.: A two-stage logarithmic goal programming method for generating weights from interval comparison matrices. Fuzzy Sets Syst. **152**(3), 475–498 (2005)
11. Samvedi, A., Jain, V., Chan, F.T.S.: An integrated approach for machine tool selection using fuzzy analytical hierarchy process and grey relational analysis. Int. J. Prod. Res. **50**(12), 3211–3221 (2012)
12. Gumus, A.T., Yesim Yayla, A., Çelik, E., Yildiz, A.: A combined fuzzy-AHP and fuzzy-GRA methodology for hydrogen energy storage method selection in Turkey. Energies **6**(6), 3017–3032 (2013)
13. Hatefi, S.M., Tamošaitiene, J.: Construction projects assessment based on the sustainable development criteria by an integrated fuzzy AHP and improved GRA model. Sustainability **10**(4), Article no. 991 (2018)
14. Güngör, Z., Serhadlioğlu, G., Kesen, S.E.: A fuzzy AHP approach to personnel selection problem. Appl. Soft Comput. J. **9**(2), 641–646 (2009)

15. Güngör, Z., Arikan, F.: Application of fuzzy decision making in part-machine grouping. Int. J. Prod.'Econ. **63**(2), 181–193 (2000)
16. Dağdeviren, M., Yavuz, S., Kilinç, N.: Weapon selection using the AHP and TOPSIS methods under fuzzy environment. Expert Syst. Appl. **36**(4), 8143–8151 (2009)
17. Kahraman, C., Kaya, I.: A fuzzy multicriteria methodology for selection among energy alternatives. Expert Syst. Appl. **37**(9), 6270–6281 (2010)
18. Ganguly, K.K., Kumar, G.: Supply chain risk assessment: a fuzzy AHP approach. Oper. Supply Chain Manag. **12**(1), 1–13 (2019)
19. Dožić, S., Lutovac, T., Kalić, M.: Fuzzy AHP approach to passenger aircraft type selection. J. Air Transp. Manag. **68**, 165–175 (2018)
20. Chan, F.T.S., Kumar, N., Tiwari, M.K., Lau, H.C.W., Choy, K.L.: Global supplier selection: a fuzzy - AHP approach. Int. J. Prod. Res. **46**(14), 3825–3857 (2008)
21. Shaw, K., Shankar, R., Yadav, S.S., Thakur, L.S.: Supplier selection using fuzzy AHP and fuzzy multi-objective linear programming for developing low carbon supply chain. Expert Syst. Appl. **39**(9), 8182–8192 (2012)
22. Jain, V., Sangaiah, A.K., Sakhuja, S., Thoduka, N., Aggarwal, R.: Supplier selection using fuzzy AHP and TOPSIS: a case study in the Indian automotive industry. Neural Comput. Appl. **29**(7), 555–564 (2018)

Analysis of Humanware Readiness Level for a Technology Transfer Process: Case Study in Arms Manufacturing Industry

Lucky Apriandi[1], Praditya Ajidarma[2(✉)], Fariz Muharram Hasby[2], and Dradjad Irianto[2]

[1] Engineering Management Program Study, Bandung Institute of Technology, Jl. Ganesha 10, Bandung, West Java 40132, Indonesia
[2] Industrial Engineering Program Study, Bandung Institute of Technology, Jl. Ganesha 10, Bandung, West Java 40132, Indonesia
ajidarmap@itb.ac.id

Abstract. Humanware plays a key role during a technology transfer within a manufacturing company. This research aims to measure humanware readiness level during a technology transfer process on department level and observe its implementation on corporate level. Information gained from the result of the measurement is to be used for developing humanware competency improvement training programs based on Bloom's Taxonomy for its material design. The object of this study is an Indonesian manufacturer of military products. In 2018, the company underwent a technology transfer process through its Quality, Occupational Health, Safety, and Environment System (QOHSE) Department. They aim to accommodate three management system standards: quality, environmental, and occupational health and safety into its management system. The methodology used in this research is referenced from humanware readiness model. The result shows the limit values from each of the four technology transfer phases, which are 2300, 3314, 3927, and 4547, correspondingly. Out of four subsequent phases of technology transfer with regard to the IMS implementation, each respondent manages to reach at least the first phase. The result indicates the necessity of further training program related to the three remaining humanware readiness subcriteria: empathy, work values and ethics, and external awareness.

Keywords: Humanware readiness model · Technology transfer · Integrated Management System · Humanware training program

1 Introduction

An Indonesian arms manufacturing company aims to implement a comprehensive system of quality management, environmental management, and occupational health and safety management, from its upstream operation to its downstream production. The system is desired to regulate work procedure and instruction as a necessary to meet the market's demand. The fulfillment of three management system, for instance ISO

M. N. Osman Zahid et al. (Eds.): iMEC-APCOMS 2019, LNME, pp. 60–66, 2020.
https://doi.org/10.1007/978-981-15-0950-6_10

9001:2015 for Quality Management System, is imperative. The company would not be able to acquire any clients if any of such standards is not met.

The joint implementation of Quality, Occupational Safety and Health, and Environmental Safety Management, or commonly abbreviated as QOHSE, is a process of technology transfer. Currently there is only one division responsible for QOHSE implementation in the arms manufacturing company. As the company aims to implement such policy to the whole company, including its upstream and downstream operation, the technology needs to be transferred from the division onto the whole corporation. According to the International Code of Conduct on The Transfer of Technology, released by United Nations Conference of Trade and Development (UNCTAD) [1], technology is a systematic knowledge used to design a process to make a product or service. QOHSE is a technology because the knowledge is mostly used as benchmark on the corporate level. The implementation of QOHSE involves a series of activity that promotes consistent practice of the knowledge in every division of the company.

This research aims to study how a transfer of technology is supported by a certain technological component, humanware. Recommendations and additional programs are also proposed to further enable the practice of technology transfer. The scope of this research is limited to the business practice in Indonesian arms manufacturing industry.

2 Literature Review

Mittleman and Pasha [2] proposed that technology transfer is a transfer of knowledge, skill, organization, values, and other assets from a certain source to be implemented in another place with a purpose of problem solving. The level of humanware readiness in a technology transfer is a condition that must be fulfilled by every humanware involved in the process, to ensure that the transfer goes accordingly [3]. It is crucial to assess its readiness level because strategic use of humanware aspects will improve competitive advantages of a manufacturing firm, as proven by various case studies [4, 5]. Training program is an organizational attempt planned to help employees in acquiring sufficient skill sets, expertise, attitude, and knowledge related to their work descriptions, so that the result could be implemented in the workplace [6].

Syafriana [7] has developed a humanware readiness model that is based on generic 1989 UNESCAP technological advancement model, complemented by competence dictionary which is developed by Spencer and Spencer [8], University of Guelph and Georgia Merit System. Furthermore, competence is defined as every basic nature of an individual human being which determine the overall upbringing of such individual. The humanware readiness model [7] measures the extent of humanware capability in terms of 6 criteria: creativity, performance orientation, teamwork orientation, efficiency orientation, risk-facing skills, and time orientation.

This study also develops a training program, previously developed by Noe [6], and base its material from Bloom's Taxonomy, developed by Anderson dan Krathwohl [9]. Prior designing such program, a survey of actual work descriptions in arms manufacturing industry is studied [10]. Furthermore, the majority of data collection in this study is based on expert insights gathered using Delphi technique [11]. Additional implementation of humanware readiness model in electrical machinery company is studied [12].

3 Methodology

This research follows the methodology of humanware readiness model [5]. By implementing such model, this research aims to calculate the threshold score for humanware readiness, and the actual humanware readiness level of current employees who are involved in the technology transfer process. To ensure the compatibility of research model within the research boundary, this study extract the necessary data from on expert respondents. The data collection uses similar approach to Delphi method, with four main respondents, as follows: Department Head of Quality System, Group Head of Quality System, Group Head of Internal Quality Audit, and the Group Head of Occupational Safety, Health, and Environment.

Determining threshold score of humanware readiness in technology transfer requires a weighting process, which should be tailored specifically to the company. The weighting governs coupling improvement technique that compares one humanware readiness criterion with other criteria, and also one subcriterion with other subcriteria. The data is further processed using Analytic Hierarchy Process. After the weight for each criterion and subcriteria of is acquired, the methodology proceeds by conducting interview with the four expert respondents. The interview gives insight on the current state of technology transfer process in the arms manufacturing industry, specifically in the context of the newly-designed QOHSE. The four defined stages of the technology transfer implementation are Technology Utilization (Phase I), Technology Integration (Phase II), Technology Development (Phase III), and Basic Research (Phase IV).

The calculation of humanware readiness threshold score aims to define the scores associated with each stage of technology transfer. The information required to determine the threshold value of humanware readiness is equal to the weight of each criterion and subcriterion, that is previously calculated, and also the degree of assessment of minimum humanware readiness for each technology transfer stage – both data are acquired from the expert respondents. The detailed calculation results of the humanware readiness threshold score is shown on Table 1.

Table 1. Humanware readiness threshold scores for every criteria.

Criteria		Subcriteria	Weight		Stages			
			Local	Global	I	II	III	IV
1	Creativity	Technical skills	0.240	0.040	2	3	4	4
		Analytical thinking	0.459	0.077	2	3	4	5
		Conceptual thinking	0.158	0.026	2	4	4	4
		Creative thinking	0.143	0.024	2	4	4	5
Creativity-oriented criterion score					**0.33**	**0.55**	**0.67**	**0.77**
2	Performance	Will to achieve	0.457	0.114	2	3	4	4
		Customer orientation	0.220	0.055	2	4	4	5
		Continuous learning	0.214	0.053	2	3	4	4
		Individual flexibility	0.110	0.027	2	3	4	4
Performance-oriented criterion score					**0.49**	**0.80**	**0.99**	**1.05**
3	Teamwork	Teamwork ability	0.348	0.066	2	3	4	5
		Relationship building	0.325	0.061	3	4	4	5
		Empathy	0.134	0.025	3	4	4	4
		Work ethics & value	0.193	0.036	2	3	4	4
Teamwork-oriented criterion score					**0.46**	**0.65**	**0.75**	**0.88**
4	Efficiency	Attention to quality, clarity, & performance	0.586	0.073	2	3	3	4
		Resource management	0.414	0.051	3	4	4	5
Efficiency-oriented criterion score					**0.30**	**0.42**	**0.42**	**0.55**
5	Risk facing skills	Risk management	0.302	0.058	2	2	4	4
		Adaptability	0.197	0.038	2	3	4	5
		External awareness	0.286	0.055	3	4	4	5
		Initiative	0.215	0.041	3	4	4	5
Risk-managing skill criterion score					**0.48**	**0.62**	**0.77**	**0.90**
6	Time management	Goal and objective	0.584	0.046	2	3	4	5
		Strategic approach	0.416	0.033	4	4	4	5
Time management-oriented criterion score					**0.22**	**0.27**	**0.32**	**0.40**
Humanware readiness threshold score					**2.30**	**3.31**	**3.93**	**4.55**

The next step of the methodology is determining the actual humanware readiness level of the current employees involved in the technology transfer process. The data is gathered in a series of interview that ask respondents to assess the extent of fulfillment in each subcriteria of humanware readiness using Guttman scale. The degree of sub-criteria assessment is calculated cumulatively, meaning that higher degree of assessment on a certain subcriteria is meaningless as long as there is a lower degree subcriteria that is left unfulfilled. On this phase, the validity test comprises the calculation of Guttman scale reproducibility coefficient and Spearman-rank coefficient of correlation, while reliability testing is conducted using split-half method. The calculation result of humanware readiness level is shown on Table 2.

Table 2. Sampling on the actual humanware readiness level.

Respondent	1	2	3	4	5	6	7	8	9
Humanware readiness score	4.86	4.48	4.78	4.63	4.55	5.00	4.95	2.58	3.84
Stage	IV	III	IV	IV	IV	IV	IV	I	II

4 Analysis

4.1 Analysis of Humanware Readiness Level Analysis Calculation Results

According to Table 2, it is apparent that the lowest score of employees being sampled is equal to 2.58. Based on the calculation on Table 1, which is formally summarized in Table 3, the score of 2.58 fulfill the first stage of technology transfer, or the Technology Utilization stage. Thus, as the sampling is assumed to represent the whole employees involved, it is objectively conclusive to state that all employees in the Departments related to QOHSE in the arms manufacturing company are ready to start delivering the technology transfer process, as most of them already passed the first stage.

Table 3. Humanware readiness threshold score (summary).

Technological transfer stage		Humanware readiness threshold score
I	Technology utilization	Score between 2.300 and 3.314
II	Technology integration	Score between 3.314 and 3.927
III	Technology development	Score between 3.927 and 4.547
IV	Basic research	Score larger than 4.547

4.2 The Development of Humanware Training Program

Additional training program is necessary to ensure the full implementation of technology transfer. The design and development of the training program is referred from

Noe [4], which involves a series of requirement assessment and program planning. The participant of this training program is the non-respondent expert employees which are involved in the technology transfer currently occurring in the QOHSE-related departments in the arms manufacturing company.

Requirement assessment for the program has been analyzed in this study. The assessments comprise of different level of the company: organization as a whole, employees, and job description, which are further described as follows:

1. Organizational level assessment is conducted to achieve a training program that conforms with regard to the arms manufacturing company's goal and objective.
2. Employee-level assessment is conducted to observe the assessment degree of humanware readiness subcriteria that is acquired from non-respondent expert employees; subcriteria which are subject to improvement are empathy, work ethics and value, and external awareness.
3. Assessment on job description aims to determine a variety of training program that are well-suited to each employee's tasks; the assessment shows that there are at least 28 core activities that are related to QOHSE technology transfer, and thus, each of them must be included in the training program.

Training program planning is conducted to determine the direction of the activities, material, methods, and technique being used to deliver the program. The explanation related to the training program planning is as follows:

1. The objective of each training program is defined on a 5-degree scale for each subcriteria of humanware readiness; for instance, in terms of work ethics, a score of 5 will have an objective criterion of "able to give example and demonstrate the implementation of organizational value and ethics", while a lower score of 3 only requires employee to be "able identify the implication of ethics".
2. The training program is designed based on cognitive process in Bloom's Taxonomy, which are revised by Anderson dan Krathwohl [7], with a total of 20 modules being recommended to be included in the proposed training program.

5 Conclusion

This study conducts a measurement of humanware readiness level on the technology transfer process in an Indonesian arms manufacturing industry. The analysis shows that every employee in the departments of Quality, Occupational Safety and Health, and Environmental Safety Management (QOHSE) has achieved the first stage of technology transfer, which is Technology Utilization. Thus, the humanware in the company is determined as capable of transferring the technology required for the implementation of Integrated Management System (IMS). There are three humanware readiness subcriteria that are subject to improvement: empathy, work ethics and values, and external awareness. These subcriteria could be improved using a series of training program which includes a total of 20 modules that govern all the necessary aspects to support the technology transfer process in the arms manufacturing industry.

References

1. United Nations Conference on Trade and Development (UNCTAD): Transfer of Technology: IIA Issues Paper Series. United Nations, Geneva (2001)
2. Ramanathan, K.: An overview of technology transfer and technology transfer models (2008)
3. Gandjar, R., et al.: Measurement of humanware readiness level in the technology transfer process at motorcycle component manufacturing company. In: Proceedings on the 11th Asia Pacific Industrial Engineering and Management Systems Conference (2010)
4. Anagnostopoulos, K., Sioutis, P.: Performance measurement of technology-production base of the firms: ascertaining their strategic competitive advantage. J. Knowl. Econ. **7**, 694–719 (2016)
5. Soosay, C., Nunes, B., Bennett, D., Sohal, A.: Strategies for sustaining manufacturing competitiveness: comparative case studies in Australia and Sweden. J. Manuf. Technol. Manag. **27**(1), 6–37 (2016)
6. Noe, R., et al.: Fundamentals of Human Resource Management, 4th edn. McGraw-Hill/Irwin, NewYork (2011)
7. Syafriana, R.N., Wiratmadja, I.I., Sunaryo, I., Govindaraju, R.: Pengukuran Tingkat Kesiapan Humanware pada Proses Alih Teknologi Untuk Mengembangkan Program 6 Peningkatan Kompetensi Nonteknis Humanware. Tugas Akhir Manajemen Rekayasa Industri ITB (2013)
8. Spencer, L.M., Spencer, S.M.: Competence at Work. Wiley, New York (1993)
9. Anderson, L.W., Krathwohl, D.R.: A Taxonomy for Learning, Teaching, and Assessing: A Revision of Bloom's Taxonomy of Educational Objectives. A.W. Longman, Inc., New York (2001)
10. Departemen Sistem Mutu dan K3LH PT Pindad (Persero): Deskripsi Kerja (2018)
11. Hsu, C.: The Delphi technique: making sense of consensus. Pract. Assess. Res. Eval. **12**(10), 1–8 (2007)
12. Syafriana, R.N., Wiratmadja, I.I., Sunaryo, I., Govindaraju, R.: The measurement of humanware readiness in a technology transfer process (case study in an electrical machinery company). In: Proceedings on the 2nd International Conference on Technology, Informatics, Management, Engineering and Environment, pp. 321–325 (2014)

Inventory Policy for Cross Selling Item

Nadia Laksita Devy, The Jin Ai, and Ririn Diar Astanti(✉)

Department of Industrial Engineering, Universitas Atma Jaya Yogyakarta,
Sleman 55281, Indonesia
ririn.astanti@uajy.ac.id

Abstract. There exists a situation in a retail where the demand of one product generate the demand for other products or it is called as cross selling item. Therefore, in order to minimize total inventory cost, inventory policy needs to be planned according to that. When two products that are correlated are supplied from the same vendor, then the joint replenishment policy need to be applied. The research presented in this paper tries to model the inventory policy for cross selling item.

Keywords: Inventory policy · Total inventory cost · Cross selling item

1 Introduction and Literature Review

The characteristics of products used to meet human needs can vary. One of them is that there is a product that usually uses another product. For example: computer and mouse, shoes and socks, toothpaste and toothbrushes, razor and shaving cream, and much more. These type of products then are said as complementary product. This characteristic will lead to the situation where the demand of one product (i.e., major product) generate the demand for other product (i.e. minor product), not vice versa. These type products are called as cross selling item [1, 2].

The existence of cross selling items raises the consequence that in order to minimize the total inventory cost, it is necessary to pay attention to the demand of both during the inventory planning [3, 4]. Inventory policy by paying attention to more than one item, where both items are supplied from the same source called the joint replenishment policy [5–7].

Research on joint replenishment policy itself can be classified into two types which are Direct Group Strategies (DGS) and Indirect Group Strategy (IGS) [5]. For DGS strategy products are replenished within the same cycle while for IGS strategy products are replenishment with regular time interval. In addition, for IGS strategy the quantity of replenishment is assumed to be sufficient exactly multiple integer of time interval [5].

Researches on joint replenishment policy consider the demand as deterministic static [5, 8], deterministic dynamic [9], and stochastic demand [10, 11] among others. The research presented in this paper consider joint replenishment policy where the demand is deterministic. However, unlike other model, due to cross selling situation,

M. N. Osman Zahid et al. (Eds.): iMEC-APCOMS 2019, LNME, pp. 67–73, 2020.
https://doi.org/10.1007/978-981-15-0950-6_11

the inventory model is developed based on assumption that demand of one item is divided into two parts. They are: (1) static independent deterministic demand and (2) static dependent on other demand.

2 Model Development

The inventory model purposed in this paper uses IGS strategy. The purpose of the joint replenishment policy 2 item model by considering cross selling item is to minimize the total inventory cost per unit of time (total relevant cost per unit time) at 2 cross selling items. Decision variables in the joint replenishment policy model 2 items taking into account the demand dependence phenomenon are: (1) optimum cycle time; (2) optimum item sales lost time interval 1; (3) optimum item lost sales interval time 2.

Notation:

R_1: demand item 1 (unit/time)
R_2: demand item 2 (unit/time)
T: cycle time
t_1: time interval during lost sales item 1 (time)
t_2: time interval for lost sales item 2 (time)
Q_1: item inventory level 1 (unit)
Q_2: item inventory level 2 (unit)
Q_1': item 1 inventory level when lost sales item 1 (unit) starts
Q_2': item 2 inventory level when lost sales item 2 (unit) starts
S: major setup/ordering cost ($)
s_1: minor setup/ordering cost item 1 ($)
s_2: minor setup/ordering cost item 2 ($)
p_1: lost sales cost item 1 ($/unit)
p_2: lost sales cost item 2 ($/unit)
r_1: carrying/holding cost per item 1 per unit of time ($/unit/time)
r_2: carrying/holding cost per item 2 per unit of time ($/unit/time)
d: proportion of demand dependence item 1 and item 2
C_c: total carrying/holding costs per unit time ($)
C_s: total setup/ordering costs per unit time ($)
C_{ls}: total lost sales costs per unit time ($)
TRC: total inventory costs per unit time/Total relevant costs per unit time ($)

Unlike previous research on joint replenishment policy, the demand process for this model is following the assumption as presented below:

$$R_1 = R_1 dependence R_2 + R_1 independence \tag{1}$$

$$R_2 = R_2 dependence R_1 + R_2 independence \tag{2}$$

Therefore, joint replenishment policy proposed in this paper is presented in Fig. 1.

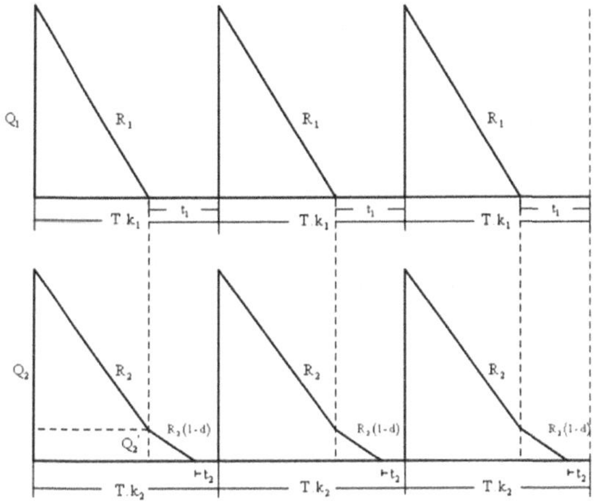

Fig. 1. Joint replenishment policy for cross selling item when $t_1 > t_2$

Total inventory cost per unit time (TRC) consists of carrying/holding cost per unit time, setup/ordering cost per unit time and lost sales cost per unit time. Detail expression of each term of total inventory cost per unit time is presented in the following subsection:

2.1 Carrying/Holding Cost Per Unit Time ($/Time)

Item 1

Carrying/holding cost per unit for item 1 is presented in the formula below.

$$Cc_1 = \frac{\frac{1}{2}.Q_1.(T.k_1 - t_1).r_1}{T.k_1} = \frac{\frac{1}{2}.R_1.(T.k_1 - t_1)(T.k_1 - t_1).r_1}{T.k_1} = \frac{\frac{1}{2}.\mathrm{R}_1.(\mathrm{T.k}_1 - \mathrm{t}_1)^2.\mathrm{r}_1}{\mathrm{T.k}_1} \tag{3}$$

Item 2

Carrying/holding cost per unit for item 2 is determined using a certain approach that is presented in Fig. 2 below. Carrying/holding cost item 2 is the sum of costs for area I and area II.

$$Cc_2 = \frac{(Area\ I + Area\ II).r_2}{T.k_2} \tag{4}$$

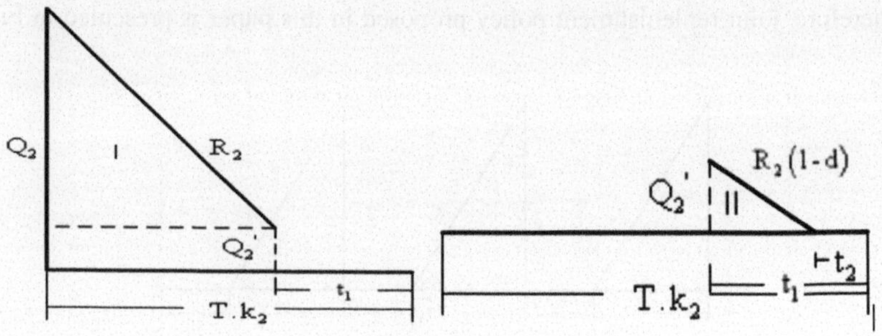

Fig. 2. Area I and Area II for condition $t_1 > t_2$

Carrying/holding cost item 2 can be expressed as:

$$Cc_2 = \frac{\left(\frac{1}{2}.R_2.(T.k_2 - t_1)^2\right) + (R_2.(1-d).(t_1 - t_2).(T.k_2 - t_1)) + \left(\frac{1}{2}.R_2.(1-d).(t_1 - t_2)^2\right).r_2}{T.k_2} \quad (5)$$

Hence, total carrying/holding cost per unit time is the sum of (3) and (5).

$$Cc = \frac{\frac{1}{2}.R_1.(T.k_1 - t_1)^2.r_1}{T.k_1} + \frac{\left(\frac{1}{2}.R_2.(T.k_2 - t_1)^2\right) + (R_2.(1-d).(t_1 - t_2).(T.k_2 - t_1)) + \left(\frac{1}{2}.R_2.(1-d).(t_1 - t_2)^2\right).r_2}{T.k_2} \quad (6)$$

2.2 Setup/Ordering Cost Per Unit Time ($/Time)

Total setup/ordering cost per unit time ($/unit time) item 1 and item 2 can be expressed as follows.

$$Cs = \frac{S}{T} + \frac{s_1}{T.k_1} + \frac{s_2}{T.k_2} = \left(S + \frac{s_1}{k_1} + \frac{s_2}{k_2}\right)/T \quad (7)$$

2.3 Lost Sales Cost Per Unit Time ($/Unit Time)

$$Cls = \frac{\mathrm{R}_1.\mathrm{t}_1.\mathrm{p}_1}{\mathrm{T.k}_1} + \frac{R_2.(1-d).t_2.p_2}{T.k_2} \quad (8)$$

Based on (3), (6), (7) and (8), then total inventory costs per unit time ($) (*TRC*) can be expressed as:

$$
\begin{aligned}
&= \frac{1}{2}.R_1.T.k_1.r_1 - R_1.t_1.r_1 + \frac{R_1.t_1^2.r_1}{2.T.k_1} + \frac{1}{2}.R_2.T.k_2.r_2 - R_2.t_1.r_2 + \frac{R_2.t_1^2.r_2}{2.T.k_2} \\
&+ R_2(1-d)(t_1-t_2)r_2 - \frac{R_2(1-d)(t_1-t_2)t_1.r_2}{T.k_2} + \frac{R_2(1-d)t_1^2.r_2}{2.T.k_2} - \frac{R_2(1-d)t_1.t_2.r_2}{T.k_2} \\
&+ \frac{R_2(1-d)t_2^2.r_2}{2.T.k_2} + \left(\frac{R_1.t_1.p_1}{k_1} + \frac{R_2(1-d).t_2.p_2}{k_2}\right)T^{-1} + \left(S + \frac{s_1}{k_1} + \frac{s_2}{k_2}\right)T^{-1}
\end{aligned}
\tag{9}
$$

Using standard optimization condition for Eq. (9) (i.e., first derivatives are zero and the Hessian matrix is definite positive), the optimum values of the decision variables can be obtained as follow:

$$
T^* = \left(\frac{\frac{R_2.r_2}{k_2}\left(\frac{1}{2}t_1^2 - (1-d)(t_1-t_2)t_1 + \frac{(1-d)t_1^2}{2} - (1-d)t_1.t_2 + \frac{(1-d)t_2^2}{2}\right) + \left(\frac{R_1.t_1^2.r_1}{2.k_1}\right) + \frac{R_2(1-d).t_2.p_2}{k_2} + \frac{R_1.t_1.p_1}{k_1} + \left(S + \frac{s_1}{k_1} + \frac{s_2}{k_2}\right)}{\frac{1}{2}(R_1.k_1.r_1 + R_2.k_2.r_2)}\right)^{\frac{1}{2}}
\tag{10}
$$

$$
t_1^* = \frac{R_1.r_1 + R_2.r_2 - R_2(1-d)r_2 - \frac{R_1.p_1}{k_1.T}}{\frac{R_1.r_1}{T.k_1} + \frac{R_2.r_2}{T.k_2}(1 - 2(1-d) + (1-d))}
\tag{11}
$$

$$
t_2^* = \frac{R_2(1-d)r_2 - \frac{R_2(1-d).p_2}{k_2.T}}{\frac{R_2(1-d).r_2}{T.k_2}}
\tag{12}
$$

3 Procedure for Obtaining for Finding the Solutions

Based on the equations presented in Sect. 2, it is known that the optimum T^* value can be obtained from the optimum t_1^* and t_2^* values. Because decision variables are interrelated and it is set that $k_1 = k_2 = 1$, the steps of the iterative procedure applied are as follows:

Step 1: Perform the first iteration by setting item 1 and item arbitrarily.
Step 2: Set the initial value $t_1^* = 0$, and $t_2^* = 0$
Step 3: Calculate the optimum T^* value with the values $k_1 = k_2 = 1$, and the initial value t_1^* and t_2^* from step 2 using Eq. 10. When calculating T^*, following conditions should be satisfied:

$T^* \geq R_1.p_1/(k_1.\ (R_1.r_1 + R_2/r_2.d))$ so that the value $t_1^* \geq 0$
$T^* \geq P_2/(r_2.\ k_2)$ so that the value $t_2^* \geq 0$.
If the T^* value calculated from (10) does not meet these conditions, then set the T^* with the biggest value from these conditions.
Step 4: Update the T^* value from step 3 to calculate the values t_1 using Eq. 11 and t_2^* using Eq. 12.
Step 5: Repeat steps 3 and 4 until convergence reached, i.e. the values of T^*, t_1^* and t_2^* are similar with the values of the previous calculation.
Step 6: Perform the second iteration by changing the setting items in first iteration, i.e. item 1 become item 2, and item 2 become item 1.
Step 7: Repeat steps 2 to step 5 with the setting in step 6.
Step 8: Select the decision variable from the first and second iterations that produces the smallest *TRC* value and has the decision variables with the pattern $t_1^* > t_2^*$.

4 Numerical Example

To illustrate the applicability of the proposed model, then a set of hypothetical data is used as follows: $R_1 = 5000$, $R_2 = 200$, $r_1 = 4$, $r_2 = 2$, $s_1 = 15$, $s_2 = 16$, $S = 40$, $p_1 = 3$, $p_2 = 2$, $d = 0.3$. The result is presented in Table 1.

Table 1. Results of iteration procedure

1st iteration	T^*	t_1^*	t_2^*	*TRC*
0	1.0000	0.0000	0.0000	10271.0000
1	1.0000	0.2544	0.0000	9619.5487
2	1.0000	0.2544	0.0000	9619.5487
2nd iteration	T^*	t_1^*	t_2^*	*TRC*
0	0.7500	0.0000	0.0000	7744.6667
1	0.7500	0.6875	0.0000	5728.0000
2	0.7500	0.6875	0.0000	5728.0000

From Table 1 it can be seen that both iterations have value of $t_1^* > t_2^*$, but 2nd iteration has *TRC* smaller than *TRC* 1st iteration. Therefore, the decision variable chosen is the result of the calculation of the 2nd iteration.

5 Conclusion

This paper successfully proposed a mathematical model for managing inventory in the case of cross selling item in order to minimize the total inventory cost. The procedure to find the optimum solution also successfully presented in this paper.

References

1. Park, C., Seo, J.: Consideration of purchase dependence in inventory management. Comput. Ind. Eng. **66**(2), 274–285 (2013)
2. Zhang, R., Kaku, I., Xiao, Y.: Model and heuristic algorithm of the joint replenishment problem with complete backordering and correlated demand. Int. J. Prod. Econ. **139**(1), 33–41 (2012)
3. Zhang, R.Q., Yi, M., Wang, Q.Q., Xiang, C.: Polynomial algorithm of inventory model with complete backordering and correlated demand caused by cross-selling. Int. J. Prod. Econ. **199**, 193–198 (2018)
4. Park, C.: Iterative approach to calculating the order fill rate under purchase dependence. Ind. Eng. Manage. Syst. **16**(3), 363–374 (2017)
5. Khouja, M., Goyal, S.: A review of the joint replenishment problem literature: 1989–2005. Eur. J. Oper. Res. **186**(1), 1–16 (2008)
6. Moon, I.K., Goyal, S., Cha, B.C.: The joint replenishment problem involving multiple suppliers offering quantity discounts. Int. J. Syst. Sci. **39**(6), 629–637 (2008)
7. Tsai, C.-Y., Tsai, C.-Y., Huang, P.W.: An association clustering algorithm for ca-order policies in the joint replenishment problem. Int. J. Prod. Econ. **117**, 30–41 (2009)
8. Devy, N.L., Ai, T.J., Astanti, R.D.: A Joint replenishment inventory model with lost sales. In: IOP Conference Series: Materials Science and Engineering, vol. 337, p. 012018 (2018)
9. Lee, L.H., Chew, E.P.: A dynamic joint replenishment policy with auto-correlated demand. Eur. J. Oper. Res. **123**(3), 490–503 (2005)
10. Liu, L., Yuan, X.-M.: Coordinated replenishments in inventory systems with correlated demands. Eur. J. Oper. Res. **123**, 490–503 (2000)
11. Larsen, C.: The Q(s, S) control policy for the joint replenishment problem extended to the case of correlation among item-demands. Int. J. Prod. Econ. **118**(1), 292–297 (2009)

Analysis of Magnetic Component Manufacturing Cost Through the Application of Time-Driven Activity-Based Costing

Nik Nurharyantie Nik Mohd Kamil, Mohd Yazid Abu(✉), Nurul Farahin Zamrud, and Filzah Lina Mohd Safeiee

Faculty of Manufacturing Engineering, Universiti Malaysia Pahang, 26600 Pekan, Pahang, Malaysia
niknurharyantie@gmail.com, myazid@ump.edu.my

Abstract. With a continuous innovation of Electrical & Electronic (E&E) industry, the magnetic component gets increasing demand from customer and their target in production is 5000 units in one shift. Hence, to fix a product cost and to achieve the target finished product are the case study on E&E industry. In other words, the company's product cost is calculated using traditional cost accounting based on volume product of magnetic component. This cost is unable to record accurately the changes of resources used to manufacture the product. The aim of this work to analyze the manufacturing cost of magnetic component incurred on production of E&E industry and time for good by Time-Driven Activity-Based Costing system (TDABC). In methodology, TDABC will be developed by using seven stages which are identify resources group and service processes of all activity and sub-activity, estimate cost of all resources supplied, estimate practical capacity, calculate capacity cost rates, develop time equation, determine time estimates for each sub-activity, and estimate capacity cost required. By having the analysis, loss manufacturing cost winding toroid core can be identified at −MYR2967504.12 and identifying of unused capacity (−889200.12 min) is capable to improve the time efficiency.

Keywords: Magnetic component · Manufacturing cost · Time-Driven Activity-Based Costing

1 Introduction

Due to today's competitive market condition in E&E industry, consumers have more choice to purchase a product. So, the company must have strategy in marketing and cheap selling price of product to stay competitive in the market. In recent years, manufacturing cost also increase due to increases of raw material cost, maintenance cost, and labor cost. However, to raise the price of the component, a company takes a risk of losing customers, inability to compete, and losing market position. Consequently, if the selling price is cheapest, it will impact of a company to generate profit margin. Therefore, an accurate cost of a component is very important for a company to sell the optimal price in market.

M. N. Osman Zahid et al. (Eds.): iMEC-APCOMS 2019, LNME, pp. 74–80, 2020.
https://doi.org/10.1007/978-981-15-0950-6_12

Generally, a company uses Traditional Cost Accounting (TCA) to calculate their product cost. Then, Activity base Costing (ABC) has been proposed which provides better accuracy. Abu [10] estimated the cost of remanufactured crankshaft using ABC and Kamil [11] applied ABC as a method of estimation for the remanufacturing cost of crankshaft. Abu [12] integrated optimization method with ABC for estimating final cost of product and Zheng [13] applied the ABC as a method of cost estimation for a palm oil plantation.

Although, Time-Driven Activity-Based Costing (TDABC) is an innovative costing method that provide more accurate cost than conventional methods. TDABC that can define as a costing model that consider a time which known as inducer time. This method is provided the cost the cost of activities with base that consume of time per activities. According to Bagherpour [1], TDABC is a costing that show the differences between the total time needed to carry all activities that performed in each department. Consequently, this method also provided the total amount time of employees that performed in the department. The advantages of TDABC are providing more reliable information for decision making process [2], more accurate costing [3], and need less cost implementation because the process is simple and easy to apply for a company [4]. According to Zaini [5] found that the percentage of TDABC application belong to 66% of healthcare service, 23% of industrial service, and 11% of library service. The application of TDABC can be categorized into nine applications which are manufacturing, healthcare, automotive, retail trade, information technology, agriculture, academic, and others [6]. The highest percentage of TDABC belongs to 66% of healthcare and lowest percentage belong to 2% of information technology. Although, in TDABC is famous method to apply in healthcare application, this method also suitable to apply in manufacturing sector.

2 Research Methodology and Results

In this section, this research started by collecting data of magnetic component and record time to finish a product in each workstation. Observation were made and data collected from actual 12 workstations in the production. Eventually, the data were recorded in relation to the consumption cost of labor, consumable, material, and maintenance. The data was collected in January 2018 until March 2019. As mention by Erhun [7], there are seven steps for applying TDABC to produce magnetic component. Firstly, the production activities must be identified activity and sub-activities. Besides, determine the cost of all resources used which consist of labor cost, material cost, maintenance cost, and consumable. Next, estimate the practical capacity that are required for the calculation of capacity cost rate. Then, develop time equation for each activity to calculate the production costs of magnetic components. After that, determine the time estimation for each activity to produce the product and lastly, calculate the capacity required of each activity.

2.1 Production Activities Used (Activity Center and Sub-activities)

The first step in the manufacturing process is a two wires are winding to the right and left at cores for 5 turns. Then, the wound core is put on the lead forming machine to cut the excessive length of leadout wire. Next, assemble the wound core with the header with the position of part number must be in front during this process. After that, use the holder to prevent leadout from affecting of epoxy. Put an epoxy at leadouts between core and header. The units were arranged into curing jug to transfer into curing oven. In curing condition, it takes 30 min, and the temperature set 130 °C with tolerance ±5. Then, cool the component under a cooling fan around 5–15 min. Insert the component into chopper base to cut the leadout and transfer to conveyor for marking process. Pass or fail of quality are depend on visual mechanical inspection requirement and final test. Lastly, packaging the component after pass with quality check by both process before.

2.2 Estimated Total Cost of the Resources Supplied

If the resources sharing the same activity, costs can be allocated directly to that activity while different activity, the cost-driver has to be used to allocate the cost. There are four types of resources which are labor, maintenance, material, and consumable.

This work only considers main activity which is winding toroid core that requires two employees at a cost MYR79200. 1,200,000 pieces of cores incurs material cost at MYR1008000 per year. Due to private and confidential to access the data, no costing listed for consumable cost and maintenance cost. So, the total cost is MYR1087200.

Table 1 shows cost of all resources supplied of main activity to produce magnetic component.

2.3 Estimated Practical Capacity

The E&E company's working hours are Monday to Saturday, 7.30 a.m. to 5.30 p.m. within 2 shift. An employee works of eight hours 35 min a day, for 20 days for a month and 240 days for a year. Deduction for breaks is 45. So, an employee has acceptable practical capacity of 10,300 min each per month and 123,600 min each per year.

2.4 Calculation of the Capacity Cost Rate

Equation 1 is used to calculate the capacity cost rate of main sub-activities.

$$\textit{Capacity cost rate} = \frac{\textit{Cost of all resources supplied}}{\textit{Practical capacity}} \tag{1}$$

By using this equation, sub-activity for winding toroid core which is transfer the toroid core on the winding is MYR362,400/123,600 min is equal MYR2.93 per minutes. Table 2 shows the summary of capacity cost rate of main sub-activity.

Table 1. Labor, maintenance, material, and consumable cost of main activity

Workstation	Sub-activities	Labor cost (MYR)	Maintenance cost (MYR)	Material cost (MYR)	Consumable cost (MYR)	Cost of all resources supplied (MYR)
Winding toroid core	1. Transfer the core to the winding machine	26,400	Nil	336,000	Nil	362,400
	2. Put a wire at the right core for winding	26,400	Nil	336,000	Nil	362,400
	3. Put a wire at the left core for winding	26,400	Nil	336,000	Nil	362,400
	Total	**79,200**	**Nil**	**1,008,000**	**Nil**	**1,087,200**

Table 2. Capacity cost rate for main sub-activity to produce magnetic component

Workstation	Sub-activities	Cost of all resources supplied (MYR/year)	Practical capacity (min/year)	Capacity cost rate (MYR/min)
Winding toroid core	1. Transfer the core to the winding machine	362,400	123,600	2.93
	2. Put a wire at the right core for winding	362,400	123,600	2.93
	3. Put a wire at the left core for winding	362,400	123,600	2.93
	Total	**1,087,200**	**370,800**	

2.5 Development of Sub-activity Time Equation

According to [8], TDABC time equation must be develop of all sub-activity for all the workstation to incorporate all the time needed to produce the magnetic component. Mathematically, the TDABC time equation can be explained by using Eq. 2 [9].

$$T_t = \beta_0 + \beta_i X_i \tag{2}$$

Where:

T_t = the time needed to perform an activity (minute)
β_0 = the standard time needed to perform the basic activity (minute)

β_i = the estimated time to perform the incremental activity (minute)
X_i = the quantity of the incremental activity (time).

2.6 Determination of the Estimated Time for Each Activity

Determination of time equation for each activity is needed to calculate the estimated production time. Time equation of each main sub-activities is developed by taking time taken multiplied by the relevant cost driver that summarized Table 3 below.

As an example, time equation is developed as shown in Eq. 3 below.

$$T_{\text{winding toroid core}} = 1.05X_1 + 0.02X_2 + 0.04X_3 \tag{3}$$

Table 3. Time equations of main sub-activities

Workstation	Sub-activities	Time equation
Winding toroid core	1. Transfer the core to the winding machine	$1.05X_1$
	2. Put a wire at the right core for winding	$0.02X_2$
	3. Put a wire at the left core for winding	$0.04X_3$

2.7 Estimated Capacity Required by Each Activity

The estimated capacity required by each activity was determined by quantifying the frequency of the activity in a year. The total time spent on the activity can be calculated by multiplying the amount of a given activity by the time spent.

As an example, the actual time on winding toroid core per year was determined by substituting the volume of cost-drivers from Table 4 into Eq. 3, as shown below.

The actual time spent = (1.05 × 1200000) + (0.02 × 2) + (0.04 × 2) = 60,000.12 min. Table 5 refers the production cost of main activity to produce magnetic component.

Table 4. Volume of cost driver for main workstation

Variable	Driver	Quantity/year
X1	Number of core (pieces/year)	1200000
X2	Number of winding machine operating (frequency/year)	2
X3	Number of winding machine operating (frequency/year)	2

Table 5. Elapsed time and total production costs of sub-activities

Workstation	Sub-activities	Used time (min)	Capacity cost rate (MYR/min)	Total cost (MYR/year)
Winding toroid core	Transfer the core to the winding machine	1260000	2.93	3691800
	Put a wire at the right core for winding	0.04	2.93	0.12
	Put a wire at the left core for winding	0.08	2.93	0.23
	Total	**1260000.12**		**3691800.35**

3 Conclusion and Discussion

As a result, analysis of capacity utilization is proposed to reduce production cost. Table 6 summarized the analysis of capacity utilization of main workstation.

Table 6. Analysis of capacity utilization of sub-activities to produce magnetic component

Workstation	Sub-activities	Practical capacity (min/year)	Used time (min)	Un-used capacity (min)	Capacity cost rate (MYR/min)	Lost manufacturing cost (MYR)
Winding toroid core	1. Transfer the core to the winding machine	123,600	1260000	−1136400	2.93	-3329652
	2. Put a wire at the right core for winding	123,600	0.04	123599.96	2.93	362147.88
	3. Put a wire at the left core for winding	123,600	0.08	123599.92	2.93	362147.77
	Total	**370,800**	**1260000.12**	**−889200.12**		**−2967504.12**

The sub-activity 2 incurs a lot of waste, at MYR362147.88, followed sub-activity 3 at MYR362147.88 and sub-activity 1 at –MYR3329652 of waste. This work found the sub-activity 2 and 3 are spent more at labor cost.

By having the analysis, we have been able to identify an un-used capacity and loss manufacturing cost in each workstation. This will lead the resource waste can be reduced and time efficiency will be more accurate to manufacture the magnetic component.

Acknowledgement. This research is fully supported by RDU190156 and FRGS/1/2018/TK03/UMP/02/34. The authors fully acknowledge University Malaysia Pahang for the approved fund which makes this important research viable and effective.

References

1. Bagherpour, M., Nia, A.K, Sharifian, M.: Time-driven activity-based costing in a production planning environment. **227**(2), 333–337 (2013)
2. Medeiros, H.D.S., Santana, A.F.B., Guimaraes, S.: The use of costing methods in lean manufacturing industries; a literature review. 395–406 (2017)
3. Kaplan, A.L., Agarwal, N., Setlur, N.P., Tan, H.J., Niedzwiecki, D., Mclaughlin, N., Saigal, C.S.: Measuring the cost of care in benign prostatic hyperplasia using time-driven activity-based costing. **3**(1), 43–48 (2015)
4. Barros, R.S., Maria, A., Simões, D.: Time-driven activity-based costing Designing a model in a Portuguese. **47**(8), 712–735 (2017)
5. Zaini, S.N.A.M., Abu, M.Y.: A review on time-driven activity based costing system in various sectors. J. Mod. Manuf. Syst. Technol. **2**, 15–22 (2019)
6. Gonzalez, M., Nachtmann, H., Pohl, E.: Time-driven activity-based costing for health care provider supply chains. Eng. Econ. **62**(2), 161–179 (2017)
7. Erhun, F., Mistry, B., Platchek, T., Milstein, A., Narayanan, V. G., Kaplan, R.S.: Time-driven activity-based costing of multivessel coronary artery bypass grafting across national boundaries to identify improvement opportunities: study protocol, pp. 1–8 (2018)
8. Stout, D.E., Propri, J.M.: Implementing time-driven activity-based costing at a medium-sized electronics company. Manage. Acc. Q. **12**(3), 1–11 (2011)
9. Bruggeman, W., Everaert, P.: Time-driven activity-based costing: exploring the underlying model. Cost Manage. **21**(2), 16–19 (2007)
10. Abu, M.Y., Jamaludin, K.R., Zakaria, M.A.: Characterisation of activity based costing on remanufacturing crankshaft. Int. J. Autom. Mech. Eng. **14**(2), 4211–4224 (2017)
11. Kamil, N.N.N.M., Abu, M.Y.: Integration of Mahalanobis-Taguchi System and activity based costing for remanufacturing decision. J. Mod. Manuf. Syst. Technol. **1**, 39–51 (2018)
12. Abu, M.Y., Nor, E.E.M., Rahman, M.S.A.: Costing improvement of remanufacturing crankshaft by integrating Mahalanobis-Taguchi system and activity based costing. In: IOP Conference Series: Materials Science and Engineering, vol. 342, pp. 1–10 (2018)
13. Zheng, C.W., Abu, M.Y.: Application of Activity based Costing for Palm Oil Plantation. J. Mod. Manuf. Syst. Technol. **2**, 1–14 (2019)

The Impact of Capacity Cost Rate and Time Equation of Time-Driven Activity-Based Costing (TDABC) on Electric Component

Nurul Farahin Zamrud, Mohd Yazid Abu(✉), Nik Nurharyantie Nik Mohd Kamil, and Filzah Lina Mohd Safeiee

Faculty of Manufacturing Engineering, Universiti Malaysia Pahang, 26600 Pekan, Pahang, Malaysia
nurulfarahinzamrud@gmail.com, myazid@ump.edu.my

Abstract. Today, in this big business competition, it is important for the company to incorporate an accurate cost estimation to decide the best price for products to gain profits. This research study revolves around a case study company, which manufactures electric and electronics products. The objective of this study is to analyze the manufacturing process cost at the case study company using Time-Driven Activity-Based Costing (TDABC). This research began by collecting data related to the production costs of a selected product. The product has been selected based on the volume of production which is high and continuous production. There are two parameters of TDABC to be considered. Firstly, the capacity cost rate and secondly, the time required to perform activities. Both parameters can be estimated easily and objectively. Eventually, the results showed that by applying the TDABC method, the capacity utilization and time efficiency can be clearly viewed. The company gained information on manufacturing costs and time utilization for those activities and could generate further action to increase efficiency and have the opportunity to use this method to enhance the accuracy in determining the appropriate process for any kind of product.

Keywords: Time-Driven Activity-Based Costing · Capacity cost rate · Time equation

1 Introduction

Nowadays, a consumer has a variety of choices of their desired products, whether as they walk into a physical store or as they scroll at the website. Not to mention that there is a wide competition of price range and quality of the same product in the market. Whoever comes out with a reasonable price and product quality wins the competition. Furthermore, there are a lot of factors that contribute to product pricing such as raw materials, manufacturing costs, and labor costs. Therefore, it is important for the company to decide the best price so that they could gain profit, not loss. So, to stay in the competition in the industry, manufacturers would need accurate cost estimation. This will help them to sell the product at an optimal price. Normally company uses traditional cost system to calculate their product cost. Then, activity base costing

M. N. Osman Zahid et al. (Eds.): iMEC-APCOMS 2019, LNME, pp. 81–87, 2020.
https://doi.org/10.1007/978-981-15-0950-6_13

(ABC) has been proposed which provides better accuracy. Abu [7] estimated the cost of remanufactured crankshaft using ABC and Kamil [8] applied ABC as a method of estimation for the remanufacturing cost of crankshaft. Abu [9] integrated optimization method with ABC for estimating final cost of product and Zheng [10] applied the ABC as a method of cost estimation for a palm oil plantation. A costing model known as Time-driven Activity-based Costing (TDABC) uses time as an inducer in costing. TDABC provides a cost of activities by calculating time consumption per activities [1]. The advantage of this method is it provides accurate estimates of care cycle cost and more transparency into the cost drivers [2]. Moreover, significant information on idle capacities is delivered and unused resources recognized by TDABC [3]. According to Zaini [4] it is found that the percentage of TDABC application belongs to 66% of healthcare service, 23% of industrial service, and 11% of library service.

2 Methodology

In this research, the data is collected from 1 line of production which consists of 13 work stations. The data which are based on 2018 was recorded in relation to the labor costs, maintenance, cost of raw materials and consumables. According to existing literature, there are 8 stages to calculate the cost using TDABC [5]. The first step is to identify and analyze activities to understand the scope and sequence of events. Secondly, the cost of resources supplied in each activity and sub-activities are listed and estimated. Practical capacity is the details about employees working hours which to be used for calculation of capacity cost rates. Next, TDABC time equation is developed to calculate the estimated production time. The estimated capacity required by each sub-activity is calculated and finally, the unit cost is determined.

2.1 Production Activities – Activity Centers/Sub-activities

The production for the magnetic inductor is mainly on the first floor while there is one workstation located on the ground floor. The production process of the magnetic inductor starts with winding workstation and ends with packaging workstation. This process is shown in Fig. 1. Note that the flattening workstation is on the ground floor.

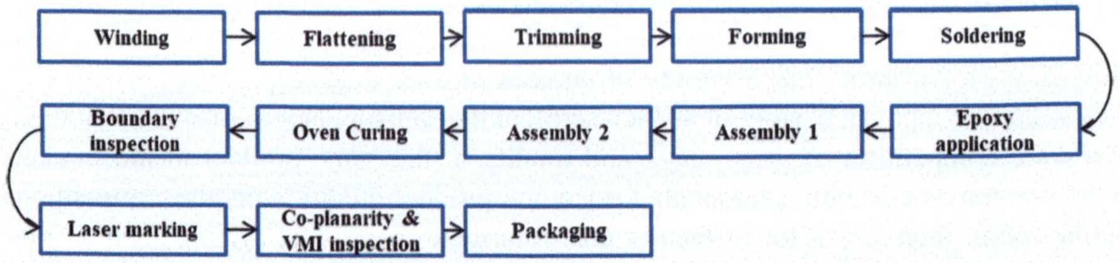

Fig. 1. Production process flow of the magnetic inductor

This paper explains in detail on how to apply the TDABC method for the cost calculation in activity 1 and 2 – winding and flattening only. The activity 1 has 1 sub-activity: (i) wind the wires using CNC machine while activity 2 has 2 sub-activities: (i) pick up the coils from winding station. (ii) flatten the coils by using a hydraulic press machine.

2.2 Estimated Total Cost of Resources Used

In this study, the resources allocated are based on 4 groups: (i) labor costs, (ii) maintenance cost for the machinery (iii) raw materials costs and (iv) consumable material costs. The costs incurred are detailed in Table 1.

Table 1. Labor, maintenance, material and consumable cost for 13 workstations (unit: MYR/year)

Activity	Sub-activities	Labor	Maintenance	Material	Consumable	Cost of all resources supplied
Winding	The wires are winded using CNC machine	24,000	26,449.89	384,000	Nil	434,449.89
Flattening	Pick up the coils from winding station	12,000	Nil	Nil	Nil	12,000
	Flatten the coils by using hydraulic press machines	36,000	70	Nil	144	36,214
Total		**72,000**	**26,519.89**	**384,000**	**144**	**482,663.89**

2.3 Estimated Practical Capacity

The factory's working hours are from Monday to Friday, and from 8 a.m. to 5.35 p.m. daily. The operators work on average 8 h per day, for 20 days a month. Every operator has a 1-hour break daily. Each operator has an acceptable capacity of 9,100 min every month. Therefore, the practical capacity per year is 109,200 min.

2.4 Calculation of the Capacity Cost Rates

The capacity cost rate (MYR per minute) can be obtained using the following Eq. (1).

$$\text{Capacity cost rate} = \frac{cost\ of\ capacity\ supplied}{practical\ capacity\ of\ resources\ supplied} \quad (1)$$

The equation is used in Table 2. Table 2 listed out the summary of capacity cost rate of each sub-activity. For sub-activity winding the wire using CNC machine, the capacity cost rate is MYR 434,449.89/218,400 min or MYR 1.99 per minute. As for sub-activity pick up the coils and flatten the coils is MYR 0.05 per minute and MYR 0.11 per minute respectively.

Table 2. Capacity cost rate of each sub-activity

Activity	Sub-activities	Cost of all resources supplied (MYR/year) [1]	Practical capacity (min/year) [2]	Capacity cost rate (MYR/year) [1]/[2] = [3]
Winding	The wires are winded using CNC machine	434,449.89	218,400	1.99
Flattening	Pick up the coils from the winding station	12,000	109,200	0.05
	Flatten the coils by using hydraulic press machines	36,214	327,600	0.11
Total		**482,663.89**	**655,200**	

2.5 Development of Time Equation

TDABC time equation is used to incorporate all the time needed for all sub-activities in one single equation [6]. The mathematical model used to establish the TDABC time equation is shown below [9]:

$$T_t = \beta_0 + \beta_i X_i \quad (2)$$

Where

T_t = the time needed to perform an activity (minute)
β_0 = the standard time to perform the basic activity (minute)
β_i = the estimated time to perform the incremental activity (minute)
X_i = the quantity of the incremental activity (time).

2.6 Determination of the Estimated Time for Each Activity

In order to calculate the estimated production time, it is necessary to develop a time equation. The estimated time for each activity is obtained by observation of the operators doing their tasks. For instance, the average time taken for the wire to be wind by using a CNC machine to become a single coil is 0.12 min. This figure is multiplied by the relevant variables or cost drivers to develop the time equation, as shown in Table 3. Each variable in the time equation is defined in Table 4.

Table 3. Time equations for all sub-activities

Activity	Sub-activities	Time equations
Winding	The wires are winded using CNC machine	$0.12X_1$
Flattening	Pick up the coils from the winding station	$10.00X_2$
	Flatten the coils by using hydraulic press machines	$0.22X_3$

From Table 3, the time equation for all the sub-activities was developed as shown in Eq. (3)

$$T\ \text{sub-activities} = 0.12X_1 + 10X_2 + 0.22X_3 \tag{3}$$

2.7 Estimated Capacity Required by Each Activity

The estimated capacity required by each activity was determined by quantifying the frequency of the activity in a year. By multiplying the amount of a given activity by the time spent doing it, one could calculate the total time spent on the activity. The volume of cost drivers for the wood preparation activity center is summarized in Table 4.

Table 4. The volume of cost drivers for the magnetic inductor (unit: quantity/year)

Var.	Driver	Quantity/year
X_1	Amount of coils produced	4,800,000
X_2	Pick up the coils from the winding station (rounds)	480
X_3	Number of hydraulic press machine operating (frequency/month)	2

The actual time spent on this activity center per year was 1,351,293.63 min determined by substituting the volume of cost-drivers from Table 4 into Eq. (3). The total time for the wire to be winded using CNC machine in a year can be represented by X_1 equals 4,800,000 in $0.12X_1$, so that $0.12 \times 4{,}800{,}000 = 576{,}000$ min. When multiplied by capacity city cost of MYR 1.99 per minute, it can be determined that the total production cost of this activity is MYR 1,146,240 per year. Based on the same approach, the production cost for the work station is shown in Table 5.

Table 5. Elapsed times and total production costs for the magnetic inductor sub-activities

Activity	Sub-activities	Used time (min)	Capacity cost rate (MYR/min)	Total cost (MYR/year)
Winding	Wind the wires using CNC machine	576,000	1.99	1,146,240
Flattening	Pick up the coils from winding station	4,800	0.05	240
	Flatten the coils by using hydraulic press machines	0.44	0.11	0.05

3 Conclusion and Discussion

Consequently, based on the analysis of utilization, cost of production can be reduced. By using TDABC, the unused capacities are identified as in Table 6.

Table 6. Analysis of capacity utilization in the magnetic inductor production

Activity	Sub-activities	Practical capacity (min/year) [1]	Used time (min) [2]	Unused capacity (min) [3] = [1] − [2]	Capacity cost rate (MYR/min) [4]	Loss of manufacturing costs (MYR) [3] × [4]
Winding	The wires are winded using CNC machine	218,400	576,000	−357,600	1.99	−711,352.02
Flattening	Pick up coils from winding station	109,200	4,800	104,400	0.11	11,472.53
	Flatten the coils by using hydraulic press machine	327,600	0.44	327,599.56	0.11	36,213.95
Total		**655,200**	**580,800.44**	**74,399.56**		**−663,665.54**

From Table 6, sub-activity 3 has the most waste of resources followed by sub-activity 2 and 1. Among these three sub-activities, it can be concluded that waste is related to labor costs. Therefore, it is suggested that the workload of each work station should be revised in order to reduce waste of resources and increase the time efficiency of the operators.

The results of this study have given the case study company a clear view of their unused capacity and time efficiency, and also identified processes that contribute to high waste. Therefore, future action should be taken in order to gain more profit and having less waste in the production of magnetic inductor.

Acknowledgment. This research is fully supported by RDU190156 and FRGS/1/2018/TK03/UMP/02/34. The authors are fully acknowledged Universiti Malaysia Pahang for the approved fund which makes this important research viable and effective.

References

1. Akhavan, S., Ward, L., Bozic, K.J.: Time-driven activity-based costing more accurately reflects costs in arthroplasty surgery. Clin. Orthop. Relat. Res. **474**(1), 8–15 (2016)
2. Helmers, R.A., Dilling, J.A., Chaffee, C.R., Larson, M.V., Narr, B.J., Haas, D.A., Kaplan, R.S.: Overall cost comparison of gastrointestinal endoscopic procedures with endoscopist- or anesthesia-supported sedation by activity-based costing techniques. Mayo Clin. Proc.: Innov. Qual. Outcomes **1**(3), 234–241 (2017)
3. Zhuang, Z.Y., Chang, S.C.: Deciding product mix based on time-driven activity-based costing by mixed-integer programming. J. Intell. Manuf. **28**(4), 959–974 (2017)
4. Zaini, S.N.A.M., Abu, M.Y.: A review on time-driven activity based costing system in various sectors. J. Mod. Manuf. Syst. Technol. **2**, 15–22 (2019)
5. Somapa, S., Cools, M., Dullaert, W.: Unlocking the potential of time-driven activity-based costing for small logistics companies. Int. J. Logist. Res. Appl. **15**(5), 303–322 (2012)
6. Pongwasit, R., Chompu-inwai, R.: Analysis of wooden toy manufacturing costs through the application of a time-driven activity-based costing system. Mem. Muroran Inst. Tech. **65**, 7–14 (2015)
7. Abu, M.Y., Jamaludin, K.R., Zakaria, M.A.: Characterisation of activity based costing on remanufacturing crankshaft. Int. J. Autom. Mech. Eng. **14**(2), 4211–4224 (2017)
8. Kamil, N.N.N.M., Abu, M.Y.: Integration of Mahalanobis-Taguchi system and activity based costing for remanufacturing decision. J. Mod. Manuf. Syst. Technol. **1**, 39–51 (2018)
9. Abu, M.Y., Nor, E.E.M., Rahman, M.S.A.: Costing improvement of remanufacturing crankshaft by integrating Mahalanobis-Taguchi system and activity based costing. In: IOP Conference Series: Materials Science and Engineering, vol. 342, pp. 1–10 (2018)
10. Zheng, C.W., Abu, M.Y.: Application of activity based costing for palm oil plantation. J. Mod. Manuf. Syst. Technol. **2**, 1–14 (2019)

The Application of Time-Driven Activity Based Costing System on Inductors in Electrics and Electronics Industry

Filzah Lina Mohd Safeiee, Mohd Yazid Abu(✉), Nik Nurharyantie Nik Mohd Kamil, and Nurul Farahin Zamrud

Faculty of Mechanical and Manufacturing Engineering, Universiti Malaysia Pahang, 26600 Pekan, Pahang, Malaysia
filzahlinasafeiee@gmail.com, myazid@ump.edu.my

Abstract. The adoption of traditional costing system (TCS) in manufacturing industry has motivated the need for improvements in the accounting system. The main problem of TCS is the inability to provide useful feedback to understand and allocate overhead costs and inaccurate to forecast on unused capacity. To perform this study, Time-driven activity-based costing (TDABC) is well suited for manufacturing industry, involving many activities with complex time drivers. TDABC is a variation and expansion of activity based costing where by process costs are analyzed based on time and resource consumption. TDABC seems to be one of the best tools for understanding cost behavior and for refining a cost system. This study is to apply TDABC method on electrics and electronics industry system. This improvement is to maximize the efficiency and effectiveness throughout the production system, cost and eventually increase the net income. From this work the information about cost and profitability quickly get and inexpensively. Several generally accepted methods of cost accounting have been described, of which TDABC is considered the most sophisticated and precise.

Keywords: Time-driven activity-based costing · Activity-based costing · Traditional cost system · Capacity cost rate · Cost analysis

1 Introduction

Electrics and electronics industry has become the leading sector in manufacturing industry in our country, Malaysia since 2010s. It accounted for 44.6% of the total investment approved in the industry [1]. In this paper, will be focusing on one of the electrical components in electric and electronics sector which is inductors. Inductor is a component in an electric and electronic circuit which possesses inductance. Due to increasing competition and customer requirement in global, forced many industrial company firms to continuously modify and optimize their strategic and tactical levels [2]. Until this date, TCA has been widely used among industrial practitioners but TCA has been criticized for not facilitating the accurate true cost and very outdated and unable to forecast [3]. There are studies argues that the primary failing of traditional

M. N. Osman Zahid et al. (Eds.): iMEC-APCOMS 2019, LNME, pp. 88–95, 2020.
https://doi.org/10.1007/978-981-15-0950-6_14

costing systems [4]. After failing in TCA, activity-based costing method (ABC) is created. ABC model built up as a single-faceted model initially and provides the accurate cost information which is based on the production process and activities [5]. Abu [10] estimated the cost of remanufactured crankshaft using ABC and Kamil [11] applied ABC as a method of estimation for the remanufacturing cost of crankshaft. Abu [12] integrated optimization method with ABC for estimating final cost of product and Zheng [13] applied the ABC as a method of cost estimation for a palm oil plantation. TD-ABC further evolved from the ABC [6], and has been stated that many industries have successfully applied the concept of TDABC as first introduced by Kaplan and Anderson in 2004. The TDABC approach is mainly based on the time drivers spent on cost pools; however, TDABC poses some difficulties in calculations of the assigned costs [7]. One example for the application of TDABC comes from palm oil plantation area. They applied TDABC because there are no previous researches on this sector and the organization has accurate costing system and operate smooth process in the workstation [8]. This paper also gave reasons that TDABC supported operational improvement, inform reimbursement policy, can capture the cost of care accurately, able to manage the complexity inherent, more efficient and lastly simple than ABC.

2 Methodology

This research study started by collecting the data from the electrics and electronics company and the related data on for this work which are detail in production process, orders, costing and time taken for each detail process. Observations and interview were made in process of collecting the data from actual workstation. The data were noted contains of cost of raw materials, labor costs, maintenance cost, consumable cost, as well as the time taken for each process. The data was collected in October 2018. To start the study, the production activities (activity centers/sub-activity) have to be determined. Then, the cost is estimated for all resources used and the calculation of capacity cost rate. To calculate the production costs, the time equation has to be created for each activity center. After that, the estimate time for each activity is determined, and the capacity demand of each activity center also has to be calculated. Finally, the cost per product unit is derived.

2.1 Production Activities Used (Activity Center/Sub-activities)

TDABC method is going to take the first step with analyze and identify the manufacturing activities related in order to understand their problem, scope, and the detail events that takes place, because all activities vary but have their own purpose. In the case study, the first step in manufacturing process on inductor component is wind the drum core with auto epoxy by using winding machine. Then, the drum core has to be heated and cool it down. After go through many processes, the core has to be inspect for pass or fail characteristics at co-planarity check station and the final test is the inductance test before packaging. The full process has been shown in Fig. 1 below.

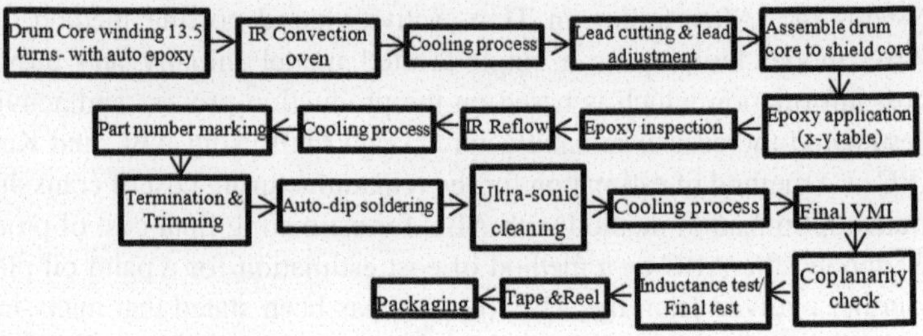

Fig. 1. Production process on one of the line production at the company

2.2 Estimated Total Cost of the Resources Used Cost of All Resources Supplied

From this research study, the resources have been allocated based on the types of resource used for each activity. About four main groups have been decided: (1) labor cost (2) material cost (3) maintenance cost (4) consumable cost. The details of the costs incurred in each study on sub-activity are as follows (Table 1):

Table 1. Labor, material, maintenance and consumable costs within the inductor preparation activity center (MYR/year)

No.	Sub-activities	Labor Cost (MYR)	Maintenance cost (MYR)	Material cost (MYR)	Consumable cost (MYR)	Cost of all resources applied (MYR)
1	Insert the core to the winding machine	792,000	Nil	16,200	Nil	808,200
2	Put the wire on the right core for winding	792,00	Nil	39,235.20	Nil	831,235.20
3	Put the wire at the left core for winding	792,00	Nil	39,235.20	Nil	831,235.20
4	Cut the finish wire	792,000	Nil	39,235.20	Nil	831,235.20
5	Hang up the finish unit at the curing fixture	792,000	Nil	Nil	Nil	792,000
6	Bring the curing fixture into the curing oven to be heated	792,000	Nil	Nil	Nil	792,000
Total		**4,752,000**	**Nil**	**133,905.60**	**Nil**	**5,717,140.80**

2.3 Estimated Acceptable Capacity (Practical Capacity)

The company's working hours are Monday to Saturday, with 2 shifts. Each shift will take about 9 h and 35 min which is 8 a.m. to 5.00 p.m and 8 p.m until 5 a.m for 20 days a month.

2.4 Calculation of the Capacity Cost Rates

The capacity cost rate (Ringgit Malaysia per minutes) can be obtained using the following equation (Table 2).

$$Capacity\ cost\ rate = \frac{Cost\ of\ all\ resources\ supplied}{Practical\ capacity} \quad (1)$$

Table 2. Capacity cost rate of each sub-activity for the inductor component preparation activity center

No.	Sub-Activities	Cost of all resources applied (MYR/Month)	Practical capacity (min/month)	Capacity cost rate (MYR/min)
1	Insert the core to the winding machine	808,200	10,300	78.47
2	Put the wire on the right core for winding	831,235.20	10,300	80.70
3	Put the wire at the left core for winding	831,235.20	10,300	80.70
4	Cut the finish wire	831,235.20	10,300	80.70
5	Hang up the finish unit at the curing fixture	792,000	10,300	76.89
6	Bring the curing fixture into the curing oven to be heated	792,000	10,300	76.89
Total		**4,885,905.60**	**61,800**	**474.35**

2.5 Development of Activity Center Time Equation

The time equation is able to fit in all the time needed to undertake all sub-activities in each activity center within single equation [9] and the mathematical model used to establish TDABC; time equation is as shown below.

$$T_t = \beta_0 + \beta_i X_i \quad (2)$$

Where,

T_t = *the time needed to perorm an activity (minute)*
β_0 = *the standard time to perform the basic activity (minute)*
β_i = *the estimated time to perform the incremental activity (minute)*
X_i = *the quantity of the incremental activity (time)*

2.6 Determination of the Estimated Time for Each Activity

A time equation is needed to be developed to calculate the estimated production time. For example, the average time taken for the first process on this production line, winding core with auto epoxy is 5 s or 0.0833 min per session. The figure was then multiplied by the relevant variables or cost-drivers to develop the time equation, as shown in Table 3. Each variable in the time equation is defined in Table 4.

Table 3. Time equations for sub-activities of the inductor components preparation activity center

No.	Sub-activities	Time equations
1	Insert the core to the winding machine	$0.035X_1$
2	Put the wire on the right core for winding	$0.0567X_2$
3	Put the wire at the left core for winding	$0.054X_2$
4	Cut the finish wire	$0.036X_2$
5	Hang up the finish unit at the curing fixture	$0.06X_2 + 0.036X_3$
6	Bring the curing fixture into the curing oven to be heated	$0.036X_3 + 0.086X_4$

2.7 Estimated Capacity Required by Each Activity Center

The estimated capacity required by each activity was determined by quantifying the frequency of the activity per year. By multiplying the amount of a given activity by the time spent doing it, one could calculate the total time spent on the activity. The volumes of cost-drivers for the inductors preparation activity center are summarized in Table 4.

Table 4. Volume of cost-drivers for the inductors preparation activity center

Var.	Driver	Quantity/year
X_1	Transfer the wound drum core from storage to the winding machine	2,400,000
X_2	Number of turns of the machine to wind the lead on the core	31,200,000
X_3	Transfer the finish wound drum core to the curing fixture	2,400,000
X_4	Full up the curing fixture with the units to be heated	2,400,000

The actual time spent on this activity center per month was 6904.8 min determined by substituting the volume of cost-drivers from Table 4 into Eq. 3, as shown below. The total time for the transfer the wound drum core from storage to the winding machine can represent by X_1 equals to 2,400,000 in $0.035X_1$ so that $0.035 \times 2{,}400{,}000 = 84{,}000$ min. When the value multiplied by the capacity cost of 78.47 Ringgit Malaysia per minute, it can be determined the total production cost of this activity comes out as 6,591,480 Ringgit per year. Based on the same approach, the total production costs for each of the inductor components preparation sub-activities are shown in Table 5. This shows that the total production cost for the inductor preparation activity center is 549,456,876 Ringgit Malaysia.

Table 5. Elapsed time and total production costs for the inductor preparation sub-activities.

No.	Sub-activities	Used time (min)	Capacity cost rate (MYR/min)	Total cost (MYR/month)
1	Insert the core to the winding machine	84,000	78.47	6,591,480
2	Put the wire on the right core for winding	1,769,040	80.70	142,761,528
3	Put the wire at the left core for winding	1,684,800	80.70	135,963,360
4	Cut the finish wire	1,123,200	80.70	91,045,740
5	Hang up the finish unit at the curing fixture	1,958,400	76.89	150,581,376
6	Bring the curing fixture into the curing oven to be heated	292,800	76.89	22,513,392
Total		**6,912,240**		**549,456,876**

3 Conclusion

From the Table 6, among these six sub-activities it was found that all the wasted costs are related to the time. By using TDABC can identify the production cost.

Table 6. Analysis of capacity utilization in inductor preparation activity center

No.	Subs-activities	Practical capacity (min/month)	Used time (min)	Un-used capacity (min)	Capacity cost rate (MYR/min)	Loss of manufacturing costs (MYR)
1	Insert the core to the winding machine	10,300	84,000	−73,700	78.47	−5,783,239
2	Put the wire on the right core for winding	10,300	1,769,040	−1,758,740	80.70	−141,930,318
3	Put the wire at the left core for winding	10,300	1,684,800	−1,674,500	80.70	−135,132,150
4	Cut the finish wire	10,300	1,123,200	−1,112,900	80.70	−89,811,030
5	Hang up the finish unit at the curing fixture	10,300	1,958,400	−1,948,100	76.89	−149,789,409
6	Bring the curing fixture into the curing oven to be heated	10,300	292,800	−282,500	76.89	−21,721,425
Total		**61,800**	**6,912,240**	**−6,850,440**		**−544,167,571**

Acknowledgement. This research is fully supported by Ministry of Higher Education through RDU190156 and FRGS/1/2018/TK03/UMP/02/34. The authors fill acknowledged Universiti Malaysia Pahang for the approved fund which makes this important research viable and effective.

References

1. Essays, UK: Electric and Electronic Industry in Malaysia, November 2018. https://www.ukessays.com/essays/marketing/performance-of-electric-and-electronic-industry-in-malaysia-marketing-essay.php?vref=1
2. Zhou, B.: Lean principles, practices, and impacts: a study on small and medium-sized enterprises (SMEs). Ann. Oper. Res. **241**, 457–474 (2012)
3. Schulze, M., Seuring, S., Ewering, C.: Applying activity-based costing in a supply chain environment. Intern. J. Prod. Econ. **135**(2), 716–725 (2012)
4. Maiga, A.S.: Advances in accounting, incorporating advances in international accounting assessing self-selection and endogeneity issues in the relation between activity-based costing and performance. Int. J. Cardiol. **30**(2), 251–262 (2014)
5. Zhang, X., Lee, C.K.M., Chen, S.: Supplier evaluation and selection: a hybrid model based on DEAHP and ABC. Int. J. Prod. Res. **50**(7), 1877–1889 (2012)
6. Dyk, J.V., Zubizarreta, E., Lievens, Y.: Cost evaluation to optimise radiation therapy implementation in different income settings: a time-driven activity-based analysis. Radiother. Oncol. **125**(2), 178–185 (2017)
7. Taha, S., Mortaji, H., Bagherpour, M., Mahdavi, M.: Fuzzy time-driven activity-based costing. Eng. Manage. J. **25**(3), 63–73 (2018)

8. Zaini, S.N.A.M., Abu, M.Y.: A review on time-driven activity based costing system in various sectors. J. Mod. Manuf. Syst. Technol. **2**, 15–22 (2019)
9. Stout, D.E., Propri, J.M.: Implementing time-driven activity-based costing at a medium-sized electronics company. Manage. Acc. Q. **12**(3), 1–11 (2011)
10. Abu, M.Y., Jamaludin, K.R., Zakaria, M.A.: Characterisation of activity based costing on remanufacturing crankshaft. Int. J. Autom. Mech. Eng. **14**(2), 4211–4224 (2017)
11. Kamil, N.N.N.M., Abu, M.Y.: Integration of Mahalanobis-Taguchi System and activity based costing for remanufacturing decision. J. Mod. Manuf. Syst. Technol. **1**, 39–51 (2018)
12. Abu, M.Y., Nor, E.E.M., Rahman, M.S.A.: Costing improvement of remanufacturing crankshaft by integrating Mahalanobis-Taguchi system and activity based costing. In: IOP Conference Series: Materials Science and Engineering, vol. 342, pp. 1–10 (2018)
13. Zheng, C.W., Abu, M.Y.: Application of Activity based Costing for Palm Oil Plantation. J. Mod. Manuf. Syst. Technol. **2**, 1–14 (2019)

Review on the Prominence of SMEs in Malaysia and Its' Imprint on University Industry Collaboration

Darshana Kumari Ragupathy(✉), Shamsuddin Baharin, and Faiz Bin Mohd Turan

Faculty of Mechanical and Manufacturing Engineering, University Malaysia Pahang, Pekan, Malaysia
Darshanamohe@gmail.com

Abstract. Small and Medium Enterprises (SMEs) represent a very large percentage in the Malaysian economy landscape and has a huge impact on University Industry Collaborations (UIC). In many developed and developing countries, UIC have clearly benefited both the industry and university on a large scale. However, Malaysia faces a different set of challenges to overcome in establishing intensified UIC. The representation of SMEs must be considered and amplified in a collaboration between industries and universities to ensure sustainability in the collaboration and continuous innovative development of both the industry and university. Herein, this paper explores the prominence of SMEs within the Malaysian economic landscape. It further examines the differences in policy direction and supply chain management for SMEs by developed economies. Based on that parameters, this paper also discusses the imprint SMEs has on UIC.

Keywords: SMEs · University · University industry collaboration

1 Introduction

SMEs are prominent in the Malaysian economy landscape. In total there are 920,624 companies and of this a huge 907,065 (98.5%) companies are SME establishments; only 5.3% are in the manufacturing sector. Although this percentage is very small comparatively to the services sector (89.2%), foreign direct investments in the manufacturing sector supersedes the services sector establishing the SMEs in the manufacturing sector significant not only in the economy but also in fostering sustainable University Industry Collaborations (UIC). For a developing economy like Malaysia, UIC has to be at a high scale, as it is critical for skills development, developing, acquiring and adopting knowledge as well as to promote entrepreneurship [1]. Government research policies have strongly emphasized cooperation between universities and industries as a key public policy in fostering innovation across the country. Amongst the many policies and initiatives, the 11th Malaysia Plan (2016–2020) gives preeminent importance on improving collaborations between Government, industries and universities on industry driven research initiative by outlining prerequisites for

M. N. Osman Zahid et al. (Eds.): iMEC-APCOMS 2019, LNME, pp. 96–100, 2020.
https://doi.org/10.1007/978-981-15-0950-6_15

innovation culture and supportive initiatives. However, despite such progressive policies and initiatives, sustainability in UIC is an issue faced by both university and the industry. While other countries have developed initiatives based on their economic strengths to bridge the gap between industry and university leading to demand driven innovations, Malaysia continues to struggle to fortify the collaborations between these entities. Herein, the question arises – what is the economic strength Malaysia possesses to foster a fortified UIC? While many researched the availability of infrastructure, human capital to do research, industry willingness and incentives to collaborate as variables to foster UIC, this paper is interested in exploring the prominence of SMEs within the Malaysian economic landscape as a variable to create, operate, foster and sustain university industry collaborations. It further examines the differences in policy direction and supply chain management by SMEs in developed economies. This paper also provides discussions on the imprint SMEs has on UIC.

2 SMEs in Malaysia - Landscape and Contribution to the Economy

In 2011, there was a total of 648,260 companies in the country. Of this, 638,790 (98.5%) were SMEs and 9,470 (1.5%) were large companies. The revision of the SME classification in 2014 saw an increment of more than 260,000 SME establishments in 2016. Based on the SME Annual Report 2016–2017, SMEs contributed 36.6% to the GDP and has an average annual growth rate of 6.5% (2011–2016). This growth was largely contributed by the services sector and it has the largest portion of microenterprises. On total exports, SMEs contributed 18.6%, where 9.4% came from services sector, 8.8% from manufacturing and 0.4% from agriculture. However, the manufacturing sector recorded the highest percentage of foreign investments (47%) compared to the services sector (20%). This shows that the foreign investors prefer the manufacturing sector. Hence, to attract more foreign investments, current policies on SMEs development and improvement on supply chain management should be focused upon (Fig. 1).

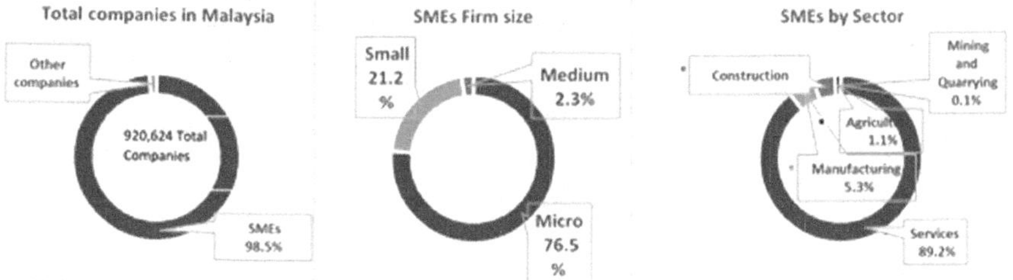

Fig. 1. A snapshot SMEs characteristic in Malaysia for year 2016 (Source: SME Annual Report 2016–2017)

3 SMEs in Malaysia - Landscape and Contribution to the Economy

SMEs in developed countries are well established and are a major entity in contributing to large firms' sustainability and profitability as well as the export market. Figure 2, compares basic SME demographics of Japan, South Korea and Germany respectively. The services sector is the largest sector in these economies, which is similar to the Malaysian landscape. SMEs in the developed economies contribute on an average of 50% to the GDP and total exports, contrary to the SMEs in the Malaysian economy. Nonetheless, for Malaysia, this contribution is commendable compared to the contribution by developing economies in the Organization for Economic Co-operation and Development (OECD) that stands at 33% (Meeting of the OECD Council at Ministerial Level, 2017). From a policy point of view, while Malaysia draws policies towards creating a contributive business environment for SMEs and entrepreneurship development, Japan on the contrary, draws policies for SMEs to specifically prevent the concentration of economic power through developing sound SMEs and creating jobs for the locals. In South Korea, SME policies are designed with an aim to diversify the economy ownership by concentrating more on small industries as compared to big conglomerates (known as *chaebols*). This was a necessary strategy policy to cope with some of the major economic and social restructuring issues facing the country. In Germany, SME policy outlines development for key technologies, promoting investments and utilizing existing skilled labour pool for SMEs. With these policies implemented, developed economies have managed to shape the objectives of SMEs growth for sustainable development. Moreover, SMEs in developed economies have grown to support the functions of each other. This is attributed to the good supply chain that has been established [2]. By meeting the customer requirement, close integration of internal functions within the firm and external linking with suppliers, customers and other stakeholders is established. This keeps the SMEs competitive in the global arena and grow profitability. However, in Malaysia, literature shows that SMEs generally lack of knowledge and information [3] on many aspects such supply chain, management practices [4], potential markets and customers, and application of ICT.

Fig. 2. Comparison on SMEs in Japan, South Korea and Germany (Source: SME Agency, Ministry of Economy, Trade and Industry, Japan 2013, National Statistical Office, Ministry of Strategy and Finance, South Korea 2013, KfW SME Panel 2003 to 2015)

4 SMEs and Its' Imprint on University Industry Collaboration

At this juncture, thorough understanding about SMEs, how they function in the economy and what more initiatives and incentives are needed to increase their contribution is pertinent for the Government to make strategic policy decisions. Equally and essential, Universities need to comprehend the economic landscape as a prerequisite to industry collaboration, considering the influence it has on UIC. SMEs focus essentially on business - profit, and desire the production and expansion of high-quality products/services. In contrary to the universities, SMEs prefer very limited knowledge sharing and respond reactively to market competition. This is because in business, sharing of knowledge/technology and response rate can directly affect the profit and market control of the enterprise. Also, SMEs are self-funded and even if they do provide funding, it is a very small proportion of academic research funds. This contribution is significant because it represents an area of growth and does become a prerequisite in attracting state and federal funds offered for innovation and local economic development [5]. Besides this, SMEs are regularly challenged and at times are at the mercy of larger companies which impedes their drive towards innovative transformation [6]. Nevertheless, SMEs has had its fair share of great innovations especially in industrious sectors, but these too is often indebted to larger firms. In addition to this, SMEs especially the microenterprises have been mushrooming excessively in the last number of years (2011: 496,458 (74%); 2016: 693,670 (76%)) with a failure rate of 60% after 5 years of establishment [7]. Investing in new technology developed in-house would require more resources and high risk as SMEs would have to venture into new markets and withstand well-established competitors. This lack, contributes to the high rate of bankruptcy within the first five years of operations. With low capital investments, they lack the ability to take risk or even take advantage on an opportunity of expansion. Ratification to this common situation within the SME sector would alternatively be to outsource R&D to universities. The ability to obtain a few new technologies for products and processes at a much-lowered cost would benefit SMEs long term.

5 Conclusion

In recent years, many initiatives have been put in place to allow SMEs contribution as part of the Malaysian economic strength in establishing strong UIC. The development dimensions of SMEs have particular relevance on the intensification and sustainability of UIC, hence executing pertinent measures primarily on policy direction and supply chain network will help develop SMEs further and set the momentum for university industry collaboration. The main contribution of this paper is that in a university industry collaboration equal amount of consideration and importance must be given to understanding of the economic landscape. In the collaboration, strategies and policies must be aligned towards improving the contributions of both SMEs and universities to the economic wellbeing.

References

1. Vaaland, T., Ishengoma, E.: University-industry linkages in developing countries: perceived effect on innovation. Educ. + Train. **58**, 1014–1040 (2016)
2. Sundram, V., Ibrahim, A.R., Govindaraju, C.: Supply chain management practices in the electronics industry in Malaysia. Benchmark. Int. J. **18**, 834–855 (2011)
3. Saleh, A.S., Ndubisi, N.O.: An evaluation of SME development in Malaysia. Int. Rev. Bus. Res. **2**(1), 1–14 (2006)
4. Zailani, S., Jeyaraman, K., Vengadasan, G., Premkumar, R.: Sustainable supply chain management (SSCM) in Malaysia: a survey. Int. J. Prod. Econ. **140**, 330–340 (2012)
5. Etzkowitz, H., Leydesdorff, L.: Introduction to special issue on science policy university-industry-government relations. J. Sci. Public Policy **24**, 2–5 (1997)
6. Ranga, M., Etzkowitz, H.: Triple Helix systems: an analytical framework for innovation policy and practice in the knowledge society. J. Ind. Higher Education **27**, 237–262 (2013)
7. Chong, W.Y.: Critical success factors for small and medium enterprises: perceptions of entrepreneurs in urban Malaysia. J. Bus. Policy Res. **7**(4), 204–215 (2012)

Critical Success Factors that Affect Implementation of Construction Project in Improving Project Performance: A Case of Cement Plant Construction Industry

Ahmad Subekti[1(✉)], Nilda Tri Putri[2], and Henmaidi[2]

[1] Semen Padang Co., 25237 Padang, Indonesia
ahmadsubekti.ptsp@gmail.com
[2] Department of Industrial Engineering, Andalas University, 25163 Padang, Indonesia
{nilda,henmaidi}@ft.unand.ac.id

Abstract. Construction projects are often completed with large cost overruns, late schedules, and problems related to quality. The success rate of the current project in Semen Padang Co. cement plant construction industry is very low. The purpose of this study is to identify the delay factors that exist by revealing fundamental problems that affect project completion and determine the most influential CSF (Critical Success Factors) in improving the performance of cement plant construction projects from the perspective of the owner, contractor, and consultant. The research methodology uses Delphi interviews to identify CSF and the AHP (Analytical Hierarchy Process) method by pairwise comparisons to prioritize CSF rankings in the implementation of a cement plant construction project at Semen Padang Co. The results of the 3 (three) rounds of Delphi interview, the expert panel identified that the most dominant category of CSF affecting were project management, procedures, commitments, and implementation methods. While for factors include procedures and bureaucracy, plans and schedules used, project monitoring, commitment of all parties, accuracy of implementation methods, control mechanisms, placement of workers in accordance with experience, and the implementation of procurement systems. This was identified as the most important CSF to be taken into account when considering how to improve the performance of cement plant construction project. Future research will create a tool to control risk and evaluate the next cement plant construction projects based on CSF that we obtained in this study.

Keywords: Project management · Critical success factors · Delphi · AHP approach · Cement plant construction industry

1 Introduction

Large and complex construction projects are more difficult to succeed in developing countries [15]. Research on the current relevant literature shows that construction projects are often completed with large cost overruns, late schedules, and quality problems [13]. Delays in construction projects have a major impact on the financial

M. N. Osman Zahid et al. (Eds.): iMEC-APCOMS 2019, LNME, pp. 101–107, 2020.
https://doi.org/10.1007/978-981-15-0950-6_16

results of the project concerned, therefore it is important to address the causes of the delay [9]. Some studies directly investigate delays and try to identify the causes and also ways to avoid those [1].

The study of project success and critical success factors (CSF) is often considered as one of the important ways to increase the effectiveness of project delivery [2]. The concept of CSF provides an astute way of identifying certain factors that tend to make the project successful. Several studies have directly tried to identify critical success factors in construction project management [5, 8].

The construction project of the cement plant at Semen Padang Co. is often not completed according to the planned schedule. Settlement of civil construction is a sequence of mechanical and electrical work that affects the completion of the overall project time. The purpose of this study is to identify the delay factors that exist in the cement plant construction industry by revealing fundamental problems that affect project completion and determine the most influential critical success factor (CSF) in improving the performance of cement plant construction projects.

2 CSF of Implementation Construction Project: DHP Methodology

In this study, the researchers used the DHP methodology to evaluate the success of CSF in implementing the cement plant construction project at Semen Padang Co. DHP is a combination of the Delphi and AHP techniques. According to Dilworth [4], the Delphi Method is a systematic tool for obtaining consensus from a group of experts (panel). The Delphi method is expected to get opinions, consensus or problems qualitatively.

Ciptomulyo [3], the Delphi Method is used as a group opinion polling whose participation consists of experts who have competence in their fields. The Delphi method is considered appropriate for obtaining opinions to formulate an objective vision with the condition when the expected subjective factors are very important or when accurate quantitative data is difficult to obtain. The aim is to reduce the negative effects of interaction groups and to get the most reliable consensus from the opinion of a group of experts [7]. The characteristic of the Delphi Method is that all participants are treated as strangers [14]. This aims to reduce the influence of shy or offensive feeling to other participants who are considered to have a higher influence or position. The approach can be done by interviewing directly or using electronic mail and then comparing and analyzing the responses of the experts and then reporting back to the participants to get a response [6].

AHP (Analytical Hierarchy Process) is a functional hierarchy with the main input of human perception. This method was developed by Prof. Thomas Lorie Saaty from Wharton Business School in the early 1970s [11], which was used to search rankings or priority sequences of various alternatives in solving a problem. AHP describes multi-factor problems or complex multi criteria into a hierarchy. With hierarchy, a complex problem can be broken down into groups which are then organized into a hierarchical form so that the problem will appear more structured and systematic.

The chosen respondents consisted of experts with a minimum of five years of experience in the field of construction. Respondents in the study include owner, contractor and consultant. The job levels chosen were starting from the manager and from various disciplines includes civil, mechanical and electrical.

For the first round, the questionnaires were sent the correspondents' email addresses and also through direct interviews. The questionnaire was also designed to allow additional factor suggestions. If there was any suggestion of additional factor, it would be printed for the second round. After the respondent returned the first round questionnaire, the results were analyzed, tabulated and returned to the respondent for further consideration, along with the questionnaire for the second round. From the first round of Delphi, the experts identified 20 sources of factors (categories) and identified the initial factors with 122 dominant factors affecting the implementation of cement plant construction projects. The first round of responses to the Delphi interview were tabulated and analyzed using a distribution ranking table to explain the average rank, top and bottom ranks of the importance of each factor. From the first round of Delphi evaluation, 20 sources of factors (categories) and 67 factors were produced which later became the second round of the Delphi questionnaire.

In the second round of questionnaires, each respondent was asked to review the original response about the rank of critical success factors, and compare them with the factors from the entire group, before making a final decision. The aim of the second round was to re-determine and confirm the critical success factors identified by the experts in the first round. In addition, the second round increased the chances of getting a deeper perspective on the evaluation process.

The strategy ranking responses from the second round of the Delphi interview were tabulated, averaged, and used to validate each ranking of importance strategies. The ranking results above the average in the second round of Delphi were from 20 sources of factors (categories) sorted into 13 sources of factors (categories), and from the second round factors, as many as 67 factors were sorted into 46 dominant factors that affect the implementation of cement plant construction projects. The evaluation results from 13 categories were combined into 9 categories (sources of factors), and the 46 factors were combined into 35 factors. The results of the second round Delphi evaluation were then used for the third round of the Delphi process.

The purpose of the third round was to create consensus or agreement among experts based on evaluation in the second round. The conclusion of the evaluation of the results of the third round was that all respondents agreed on the 9 (nine) sources of factors (categories) and the 35 (thirty five) factors, except for 1 (one) respondent who did not provide feedback.

From the results of the expert panel consensus on the third round of Delphi, then a pairwise comparison of the critical success factors was conducted. With pairwise comparison data, we can get priority and ranking for each category and the dominant factors influencing the implementation of effective and successful cement plant construction projects. In this AHP process, researchers can obtain the weight of local and global priorities. Local weight is the priority of an element associated with the previous and first element calculated. Meanwhile the global weight of each element related to the goal of successful CSF implementation on construction projects is calculated by multiplying the local weight of an element by the weight of the previous element.

The next stage of the AHP approach was to calculate the consistency ratio (CR) to measure how consistent the assessment was by the expert panel. Because comparisons were made through personal or subjective judgments, some inconsistency can occur. The results of the study indicated that the overall consistency of evaluator evaluations was in an acceptable ratio of 0.10 as suggested by Saaty [12].

After all pairwise comparisons were carried out at each level, and proven to be consistent, then the next step was to synthesize evaluations from evaluators to rank each category and factor. The geometric mean approach as suggested by Saaty and Vargas [10] was by synthesizing the evaluations of each evaluator. The results of the geometric mean of evaluators were combined into a pairwise comparison matrix.

3 Result and Discussion

Using the DHP methodology, which is a combination of Delphi and AHP techniques, the expert panel identified CSF which consisted of nine categories (source of factors) and 35 factors that influenced the implementation of the cement plant construction project at Semen Padang Co., which was validated through three Delphi method rounds.

Table 1. Local and global normalized weights of judgments from panel of expert

Category/main factors	Local weight	Factors	Local weight	Global weight
Project management	22.75%	Communication system (project coordination meeting)	9,04%	2,06%
		Control mechanism (project control system)	24,30%	5,53%
		Project monitoring (monitoring during project implementation)	25,53%	5,81%
		Plans and schedules used (time or duration of work)	27,92%	6,35%
		Risk identification and allocation	13,22%	3,01%
Contractor	10.58%	Contractor experience in construction projects	18,35%	1,94%
		Technical and professional abilities	24,59%	2,60%
		Field management including materials and equipment	15,48%	1,64%
		Supervision of construction projects	8,13%	0,86%
		Timeliness of ordering material	6,73%	0,71%
		Planning material procurement schedule	8,42%	0,89%
		Good flow of funds (cash flow)	13,92%	1,47%
		Speed of information flow and coordination	4,39%	0,46%

(continued)

Table 1. (*continued*)

Category/main factors	Local weight	Factors	Local weight	Global weight
Project manager	6.72%	Project manager competency	38,02%	2,56%
		Project manager experience	11,37%	0,76%
		Expertise in leading project	18,01%	1,21%
		The ability of the project manager to delegate authority	7,98%	0,54%
		Accuracy of decisions made by project managers	14,72%	0,99%
		Speed of making decisions from the project manager	9,90%	0,67%
Specification	7.04%	Achievement of specifications	36,60%	2,58%
		Complete design and detailed design	63,40%	4,46%
Manpower (human resource)	9.27%	Worker placement in accordance with experience in their field	56,82%	5,27%
		Use of skilled workers in their fields	43,18%	4,00%
Equipment	6.51%	Quality of equipment used	18,99%	1,24%
		Compatibility of specifications of equipment used	50,99%	3,32%
		Number of equipment used	30,02%	1,95%
Construction method	10.65%	The accuracy of the method in carrying out the work	53,03%	5,65%
		Physical work environment (natural disasters, weather, pollution)	14,48%	1,54%
		Interface (overlapping) with other units (civil, mechanical, electrical)	19,25%	2,05%
		Access road conditions	13,24%	1,41%
Commitment	13.18%	Commitment of all parties to the project	42,97%	5,66%
		Top management support	29,82%	3,93%
		The purpose and scope of the project	27,21%	3,59%
Procedure	13.29%	System for implementing project procurement	38,99%	5,18%
		Procedure and bureaucratic of the construction project	61,01%	8,11%
Total category	**100%**	**Total Global Weight**		**100%**

From Table 1, these were identified as the most important CSF to take into account when considering how to improve the performance of a cement plant construction project.

4 Conclusion and Further Research

The conclusion of this study is that the most dominant categories affecting CSF include project management, procedures, commitments, and implementation methods. While for factors include procedures and bureaucracy, plans and schedules used, project monitoring, commitment of all parties, accuracy of implementation methods, control mechanisms, placement of workers in accordance with experience, and the implementation of procurement systems.

Finally, the critical success factors in the construction process investigated in this study forms an empirical study for future research on the critical success factors in the cement plant construction projects. Further research can be done by making applications to control risk and evaluate the next cement plant construction project based on CSF results in this study.

References

1. Andresen, P.E., Landmark, A.D., Hajikazemi, S., Johansen, A., Andersen, B.: Remedies for managing bottlenecks and time thieves in Norwegian construction projects – public vs private sector. International Project Management Association. Westin Playa Bonita, Panama (2015)
2. Chan, A.P.C., Scott, D., Chan, A.P.L.: Factors affecting the success of a construction project. J. Constr. Eng. Manage. **130**(1), 153–155 (2004)
3. Ciptomulyo, U.: Integrasi metode Delphi dan Prosedur Analisis Hierarkhis (AHP) untuk identifikasi dan penetapan prioritas objektif/kriteria keputusan. J. IPTEK **12**, 42–52 (2001)
4. Dilworth, J.B.: Operation Management: Design, Planning and Control for Manufacturing and Services. McGraw Hill, New York (1992)
5. Garbharran, H., Govender, J., Msani, T.: Critical success factors influencing project success in the construction industry. Durban University of Technology, Durban, South Africa (2012
6. Graham, B., Regher, G., dan Wright, J.G.: Delphi as a method to establish consensus for diagnostic criteria. J. Clin. Epidemiol. **56**, 1150–1156 (2003)
7. Gupta, U.G., dan Clarke, R.E.: Theory and application of the delphi technique: a bibliography (1975–1994). Technol. Forecast. Soc. Change **53**(2), 185–211 (1996)
8. Ihuah, P.W., Kakulu, I.I., Eaton, D.: A review of critical project management success factors (CPMSF) for sustainable social housing in Nigeria. Int. J. Sustain. Built Environ. **3**(1), 62–71 (2014)
9. Khoshgoftar, M., Bakar, A.H.A., Osman, O.: Causes of delays in Iranian construction projects. Int. J. Constr. Manage. **10**, 53–69 (2010)
10. Saaty, T.L., Vargas, L.: Decision Making for Leaders. Lifetime Learning Publications, Pittsburgh (1982)
11. Saaty, T.L.: Decision Making For Leaders (The Analytical Hierarchy Process for Decision in a Complex World). University of Pittsburgh, Pittsburgh (1970)
12. Saaty, T.L.: Multi Criteria Decision Making: The Analytic Hierarchy Process. RWS Publications, Pittsburgh (1988)

13. Salleh, R.: Critical success factors of project management for brunei construction projects: improving project performance. School of Urban Development, Faculty of Built Environment and Engineering, Quensland University of Technology, Brunei (2009)
14. Schroeder, R.G.: Operation Management: Contemporary Concepts and Cases. Mc Graw-Hill, New York (2000)
15. Swan, W., Khalfan, M.M.A.: Mutual objective setting for partnering projects in the public sector. Eng. Constr. Archit. Manage. **14**(2), 119–130 (2007)

Proposing of Mahalanobis-Taguchi System and Time-Driven Activity-Based Costing on Magnetic Component of Electrical & Electronic Industry

Nik Nurharyantie Nik Mohd Kamil, Mohd Yazid Abu(✉), Nurul Farahin Zamrud, and Filzah Lina Mohd Safeiee

Faculty of Manufacturing Engineering, Universiti Malaysia Pahang, 26600 Pekan, Pahang, Malaysia
niknurharyantie@gmail.com, myazid@ump.edu.my

Abstract. In the complex and dynamic environment today, it is very important for Electrical & Electronic (E&E) industry to know the better quality and profitability of magnetic component. The objective of this work is proposing of Mahalanobis-Taguchi System (MTS) and Time-Driven Activity-Based Costing (TDABC) on production of E&E industry. In general, MTS is used to evaluates the parameters in workstation either critical or non-critical parameter while TDABC requires estimate two key parameters: (1) develop time equation to determine the estimated time for each activity and (2) calculate capacity cost rate of each sub-activity. Furthermore, based on collecting the data for a workstation visual mechanical inspection, MTS can identify the diagnosis parameters while TDABC can measure the unused capacity in term of resources and time for each process in a workstation. Eventually, MTS found that Mahalanobis Distance (MD), 7 and TDABC predicted MYR62659.55.

Keywords: Diagnosis · Mahalanobis-Taguchi System · Time-Driven Activity-Based Costing

1 Introduction

MTS and TDABC are proposed on production of E&E industry. Due to today's competitive market condition, the quality and profitability of component is very important to stay competitive in the market. The high contribution of diagnosis parameters at VMI must be focus to avoid many rejected component and profit loss at E&E industry.

MTS is a method to diagnosis and predict the system performance that use multivariate data to make a quantitative decision with construction [1]. In multivariate system, decision making can be analyzed when information provided one or more variable. The advantages of MTS are to select the important parameters to improve the quality of product and process [2], easily determined the abnormality of observation, and considered correlation variables of combine all the variables within Mahalanobis Distance [3]. Yazid [11] provided a systematic pattern recognition using MTS by

M. N. Osman Zahid et al. (Eds.): iMEC-APCOMS 2019, LNME, pp. 108–114, 2020.
https://doi.org/10.1007/978-981-15-0950-6_17

constructing a scatter diagram while Abu [12] classified crankshafts' end life into recovery operations. Kamil [13] developed a distinctive pattern of crankshaft and identify the critical and non-critical parameter of crankshaft based on the MTS while Abu [14] identified the critical and non-critical variables during remanufacturing process. Abu [15] evaluated the criticality of parameters on the end of life crankshaft and finally estimate the true cost. As mention by Gonzalez [4], there are applications and percentages of MTS which are manufacturing, automotive, information technology and others, healthcare, academic and agriculture. TDABC is a costing model that consider a time that provided the cost of activities with base that consume of time per activities that shows the differences between the total time needed to carry all activities that performed in departments [5]. The advantages of TDABC are can be used quickly and easily, can be applied in budgeting activities and planning capacity, and can be used with complex processes, systems or organizational structures of different industries [6]. According to Zaini [7] found that the percentage of TDABC application belong to 66% of healthcare service, 23% of industrial service, and 11% of library service. The applications and percentages of TDABC are healthcare (66%), manufacturing and academic (14%), retail trade (3%), others (3%), and information technology (2%) [4].

2 Research Methodology and Results

There are 4 stages for applying MTS which are construction of (1) Mahalanobis Scale (MS) which determine and define the main characteristics and need to identify the target by collecting the data, (2) Validation of MS that identify off-target group, (3) Identify the useful variables to get best criteria, and (4) Prediction of future diagnosis which obtain new data sets of useful variables [8]. There are seven steps for applying TDABC to produce magnetic component which are identify activity and sub-activity of production, determine all resources cost used including labor, material, maintenance, and consumable, calculate capacity cost rate by estimation of practical capacity, develop time equation to calculate production cost, determine time estimation for each activity, and calculate the capacity required of each activity [9]. The research methodology and results as follows by collecting data at main workstation of rejected component which is VMI workstation.

2.1 MTS

MTS is used to diagnosis parameters of VMI workstation at E&E industry. There are 10 parameters of rejected component in VMI which are number of scratches, height of insulator wire, condition of stripping, number of epoxy at header corner, number of epoxy spot, condition of marking, distance of winding gap, type of coating damage, height of excess plastic header, and level of tinning. The data was considered on 24 January on target 5000 pieces/day. The valid percentage belongs 99.89% while the outliers belong 0.11%. If the outliers over 25%, samples data may need to be reevaluated. Figure 1 illustrates MD for off target participants. In this work, 411 off target participants were identified from 5000 participants. The threshold is

(0.60117,6.733753). Sample number 202 shows the lowest MD off target while sample number 1019 shows the highest MD off target.

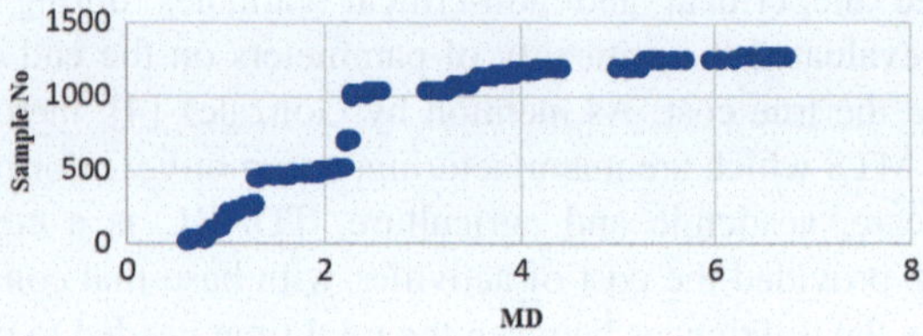

Fig. 1. Scatter diagram of MD for off target participants

All the parameters are optimized to screen the critical parameters. Table 1 shows the optimized parameters which are number of scratches (1), height of insulator wire (2), condition of stripping (3), number of epoxy spot (4), condition of marking (5), distance of winding gap (6), type of coating damage (7), and height of excess plastic header (8).

Table 1. Optimization parameters

Impact	1	2	3	4	5	6	7	8
In-control	56	46		75	4	44	66	48
Above 3 standard deviations of target	29	34	11	7	12		4	2
Below 3 standard deviations of target						41		
Total	85	80	11	82	16	85	70	50

Table 2 shows the variable and distance contribution (percentage) for diagnosis and prediction. As an example, sample number 1046 belongs height of insulator drives only 0% with distance of winding gap being key driver, 92%. It means, for this sample, distance of winding gives high contribution of rejected parameter and the deviation is below acceptable range of target group.

Table 2. Variable and distance contribution (percent)

No of samples	Target Criteria Value: Mechanical yield	Distance ▲	Number of scratches	Height of insulator wire	Condition of stripping	Number of epoxy spot on winding coil	Condition of marking	Distance of winding gap	Type of coating damage	Height of excess plastic header
1046	92	7	1	0	1	2	1	92	3	3
1045	90	7	3	0	1	2	1	92	2	3
200	90	1	29	2	1	1	2	11	56	1
199	92	1	44	3	4	4	2	13	32	3

2.2 TDABC

There are seven stages to analyze result of TDABC. Generally, TDABC is used to analyze and identify the manufacturing activities.

Initial stage is identification of production activity with sub-activity started from winding process until the packaging as shown in Fig. 2.

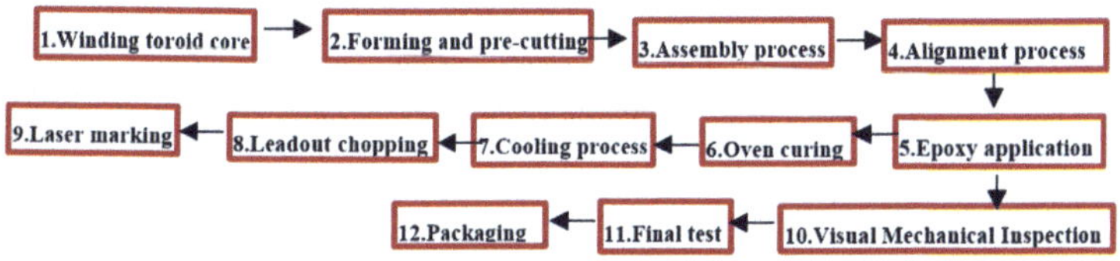

Fig. 2. Production process of magnetic component at E&E industry

Secondly, estimated the total cost of the resource supplied. The cost can be allocated directly when the resources is sharing the same activity if not, the cost-driver is considered to allocate the cost. Table 3 shows the labor costs of five employees at a cost MYR 66,000. Due to private and confidential, no costing listed for maintenance, material, and consumable cost.

Table 3. Labor, maintenance, material, and consumable cost of VMI

Workstation	Sub-activities	Labor cost (MYR)	Maintenance cost (MYR)	Material cost (MYR)	Consumable cost (MYR)	Cost of all resources supplied (MYR)
VMI	1. Inspect the winding gap by using gauge	13,200	Nil	Nil	Nil	13,200
	2. Inspect leadout pitch and length by using gauge	13,200	Nil	Nil	Nil	13,200
	3. Checking the unit by using magnifying glass based on VMI requirement	39,600	Nil	Nil	Nil	39,600
	Total	**66,000**	**Nil**	**Nil**	**Nil**	**66,000**

Thirdly, estimate practical capacity of production. The E&E company's working hours are Monday to Saturday, 7.30 a.m. to 5.30 p.m. within 2 shift. An employee works of eight hours 35 min a day, for 20 days for a month and 240 days for a year.

Deduction for breaks is 45. So, an employee has acceptable practical capacity of 10,300 min each per month and 123,600 min each per year.

Subsequently, this work needs to determine the capacity cost rate for VMI. For sub-activity 1 is MYR13,200/123,600 min is equal MYR0.11 per minutes. Table 4 shows the summary of capacity cost rate of VMI sub-activity.

Table 4. Capacity cost rate of VMI sub-activity to produce magnetic component

Workstation	Sub-activities	Cost of all resources supplied (MYR/year)	Practical capacity (min/year)	Capacity cost rate (MYR/min)
VMI	1. Inspect the winding gap using gauge	13,200	123,600	0.11
	2. Inspect the leadout pitch and length by using gauge	13,200	123,600	0.11
	3. Checking the unit by using magnifying glass based on VMI requirement	39,600	370,800	0.11
	Total	**66,000**	**618,000**	

After that, calculate estimated production time and then estimate capacity required that determined by quantifying the frequency of the activity in a year. Table 5 summarized time equation multiplied by relevant cost driver for VMI.

Table 5. Time equations and volume of cost driver for VMI workstation

Variable	Sub-activities	Driver	Time equation	Quantity/year
X1	1. Inspect the winding gap by using gauge	Number of boundary jig gauge (unit/year)	$1.44X_1$	2
X2	2. Inspect the leadout pitch and length by using gauge	Number of leadout and pitch jig gauge (unit/year)	$0.57X_2$	2
X3	3. Checking the unit by using magnifying glass based on VMI requirement	Number of units pass with VMI Requirement (pieces/year)	$0.01X_3$	1200000

As an example, time equation is developed as shown in Eq. 1 below.

$$T_{VMI} = 1.44X_1 + 0.57X_2 + 0.01X_3 \quad (1)$$

As an example, the actual time of VMI per year was determined by substituting the volume of cost-drivers from Table 5. The actual time spent = (1.44x2) + (0.57x2) + (0.01x1200000) = 12004.02 min. Table 6 refers the production cost of VMI activity that inspect the component.

Table 6. Elapsed time and total production costs of sub-activities

Workstation	Sub-activities	Used time (min)	Capacity cost rate (MYR/min)	Total cost (MYR/year)
VMI	1. Inspect the winding gap by using gauge	2.88	0.11	0.32
	2. Inspect the leadout pitch and length by using gauge	1.14	0.11	0.13
	3. Checking the unit by using magnifying glass based on VMI requirement	12,000	0.11	1320
	Total	**12004.02**		**1320.45**

Equation 2 is presented the time equation of all the total time spent in the production.

$$T_{\text{production of magnetic component}}$$

$$= 1.44X_1 + 0.02X_2 + 0.04X_3 + 1.15X_4 + 0.13X_5 + 0.28X_6 + 0.03X_7 + 0.03X_8 + 0.07X_9 + 0.02X_{10} + 0.03X_{11} + 0.87X_{12} + 0.67X_{13} + 0.65X_{14} + 5X_{15} + 0.99X_{16} + 0.08X_{17} + 0.84X_{18} + 0.84X_{19} + 1.05X_{20} + 0.02X_{21} + 0.04X_{22} + 0.02X_{23} + 1.34X2_4 + 1.34X2_5 + 1.54X_{26} \quad (2)$$

3 Conclusion and Discussion

The sub-activity 3 incurs a lot of waste, MYR35468, followed sub-activity 2 at MYR13595.87 and sub-activity 1 at MYR13598.68 of waste. Based on analysis, TDABC can forecast capacity either the cost is increased or decreased for the next year. Analysis as shown in Table 2 found that MTS can diagnosis and prediction for each sample number either the parameter is contributed or not for rejected the component. The bigger value of key driver, the higher contribution to reject the sample number.

Acknowledgement. This research is fully supported by RDU190156 and FRGS/1/2018/TK03/UMP/02/34. The authors fully acknowledge University Malaysia Pahang for the approved fund which makes this important research viable and effective.

References

1. Cudney, E.A., Paryani, K., Ragsdell, K.M.: Applying the Mahalanobis-Taguchi system to vehicle handling. Concurr. Eng. **14**, 343–354 (2006)
2. Peng, C., Tsai, S.: Applying the Mahalanobis-Taguchi System to improve tablet PC production processes, pp. 1–17 (2017)
3. Resendiz-Flores, E.O., Navarro-Acosta, J.A., Mota-Gutierrez C.G., Reyes-Carlos, Y.I.: Fault detection and optimal feature selection in automobile motor-head machining process, pp. 2613–2622 (2018)
4. Gonzalez, M., Nachtmann, H., Pohl, E.: Time-driven activity-based costing for health care provider supply chains. Eng. Econ. **62**(2), 161–179 (2017)
5. Bagherpour, M., Nia, A.K., Sharifian, M.: Time-driven activity-based costing in a production planning environment. **227**(2), 333–337 (2013)
6. Kaplan, R.S., Anderson, S.R.: Time-Driven Activity Based Costing: A Simpler and More Powerful Path to Higher Profits. Harvard Business School Publishing Co., Brighton (2007)
7. Zaini, S.N.A.M., Abu, M.Y.: A review on time-driven activity based costing system in various sectors. J. Mod. Manuf. Syst. Technol. **2**, 15–22 (2019)
8. Ghasemi, E., Aaghaei, A., Cudney, A.: Mahalanobis Taguchi system: a review. Int. J. Qual. Reliabil. Manage. **32**(3), 291–307 (2015)
9. Erhun, F., Mistry, B., Platchek, T., Milstein, A., Narayanan, V. G., Kaplan, R.S.: Time-driven activity-based costing of multivessel coronary artery bypass grafting across national boundaries to identify improvement opportunities: study protocol, pp. 1–8 (2018)
10. Pongwasit, R., Chompu, R.: Analysis of wooden toy manufacturing costs through the application of a time-driven activity-based costing system. Mem. Muroran Inst. Tech. **65**, 7–14 (2015)
11. Yazid, A.M., Rijal, J.K., Awaluddin, M.S., Sari, E.: Pattern recognition on remanufacturing automotive component as support decision making using Mahalanobis-Taguchi system. In: 12th Global Conference on Sustainable Manufacturing, vol. 26, pp. 258–263 (2015)
12. Abu, M.Y., Jamaluddin, K.R., Zakaria, M.A.: Classification of crankshaft remanufacturing using Mahalanobis-Taguchi system. Int. J. Autom. Mech. Eng. **13**(2), 3413–3422 (2017)
13. Kamil, N.N.N.M., Abu, M.Y.: Integration of Mahalanobis-Taguchi system and activity based costing for remanufacturing decision. J. Mod. Manuf. Syst. Technol. **1**, 39–51 (2018)
14. Abu, M.Y., Nor, E.E.M., Rahman, M.S.A.: Costing improvement of remanufacturing crankshaft by integrating Mahalanobis-Taguchi system and activity based costing. In: IOP Conference Series: Materials Science and Engineering, vol. 342, pp. 1–10 (2018)
15. Abu, M.Y., Norizan, N.S., Rahman, M.S.A.: Integration of Mahalanobis-Taguchi system and traditional cost accounting for remanufacturing crankshaft. In: IOP Conference Series: Materials Science and Engineering, vol. 342, pp. 1–9 (2018)

Optimization of Utilities Capacity at Aircraft Heavy Maintenance Center Using Linear Programming Models

M. Johny Ali Firdaus, Muhammad Gharutha, and Rachmawati Wangsaputra(✉)

Industrial Engineering Program Study, Bandung Institute of Technology, Jl. Ganesha 10, Bandung, West Java 40132, Indonesia
rachmawati_wangsaputra@yahoo.com

Abstract. In aviation industry, aircraft maintenance is critical in ensuring safe flight to end customers. This study evaluates the aircraft maintenance services performance in an Indonesian aircraft manufacturer and services company with the intention of developing a model for maintenance optimization. The optimization model was build based on Mixed Integer Programming model and solved using Microsoft Excel's Solver feature. The model calculates optimum profit based on two main factors, the amount of aircraft serviced per type and maintenance resources utilization. Aircraft serviced in this company can be divided in two type, fixed wing and rotary wing, with each type having its own aircraft models with their own variations of maintenance service inspections. Resources utilization are measured by the following factors, manpower, serviceable area (hangar size) and capital. The result shows that the optimum model able to attain an average of 90% resource utilization with total revenue around IDR 64 billion (about USD 4.5 million with exchange rate of IDR 14.000 per dollar).

Keywords: Mixed Integer Programming · Linear Programming · Aircraft maintenance · Maintenance utility · Profit optimization · Resource utilization · Indonesia aircraft maintenance services

1 Introduction

Aircraft maintenance service in an Indonesian aircraft manufacturing company has been under performing since 2015 to 2017. It was indicated by low number of planes serviced with the average of four planes a year, high operator idle time, and low hangar utilization. This situation was further aggravated by management unable to plot a clear target for maintenance service. In order to solve this issue a model that can state optimum aircraft servicing configuration is required in order to assist management team to plan maintenance target depending on available resource and profitability.

M. N. Osman Zahid et al. (Eds.): iMEC-APCOMS 2019, LNME, pp. 115–120, 2020.
https://doi.org/10.1007/978-981-15-0950-6_18

2 Literature Review

The basic model used in this research is the Mixed Integer Programming by Dakin [1]. This Mixed Integer Programming basic model can be seen on Eq. (1).

Objective function

$$Maximize\ Z = \sum\nolimits_{j=1}^{n} (c_j)(x_j) \tag{1}$$

Subject to

$$\sum\nolimits_{j=1}^{n} (a_{ij})(x_j) \le (b_i)$$
$$\text{for } i = 1, 2\ldots\text{n},\ X_j \ge 0,\ \text{integer for j} = 1, 2\ldots\text{m} \tag{2}$$

3 Mathematical Model Formulation

The next step is the development of these models that relate to the conditions in the research. Within the scope of this research the notations used in linear programming modeling are as follows:

3.1 Parameters

T_{ijk}: time in days, for i aircraft type at j aircraft variation and k inspection type
To: additional work in hour
Dm: time in days per month
Ak: accommodation cost per hour additional work
A_{ij}: area dimension, for i aircraft type at j aircraft variation
P_{ijk}: manpower required, for i aircraft type at j aircraft variation and k inspection type
M_{ijk}: material cost, for i aircraft type at j aircraft variation and k inspection type (IDR)
PX_i: total manpower, for i aircraft type
PX_{ij}: total manpower in accordance with i aircraft type and j aircraft variation
P_x: amount of manpower more than one personal rosters in accordance with i aircraft type and j aircraft variation
Pt_{ijk}: revenue, for i aircraft type at j aircraft variation and k inspection type (IDR)
Lef: hangar area dimension per day
Rl: average hangar unused per day
TF: total time to be needed in days
GP: manpower cost per month
M: total capital for material (IDR)
N: constant based on personal roster N:1 for two personal rosters, N:2 for three personal rosters and N:(n-1) for n personal rosters.

3.2 Decision Variable

X_{ijk}: amount of aircraft based on i aircraft type at j aircraft variation and k inspection type

$X_{ijk} \geq 0$, integer

3.3 Objective Function Formulation

$$Zmax = \sum_{i=1}^{n}\sum_{j=1}^{m}\sum_{k=1}^{q}(Pt_{ijk})(X_{ijk}) - \left(\left(\sum_{i=1}^{n}\sum_{j=1}^{m}\sum_{k=1}^{q}(To)\left(\frac{GP}{8(Dm)}+(Ak)\right)(P_{ijk})(T_{ijk})(X_{ijk})\right)+(TF)\left(\frac{GP}{Dm}\right)\sum_{i=1}^{n}(PX_i)\right) \quad (3)$$

Subject to

$$\sum_{i=1}^{n}\sum_{j=1}^{m}\sum_{k=1}^{q}\frac{A_{ij}}{Day}(T_{ijk})(X_{ijk}) \leq (TF)\frac{(Lef - Rl)}{Day} \quad (4)$$

$$\sum_{i=1}^{n}\sum_{j=1}^{m}\sum_{k=1}^{q}(P_{ijk})(T_{ijk})(X_{ijk}) \leq (TF)\left(\sum_{i=1}^{n}PX_i\right) \quad (5)$$

$$\sum_{i=1}^{n}\sum_{j=1}^{m}\sum_{k=1}^{q}(P_{ijk})(T_{ijk})(X_{ijk}) \leq (TF)\left(\sum_{i=1}^{n}\sum_{j=1}^{m}PX_{ij}\right) \quad (6)$$

$$\sum_{i=1}^{n}\sum_{j=1}^{m}\sum_{k=1}^{q}(P_{ijk})(T_{ijk})(X_{ijk}) \leq (TF)\left(\sum_{i=1}^{n}\sum_{j=1}^{m}(PX_{ij}) - N(P_x)\right) \quad (7)$$

$$\sum_{i=1}^{n}\sum_{j=1}^{m}\sum_{k=1}^{q}(M_{ijk})(X_{ijk}) \leq (M) \quad (8)$$

$$\sum_{i=1}^{n}\sum_{j=1}^{m}\sum_{k=1}^{q}\left(\frac{GP}{Dm}\right)(P_{ijk})(T_{ijk})(X_{ijk}) \leq (TF)\left(\frac{GP}{Dm}\right)\sum_{i=1}^{n}PX_i \quad (9)$$

The objective function (3) is the optimization of the total profit, which is the through value (revenue) minus total manpower cost available and overtime cost if any. Constraint (4) is indicating sum of aircraft areas per day can be able to maintenance based on hangar areas minus possible of aircraft area based on manpower and time to aircraft maintenance per day available. Constraint (5) is indicating manpower used in aircraft maintenance based on total manpower available and total manpower aircraft type (fixed wing or rotary wing). Constraint (6) is indicating manpower available based on personal a roster. Constraint (7) indicating manpower available based on more than personal a roster. Constraint (8) is indicating material used in aircraft maintenance based on capital available. Constraint (9) is indicating manpower cost contribution in aircraft maintenance based on total manpower cost.

4 Implementation Model

The development of a Mixed Integer Linear Programming (MILP) model in this research was carried out at the aircraft maintenance center which can maintain 2 (two) types of aircraft. The First is fixed wing aircraft three variations, namely B737-s, CN235, and NC212. The other is rotary wing aircraft, namely AS332, Bell412 and BO105 helicopters. Each type has their own service inspection type that can be seen on Table 2 under inspection type column.

4.1 Manpower Data

In this research aircraft maintenance center has a total 70 manpower. Of the 70-manpower divided into several groups according to personnel rosters based on the type and variety of aircraft. Compositions of manpower based on personnel rosters for fixed and rotary wing service can be seen in Fig. 1.

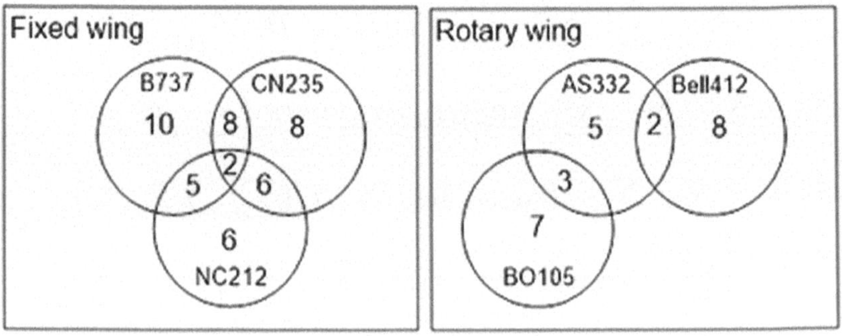

Fig. 1. Manpower composition for fixed and rotary wing service

4.2 Area Data

The effective hangar area in this study was 72 m in length and 48 m in width, so the total hangar area was 3456 m^2. The data labeled in Table 1 is the required service area for each serviceable aircraft.

Table 1. Required service area of each aircraft

Aircraft type	Length (m)	Wide (m)	Area (m^2)
B737-s	40	25	**1000**
AS332	20	23	**460**
Bell 412	18	21	**378**
CN235	31	23	**713**
NC212	23	19	**437**
BO 105	15	14	**220**

4.3 Maintenance Data

Maintenance data for each inspection type includes manpower requirement, maintenance duration, labor cost, material cost and the revenue for each type of maintenance in the USD ($) exchange rate IDR 14,000, - can be seen in Table 2.

Table 2. Inspection type data

Type	insp. type	Pers. Req	Workdays	Lab cost	Material cost	Revenue
B737-s	C1 (1)	12	14	76.363.636	202.300.000	473.200.000
	C2 (2)	12	15	81.818.182	202.300.000	532.700.000
	C4 (3)	14	14	89.090.909	202.300.000	562.450.000
	C6 (4)	20	20	181.818.182	297.500.000	1.172.500.000
	C8 (5)	14	14	89.090.909	202.300.000	574.700.000
CN235	1C (1)	14	14	89.090.909	226.100.000	564.900.000
	2C (2)	16	19	138.181.818	297.500.000	892.500.000
	3C (3)	16	19	138.181.818	892.500.000	997.500.000
NC212	1C (1)	8	7	25.454.545	59.500.000	150.500.000
	2C (2)	8	8	29.090.909	61.880.000	168.420.000
	3C (3)	10	8	36.363.636	119.000.000	231.000.000
	4C (4)	8	8	29.090.909	66.640.000	169.260.000
	6C (5)	10	9	40.909.091	368.900.000	310.100.000
AS-332	PI600 (1)	8	6	21.818.182	1.915.900.000	467.600.000
BL-412	PI600 (1)	10	8	36.363.636	214.200.000	247.800.000
BO-105	PI600 (1)	3	5	6.818.182	107.100.000	54.600.000
	PI1200 (2)	5	5	11.363.636	107.100.000	81.900.000

4.4 Research Data

These data are then incorporated into the Mixed Integer Linear Programming model formulation that has been developed and implemented by using Microsoft Excel's solver. Assuming all types of aircraft maintenance are available, the results can be seen in Table 3 with utility rate 25$ and IDR 14000 per USD.

Table 3. Optimization result

To = 0, Ad = 0		Aircraft type					
		Fixed wing			Rotary wing		
		B737-s	CN235	NC212	AS332	Bell412	BO105
Aircraft Result		4	13	55	55	26	73
Utilities	Manpower	24,24%	62,37%	98,68%	100,00%	78,79%	69,13%
		88,40%			99,17%		
		92,25%					
	Hangar	99,89%					
	Capital	63,27%					
	Manpower cost	92,25%					
Profit		64.452.500.000					

5 Conclusions

The main result of this study is an optimized model for an Indonesian aircraft maintenance services with utilization averaging over 90%, occurring in exchange rate of IDR 14,000, with optimal profits of IDR 64,452,500,000 (around USD 45 Million). This result is achieved with no additional works and overtime because no resources exceeded 100% utilization. The optimum combination of aircraft maintenance per year (264 days) is B737-s = 4, CN235 = 13, NC212 = 55, AS332 = 55, Bell412 = 26 and BO105 = 73 based on aircraft type, variations and types of inspection available. This model will allow the object company to have a clear target for aircraft service sales and can also be used as a reference model for other aircraft service company.

References

1. Dakin, R.J.: A Tree Search Algorithm for Mixed Integer Programming Problem. Comput. J. **8**, 250–255 (1965)
2. Qin, Y., Chan, F.T.S., Chung, S.H., Qu, T., Wang, X.P., Ruan, J.H.: MIP models for the hangar space utilization problem with safety consideration. In: Proceedings of the 3rd World Congress on Mechanical, Chemical, and Material Engineering, pp. 101-1–101-2. International ASET Inc., Rome (2017)
3. Haghani, A., Sriram, C.: An optimization model for aircraft maintenance scheduling and re-assignment. Transp. Res. Part A: Policy Pract. **37**, 29–48 (2003)
4. Patel, P., Ali, A.: Budget airline operations optimization using linear programming. In: Proceedings of the International Conference on Industrial Engineering and Operations Management, pp. 3496–3504. IEOM Society International, Bandung (2018)

Diagnosis and Costing Optimization on Inductors in Electrics and Electronics Industry

Filzah Lina Mohd Safeiee, Mohd Yazid Abu(✉), Nik Nurharyantie Nik Mohd Kamil, and Nurul Farahin Zamrud

Faculty of Mechanical and Manufacturing Engineering, Universiti Malaysia Pahang, 26600 Pekan, Pahang, Malaysia
filzahlinasafeiee@gmail.com, myazid@ump.edu.my

Abstract. The issue of quality and cost of product will be the important aspects in manufacturing industry. But generally, most organization will separate between these two but it is good chance to combine between quality and cost during production process especially for costing improvement. This work presents the application of the Mahalanobis–Taguchi system (MTS) with Time-driven activity-based costing (TDABC). Therefore, the aim of this research is to do the diagnosis process on production line in electric and electronics Company using MTS and TDABC. Mahalanobis distance (MD), which is well known in multivariate statistics through MTS which is a data analytic method for diagnosis and pattern recognition of multivariate data. From this work can state that, the higher the MD, the higher the impact of the parameter to the normal condition. TDABC is a method used to calculate the capacity cost rate multiplied by the time activity, and then MTS is the method to find out the abnormal condition of the data. By using diagnosis MTS method with ABC and TDABC the analysis of electronics manufacturing industry performance indicators, it is possible to produce accurate analysis for managerial decisions.

Keywords: Mahalanobis Taguchi method · Time-driven activity-based costing · Mahalanobis distance

1 Introduction

As a short brief on the electronic sector, it can be said to be shifted into a higher value-added products and activities accordingly to the rapid globalization and increasing cost pressures. Rather than considering the cost, quality has been considered as significant driver for success for manufacturing sector especially in the era of global competition [1]. Mahalanobis–Taguchi system (MTS), a kind of supervised technique, uses Mahalanobis distance (MD) as a multivariate measure for prediction, diagnosis and pattern recognition in multi-dimensional system without any assumption of statistical, and attempts to find out the significant features for generalization [2]. Yazid [8] provided a systematic pattern recognition using MTS by constructing a scatter diagram while Abu [9] classified crankshafts' end life into recovery operations. Yazid [10] developed a distinctive pattern of crankshaft and identify the critical and non-critical

M. N. Osman Zahid et al. (Eds.): iMEC-APCOMS 2019, LNME, pp. 121–127, 2020.
https://doi.org/10.1007/978-981-15-0950-6_19

parameter of crankshaft based on the MTS while Abu [11] identified the critical and non-critical variables during remanufacturing process. Abu [12] evaluated the criticality of parameters on the end of life crankshaft and finally estimate the true cost. While Time-driven activity-based costing (TDABC) is a costing model that considers the time as the only inducer costing. Its purpose is to provide costs of activities with base in consume of time per activities [3]. According to Zaini [13] found that the percentage of TDABC application belong to 66% of healthcare service, 23% of industrial service, and 11% of library service. Therefore, MTS is a method that starts with collecting considerable observations from the investigated dataset, tailed by separating of the unhealthy dataset from the healthy [4] and is a useful tool to optimize the delivery of value-based care [5].

2 Methodology

In this research study, started by collect the data which are process, orders, costing and time taken for each activity and sub-activity during production process. MTS uses procedures that there are data analytic and are independent of the distribution of the characteristics that define the system [6] with good classification results for imbalance data without resampling and it can be used to measure the abnormality of the data [4]. Generally accepted methods of cost accounting have been described, of which TDABC is considered the most sophisticated and precise [7] also fast and simple method that only requires two parameters, an estimation of time required to perform an activity and the unit cost per time of supplying capacity.

3 Result and Discussion

3.1 Mahalanobis-Taguchi System (MTS)

There are 11 parameters of rejected type in VMI workstation that has to be diagnose by MTS which are solder ball, wire defect, core misalignment, chip core, wrong P/N marking OR, solder defect, crack core, insufficient solder, excess epoxy, insufficient epoxy and epoxy on soldering pad. The data considered on 10 December on target 5000 pieces/day. Then the valid percentage belongs 99.99% while the outliers belong 0.10%. If the outliers over 25% the data samples need to reevaluate. From the Fig. 1, can be seen demonstration of MD for off target participants. In this work, there are 470

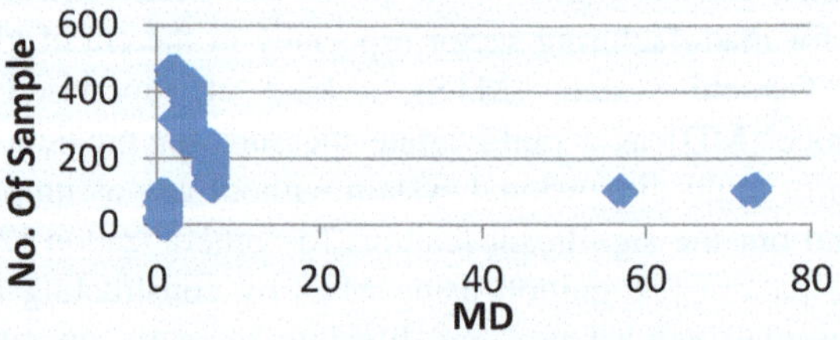

Fig. 1. Scatter diagram of MD for off target participants

off target participants were identified from 5000 participants. Then, the threshold is (0.247722578, 73.4726551). Sample no 24 shows the lowest MD off target while sample number 104 shows the highest MD off target.

Therefore, all the parameters are being optimized to screen the critical parameters. Table 1 shows the optimized parameters which are (1), condition of core alignment (2), condition of soldering defect (3), condition of wire defect and (4), condition of epoxy.

Table 1. The parameter with optimization

Impact	1	2	3	4
In-control	47	23	25	65
Above 3 standard deviations of target				
Below 3 standard deviations of target	15	201		112
Total	62	224	25	177

Table 2 illustrates the variables and distances contribution (percentage) from the diagnosis and prediction process. For example, in the sample below number 104 belongs to condition of epoxy drives only 2% with condition of core alignment being key driver, 88%. It means that, for this sample, the condition of core alignment gives high contribution of rejected parameter and the deviation is below acceptable range of target group.

Table 2. Variable and Distance contribution (percent)

No. of sample	Target criteria value: mechanical yield	Distance	Condition of core alignment	Condition of soldering defect	Condition of wire defect	Condition of epoxy
104	88	74	88	6	6	2
114	90	74	88	6	6	2
57	88	1	8	48	29	17
65	88	1	8	48	29	17

3.2 TDABC

3.2.1 Production Activities Used in VMI (Activity Center/Sub-activities)

The studies are going to start by ABC and TDABC method by analyze and identify the manufacturing activities related to understand the detail process and find out each of rejected type in final VMI station.

3.2.2 Total Cost Estimated of the Resources Used in VMI

From this research, the resources have been determined based on four groups which are labor cost, maintenance cost, material cost and consumable cost. The detail of the resources cost of each sub-activity are as follow (Table 3):

Table 3. Labor, material, maintenance and consumable costs on VMI Final test workstation (MYR/Year)

No.	Sub-Activities	Labor Cost (MYR)	Maintenance Cost (MYR)	Material Cost (MYR)	Consumable Cost (MYR)	Cost of all resources applied (MYR)
1	Cooling down the unit after ultrasonic cleaning process	1,056,000	Nil	16,200	Nil	1,072,200
2	Observe the unit using magnifying glass for final test	1,056,000	Nil	16,200	Nil	1,072,200
3	Transfer the unit into the vacuum tray for final test	1,056,000	Nil	16,200	Nil	1,072,200
Total		**3,168,000**	**Nil**	**48,600**	**Nil**	**3,216,600**

3.2.3 Estimated Acceptable Capacity (Practical Capacity)

The company's working hours are Monday to Saturday, with 2 shifts. Each shift will take about 9 h and 35 min which is 8a.m. to 5.00 p.m and 8 p.m until 5 a.m for 20 days a month.

3.2.4 Calculation of the Capacity Cost Rates

The capacity cost rate (Ringgit Malaysia per minutes) can be obtained using the following equation (Table 4).

$$Capacity\ cost\ rate = \frac{Cost\ of\ all\ resources\ supplied}{Practical\ capacity} \quad (1)$$

Table 4. Capacity cost rate of each sub-activity for the inductor component preparation activity center

No.	Sub-activities	Cost of all resources applied (MYR/Month)	Practical capacity (min/month)	Capacity cost rate (MYR/min)
1	Cooling down the unit after ultrasonic cleaning process	1,072,200	10,300	104.1
2	Observe the unit using magnifying glass for final test	1,072,200	10,300	104.1
3	Transfer the unit into the vacuum tray for final test	1,072,200	10,300	104.1
Total		**3,216,600**	**30,900**	**312.3**

3.2.5 Development of Activity Center Time Equation

TDABC time equation is able to incorporate all the time needed to undertake all sub-activities in each activity center within single equation [5] and the mathematical model used to establish TDABC, time equation is as shown below.

$$T_t = \beta_0 + \beta_i X_i \tag{2}$$

Where,

T_t = *the time needed to perorm an activity* (*minute*)
β_0 = *the standard time to perform the basic activity* (*minute*)
β_i = *the estimated time to perform the incremental activity* (*minute*)
X_i = *the quantity of the incremental activity* (*time*)

3.2.6 Determination of the Estimated Time for Each Activity

A time equation is needed to be developed to calculate the estimated production time.

From Table 5, the time equation for the inductor preparation activity center developed as shown in Eq. 3.

Table 5. Time equations for sub-activities of the inductor components preparation activity center

No.	Sub-activities	Time equations
1	Cooling down the unit after ultrasonic cleaning process	$30X_{32}$
2	Observe the unit using magnifying glass for final test	$3.11X_2$
3	Transfer the unit into the vacuum tray for final test	$1.17X_2$

$$T_{Inductor\ preparation\ activity\ center} = 30X_{32} + 3.11X_{33} + 1,17X_{34} \tag{3}$$

3.2.7 Estimated Capacity Required by Each Activity Center

The estimated capacity required by each activity was determined by quantifying the frequency of the activity per year. By multiplying the amount of a given activity by the time spent doing it, one could calculate the total time spent on the activity. The volumes of cost-drivers for the inductors preparation activity center are summarized in Table 6.

Table 6. Volume of cost-drivers for the inductors preparation activity center.

Var.	Driver	Quantity/year
X_{32}	Transfer the unit from the ultrasonic cleaning process to VMI	2,400,000
X_{33}	Put the unit under the magnifying glass to check the condition of units	2,400,000
X_{34}	Transfer the unit to final test using vacuum tray	2,400,000

The actual time spent on this activity center per month was 82,272,000 min determined by substituting the volume of cost-drivers from Table 6 into Eq. 3. The total time for the transfer the unit from the ultrasonic cleaning process to VMI process represent by X_1 equals to 2,400,000 in $30X_{32}$, so that $30 \times 2{,}400{,}000 = 72{,}000{,}000$ min.

By having the analysis of the production cost using TDABC, this work able to identify the unused capacity, which results in waste costs as summarized in Table 7. But the value showed negative (−) sign for each sub-activity with meaning that there are no unused capacity occur during this time by using TDABC then, there are no loss on manufacturing costs.

Table 7. Analysis of capacity utilization in inductor preparation activity center

No.	Subs-activities	Practical capacity (min/month)	Used time (min)	Un-used capacity (min)	Capacity cost rate (MYR/min)	Loss of manufacturing costs (MYR)
1	Cooling down the unit after ultrasonic cleaning process	10,300	72,000,000	−71,989,700	104.1	−7,494,127,770
2	Observe the unit using magnifying glass for final test	10,300	7,464,000	−7,453,700	104.1	−775,930,170
3	Transfer the unit into the vacuum tray for final test	10,300	2,808,000	−2,797,700	104.1	−291,240,570
Total		**61,800**		**−82,251,400**		**−8,561,298,510**

4 Conclusion

In conclusion, the research study on MTS is it can diagnose the database with numerical evidence while TDABC can interpret the costing matter with unused capacity value.

Acknowledgements. This research is fully supported by Ministry of Higher Education through RDU190156 and FRGS/1/2018/TK03/UMP/02/34. The authors fill acknowledged Universiti Malaysia Pahang for the approved fund which makes this important research viable and effective.

References

1. Herzallah, A.M., Gutiérrez, L., Rosas, J.F.M.: Total quality management practices, competitive strategies and financial performance: the case of the Palestinian industrial SMEs. Total Qual. Manage. Bus. Excell. **25**(5–6), 635–649 (2014)
2. Das, P., Datta, S.: Developing an unsupervised classification algorithm for characterization of steel properties (2012)
3. Da Silva Medeiros, H., Santana, A.F.B., Da Silva Guimarães, L.: The use of costing methods in lean manufacturing industries: a literature review. Gest. e Prod. **24**(2), 395–406 (2017)
4. El-banna, M.: Modified mahalanobis taguchi system for imbalance data classification, vol. 2017 (2017)
5. Yu, Y.R., et al.: Time-driven activity-based costing: a dynamic value assessment model in pediatric appendicitis. J. Pediatr. Surg. **52**(6), 1045–1049 (2017)
6. Deepa, N., Ganesan, K.: Mahalanobis Taguchi system based criteria selection tool for agriculture crops. Sādhanā **41**(12), 1407–1414 (2016)
7. Hamid, K.S., Matson, A.P., Nwachukwu, B.U., Scott, D.J., Iii, R.C.M., Deorio, J.K.: Determining the cost-savings threshold and alignment accuracy of patient-specific instrumentation in total ankle replacements (2017)
8. Yazid, A.M., Rijal, J.K., Awaluddin, M.S., Sari, E.: Pattern recognition on remanufacturing automotive component as support decision making using Mahalanobis-Taguchi system. In: 12th Global Conference on Sustainable Manufacturing, vol. 26, pp. 258–263 (2015)
9. Abu, M.Y., Jamaluddin, K.R., Zakaria, M.A.: Classification of crankshaft remanufacturing using Mahalanobis-Taguchi system. Int. J. Autom. Mech. Eng. **13**(2), 3413–3422 (2017)
10. Kamil, N.N.N.M., Abu, M.Y.: Integration of Mahalanobis-Taguchi system and activity based costing for remanufacturing decision. J. Mod. Manuf. Syst. Technol. **1**, 39–51 (2018)
11. Abu, M.Y., Nor, E.E.M., Rahman, M.S.A.: Costing improvement of remanufacturing crankshaft by integrating Mahalanobis-Taguchi system and activity based costing. In: IOP Conference Series: Materials Science and Engineering, vol. 342, pp.1–10 (2018)
12. Abu, M.Y., Norizan, N.S., Rahman, M.S.A.: Integration of Mahalanobis-Taguchi system and traditional cost accounting for remanufacturing crankshaft. In: IOP Conference Series: Materials Science and Engineering, vol. 342, pp. 1–9 (2018)
13. Zaini, S.N.A.M., Abu, M.Y.: A review on time-driven activity based costing system in various sectors. J. Mod. Manuf. Syst. Technol. **2**, 15–22 (2019)

A Quality Improvement Model Based on Taguchi's Loss Function Considering Imperfect Quality Inspection

Maghfira Devi Ramadhanty(✉), Cucuk Nur Rosyidi, and Wakhid Ahmad Jauhari

Department of Industrial Engineering, Sebelas Maret University, Surakarta, Indonesia
devimaghfira@student.uns.ac.id

Abstract. Quality cost is not only the cost that manufacturing have to pay for poor quality, but also includes cost that has to be paid by the customers. In manufacturing, customers have a high quality product requirement. In producing high quality products, the manufacturer must improve the production process continuosly. Taguchi's loss function is a common model to quantify the quality loss. Taguchi's loss function can be applied to nominal the best, smaller the better, and larger the better quality characteristic. In this study, we developed an optimization model to determine the optimal process parameters by taking into consideration of the imperfection of quality inspection. The imperfect quality inspection involves two kind of errors, namely Type I and Type II errors. Inspectors may make an error in categorizing non-defective items as defectives and some defective items as non-defectives. The inspection errors will increase the quality cost as one components of total cost considered in this study. From the numerical example, the calculation results are the expected cost for 100% inspection is \$ 3.1395, and the expected cost for proposed model is \$ 3.0898 with the optimal process mean for the proposed model is μ_n = 15.411 mm, the optimal standard deviation is σ_n = 0.20076 mm and the optimal multiplier of standard deviation is r = 3.996.

Keywords: Quality cost · Taguchi's loss function · Imperfect quality inspection · Inspection cost

1 Introduction

The development of knowledge, and technology among companies is has made competition getting tighter. The company is required to produce high-quality with low-cost products. Five manufacturing competitive priorities - quality, flexibility, cost, delivery and innovation [1]. It becomes a challenge for companies to produce products with the higher quality, lower selling price as other competitors butat the same time can minimize the manufacturing cost. Quality cost has to be reduced by the companies mean that they have to give more attention to their quality problems. Hence, quality improvement must be carried out continuously to reduce the quality costs.

M. N. Osman Zahid et al. (Eds.): iMEC-APCOMS 2019, LNME, pp. 128–133, 2020.
https://doi.org/10.1007/978-981-15-0950-6_20

Quality improvement is the reduction of variability in processes and products [2]. Quality costs or cost of quality is a means to quantify the total cost of quality-related efforts and deficiencies [3]. Quality costs are often a direct result of the hidden factory that is, the portion of the business that deals with waste, scrap, rework, work-in-process inventories, delays, and other business inefficiencies [2]. Many manufacturing and service organizations use four categories of quality costs: prevention costs, appraisal costs, internal failure costs, and external failure costs [2]. In 1987, Kapur [4] pointed out that the present process can be improved, then a short term approach to decrease variance of the items shipped to the customer is to put spesification limits on the process and truncate the distribution by inspection.In their research, [4] applied the Taguchi's quadratic quality lossfunction for designing the economic specification limits of the qualitycharacteristic with normal distribution. Taguchi's idea is to replace the concept of "Quality" by its complement "Quality Loss". This means that parts of acceptable quality inherent the lowest quality loss, particularly zero. On the other hand, the engineering experience shows that quality degrades, with some exceptions, continuously [5]. Taguchi's main objectives are to improve process and product design through the identificationof controllable factors and their settings, which minimize the variation of a product around atarget response [6].

Reference [7] developed a model based on the research of [4] by determining the optimal value of spesification limits and the adjusment process parameters. The Model of [7] aims to determine the total quality loss by adjusment of product's process parameter i.e. mean, standard deviation, and the spesification limits. The model involved quality inspection activities but the inspection is assumed to be perfect. Inreality, the inspection carried out by an inspector must have inspection errors. The inspector may make an error in an his inspection which has two types of error, namely Type I and type II errors [8]. In this research, we extend the research of Kao [7] by incorporating imperfect inspection to determine the optimal value of certain quality characteristic process parameters with the objective function to minimize total quality cost.

2 Assumption and Notation

This paper has the following three assumptions, there are:initial process mean and standar deviation are known, product quality characteristic are normally distributed and scrap cost, adjusted mean cost, adjusted standard deviation cost, cost of falsely accepting a defective products and cost of falsely rejecting a non-defective products are known

The following notations are used in this paper:

γ: percentage of defective items
γ_e: percentage of defective items observed by the inspector
$m_{1:}$ random variable representing Type I error
$m_{2:}$random variable representing Type II error
Ca: cost of falsely accepting a defective products (Type I)
Cr: cost of falsely rejecting a non-defective products (Type II)

ICa: cost of falsely accepting a defective products
ICr: cost of falsely rejecting a non-defective products
MC: adjusting mean cost per unit
DC: adjusting standard deviation per unit
SC: scrap cost per unit
k: loss cost per unit
y_0: target value
σ_1: initial standard deviation
σ_2: minimum standard deviation under the ideal production
σ_n: optimal standard deviation
μ_n: optimalmean process
$\sigma_{t,n}$: standard deviation of truncated normal distribution
$\mu_{t,n}$: mean of truncated normal distribution
q_n: probabilities that a random variable falls inside the specification limits
u: $(y_0 - \mu_n)/\sigma_n + w$
v: $(y_0 - \mu_n)/\sigma_n - w$
r: multiplier of standard deviation
$\phi(.)$: PDF of the standard normal distribution
$\Phi(.)$: CDF of the standard normal distribution
y_n: random variable of normal distribution.

3 Model Development

3.1 The Basic Model

Basic Total Expected Cost

Reference [7] proposed a quality loss model using Taguchi loss function and process capability indices for normal distribution. Kao's model aimed to determine the optimal adjusted process mean, adjusted standard deviation and specification limits. The components of total cost includes inspection cost (cost false of rejection and cost false of acceptance), scrap cost, and adjusted process cost. They assumed that quality characteristic Y follows the normal distribution with known mean and variance. According to [7], the total expected cost (*TC*) can be expressed as Eq. (1).

$$TC_n = L_n + (1 - q_n)SC + IC + e^{\left(1-\left|1-\frac{\mu_n}{y_0}\right|\right)}MC + \left(e^{\left(\frac{\sigma_1-\sigma_n}{\sigma_1-\sigma_2}\right)} - 1\right)DC \quad (1)$$

The expected quality loss for nominal-the-best type quality characteristic is presented as

$$L_n = E\left[L\left(Y_{t,n}\right)\right] = k\left(\mu_{t,n} - y_0\right)^2 + \sigma_{t,n}^2, \quad LSL \le y_n \le USL \quad (2)$$

Where

$$\mu_{t,n} = \mu_n + \frac{\sigma_n}{q_n}[(\phi(v) - \phi(u))] \tag{3}$$

$$\sigma_{t,n}^2 = \sigma_n^2\left[1 + \frac{1}{q_n}(v\phi(v) - u\phi(u)) - \frac{1}{q_n^2}(\phi(v) - \phi(u))^2\right] \tag{4}$$

$$q_n = \Phi(u) - \Phi(v) \tag{5}$$

$$LSL = y_0 - r\sigma_n \tag{6}$$

$$USL = y_0 + r\sigma_n \tag{7}$$

Imperfect Inspection Cost

Errors in the inspection process will result in additional costs that must be incurred by the manufacturer. According to Khan [8], there are two types of costs arising from inspection errors, namely cost of false rejection (*Cr*) and cost of false acceptance (*Ca*). The cost of false rejection occurs when the inspectors make a type I error, which is they classify the non-defective product as a defective one. While the cost of false acceptance occurs when the inspectors make a type II error, which is they classify a defective product as a non-defective one. The fraction of defective units as perceived by the inspectors would be:

$$\gamma_e = (1 - \gamma) * m_1 + \gamma * (1 - m_2) \tag{8}$$

Where

$$\gamma = 1 - \int_{y_0 - r\sigma_n}^{y_0 + r\sigma_n} \frac{1}{\sqrt{2\pi}\sigma} e^{\frac{-1}{2}\frac{(y_n - \mu)^2}{\sigma^2}} dy_n \tag{9}$$

The total cost of false rejection (*Cr*) and cost of false acceptance (*Ca*) are shown in Eqs. (10) and (11) respectively.

$$ICr = \frac{cr(1 - \gamma)m_1}{1 - \gamma_e} \tag{10}$$

$$ICa = \frac{ca.\gamma.m_2}{1 - \gamma_e} \tag{11}$$

3.2 Proposed Model

The total expected quality loss per item in this research is expressed in Eq. (12)

$$TC = L_n + (1 - q_n)SC + ICa + ICr + e^{\left(1-\left|1-\frac{\mu_n}{y_0}\right|\right)}MC + \left(e^{\left(\frac{\sigma_1-\sigma_n}{\sigma_1-\sigma_2}\right)} - 1\right)DC \quad (12)$$

This proposed model is combined model [7] and [8] that is calculate the minimum totally expected loss model considering imperfect inspection for normal distribution

4 Numerical Example

In this research, we use backrest thread component as the case study of numerical example. The product is one of components of hospital bed of the Supramax 73004 type produced by MAK (Mega AndalanKalasan), there is backrest thread 180 mm. The process used in the backrest thread are cutting and turning. The backrest thread 180 mm divided into 2, namely minor diameter and major diameter. In this study, we will use only the minor diameter data. We collected 40 data to estimate the initial mean and standard deviation. Table 1 lists sample of minor diameter of backrest thread 180 mm.

Table 1. Minor diameter of backrest thread data (mm)

15.3	15.45	15.55	15.7	15.4	15.5	15.6	15.9
15.4	15.45	15.55	15.75	15.4	15.5	15.6	15.9
15.4	15.45	15.6	15.8	15.45	15.4	15.6	16
15.4	15.45	15.6	15.8	15.45	15.5	15.65	16.1
15.4	15.5	15.6	15.85	15.45	15.55	15.7	16.1

Backrest thread has a major diameter of 18.70 mm, minor diameter of 15.40 mm with 180 mm length. From the table, the initial mean $\mu_1 = 15.59625$ mm and the initial standard deviation $\sigma_1 = 0.20076$ mm. The process mean does not meet the target because the value of process mean is larger than the target ($y_O = 15.40$ mm).

For 100% inspection, Let $k = 8$, $IC = 0.1$, $SC = 2$, $MC = 1$, $DC = 1$ and the ideal standard deviation (σ_2) is 0.05 mm. By Kao (2010) model, the optimal process mean is $\mu_n = 15.411$ mm, the optimal standar deviation is $\sigma_n = 0.20076$ mm and the optimal multiplier of standard deviation is $r = 3.996$. The total expected cost is $TC = \$ 3.1395$, and specification limits $LSL = 14.5977$ mm and $USL = 16.2023$ mm.

For the proposed model in this research, let $k = 8$, $ICa = 0.5$, $ICr = 0.2$, $SC = 2$, $MC = 1$, $DC = 1$, $m_1 = m_2 = 0.2$, $r = 3$ and the ideal standard deviation (σ_2) is 0.05 mm. The optimal process mean for the proposed model is $\mu_n = 15.411$ mm, the optimal stand- ard deviation is $\sigma_n = 0.20076$ mm and the optimal multiplier of standard deviation is $r = 3.996$. Hence, we get the total expected cost of $TC = \$ 3.0898$ with the specification limits of $LSL = 14.5977$ mm and $USL = 16.2023$ mm. The initial value of the standard deviation is optimal, so there is no adjusment for standard deviation parameter process (σ_n).

From the results above, the proposed model has a lower total expected cost than the previous model. Total expected cost that considering imperfect inspection results is lower than considering perfect inspection. This is due to perfect inspections carried out on all products so that inspection costs are high. But it is quite realistic to account for Type I and Type II errors committed by inspectors in this process. So there will be an additional costs for Type I errors, and for Type II errors it is possible returned products from the market. With lower total expected cost, the proposed model gives benefit for both manufacturer and buyer in the form of cost and quality of the product.

5 Conclusion

In manufacturing, the buyer has a high quality product requirement. High quality means the manufacturer must spend a lot of funds. The base model [7] have presented a totally expected quality loss model using Taguchi's loss function for normal distribution. Reference [8] describe that inspection cost divided into two costs, such as cost of false rejection occurs when the inspector makes a type I error, and the cost of false acceptance occurs when the inspector makes a type II error. This study proposed modified totally expected loss model considering imperfect inspection for normal distribution. In [7] model's, involved quality inspection activities but the inspection is assumed to be perfect. In this study, inspection which has two types error. From the calculation results, the expected cost for 100% inspection is $ 3.1395, and the expected cost for proposed model is $ 3.0898. Model that considering imperfect inspection has a lower expected cost than 100% inspection. With lower total expected cost, the proposed model gives benefit for both manufacturer and buyer in the form of cost and quality of the product.

Acknowledgement. This research was fully funded by Research and Community Service (P2 M) PNBP funds 2019 budgeting year with contract number 516/UNS27.21/PM/2019.

References

1. Russel, S.N., Millar, H.H.: Competitive priorities of manufacturing firms in the Caribbean. IOSR J. Bus. Manage. **16**(10), 72–82 (2014)
2. Montgomery, D.C.: Introduction to Statistical Quality Control, 7th edn. Wiley, New York (2012)
3. Feigenbaum, A.V.: Total quality control. Harvard Bus. Rev. (1956)
4. Kapur, K.C., Wang, C.J.: Economic design of specifications based on Taguchi's concept of quality function. In: The Winter Annual Meeting of the American Society of Mechanical Engineer, Boston, pp. 23–36 (2010)
5. Academia. https://www.academia.edu/4265348/Taguchi_loss_function. Accessed 17 July 2019
6. Vacarescu, C.F., Vacarescu, V.: The Taguchi's quality loss function in development and design for bulk goods in the automotive field. In: Annals of DAAAM for 2010 and Proceedings of the 21st International DAAAM Symposium, vol. 21, no. 1 (2010)
7. Kao, S.C.: Deciding optimal spesification limits and process adjusments under quality loss function and process capability indices. Int. J. Ind. Eng. **17**(3), 212–222 (2010)
8. Khan, M., Jaber, M.Y., Ahmad, A.R.: An integrated supply chain model with errors in quality inspection and learning in production. Omega **42**, 16–24 (2014)

Identifying Bottleneck Process Using Production Time Study in Concrete Pole Manufacturing Company in Malaysia

Afiqah Alias(✉), Atiah Abdullah Sidek, Md. Yusof Ismail, and Muataz Hazza

Department of Manufacturing and Material Engineering, Faculty of Engineering, International Islamic University Malaysia, Gombak, Malaysia
fikaalias@gmail.com, atiahas@iium.edu.my

Abstract. In the production line, productivity is the key towards maximum efficiency. Time study is the technique of establishing an allowed time standard to perform a given task based on the prescribed method. This study employs the use of a time study to measure the time taken to conduct six processes in the production of concrete poles in order to identify the bottleneck in each process. This determines the best way to execute repetitive tasks and to measure the time spent by an average worker to complete a given task in the workplace. The identification of bottleneck allows for the proposal of improvements that resolve the disruptions in the production line. The data needed for this study was obtained from a concrete poles manufacturing company in Nilai, Negeri Sembilan. The findings show that the implementation of time study contributes positively towards identifying bottleneck and achieving productivity.

Keywords: Productivity · Time study · Bottleneck

1 Introduction

Time study phase is defined as a procedure to determine the amount of time required, under certain standard conditions of measurement, for tasks involving human and machines activity [1]. Many manufacturing companies utilize the Time Study technique as one of the productivity improvement techniques. This technique is a scientific analysis method that is designed to measure the time spent by an average worker to complete a given task and determine the best way to execute the repetitive task [1].

In the manufacturing operation, bottleneck refers to a specific segment in the production process that continuously disrupts and slows down the speed of production. This includes issues such as long queues, long waiting times, poor performance, and the overall inefficiency of the production [2]. The identification of the bottleneck is important as it informs about the disruptions in the flow of production and highlights the specific area where accumulations occur. Such information serves as a vital importance to the effort of improving the efficiency and productivity of the production line. As disruptions and inefficacy in production may lead to a significant financial loss,

M. N. Osman Zahid et al. (Eds.): iMEC-APCOMS 2019, LNME, pp. 134–139, 2020.
https://doi.org/10.1007/978-981-15-0950-6_21

the entire production process should be monitored as an effort for early identification of bottleneck and to reduce its impact to the minimum.

This case study involves a concrete pole manufacturing company located in Nilai, Negeri Sembilan. The main product of this company is Pre-Stress Concrete Poles. Regardless of its size, these concrete poles are manufactured according to a standardised procedure in order to maintain its quality. This is aligned with the company's mission of providing concrete based products of good quality to its customers.

2 Methodology

The required data for this time motion study includes man power, the sequence and motion of the processes, processing time for each process, and the machine involved. The study began by identifying the processes involved in the production of the concrete poles. This was achieved through semi-structured interviews with the general manager, head of quality assurance, and the operators as well as production floor observation. The observed time of each process was recorded by using a stopwatch. Five observation were taken to compute the average observed time and to calculate the standard time of the processes.

2.1 Flow Chart of the Concrete Pole Manufacturing Process

Figure 1 shows the manufacturing process of concrete poles at the company. From the interview, the demand for concrete poles ranges between 1000 to 3000 poles per month. Prior to delivery, the poles are inspected for quality assurance in order for it to satisfy customers' expectation. Although these concrete poles range between three different sizes of 10 m, 9 m, and 7.5 m, each pole is manufactured using similar process. This study focuses on production process of the 7.5 m concrete poles.

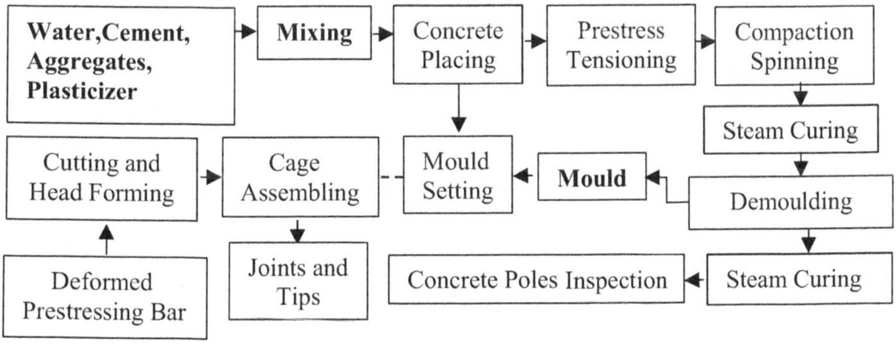

Fig. 1. Concrete poles manufacturing process.

2.2 The Process Flow by Section

In order to study the time and motion of this production, the overall processes were divided by sections.

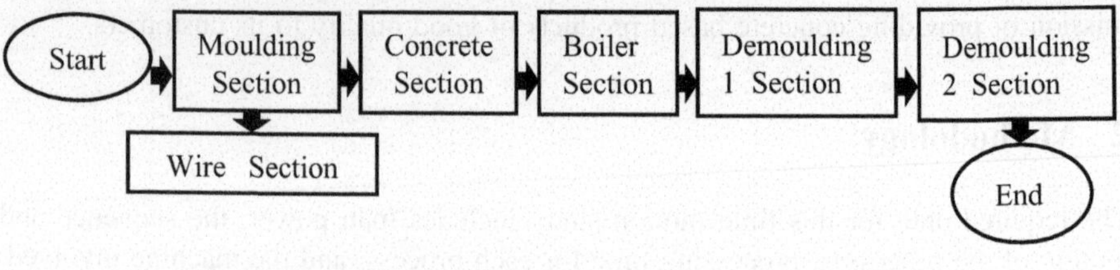

Fig. 2. Manufacturing process of concrete poles by section.

Figure 2 shows the breakdown of the concrete poles manufacturing process by section. The process begins with the wire section which is part of the moulding section. The mould is then transferred to the Concrete Section for it to be filled with concrete before being sent for the curing process in the Boiler Section. Once cured, the concrete poles will be transferred to Demoulding 1 and Demoulding 2 for inspection.

2.3 The Number of Operators by Sections

As shown in Fig. 2, the manufacturing process of concrete poles comprises six sections namely the Moulding Section, Wire Section, Concrete Section, Boiler Section, Demoulding 1 Section, and Demoulding 2 Section. A total of 51 operators were involved to operate these sections with the inclusion of 5 crane operators. All operators are trained and qualified to perform the required duties. Table 1 shows the dispersion of operators that work in every section.

Table 1. The number of operator and the total standard time by sections

Section no.	Section name	No. of operators	Total standard time (s)
1	Moulding section	11	380
2	Wire section	7	480
3	Concrete section	8	535
4	Boiler section	1	18030
5	Demoulding 1 section	12	282
6	Demoulding 2 section	7	103
Crane	Crane operator	5	–

2.4 The Average Observed Time, Normal Time and Standard Time

A stopwatch timer was used to record the processing time taken by the operators to complete a specific process. The stopwatch is chosen as the measuring equipment of this study due to its convenience and accuracy in measuring time [3]. The element of time can be entered directly into the time study sheet based on the recorded data. The data were recorded in a real-time observation at the production floor. The time taken were recorded while observing how the task was done by the operators. The data of five observations were taken to find the average observe time. The process in each section was also video recorded in order to validate the time taken using the stopwatch as well as for further observation after the real-time observation had been done. The average observed time is not the actual time taken for the operator to complete the task [4]. Standard time is defined as time taken for an average skilled operator to complete the task using the prescribed method by working at a normal pace. Equation (1) is used to determine the standard time.

$$\textit{Standard Time} = \textit{Normal Time} \times \left(\frac{100}{100 - \textit{Allowance in}\%}\right) \tag{1}$$

To determine the normal time for computing the standard time, refer Eq. (2). The performance rating factor is rated by the head of the quality assurance in the company during the observation. The allowance is calculated based on the allowances that are recommended by International Labor Organization (ILO). The observation and evulation of the allowance were done by the guidance of the head of quality assurance. Table 1 shows the total standard time for each section.

$$\textit{Normal Time} = \textit{Observe Time} \times \frac{\textit{Rating in}\%}{100} \tag{2}$$

3 Result and Discussion

3.1 Identification of Bottleneck

The identification of bottleneck in the concrete poles manufacturing is conducted by looking at the process that takes the longest processing time in each section while omitting the long processing time that is caused by machine requirement. Table 2 shows the bottleneck process for each section.

The data shows that the longest processing time in concrete poles manufacturing is the process of wet steam cooling. However, such long processing time is a standardised requirement that needs to be followed by the company in order to sustain the integrity and durability of the poles. Thus, this study focuses on the second longest processing time which is the wire section in order to identify the bottleneck in the process.

Table 2. The bottleneck process for each section.

Section	Bottleneck process	Process description	Standard time (s)
Moulding section	Wire tensioning	The wire were stretched and Ring Set were inserted	90
Wire	Straightening wire	The wire were straightened and cut	99
Concrete	Concrete injection process	The injection process of concrete based on the size of the mould which had been set up	45
Boiler	Wet steam (cooling)	The Inlet Valve were closed and The Curing Pit cover opened for cooling process	900
Demoulding 1	Impact wrench	The position of mould were set and the pole were pushed out	58
Demoulding 2	Pulling pole	The pole were pull out from the mould and were moved to the conveyor	46

Table 3. Wire section processing time

Section	Process name	Machine involved	ProcessDescription	No of operator	StandardTime (s)
Wire	Bottom caging	Wire caging	Caging process	2	90
			Spiral wire were cut and were put on the Trolley	1	15
	Straighten wire	Wire straightening	The wire were straightenand cut. Wire were tied and cut according to the measurement	2	99
	Rivet head	Upsetter	Rivet head were done on the side of P.C wire	1	35
			Rivet head process	1	24
			Pre assembly and plastic spacer were attached and were tied into the mould in the right position	1	70
	Top cage	Wire caging	Spiral wire were straightened and cut according to the measurement	1	29
			Spiral wire were inserted and were attached to the Hook	1	30
			Caging process and the cage were cut according to the measurement	1	88

3.2 Bottleneck Process in Wire Section

Table 3 contains the information on the process details, the machine involved, number of operator, and the processing time of each process in the wire section. The process of straightening the wires is identified as the bottleneck for the wire section. In such process the wires will be straightened, measured, and cut according to prefix measurements and standard. Two operators were assigned to operate the machine that straightens and cuts the wires and transfer the wires to the transverser. It was observed that the operators required a lot of time to transfer the wires manually.

4 Conclusion

The present case study investigates on the bottleneck that disrupts the production of concrete poles by a company in Nilai, Negeri Sembilan. This is achieved through the use of time study in order to determine the production process with the longest processing time. The finding indicates that the process of straightening the wires is time consuming as the operators have to transfer and move the wires manually. Such bottleneck can be minimized by re-evaluating the amount of manpower and machines required to complete the task.

Acknowledgement. This research has received financial support from the Ministry of Higher Education Malaysia (MOHE) through the Fundamental Research Grant Scheme (FGRS), to which the author is grateful. Special gratitude also goes to the participating concrete poles company, interviewed staffs and workers, and other relevant parties that have been involved with this study.

References

1. Reddy, A.S.N., Rao, P.S., Rajyalakshmi, G.: Productivity improvement using time study analysis in a small scale solar appliances industry - a case study. ARPN J. Eng. Appl. Sci **11** (1), 666–674 (2016)
2. C. Study: Bottleneck Identification Using Time Study and Simulation Modeling of Apparel Industries, pp. 321–331 (2017)
3. Chandra, P.V.: An effort to apply work and time study techniques in a manufacturing unit for enhancing productivity. Int. J. Innov. Res. Sci. Eng. Technol. **2**(8), 4050–4058 (2013)
4. Permata, L., Hartanti, S.: Work measurement approach to determine standard time in assembly line. Int. J. Manag. Appl. Sci. **10**, 192–195 (2016)

The Development of Industry 4.0 Readiness Model. Case Study in Indonesia's Priority Industrial Sector of Chemical

Idriwal Mayusda(✉) and Iwan Inrawan Wiratmadja

Industrial Engineering and Management Department, Institut Teknologi Bandung, Bandung 40132, Indonesia
idriwal.mayusda@gmail.com, iwan@lspitb.org

Abstract. Various countries prepare to face a new industrial revolution, known as Industry 4.0 by making Industry 4.0 initiative strategy. Indonesia as one of the developing countries, pursue a strategy named 'Making Indonesia 4.0'. Industry 4.0 Readiness assessment is the next stage to carry out the requirements, technology, and resources needed to implement the strategy. This study aims to identify some factors used in assessing the readiness of Making Indonesia 4.0. The results of this study indicated that there were nine dimensions of Industry 4.0 readiness such as smart factory; smart products; employee; digitization and value chain integration; strategy and organization; compliance, security, legal and tax; leadership; culture; and business model. This paper evaluates the chemical industrial sector. It was shown that this sector was at level 2 out of 4 levels of the Industry 4.0 Readiness.

Keywords: Industry 4.0 · Making Indonesia 4.0 · Readiness · Assessment · Industrial sectors

1 Introduction

The world has introduced a new industrial revolution phase, known as Industry 4.0. Various countries prepare to face the fourth industrial revolution challenge by making Industry 4.0 strategy for example 'Industrie 4.0' (Germany), "Made in China 2025" (China) [1], "Advanced Manufacturing Partnership 2.0" (USA) and Manufacturing Innovation 3.0 (South Korea) [2]. On April 2018, Indonesia pursued a strategy named 'Making Indonesia 4.0'.

Making Indonesia 4.0 places five priority industrial sectors to implement Industry 4.0 strategies, such as food and beverage; textile and apparel; automotive; electronic; and chemical. The chemical sector particularly contributes significantly to economic growth and play an important role as a producer of the raw materials for other industrial production needs. In 2017, this sector gave 10,05% of national GDP and expectantly establish a position as a leading bio speciality chemical hub in the next 15 years.

According to Schumacher et al. [3], one of the problems appear when implementing Industry 4.0 strategy is the failure of the company to assess their own

M. N. Osman Zahid et al. (Eds.): iMEC-APCOMS 2019, LNME, pp. 140–146, 2020.
https://doi.org/10.1007/978-981-15-0950-6_22

capabilities in Industry 4.0 which detains from taking any coordinated measures. In solving this problem, maturity models and assessment of Industry 4.0 maturity become decidedly important, since a lot of companies struggle to initialize Industry 4.0 transformation [4].

The purpose of this paper is to develop the Industry 4.0 readiness model to assess manufacturing companies' state of development in relation to the Industry 4.0 vision.

2 Theoretical Framework

There are two terms used in explaining the readiness of Industry 4.0 strategy, 'readiness' and 'maturity'. According to Carolis et al. [5], the term 'readiness' is defined as the capability of a manufacturing company to organize the concepts, and the term 'maturity' is defined as the capability of a manufacturing company to employ concepts or its technology. For that reason, the terms 'readiness' and 'maturity' are relative and related. Similarly, Tetlay and John [6] stated that Maturity as a part of Readiness. "Maturity" is encapsulated within the concept of "readiness". In conclusion, discussing readiness will mention maturity and vice versa; they are not mutually exclusive. This paper uses the term 'readiness' to describe the Industry 4.0 strategy readiness implementation.

According to Rajnai and Kocsis work [7], the maturity model framework consists of four domains: technology, culture, organization, and insight. Whilst, most of the recent studies were limited in two domains; technology and organization [4]. In order to reach more comprehensive results, this study extended the current models and tools through its focus on the four domains.

3 Research Methodology

3.1 The Development of Industry 4.0 Readiness Model

This study used the methodology framework of De Bruin et al. in the development of Industry 4.0 Readiness Model. It was stated that the proposed standard framework forms to guide the improvement of an existing model through several phases to enable the evolution of the model within a given domain [8].

3.2 Proposed Industry 4.0 Readiness Model

Through an extensive literature review about Industry 4.0 Maturity and Readiness Model, there are some extension domain components were identified for the development of Industry 4.0 Readiness model based on Rajnai and Kocsis' [7] maturity model framework. The resultant list validated through interviews and case studies in organizations. Table 1 provides an overview of the results achieved related to Industry 4.0 Readiness model development.

Table 1. Proposed Industry 4.0 readiness dimension and maturity item

Dimension	Exemplary maturity item
Smart factory [9]	Automation; Utilization Machine & Operation system integration (M2 M); Utilization of mobile devices; Self-optimizing processes;...
Smart products [9]	Product customization; Digital feature of products; ICT add on functionalities; The possibility to integrate products into other systems;...
Employees [9]	The openness of employees to new technology; Autonomy of employees); ICT competence of employees;...
Digitization and value chain integration [10]	Integrate IS vertically and horizontally; Standardize data interfaces; Implemented data governance; Automated data analysis;...
Strategy and organization [3]	Utilization of an Industry 4.0 roadmap, Availability of resource for Industry 4.0, Communication of Industry 4.0 activities, ...
Compliance, security, legal & tax [10]	Government regulations for Industry 4.0; Suitability of technological standards; Protection of intellectual property; Tax & Policies;...
Culture [3]	Recognize the value of mistakes, Knowledge sharing, Open innovation and cross-company collaboration, ...
Leadership [3]	The willingness of leaders, Management competence and methods, Existence of central coordination for Industry 4.0, ...
Business model [11]	As 'a service' business model, Real-time tracking of a product, Real-time and automated scheduling of maintenance activities, ...

This proposed model used a survey that integrated quantitative measurement which enabled the collection of results. It also empowered consistent statistical analysis and improved the comparability of results. The survey used Likert scales that improved the reliability and consistency of response and allowed results to be easily mapped to maturity stages. Each maturity item was graded with related survey questions by 0–4 points. After all, calculated points of maturity items were grouped under several dimensions to identify maturity levels individually and overall. Equations (1), (2), and (3) were applied to calculate maturity levels [4] (Table 2).

M: Maturity
D: Dimension
I: Maturity Item
Q: Question Number
O: Overall
n: Number of respondents
m: Number of maturity item

$$M_{DIi} = \frac{\sum_{j=1}^{n} Q_{Iij}}{n} \tag{1}$$

$$M_D = \frac{\sum_{i=1}^{m} M_{DIi}}{m} \quad (2)$$

$$M_o = \min(M_1, M_2, M_3, \ldots M_9) \quad (3)$$

Table 2. Limit values to determine the maturity level

Maturity level	Low	High
Level 0: novice	0.00	0.90
Level 1: beginner	0.90	1.80
Level 2: learner	1.80	2.70
Level 3: intermediate	2.70	3.60
Level 4: expert	3.60	5.00

4 Case Study in Making Indonesia 4.0 Chemical Industrial Priority Sector

The study was conducted in **three pioneer companies** listed in chemical industry sectors of Making Indonesia 4.0 roadmap. The companies already engaged in Industry 4.0 and had basic knowledge and understanding of Industry 4.0 basic concepts.

4.1 Validity and Reliability

The survey questions were validated by referencing existing literature and by seeking assessment within a selected group of expert and practitioners in Indonesia's priority sectors of industry. Respondents were asked to comment on survey structure, ease of survey completion, and perceived completeness of the questions.

The reliability is exhibited by the value of Cronbach's Alpha [12]. The result of the reliability test addressed in this model is reflected in Table 3. It showed that three of the dimensions were not reliable to assess the Industry 4.0 readiness. Thus, these dimensions were not used further in the analysis of Industry 4.0 readiness case study.

4.2 Result and Discussion

The result of scores corresponding to the answers given to the questions related to dimensions were calculated according to Eq. (1), (2), (3). Based on Eq. (3), the overall maturity level of a company is determined by the minimum maturity level of dimensions. The minimum maturity level score (Table 4) is 1.96 which is calculated for Smart Products dimension. Therefore, the chemical industry priority sector in Making Indonesia 4.0 strategic plan in this case study was at "Level 2: Learner" regarding Industry 4.0 readiness model. Summary of overall maturity scores is given in Table 4. Figure 1a visualized the general result of the maturity level at-a-glance.

Table 3. Reliability test of assessment survey

Dimension	Cronbach's alpha	Reliability
Smart factory	0,82	Reliable
Smart products	0,75	Reliable
Employee	0,04	**Not reliable**
Digitization & integration value chain	0,87	Reliable
Compliance, security, legal & tax	0,52	**Not reliable**
Strategy & organization	0,70	Reliable
Leadership	0,41	**Not reliable**
Culture	0,76	Reliable
Business model	0,92	Reliable

Table 4. Maturity score and level for each dimension of Industry 4.0 readiness in the chemical industry sector

Dimension	Maturity score	Maturity level
Smart factory	2,40	Level 2: learner
Smart products	1,96	Level 2: learner
Digitization & value chain integration	2,43	Level 2: learner
Strategy & organization	2,73	Level 2: learner
Culture	2,63	Level 2: learner
Business model	2,20	Level 2: learner

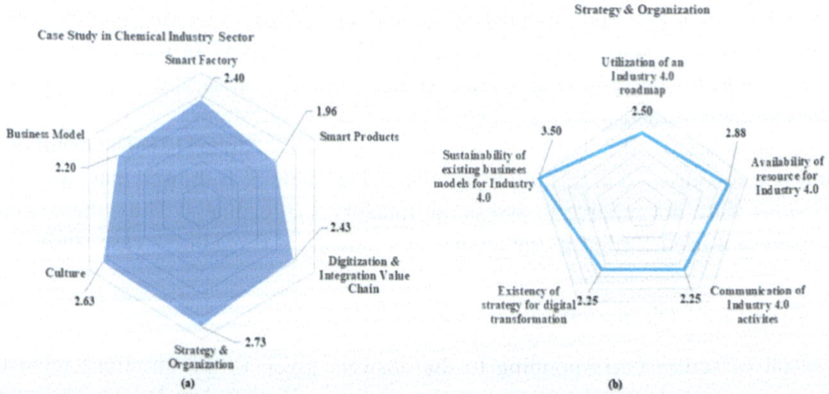

Fig. 1. Radar chart visualizing of (a) industry 4.0 readiness in chemical industry sector case study, (b) dimension of strategy & organization.

To get a better understanding of the model, the highest maturity score in the dimension of Strategy & Organization (2,73) is presented in Fig. 1b.

It can be seen that the three companies in the chemical industry sector have achieved level 3 of readiness in the utilization of Industry 4.0 strategy that sustained with the existing of the company's business model. Moreover, it is supported by the availability of resources for Industry 4.0 implementation. Whereas the companies have just initiated the Industry 4.0 strategy and did not all the companies has the digital transformation strategy before. In contrast, the communication activity to the employee is not fully applied by the companies.

5 Conclusion

The study presented the development of an Industry 4.0 readiness model with nine dimensions and five maturity levels. This study exhibited that the chemical industry sector of Making Indonesia 4.0 was at level 2 (learners) out of 4 levels of the Industry 4.0 Readiness.

Acknowledgements. This research was supported by the Indonesia Endowment Fund for Education (LPDP) Scholarships.

References

1. Li, L.: China's manufacturing locus in 2025: with a comparison of "Made-in-China" and "Industry 4.0". Technol. Forecast. Soc. Chang. **135**, 66–74 (2018)
2. Lee, M., Yun, J.J., Pyka, A., Won, D., Kodama, F., Schiuma, G., Park, H., Jeon, J., Jung, K.: How to respond to the fourth industrial revolution, or the second information technology revolution? Dynamic new combinations between technology, market, and society through open innovation. J. Open Innov. Technol. Mark. Complex **4**, 21 (2018)
3. Schumacher, A., Erol, S., Sihn, W.: A maturity model for assessing industry 4.0 readiness and maturity of manufacturing enterprises. Proc. CIRP **52**, 161–166 (2016)
4. Akdil, K.Y., Ustundag, A., Cevikcan, E.: Maturity and Readiness Model for Industry 4.0 Strategy. Springer Series in Advanced Manufacturing, pp. 61–94. Springer International Publishing, Switzerland (2018)
5. Carolis, A.D., Macchi, M., Brundag, M.P., Kulvatunyou, B., Terzi, S.: Maturity models and tools for enabling smart manufacturing systems: comparison and reflections for future developments. Springer International Publishing IFIP AICT 517, pp 23–35 (2017)
6. Tetlay, A., John, P.: Determining the lines of system maturity, system readiness and capability readiness in the system development lifecycle. In: Conference on Systems Engineering Research, pp 1–8. CSER (2009)
7. Rajnai Z., Kocsis I.: Assessing industry 4.0 readiness of enterprise. In: Symposium on Applied Machine Intelligence and Informatics, pp. 225-230. IEEE, Kosice (2018)
8. De Bruin, T., Freeze, R., Kaulkarni, U., Rosemann, M.: Understanding the main phases of developing a maturity assessment model. In: Campbell, B., Underwood, J., Bunker, D. (eds.) Conference on Information Systems. ACIS, Sydney (2005)
9. Lichtblau, K., Stich, V., Bertenrath, R., Blum, M., Bleider, M., Millack, A., Schmitt, K., Schmitz, E., Schroter, M.: Industrie 4.0 Readiness. IMPULS-Stiftung for Mechanical Engineering, Plant Engineering, and Information Technology, Frankfurt (2015)
10. Geissbauer, R., Vedso, J., Schrauf, S.: Industry 4.0: Building the Digital Enterprise. PWC 2016 Global Industry 4.0 Survey (2016)

11. Agca, O., Gibson, J., Godsell, J., Ignatius, J., Davies, C.W., Xu, O.: An Industry 4.0 Readiness Assessment Tool. WMG International Institute for Product and Service Innovation University, Coventry (2016)
12. Hair Jr., J.F., Black, W.C., Babin, B.J., Anderson, R.E.: Multivariate Data Analysis. Pearson New International Edition, London (2014)

Open Innovation Practices and Sustainability Performance in Small and Medium Industries

Amelia Kurniawati[1,2(✉)], Praditya Ajidarma[1], Iwan Inrawan Wiratmadja[1], Indryati Sunaryo[1], and T. M. A. Ari Samadhi[1]

[1] Bandung Institute of Technology, Jl. Ganesa 10, Bandung 40132, Indonesia
amelia.kurniawati@gmail.com
[2] Telkom University, Jl. Telekomunikasi no. 1, Bandung 40257, Indonesia

Abstract. Sustainability performance is the achievement of an organization in economic, environmental, and social dimensions by considering the interests of stakeholders. Achieving a high sustainability performance is a challenge for not only large industries but also Small and Medium Industries (SMIs). In order to achieve high sustainability performance, an organization needs to innovate in its operations, especially in activities related to the environment, employees, society, and ethics. Conducting innovation needs resources, both financial and nonfinancial, which are very limited in the SMIs. To overcome the problem of limited resources, SMIs can implement open innovation that utilizes both internal and external resources. Open innovation has three types of practices which are inbound, outbound, and coupled. This study aims to identify the relationship between the three open innovation practices and sustainability performance in the SMIs. The respondents of this study are 125 SMIs which produce batik in Indonesia. The model is tested using the partial least square. The result shows a significant relationship between inbound open innovation and economic and environmental performance; outbound open innovation and environmental and social performance; and coupled open innovation and social performance.

Keywords: Open Innovation · Small and medium industries · Sustainability performance

1 Introduction

Sustainability performance is the achievement of an organization in economic, environmental, and social dimensions by considering the interests of stakeholders [1–3]. Achieving a high sustainability performance is a challenge for not only large industries but also Small and Medium Industries (SMIs). In order to achieve high sustainability performance, an organization needs to innovate in its operations, especially in activities related to the environment, employees, society, and ethics [4]. Innovation is a multi-stage process conducted by an organization to transform concepts into a new better product, or process in order to gain competitive advantage [5]. Conducting innovation needs resources, both financial and nonfinancial, which are very limited in the SMIs.

M. N. Osman Zahid et al. (Eds.): iMEC-APCOMS 2019, LNME, pp. 147–152, 2020.
https://doi.org/10.1007/978-981-15-0950-6_23

To overcome the problem of limited resources, SMIs can implement open innovation that utilizes both internal and external resources [6]. Open innovation is a distributed innovation process based on knowledge inflow and outflow deliberately across organization boundaries, using various mechanisms that are in line with the organization's business model whether involving money or not, with the aim of speeding up internal innovation and widening markets for external use of innovation [7]. Open innovation has three types of practices which are inbound, outbound, and coupled [8].

The relationship among inbound, outbound, coupled open innovation practices, and organizational performance is discussed in Mazzola et al. [9]. The model is tested in biopharmaceutical industry. The organizational performance explored in Mazzola et al. [9] is in term of innovation and financial performance. Therefore, for sustainability performance, the environmental and social dimensions need to be further investigated, especially in the SMIs context. This study aims to identify the relationship between the three open innovation practices and sustainability performance in the SMIs.

This paper is organized in five sections. Following this introduction section, the second section discusses the research model and hypotheses development based on the previous research. In the third section, the methodology in conducting this study is explained. After that, the result and discussion are presented in the fourth section. Finally, the findings are summarized as the conclusion in the fifth section.

2 Research Model and Hypotheses

2.1 Open Innovation

Lopes and de Carvalho [10] conducted a systematic literature review on the scientific articles related to open innovation. The result shows that most of the article using inbound and outbound to represent the open innovation concept. Inbound and outbound as representation of open innovation in the SMIs context is used in Martinez-Conesa et al. [11], which explored the relationship between open innovation and knowledge management capability.

Mazzola et al. [9] conducted a systematic literature review on the empirical study articles which investigate the role of open innovation practices in explaining the financial and innovation performance in an organization. The result shows that inbound, outbound, and coupled open innovation practices are related to innovation and financial performance in an organization. The other empirical study which supports Mazzola et al. [9] is Oltra et al. [12] which explored the influence of inbound, outbound, and coupled open innovation toward firm performance. Therefore, this study represents open innovation practices with inbound, outbound, and coupled.

In this study, inbound is defined as the inflows of new knowledge and technology from outside sources such as competitors, suppliers, customers, government, consultants, universities, or research organizations, that support innovation in the SMIs [9, 11]. Outbound is defined as the exploitation of new internal knowledge and technology to be used by external partners so that the SMIs get monetary and non-monetary benefits [9, 11]. Coupled is defined as co-innovation with external partners in the form of structured cooperation [9].

2.2 Sustainability Performance

The measurement of organizational performance evolved from the focus on productivity and budget control into a more integrated manner with regard to environmental and social aspects, which is known as sustainability performance [13]. Sustainability performance can be measured by the standard measurement tools such as the sigma sustainability scorecard and sustainable balanced scorecard [14], or through indicators formulated according to the organizational context, as Wijethilake [15] and Sajan et al. [16]. Among the studies which formulate its own indicators, most of the studies use the economic, environmental, and social performance to represent the sustainability performance. Therefore, this study uses the same approach.

In this study, the definition of economic, environmental, and social performance are adapted from Sajan et al. [16] which discussed sustainability performance in SMIs context. Economic performance is the profitability and economic viability of the SMIs. Environmental performance is the SMIs' initiative to mitigate the environmental impact of processes and products through material selection, energy savings and waste treatment. Social performance is the SMIs' initiative to support the welfare of employees and the local community.

2.3 Hypotheses

Previous studies which explore the relationship between open innovation practices and sustainability performance are very limited. Lopes and de Carvalho [10] synthesized the previous studies on open innovation and found that the organizational performance variables which are related to open innovation are innovation performance and firm performance in term of sales growth, market share, profitability, financial indicators, customer performance, and customer. The study of Mazzola et al. [9] has similar findings that the previous studies explore the relationship among open innovation, innovation performance, and financial performance. There is no previous study which explores the social and environmental aspects of sustainability performance.

This study explores the relationship between open innovation practices and sustainability performance. In the conceptual model, open innovation practices is represented by inbound, outbound, and coupled open innovation, while the sustainability performance is represented by economic, environmental, and social performance. The conceptual model is shown in Fig. 1. The hypotheses in this study are as follow.

H1a: Inbound open innovation positively influence economic performance.
H1b: Inbound open innovation positively influence environmental performance.
H1c: Inbound open innovation positively influence social performance.
H2a: Outbound open innovation positively influence economic performance.
H2b: Outbound open innovation positively influence environmental performance.
H2c: Outbound open innovation positively influence social performance.
H3a: Coupled open innovation positively influence economic performance.
H3b: Coupled open innovation positively influence environmental performance.
H3c: Coupled open innovation positively influence social performance.

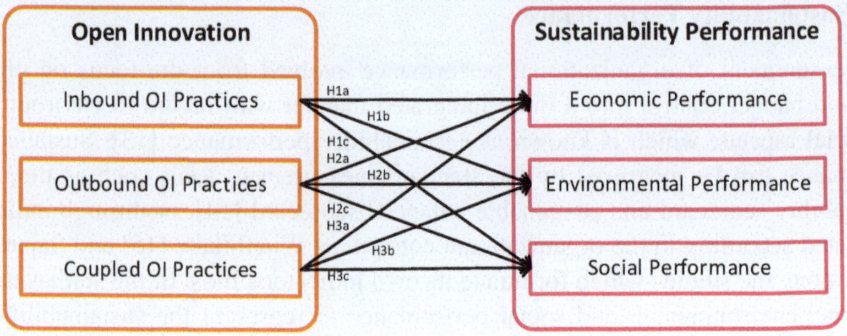

Fig. 1. Conceptual model

3 Methodology

This study starts by conducting literature review on the previous studies which explore the relationship between open innovation and organizational performance, especially sustainability performance. Based on the literature review, the studies which support the aim of this study are identified. The next step is developing the conceptual model and formulating the hypotheses.

The variables in the conceptual model are then operationalized into thirty five measurement items for the questionnaire. Some of these measurement items are adapted from Martinez-Conesa et al. [11] and Sajan et al. [16], and the others are newly developed. The questionnaire has five points of scale which are strongly disagree, disagree, neutral, agree, and strongly agree. The questionnaire is then used to test the conceptual model and hypotheses. The respondents of this study are 125SMIs which produce batik in Indonesia. The maximum indicators in the constructs of the model is seven. Therefore, the minimum sample size to identify minimum R^2 values of 0.5 in the independent variables for significant levels of 5% with statistical power of 80% is 51 [17]. The model is tested using the Partial Least Square Structural Equation Modeling (PLS-SEM) which is suitable for testing complex model with a small number of data in an exploratory study [17]. The measurement and structural model are tested using SmartPLS software [18].

4 Result and Discussion

The measurement model testing is conducted by performing the internal consistency (Composite Reliability (CR)), indicator reliability, convergent validity (Average Variance Extracted (AVE)), and discriminant validity test. The CR value of 0.6 to 0.7 and the AVE value above 0.5 are acceptable in exploratory research [17]. In order to achieve both standard values, the indicator reliability (outer loading) of each indicator needs to be observed. The indicator with outer loading value below 0.4 must be deleted. The indicator with outer loading value between 0.4 and 0.7 also needs to be deleted if the deletion increases the CR and AVE [17]. The discriminant validity is conducted by examining the cross loading of the indicators. After performing all of the test, thirteen indicators are deleted to make the CR and AVE value of the constructs acceptable.

The structural model testing is conducted by assessing the significance of the structural model relationship (using t value) and the R^2 value. Based on the t value for 5% probability of error, the significant relationship support H1a, H1b, H2b, H12c, H3c, while H1c, H2a, H3a, H3b are not supported. The R^2 values for economic, environmental, and social performance are 0.31, 0.27, and 0.16 consecutively. It means that beside the open innovation practices, there are more other variables which influence the sustainability performance.

The economic performance is only positively influenced by inbound open innovation, environmental performance is positively influenced by inbound and outbound open innovation, and social performance is positively influenced by outbound and coupled open innovation. In order to improve the economic performance, the SMIs needs to do inbound open innovation, especially for product innovation. The batik SMIs can collaborate with fashion designers to create more valuable end products. Environmental performance can be achieved by applying inbound and outbound open innovation, especially for process innovation to reduce negative environmental impact. The batik SMIs can adopt the research result from universities regarding the methods to eliminate the waste. The appropriate method can be shared among the SMIs. Social performance can be improve by contributing more in the innovation of external partner through outbound and coupled open innovation.

5 Conclusion

In SMIs context, open innovation practices are associated with sustainability performance. The significant relationships are identified between inbound open innovation and economic and environmental performance; outbound open innovation and environmental and social performance; and coupled open innovation and social performance. For future research, the model can be tested in SMIs in various industry types.

References

1. Schaltegger, S., Wagner, M.: Managing the Business Case for Sustainability: The Integration of Social, Environmental and Economic Performance, 2nd edn. Routledge, Abingdon (2017)
2. Searcy, C.: Corporate sustainability performance measurement systems: a review and research agenda. J. Bus. Ethics **107**(3), 239–253 (2012)
3. Goyal, P., Rahman, Z., Kazmi, A.A.: Corporate sustainability performance and firm performance research: literature review and future research agenda. Manag. Decis. **51**(2), 361–379 (2013)
4. Bos-Brouwers, H.E.J.: Corporate sustainability and innovation in SMEs: evidence of themes and activities in practice. Bus. Strat. Environ. **19**(7), 417–435 (2010)
5. Baregheh, A., Rowley, J., Sambrook, S.: Towards a multidisciplinary definition of innovation. Manag. Decis. **47**(8), 1323–1339 (2009)

6. Verbano, C., Crema, M., Venturini, K.: The identification and characterization of open innovation profiles in Italian small and medium-sized enterprises. J. Small Bus. Manag. **53**(4), 1052–1075 (2015)
7. Chesbrough, H., Bogers, M.: Explicating Open Innovation: Clarifying an Emerging Paradigm for Understanding Innovation, in New Frontiers in Open Innovation. Oxford University Press, Oxford (2014)
8. Huizingh, E.K.R.E.: Open innovation: state of the art and future perspectives. Technovation **31**, 2–9 (2011)
9. Mazzola, E., Bruccoleri, M., Perrone, G.: Open innovation and firms performance: state of the art and empirical evidences from the bio-pharmaceutical industry. Int. J. Technol. Manag. **70**(2–3), 109–134 (2016)
10. Martinez-Conesa, I., Soto-Acosta, P., Carayannis, E.G.: On the path towards open innovation: assessing the role of knowledge management capability and environmental dynamism in SMEs. J. Knowl. Manag. **21**(3), 553–570 (2017)
11. Lopes, A.P.V.B.V., de Carvalho, M.M.: Evolution of the open innovation paradigm: towards a contingent conceptual model. Technol. Forecast. Soc. Chang. **132**, 284–298 (2018)
12. Oltra, M.J., Flor, M.L., Alfaro, J.A.: Open innovation and firm performance: the role of organizational mechanisms. Bus. Process. Manag. J. **24**(3), 814–836 (2018)
13. Bititci, U., Garengo, P., Dörfler, V., Nudurupati, S.: Performance measurement: challenges for tomorrow. Int. J. Manag. Rev. **14**(3), 305–327 (2012)
14. Carneiro-da-Cunha, J.A., Hourneaux, F., Corrêa, H.L.: Evolution and chronology of the organisational performance measurement field. Int. J. Bus. Perform. Manag. **17**(2), 223–240 (2016)
15. Wijethilake, C.: Proactive sustainability strategy and corporate sustainability performance: the mediating effect of sustainability control systems. J. Environ. Manag. **196**, 569–582 (2017)
16. Sajan, M.P., Shalij, P.R., Ramesh, A., Augustine, B.P.: Lean manufacturing practices in Indian manufacturing SMEs and their effect on sustainability performance. J. Manuf. Technol. Manag. **28**(6), 772–793 (2017)
17. Hair, J.F., Hult, G.T.M., Ringle, C.M., Sarstedt, M.: A Primer on Partial Least Squares Structural Equation Modeling (PLS-SEM), 1st edn. SAGE Publications, Inc., California (2014)
18. Ringle, C.M., Wende, S., Becker, J.M.: SmartPLS 3, Boenningstedt: SmartPLS GmbH (2015)

Product-Service System Inventory Control for Malaysian Palm Oil Industry: A Case Study Utilizing IDEF0 Modelling

Fatkhurrahman Manani(✉) and Siti Zubaidah Ismail

Faculty of Mechanical and Manufacturing Engineering,
University Malaysia Pahang, 26600 Pekan, Pahang, Malaysia
MF018001@stdmail.ump.edu.my

Abstract. Over the last several years, the palm oil industry has multiplied in Malaysia as it is among the main contributors to the country's economic development. The government strongly emphasizes palm oil productivity in Malaysia in terms of quality and safety. The current inventory context offers a more extensive set of uncertainties that the company needs to manage due to the enhanced scope and complexity of the product and service offering. The current method of determining inventory takes an annual physical inventory by comparing the actual inventory with record levels. This method is a costly and time-consuming effort. This paper investigates Product-Service System (PSS) Inventory Control for different types of the section in the palm oil industry. In this work, an integrated and systematic methodology is presented for the assessment of inventory control. The methodology is represented as an IDEF0 activity model that rigorously defines the interconnections of information and activities with incorporating product and service elements. In general, therefore, it seems that PSS inventory control can integrate some of the elements involved in inventory control and consequently give a positive impact on all palm oil industry players.

Keywords: Product-Service System (PSS) · Inventory control · Palm oil · IDEF0

1 Introduction

The palm oil industry in Malaysia has undergone changes beginning with ornamental plants to become the most valuable commodity crops as it is among the main contributors in the country's economic development. As the economic crisis hit Malaysia around 1997, the palm oil industry played a significant role in absorbing the effects of the economic downturn through its export [1]. The palm oil industry in Malaysia is based on yearly oil planted area. In 2017 it reached 5.81 million hectares, which are increased by 1.3% compared to 5.74 million hectares from the previous year. The current method of determining inventory was done manually by takes an annual physical inventory and comparing the actual inventory with record levels. This kind of method is a costly and time-consuming effort. This research will carry out the investigation on Product-Service System (PSS) Inventory Control for different types of the

M. N. Osman Zahid et al. (Eds.): iMEC-APCOMS 2019, LNME, pp. 153–158, 2020.
https://doi.org/10.1007/978-981-15-0950-6_24

section in the palm oil industry. PSS is known as the integration between product and service and has its advantages, which will reduce the impact on the environment [2]. PSS will assist the company in terms of improving the value of products and services. In this work, an integrated and systematic methodology is presented for the assessment of inventory control. The methodology is represented as an IDEF0 activity model that rigorously defines the interconnections of information and activities with incorporating product and service elements.

Inventory control in agriculture controlling the budgets, raw materials, and finished products [3]. It will often cope with the issue of uncertainty elements, and the integration of these elements become importance in managerial decisions in order to gain efficiency and competitiveness to stay relevant in the market [4]. The theory of inventory control, especially on agriculture products, need more consideration due to its unique characteristics such as unpredictable supply and price and inconsistent demand [5]. Inventory control plays a crucial role in managing inventory so that the inventory level of palm oil product should be low but enough to meet the order of the processing unit. Inventory control provides useful information for managers to make more accurate and timely decisions.

2 Literature Review

In 2008 palm oil was one of the major commodities exports in Malaysia; the total production of 17.7 million tonnes, which contributes 41% of total world palm oil production [6]. The Malaysian palm oil industry plays an essential role as a significant producer and exporter of palm oil and palm oil products not only for local use but also for global demand for oil and fats [7]. Nowadays, palm oil production has brought countless economic benefits and is now a new emerging economic sector in Malaysia [7]. Inventory management of palm oil products explaining how palm oil industry players in Malaysia manage their inventory to keep pace with market demand without any stock surplus that will affect company operations [8].

Many manufacturers feel that developing a Product-Service System (PSS) is very important because of the trend of servitization and creating social value [9]. PSS is as an interesting business theory by creating high value-added through the integration of products and services [10]. PSS is a business model and integration concept of services and products to empower new capital business in order to increase value-added flow and meet customer demands [11]. Recent studies show that the role of producers on the sustainability of the agricultural industry network can be enhanced by adapting PSS through metric identification processes at various stages [12]. Studies also have shown that the concept of business model can be a reference to implementing PSS, but companies still need guidance in implementing it.

Inventory interpreted as a raw material used to produce the product through several processes before it can be sold. Inventory can be classified into three-component, namely raw materials, work in progress and finished goods. Inventory control described as the process of checking the supply, inventory and storage company so that it all meets the current demands and needs and avoid over-stock and other problems involving stocks. Agricultural inventory literature has not been exploring well because of its unique elements, including supply characteristics, uncertain demand, unstable prices, and selling price decision [5]. Inventory for agriculture which is based on customer-oriented will usually take into consideration the customer's request before a decision to keep the stock taken [3]. Agricultural sector faces new challenges, including managing agricultural supply chains efficiency that has become an attractive topic for researchers and practitioners.

3 IDEF0

This paper used Integration Definition for Function Modelling (IDEF0) to represent all activities and interconnections, including products and service. IDEF0 is a functional model methodology for describing processing and manufacturing. It is increasingly popular among industry and academic practitioners due to its function, which offers good and effective communication between researchers and customers by showing a simplified diagram [13]. By using IDEF0, functions are depicted via boxes, which means represent such as activities, actions, processes, and operations. Interfaces are displayed by arrows which indicate data or activity. Input arrow consists of data which entering the functions box from the left side. Output arrow exiting the box on the right side which as a result of a function that has taken place. Control-arrow entering from the top of the box and mechanism entering from the bottom of the box. Control is a constraint and direct activity in the process which fixed and seldom changed. The mechanisms are resources and tools that are needed to perform and complete the process [14]. IDEF0 helps in the implementation of the regeneration process in the agricultural sector [15]. Figure 1 below shown IDEF0 representative.

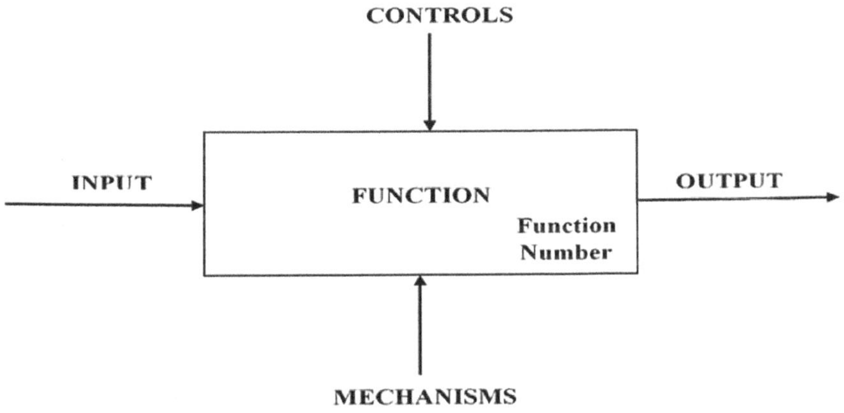

Fig. 1. IDEF0 representative

4 Result and Discussion

Figure 2 shows the palm oil agriculture characteristics with all connection be-tween parent function together with their child function in one single diagram. At the Seedling (A1), the seed is the input for the beginning of the entire process chain. Sapling is the output for A1 and also an input for A2. Controls in this stage are policy, culling process and inventory control, scheduling and quality control. Culling is a procedure to be done in order to get the best and uniform seedlings that produce high yields and is monitored by quality control. Inventory control is ensuring that is no stock dumping that will cause loss to the company. Mechanisms in this stage are sunlight, watering, weeding, manuring and pesticide application. Watering is to ensure the seedling get optimum water to reach optimum growth and avoid stagnantly. Manuring provides the growth of seedlings as additional nutrients based on the types of soil used at the nursery level. Weeding pro-gram takes places after the manuring to prevent the seedling from fungal attacks.

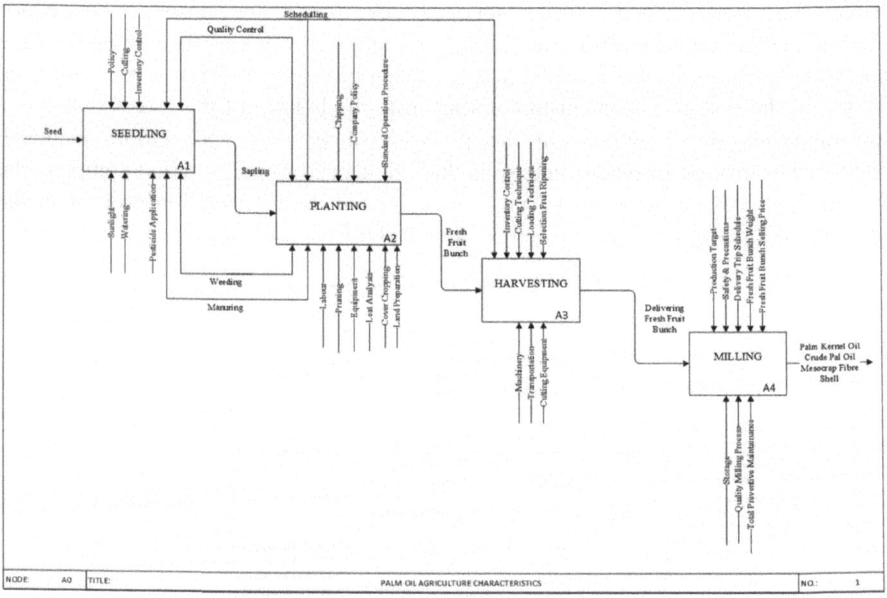

Fig. 2. Palm oil agriculture characteristics

In planting (A2), the sapling to be input and the output has recognized as Fresh Fruit Bunch. Chipping process where trunks of the palm oil trees together with its fronds will be cut and chopped. For mechanisms are like weeding, manuring, labour, pruning, equipment, leaf analysis, cover cropping, and land preparation. Leaf analysis will come out with a program, and it is essential to determine the correct fertilizer to use.

For harvesting (A3), the fresh fruit bunch has been identified as input, and the output is delivering FFB to the milling stage. Controls in this stage are like scheduling, inventory control, cutting technique, loading technique, and selection fruit ripening. Inventory control plays a role in order to secure the company's operations run smoothly and does not experience a stock surplus. Cutting technique has to be done by the specialist hired labour. Mechanisms in this stage are machinery, transportation, and cutting equipment. Machinery used in reception counters like backhoe and loading ramp. Transportation is an example or service in this study because outsiders provide it.

Delivering Fresh Fruit Bunch will be input in milling (A4) and will produce output such as crude palm oil, palm kernel oil, mesocarp fibre, and shell. To get such results, it needs to go through controls and mechanisms. Controls are like production target, safety & precautions, delivery trip schedule, FFB weight, and FFB selling price. For delivery trip schedule, it has to be inline with mill capacity in order to prevent from FFB dump at the grading area. In the case of FFB selling price, it may be slight differences from one company to another, but the government has set ceiling prices that these industry players have to follow. Grading FFB should be independent parties to perform their duties and may not be affected by any party. Every mill has to ensure that its storage well enough to stored its outputs so that the quality still in good condition. To ensure every machine works correctly and in the right conditions, the company had its maintenance schedule.

It is shown that by using IDEF0, the integration between products and service elements that happened during the operation are to facilitate any hypothesis and planning that want to be designed and action to be taken for continuous improvement. Therefore IDEF0 give a clear picture of the difference between controls and mechanisms so that the industry can identify the best process for performing its operations. In other words, Product-Service System (PSS) inventory control will give a big positive impact on the palm oil industry by interconnecting information and activities involving an inventory of the entire system indeed.

5 Summary and Conclusion

IDEF0 had been widely used for modeling and analyzing the correlation of the system elements. This paper has provided a useful understanding of overall characteristics through the developed generic model in the Malaysian palm oil industry using the IDEF0. By utilizing the IDEF0 modeling, the integration between elements indicates that Product-Service System (PSS) inventory control will benefit to this industry generally. Its a preliminary discussion for future work which will be used to develop software for efficient inventory control at the palm oil plantation.

Acknowledgment. This research would not have been possible without the financial support given by the Universiti Malaysia Pahang via RDU180293. Neither would it have been possible without the support, time and knowledge provided by the collaborating industrial company. The views expressed in this paper are those of the authors.

References

1. Vijaya, S., et al.: Life cycle inventory of the production of crude palm oil - a gate to gate case study of 12 palm oil mills. J. Oil Palm Res. **20**, 484–494 (2008)
2. Baines, T.S., et al.: State-of-the-art in product-service systems. Proc. Inst. Mech. Eng. Part B: J. Eng. Manuf. **221**(10), 1543–1552 (2007)
3. Sani, M.H.: Applicable inventory control models for agricultural business managers: Issues and concerns. Trends Agric. Econ. **7**(1), 11–25 (2014)
4. Borodin, V., et al.: Handling uncertainty in agricultural supply chain management: a state of the art. Eur. J. Oper. Res. **254**, 348–359 (2016)
5. Shi, J.J., et al.: Managing inventories for agricultural products: the optimal selling policies. Oper. Res. Manuscr. **3**, 142 (2014)
6. Umar, M.S., et al.: A policy framework and industry roadmap model for sustainable oil palm biomass electricity generation in Malaysia. Renew. Energy **128**, 275–284 (2018)
7. Alam, A.F., et al.: Malaysian oil palm industry: prospect and problem. J. Food Agric. Environ. **13**(2), 143–148 (2015)
8. Nordin, A.B.A: A study on inventory management of malaysia palm oil products. Oil Palm Ind. Econ. J. **12**(1) (2012)
9. Meier, H., et al.: Industrial product-service systems—IPS2. CIRP Ann.-Manuf. Technol. **59**, 607–627 (2010)
10. Muto, K., et al.: A guideline for product-service-systems design process. Proc. CIRP **30**, 60–65 (2015)
11. Papazoglou, M.P., et al.: Collaborative on-demand product-service systems customization lifecycle. CIRP J. Manuf. Sci. Technol. **480**, 15 (2018)
12. Ziout, A., Azab, A.: Industrial product-service system: a case study from the agriculture sector. Proc. CIRP **33**, 64–69 (2015)
13. Waissi, G.R., et al.: Automation of strategy using IDEF0 - a proof of concept. Oper. Res. Perspect. **2**, 106–113 (2015)
14. Sarkis, J., Liles, D.H.: Using IDEF and QFD to develop an organizational decision support methodology for the strategic justification of computer-integrated technologies. Int. J. Proj. Manag. **13**(3), 177–185 (1995)
15. Bevilacqua, M., et al.: A combined IDEF0 and FMEA approach to healthcare management reengineering. Int. J. Procure. Manag. **8**, 25–43 (2015)

Designing Product-Service System Inventory Control: System Requirements Analysis of Raw Material in Automotive Industry

Farah Ameelia Mohammad and Siti Zubaidah Ismail(✉)

Faculty of Mechanical & Manufacturing Engineering, Universiti Malaysia Pahang, 26600 Pekan, Pahang, Malaysia
mfo18002@stdmail.ump.edu.my, zubaidahismail@ump.edu.my

Abstract. During the past decade, growing attention to Product-Service System (PSS) is given a massively important role in development from traditional product-based business model to provision of industrial services. Every production task will incur some level of inventory, and this requires the collection and processing of both physical and information elements. Inventory control in PSS creates a more extensive set of uncertainties that the automotive industry needs to manage due to the enhanced scope and complexity of the product and service offering. There is a limited diffusion of new business models, especially by the automotive industry, and in particular in managing inventory control of raw material. To deal with this difficulty, this paper aims: firstly, suggests a generic model using IDEFØ for systematic and comprehensive analyses of how PSS inventory control characteristics in raw material integrate. Finally, this paper proposes the inventory of raw material in automotive industry characteristics to make it possible to better understand the dynamics of scope and complexity of the product and service offering.

Keywords: PSS · Inventory control · IDEFØ · Raw material inventory

1 Introduction

The industry that uses steel and aluminium as a raw material due to the fluctuation worldwide has also been affected by the progress in the world steel market. One of the affected areas is the automotive industry, where the industry needs steel to form automotive raw material used to produce a finished product. The steel industry uses large amounts of raw material, mainly scrap, aluminium, quicklime, limestone and iron ore. The industries need to enforce and execute the new method for managing stock for inventory control and adds to the comprehension of the PSS worldview in manufacturing industries. The consequences of increasing the price of steel in universal are given a significant impact on the industry to continue their market price and the product standard.

The introduction of Product-Service Systems (PSS) suggests that as a commercial system of products, services, and supporting infrastructure that is capable of satisfying customer requirements, lower environmental impact, cost-effectively and sustainably to

M. N. Osman Zahid et al. (Eds.): iMEC-APCOMS 2019, LNME, pp. 159–164, 2020.
https://doi.org/10.1007/978-981-15-0950-6_25

capture this transformation [1]. PSS development is an explication to achieve the market company and it also offers lower environmental impact by combination of products and services become into one of the solution. While the pure transaction changes to relational engagements with customers, the potential risks will increase for manufacturing companies to assume their operation responsibilities for customer processes, and engage in value co-creation [2]. By implementing a PSS business model, is that it give interactions with the customer, production operation with risk management responses [3, 4]. This trend is incredibly evident in product-centric manufacturing companies that are investing heavily in developing industrial services involving their products [5].

The problems on inventory faced by the organisation due to several factor. One of them are overproduction, stock out situations, poor documentation, delay in delivery of raw materials and discrepancy of inventory. Inventory control in PSS creates a more extensive set of uncertainties that the factory needs to manage due to the enhanced scope and complexity of the product and service offering. Inventory control mainly focuses on location, storage, and accounting for the number of inventories. Where the inventory level provides an impact to costs for storage, insurance, spoilage and interest, supporting infrastructure system for inventory control of raw material is created. Poor inventory management can create a significant influence on the capacity of productivity and market cost in an organization. Thus, planning to systemize and manage the flow of materials involved inventory management from the first purchase unit by distributing the internal operations to the service point. The availability of the resources in an organisation is the goal to achieve an optimal inventory management.

This research proposed a generic model using IDEFØ. IDEFØ is an abbreviation for "Icam Definition for Function Modelling". ICAM stands for "Integrated Computer-Aided Manufacturing. IDEFØ is a family component of IDEF modelling in software engineering and is built for Structural Analysis and Design Technique (SADT) (Press, 2001). IDEFØ is one of the graphical languages named as structured analysis and design techniques. To design a model from activities in a manufacturing organisation and analysis for the system, what is needed to perform those function and what the current situation does right, IDEFØ can be used. It often created as one of the first tasks of a system development effort. The system analysis would be achieved through modelling information, dynamics, functions, and processes.

2 Methodology

Typically, the features of IDEFØ models are the same as data flow diagrams. This tools mostly have been widely used as functional modelling in the manufacturing field because it is easy to use and cost-effective. As an analytical tool, IDEFØ can help in identifying the functions performed in the software and very suitable in systems with a higher classification of forecast.

IDEFØ views a complex system as both functions, whether implemented by using machines, people, or other means. The function is to transform the inputs into the outputs, under the influence of control, using the mechanism provided, to describe its boxes and arrows and glossary and text to define the precise meanings of diagram elements, it has diagrams through in the simple box and arrow graphics and text labels. The gradual exposition of details featuring a hierarchical structure, with the function at the top and with successive levels are not more than the limitation of subfunctions revealing well-bounded detail breakout. The benefits of IDEFØ are providing flexibility and a general format of representation of a process where it helps to allow decompose information to any detail by using inexpensive commercial software program (Fig. 1).

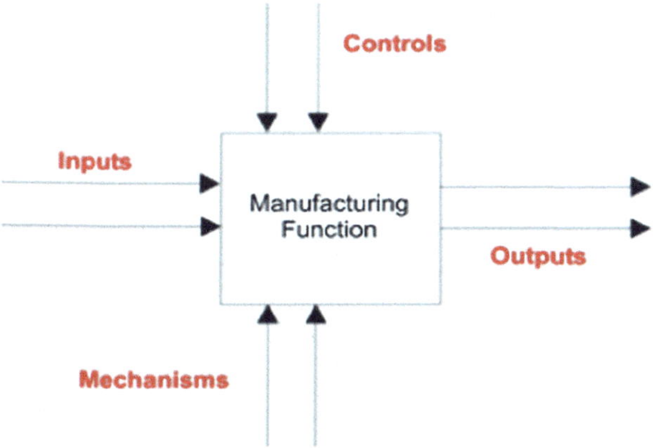

Fig. 1. IDEFØ elements box and arrow graphic

3 Results and Discussion

From the A0 diagram in Fig. 2, it shows the elements of system requirements analysis using IDEFØ, which represents the context of the model, called a context diagram. It contains the significant sub-levels of the model with each process having its specific designation in PSS inventory control. They are consist of:

A1 Market and sell; product and service

The first one tasks is where it includes all the activities that are required to identify the customer upon request. The highlights of the key input are the company policies, company procedure and standard operating procedure (SOP). The sub-activities contain all the tasks associated with the activity from negotiating with the customer, define product requirements and verify availability.

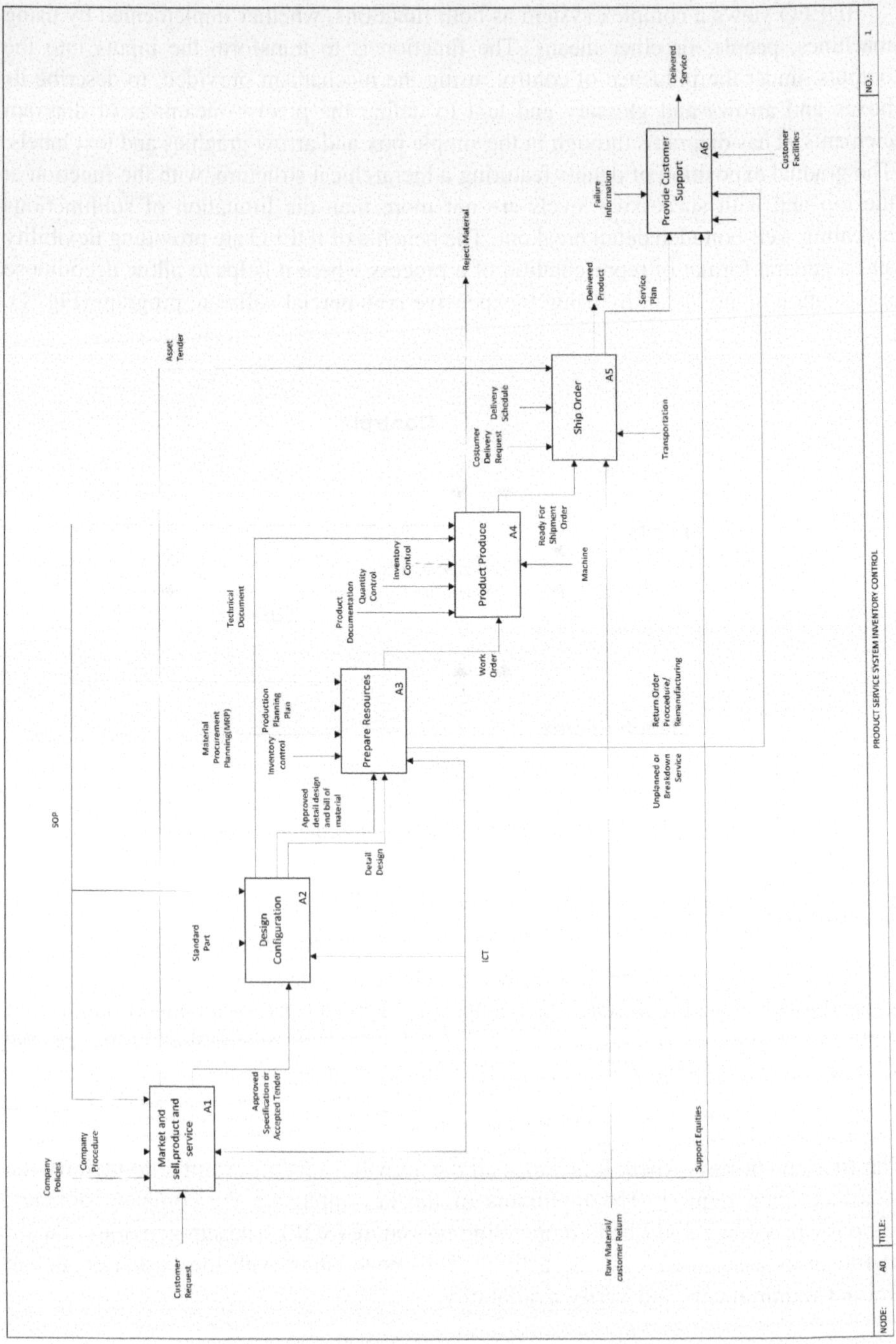

Fig. 2. Product service systems inventory control

A2 Design configuration

The second sub-activities tasks are design configuration. Before the design approved to generate planning on information to the manufacturer and the supplier, the customer needs to accept the specification and tender, overall design with all the input of budget, SOP and some scheduled will be prepared. The initial design team then will develop a detailed design to final detailed design.

A3 Prepare resources

From the detailed design and bill of material (BOM), which include all areas within the asset such as machine and labour resources, are the critical activities in A3. The fundamental mechanisms are ICT supports tools. Main production section plays a huge role when generating production schedule that it will generate some work order and documentation records to the production as they must maintain a good of spare parts, equipment, and doing some inspection and maintenance.

A4 Generate product

The task not only to create the product but also to repair/reject products. Work order, material from the supplier, rework plan and materials requisition are the key inputs. It is essential that manufacture and assembles parts activities comply with safety regulations. The mentioned product is inspected by guidelines of the QA/QC policy. The inspection is to make sure the products that go under the before the delivery process, any imperfections or reject parts to go through reworking or repairing processes until that has no flaws when delivery to the customer.

A5 Ship order

The sub-activities of A5 contain all the tasks associated with the delivery of the scheduled products. It also involves interactions of the company, customers and shipping carrier. The critical constraint is the policy, and the critical mechanism is the ICT support tools. The product or service can only be confirmed as received by the customer after the customer signs off the document. Any failure undeliverable or rejected product by the customers will be returned to the company for further checking and inspection.

A6 Provide customer support

All the tasks linked with the activity in providing customer support is shown in A6. The key inputs are the condition of QA/QC report and service plan while the key outputs are the maintenance and inspection analysis, which are the current condition documents and documentation graph. Tools and spares are used as key mechanisms. Unplanned or breakdown service and scheduled service, which have different constraints but the same mechanism are the two kinds of services.

4 Conclusions

In this paper, it was attempted to develop a generic model of Product-Service System (PSS) inventory control incorporation all the essential system requirements characteristics of the production line of raw material. By using IDEFØ, this shown the inventory control of PSS is a complex process as opportunities on market development and increase the inventory efficiency. In sum, this paper is, in essence, a preliminary discussion of an example of PSS in raw material in the automotive industry context, with a focus on inventory control.

Acknowledgement. This paper would not be possible without the financial support given by the Universiti Malaysia Pahang via RDU180385. Neither would it be possible without the support, time and knowledge provided by the collaborating industrial company.

References

1. Ismail, S.Z., Morton, S.C., Harding, J.A., Michaelides, R.: Product-service system in inventory control: a new paradigm? In: 2015 International Conference on Industrial Engineering and Operations Management (IEOM) (2015)
2. Reim, W., Parida, V., Sjödin, D.R.: Risk management for product-service system operation. Int. J. Oper. Prod. Manag. **36**(6), 665–686 (2016)
3. Mackulak, G.T.: High level planning and control: An IDEF0 analysis for airframe manufacture. J. Manuf. Syst. **3**(2), 121–133 (1984)
4. Richter, A., Sadek, T., Steven, M.: Flexibility in industrial product-service systems and use-oriented business models. CIRP J. Manuf. Sci. Technol. **3**(2), 128–134 (2010)
5. Kindström, D.: Towards a service-based business model – key aspects for future competitive advantage. Eur. Manag. J. **28**(6), 479–490 (2010)
6. Park, H., Yoon, J.: A chance discovery-based approach for new product–service system (PSS) concepts. Serv. Bus. **9**(1), 115–135 (2013)

Reliability Centered Maintenance of Mining Equipment: A Case Study in Mining of a Cement Plant Industry

Endi Alta[1(✉)], Nilda Tri Putri[2], and Henmaidi[2]

[1] Semen Padang Co., 25237 Padang, Indonesia
endialta@gmail.com
[2] Department of Industrial Engineering, Andalas University, 25163 Padang, Indonesia
{nilda, henmaidi}@ft.unand.ac.id

Abstract. Today's industrial growth is inseparable from the free market mechanism which is a consequence of economic globalization and thus cannot be avoided by any country. All this leads to a very keen business competition. This can be seen in the high competition of the Indonesian national cement industry and worsened by the oversupply of domestic cement in Indonesia. Semen Padang Co. as a national cement company sets the optimization of production capacity as one of the focus of initiatives. To achieve this, maintenance of production equipment is very important. In the mining unit, the occurrences of breakdown are very high in the crusher plant and conveyor belts having a large number of equipment and serial production activities that need to be set up for effective maintenance planning. The purpose of this study is to develop preventive maintenance planning with the implementation of the Reliability Centered Maintenance methodology and to submit a schedule of planned preventive maintenance activities with the limitation of meeting the target production hours of conveying limestone and silica to each plant. Based on the calculation results from the implementation of preventive maintenance planning, it can reduce the breakdown time by 42% and it is recommended to be applicable. The next study is to calculate the maintenance costs needed in accordance with the maintenance plan that is made with the Reliability Centered Maintenance methodology. This step is expected help the company in minimizing the cost of maintaining crusher plant and conveyor belts.

Keywords: Reliability-centered maintenance · Preventive maintenance · Mining

1 Introduction

Maintenance is an important task to keep equipment reliability to ensure that all production activities run smoothly according to the set schedule and target [1]. The high occurrences of machine breakdown is a problem that many companies face with today. This condition certainly results in the production process in the company becoming unproductive and inefficient [2]. Mining activities are the first series of work activities

M. N. Osman Zahid et al. (Eds.): iMEC-APCOMS 2019, LNME, pp. 165–170, 2020.
https://doi.org/10.1007/978-981-15-0950-6_26

at Semen Padang Co. The activities are mining limestone and silica stone as the main ingredients for making cement. Unit of mining in Semen Padang Co. also faces with a high breakdown of equipment operating conditions that have exceeding the set target. The problem of high breakdown will be explained by the Reliability Centered Maintenance (RCM) approach [3]. The application of RCM can help to decide what the different maintenance strategies required to ensure a high reliability of equipment with a reasonable cost [1, 4].

2 Reliability Centered Maintenance

RCM is the optimal mix of reactive, time-based, interval-based, condition-based, and proactive maintenance [1, 5]. RCM is used to determine what failure management strategies must be implemented to ensure that the system reaches the desired level of safety, reliability, environmental health and operational readiness in the most cost-effective way. Often, the purpose of maintenance is to prevent all possible failures, and produce an over-maintenance program and maintenance activities to be ineffective [6]. The application of RCM will produce a treatment program that is truly applicable and effective [3, 7].

RCM recognizes the value of your personnel and takes advantage of their extensive experience running the equipment [1, 8]. Run to Failure (RTF) works on the assumption that it is most cost effective to let equipment run unattended until it fails. This type of maintenance is used for the lowest priority equipment. Preventive Maintenance (PM) comprises maintenance tasks on a piece of equipment at regular intervals regardless whether the equipment needs it or not. When well implemented, PM may produce savings in excess of 25 percent [1, 9]. Even though PM is an improvement over RTF, abrupt failures that cause unscheduled downtime still occur. Predictive Maintenance (PDM) is maintenance based on real-time data collected on a piece of equipment. The data show the "health" of the equipment. Proactive Maintenance (PAM) determines the root causes of failure and implements "fixes" (e.g., redesign the equipment so that it does not break down as frequently).

3 Case Study

The case study herein was carried out in Semen Padang Co. mine conveyor belt system. The conveyors transport crushed lime stone or silica stone to storages in the plant. This study takes operating and repairs data from 2015 to 2017. The operation of the conveyor system must be able to maintain the availability of limestone and silica in each plant's storage capacity at least 50%. Totally, annual requirement for limestone is 9,850,000 tons and 750,000 tons of silica stone for all plant. Processing and transport activities of material must be carried out safely by taking into account the safety of the operator and maintenance crew and not causing pollution to the environment. In the conveying system of Semen Padang Co. mine there are 42 conveyor items. RCM applications are implemented in the most critical machines. The critical value of the machine is determined by weighing several aspects of the scale of assessment in accordance with the operational requirements of the equipment.

The weighing scale is made with low, medium and high values (with values 1, 2, 3) and the equation used is:

$$MC = 0.2 * L + 0.2 * Q + 0.1 * M + 0.2 * B + 0.3 * E \quad (1)$$

Where machine criticality (MC), conveyor length (L), storage load volume (Q), material transported (M), machine breakdown (B) and environment impact (E). As a result, we get the most critical belt conveyor as A1J12A, with a value of 3 criticality (see Table 1).

Table 1. Criticality of conveyor belt in Semen Padang. Co. Mining (partial)

No	Machine no.	Critical machine criteria										Total
		Conveyor length		Storage load volume		Material		Breakdown hours		Environment		
		Weight	20%	Weight	20%	Weight	10%	Weight	20%	Weight	30%	
		Value	Value × weight	Value	Value × weight	Value	Value × weight	Value	Value × weight	Value	Value × weight	
1	A1J12A	3	0,6	3	0,6	3	0,3	3	0,6	3	0,9	3,00
2	A5J10	3	0,6	3	0,6	2	0,2	3	0,6	3	0,9	2,90
3	A1J12P	3	0,6	3	0,6	3	0,3	2	0,4	3	0,9	2,80
4	A1J12B	3	0,6	2	0,4	3	0,3	3	0,6	3	0,9	2,80
5	A1J12C	1	0,2	3	0,6	3	0,3	1	0,2	3	0,9	2,20
6	A5J11	1	0,2	3	0,6	2	0,2	1	0,2	3	0,9	2,10
7	20104	2	0,4	2	0,4	3	0,3	1	0,2	1	0,3	1,60
8	A1J11	1	0,2	3	0,6	3	0,3	1	0,2	1	0,3	1,60
9	A4J14	2	0,4	2	0,4	3	0,3	1	0,2	1	0,3	1,60
10	15107	1	0,2	3	0,6	2	0,2	1	0,2	1	0,3	1,50

The next step is to create a Fault Tree Analysis (FTA) that aims to identify each failure and find the root cause of the failure that can be generated from each machine component. Examples of FTA on conveyor A1J12A can be seen in the Fig. 1.

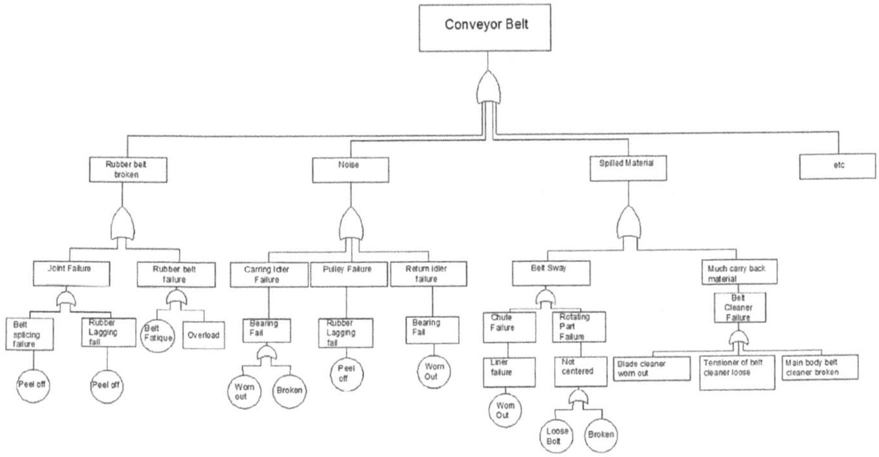

Fig. 1. Partial fault tree analysis of conveyor belt

Reliability data is needed to decide whether an item is critical. This data describes the failure process and optimizes the scheduling of preventive maintenance time. Examples of such data are those at average time between failures (MTTF), average repair time (MTTR) and failure rate function. Furthermore, an analysis of Failure Mode and Effect Analysis (FMEA) was developed which consisted of tabulation failure modes obtained from FTA and machine reliability data.

Then, an assessment is carried out so that the risk priority index (RPN) will be obtained from each failure mode. The value of RPN is the multiplication of the values of severity, occurrence and detection. Each of these parameters is made in 10 rating scales.

Table 2. Risk ranking categories

RPN	Risk level	Description
0–15	Low	Acceptable as it is
15–125	Medium	Acceptable with procedures and controls
125 < 425	High	Unacceptable, should be mitigated with specified time
>425	Extreme	Unacceptable, must be mitigated immediately

3.1 Selection of Maintenance Strategy

The next step is selection the maintenance strategy in each failure mode in the FMEA tabulation. Determination of maintenance strategies by using the RCM Logic Tree Diagram (LTA). The maintenance type determination that will be carried out is processed by answering each step of the question that is in the LTA of RCM (Fig. 2).

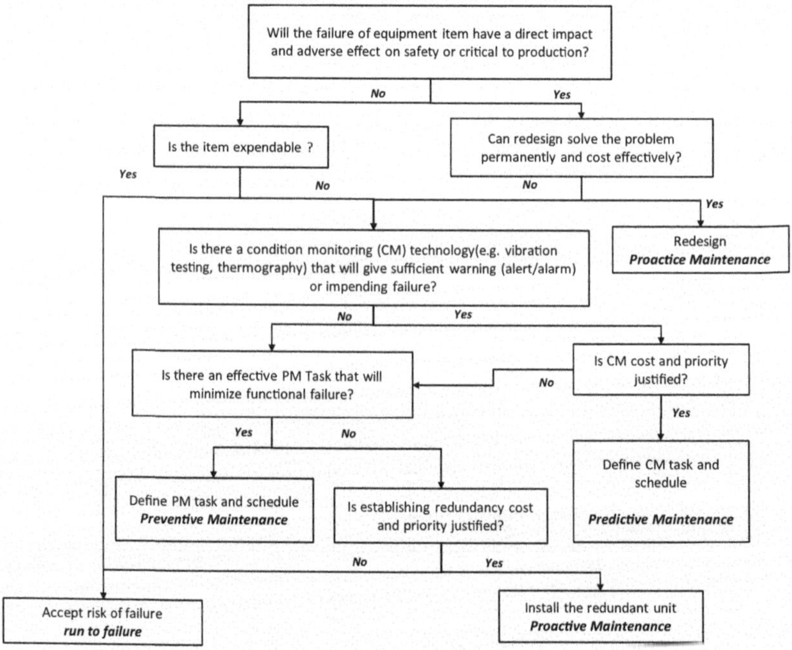

Fig. 2. RCM logic tree diagram

Next is to determine the maintenance interval from the results of the LTA. Maintenance Scheduling is performed on failure modes that have high risk as shown in Table 2.

Table 3. Failure mode and effect analysis for conveyor A1J12A (partial)

Item	Failure Mode	Failure Effect	Severity Analysis (production)	Severity Analysis (Safety)	Severity Analysis (Environment)	Severity (S)	Occurrence (O)	Detection (D)	RPN	Risk Group	LTA	Maintenance Task (Current)	Maintenance Task (proposed)
Rubber Belt	Belt Broken	Machine Stop, Downtime For Repair Or Replace 32 Hours	7	1	4	7	1	5	35	Medium	Yes	PM	PM
	Splicing Peel Off	Belt Break Apart, Downtime For Repair Or Replace 3 Hours	1	1	2	2	10	7	140	High	Yes	PM	PAM
	Belt Run Off (Sway)	Material Spilled, Downtime For Repair Or Replace 8 Hours	2	1	4	4	3	1	12	Low	No	RTF	RTF
	Belt Torn (Leaked Out)	Material Spilled, Downtime For Repair Or Replace 3 Hours	1	1	4	4	5	7	140	High	Yes	PM	PAM
	Belt Slip	Machine Stop, Downtime For Repair Or Replace 2 Hours	1	1	1	1	1	7	7	Low	No	RTF	RTF
	Belt Edge Peel Off	The Belt Has The Potential To Get Stuck And Tear Long, Downtime For Repair Or Replace 40 Hours	9	1	1	9	2	1	18	Medium	Yes	RTF	PAM
	Belt Shredded Because Of Iron Entry Or Sharp Objects	Material Spilled, Downtime For Repair Or Replace 48 Hours	10	1	4	10	1	7	70	Medium	Yes	RTF	PAM
Carrying Idler	Idler Is Not Installed Or Detached	Material Spilled, Downtime For Repair Or Replace 1 Hours	1	1	1	1	1	1	1	Low	No	PM	RTF
	Idler Not Rotate	Causing The Belt Sway Dan Material Spilled, Downtime For Repair Or Replace 1 Hours	1	1	8	8	1	1	8	Low	No	PM	RTF
	Bearing Worn Out Or Broken	Noise (Complaints From Surrounding People), Downtime For Repair Or Replace 1 Hours	1	1	6	6	3	7	126	High	Yes	PM	PAM
	Idler Broken	Material Spilled, Downtime For Repair Or Replace 1 Hours	1	1	1	1	1	7	7	Low	No	PM	RTF
	Idler Stand Fall Down	Material Spills Along The Belt, Downtime For Repair Or Replace 1 Hours	1	1	8	8	1	1	8	Low	No	PM	RTF
Return Idler	Idler Is Not Installed Or Detached	Material Spilled, Downtime For Repair Or Replace 1 Hours	1	1	1	1	1	1	1	Low	No	PM	RTF
	Can Not Rotate	Material Spills Along The Belt, Downtime For Repair Or Replace 1 Hours	1	1	8	8	1	1	8	Low	No	PM	RTF
	Bearing Worn Out Or Broken	Noise (Complaints From Surrounding People), Downtime For Repair Or Replace 1 Hours	1	1	6	6	4	7	168	High	Yes	PM	PAM

4 Result and Discussion

Base on the result of FMEA (Table 3), we can see that there are four modes of failure that belong to a "High" criticality group. In accordance with the Table 2, these four failure modes must be mitigated. The failure modes are bearing of return idler worn out or broken, rubber belt splicing peel off, belt torn (leaked out) and bearing of carrying idler worn out or broken. Based on the calculation of MTTR and MTTF using the failure mode distribution approach that occurs, the reduction in downtime is obtained by 629.8–368.8 = 261 h or 41.44% of the down-time generated by the four components that have a "high" criticality, see Table 4.

Table 4. Reduction of downtime by proposed maintenance

Component	Failure mode	MTTF (hours)	MTTR (hours)	MTBF (hours)	Reali bility	Freq. breakdown/year (current)	Down time/year (current)	Time repair (propose)	Down time/year (propose)
Carrying idler	Bearing worn out or broken	820,75	0,65	821,4	0,50	49	31,72	0,5	24,56
Return idler	Bearing worn out or broken	652,7	0,65	653,35	0,26	62	40,32	0,5	30,89
Rubber belt	Belt torn (leaked out)	1719,42	3,22	1722,64	0,14	23	75,45	2,5	58,62
Rubber belt	Splicing peel off	237,44	2,84	240,28	0,37	170	482,34	1,5	254,72
						Total	629,84	Total	368,79

5 Conclusion and Further Research

The application of the RCM approach illustrated in the case study shows an improvement in the maintenance strategy applied to maintenance of the conveyor belt at Semen Padang Co. mine. Furthermore, the critical level of failure mode with FMEA is analyzed and the MTTF, MTTR and reliability calculations are carried out by looking at the distribution of failures that occur. From these results, improvements were made which resulted in a decrease in hours of downtime by 41.44% from the current hour of downtime.

Finally, this case study to be continued on other mining equipment at Semen Padang Co. for obtaining an optimal maintenance strategy in terms of scheduling equipment maintenance and maintenance costs.

References

1. Hoseinie, S.H., Kumar, U., Behzad, G.: Reliability Centered Maintenance (RCM) for Automated Mining Machinery, Luleå University of Technology, Luleå (2016)
2. Moubray, J.: Reliability-Centred Maintenance, Biddles Ltd, Guildford and King's Lynn, Great Britain (1997)
3. Sifonte, J.R., Reyes-Picknell, J.V.: Reliability Centered Maintenance Reengineered Practical Optimization of the RCM Process with RCM-R®. CRC Press Taylor & Francis Group, Boca Raton (2017)
4. Langlo, F.: Application of reliability centered maintenance on a drilling system, Department of Mechanical and Structural engineering and Material Science, Faculty of Science and Technology, UIS, Stavanger (2014)
5. Afefy, I.H.: Reliability-centered maintenance methodology and application: a case study. Sci. Res. Eng. **2010**(2), 863–873 (2010)
6. Regan, N.: The RCM Solution: A Practical Guide to Starting and Maintaining a Successful RCM Program. Industrial Press Inc., New York (2012)
7. Bloom, N.: Reliability Centered Maintenance Implementation Made Simple. McGraw-Hill Inc, New York (2006)
8. Engineering, US Army Corps of Engineers Construction: Streamlined reliability centered maintenance (RCM): tutorial and application (RCM.ppt/R6.exe). USACERL Technical report 99/50 (1999)
9. National Aeronatics and Space Administration: Reliability Centered Maintenance Guide for Facilities and Collateral Equipment (2000)

A Comparative Study of Product Costing by Using Activity-Based Costing (ABC) and Time-Driven Activity-Based Costing (TDABC) Method

N. F. Zamrud, M. Y. Abu(✉), N. N. N. M. Kamil, and F. L. M. Safeiee

Faculty of Manufacturing Engineering, Universiti Malaysia Pahang, 26600 Pekan, Pahang, Malaysia
nurulfarahinzamrud@gmail.com, myazid@ump.edu.my

Abstract. This research study revolves around a case study company, which manufactures electric and electronics products. The objective of this work is to compare Activity-Based Costing (ABC) and Time-Driven Activity-Based Costing (TDBAC). This work began by collecting data related to the production costs of a selected product. The product has been selected based on the volume of production which is high and continuous. The next phase in this study involved analyzing costs using ABC and TDABC method. These costing methods are compared on the accuracy and the unused capacity. The results showed that TDABC improves the costing accuracy and helps reducing waste by introducing time equation on the production cost structure. Finally, it is concluded that from this comparative study, TDABC would be beneficial to the company in a competitive environment and future further studies on product costing.

Keywords: Activity-Based Costing · Time-Driven Activity-Based Costing · Comparative study

1 Introduction

Industries today seek the reduction and elimination of waste through continuous improvement projects which enable increase productivity in the production process while still maintaining the quality. One of their ways is to improve existing costing method by comparing with another method in terms of accuracy and unused capacities. For example, there are Activity-Based Costing (ABC) and Time-Driven Activity-Based Costing (TDABC). ABC is a system where it delivers accounting information that recognizes activities conducted in the organization [1]. Furthermore, it is a method for accurately costing products and ease the decision-making process by providing cost data and management information. [2] estimated the cost of remanufactured crankshaft using ABC and [3] applied ABC as a method of estimation for the remanufacturing cost of crankshaft. [4] integrated optimization method with ABC for estimating final cost of product and [5] applied the ABC as a method of cost estimation for a palm oil

M. N. Osman Zahid et al. (Eds.): iMEC-APCOMS 2019, LNME, pp. 171–178, 2020.
https://doi.org/10.1007/978-981-15-0950-6_27

plantation. A costing model known as Time-driven Activity-based Costing (TDABC) uses time as an inducer in costing. TDABC provides a cost of activities by calculating time consumption per activities [6]. The advantage of this method is it provides accurate estimates of care cycle cost and more transparency into the cost drivers [7]. Moreover, significant information on idle capacities is delivered and unused resources recognized by TDABC [8]. This paper delivers a comparative study of a magnetic inducer by using ABC and TDABC method for accuracy and unused capacity aspect.

2 Methodology

Firstly, there are six steps involved in adapting to ABC [9]. Step one is to identify a product for analysis which is a magnetic inductor, chosen based on a continuous and high volume of production. Next, step two is reviewing the company's financial information through the production report, interviews and other materials. Some of the information can also be obtained by observation of production operations, activities, machines used and equipment. Step three is to do an activity analysis. Step four is to determine the operating cost for each activity. This step includes an analysis of costs and the identification of appropriate cost drivers for each activity. Lastly, step six is to calculate product cost. Meanwhile, according to existing literature, there are eight stages to calculate the cost using TDABC [10]. The first step is to identify and analyze activities to understand the scope and sequence of events. Secondly, the cost of resources supplied in each activity and sub-activities are listed and estimated. Practical capacity details about employees working hours which to be used for calculation of capacity cost rates. Next, TDABC time equation is developed to calculate the estimated production time. The estimated capacity required by each sub-activity is calculated and finally, the unit cost is determined. In this paper, the steps above will be discussed excluding product unit cost.

3 Results and Discussion

In this study, the resources are allocated in four groups: labor costs, maintenance cost, raw materials, and consumables costs. Firstly, the ABC methodology is applied. The costs incurred are summarized in Table 1.

There are 13 activities as shown in Table 1. The percentage of time spent shows the estimated time taken for the work to be done at each activity.

The cost drivers of the activities are shown in Table 2. Each cost driver is selected from the observation and comparison of the cost that triggered the total cost of each production. To continue with the TDABC method, we estimated the practical capacity by calculating the working hours of the operator per year. Next, we calculate the capacity cost rate. The data are summarized in Table 3.

Table 1. Cost of magnetic inductor production

Activity	Operating expenses (RM)					Unit cost (RM)	Time spent (%)
	Labor	Maintenance	Material	Consumable	Total		
Winding	24,000	26,449.89	384,000	Nil	434,449.89	0.067	13.16
Flattening	48,000	70	Nil	144.00	48,214.00	0.013	1.46
Trimming	48,000	Nil	Nil	192.00	48,192.00	0.018	1.46
Forming	48,000	Nil	Nil	192.00	48,192.00	0.022	1.46
Soldering	24,000	Nil	Nil	445,064.32	469,064.32	0.087	14.21
Epoxy application	24,000	Nil	1,440,000	40,449.98	1,504,449.98	0.279	45.59
Assembly 1	72,000	Nil	480,000	403.20	552,403.20	0.102	16.74
Assembly 2	48,000	Nil	Nil	268.80	48,268.80	0.009	1.46
Oven curing	24,000	Nil	Nil	24.00	24,024.00	0.004	0.73
Boundary inspection	24,000	Nil	Nil	134.40	24,134.40	0.004	0.73
Laser marking	24,000	626.53	Nil	Nil	24,626.53	0.005	0.75
Co-planarity and VMI test	24,000	Nil	Nil	24.00	24,024.00	0.004	0.73
Packaging	24,000	nil	nil	26,028.80	50,028.80	0.009	1.52

Table 2. The cost driver of the activities.

Activity	Cost driver
Winding	Amount of material
Flattening	Amount of load
Trimming	Amount of load
Forming	Amount of load
Soldering	Amount of consumable usage
Epoxy application	Amount of consumable usage
Assembly 1	Labor time
Assembly 2	Labor time
Oven curing	Machine time
Boundary inspection	Type of jig
Laser marking	Machine time
Co-planarity and VMI test	Number of products
Packaging	Number of products

Table 3. Capacity cost rate of each sub-activity

Activity	Sub-activities	Cost of all resources supplied (RM/year) [1]	Practical capacity (min/year) [2]	Capacity cost rate, CCR (RM/year) [1]/[2] = [3]
Winding	The wires are winded using CNC machine	434,449.89	218,400	1.99
Flattening	Pick up the coils from winding station	12,000	109,200	0.11
	Flatten the coils by using hydraulic press machines	36,214	327,600	0.11
Trimming	Pick up the coils from flattening station	12,000	109,200	0.11
	Trim the coils by using pneumatic press machines	36,192	327,600	0.11
Forming	Bend the coils by using pneumatic press machines	48,192	436,800	0.11
Soldering	Dip the coils into flux	4,960	109,200	0.23
	Then, dip the coils into the solder.	444,104	109,200	4.07
Epoxy application	Arrange core on the magnetic strip	1,452,000	109,200	13.30
	Put the magnetic strip into the epoxy machine.	52,449.98	109,200	0.48
Assembly 1	Assemble coil to the I-core.	552,403.20	655,200	0.84
Assembly 2	Assemble core with I-core.	48,268.80	436,800	0.11
Oven curing	Put the inductor into oven	4,024	218,400	0.11
Boundary inspection	The inductors undergo boundary inspection	24,134.40	218,400	0.11
Laser marking	Put the inductors into a laser marking machine	24,626.53	218,400	0.11
VMI and co-planarity inspection	Inspect inductors for co-planarity and then into VMI equipment.	24,024	218,400	0.11
Packaging	Pack inductors	48,028.80	218,400	0.22
	Total	**2,863,621.71**	**4,149,600**	

Practical capacity in Table 3 is gained by the multiplication of the workers working hours with the number of working days in that year. Capacity cost rate is calculated by dividing the cost of capacity supplied over the practical capacity of resources supplied. Furthermore, the next step in TDABC is the development of time equations. All the details are included in Table 4.

Table 4. Time equations for all sub-activities

Sub-activities	Time equations	Cost driver		
		Var.	Driver	Quantity/year
The wires are winded using CNC machine	$0.12X_1$	X_1	Amount of raw material (kg)	4,800,000
Pick up the coils from winding station	$10.00X_2$	X_2	Pick up the coils from the winding station (rounds)	480
Flatten the coils by using hydraulic press machines	$0.22X_3$	X_3	Number of hydraulic press machine operating (frequency/month)	2
Pick up the coils from flattening station	$10X_4$	X_4	Pick up the coils from flattening station (rounds)	480
Trim the coils by using pneumatic press machines	$0.13X_5$	X_5	Number of pneumatic press machine operating (frequency/month)	2
Bend the coils by using pneumatic press machines	$0.17X_6$	X_6	Number of pneumatic press machine operating (frequency/month)	2
Dip the coils into flux	$0.72X_7$	X_7	Amount of flux used (liter)	432
Then, dip the coils into solder.	$0.18X_8$	X_8	Amount of solder used (kg)	120
Arrange core on magnetic strip	$0.15X_9$	X_9	Arrange core on the magnetic strip (amount of core)	4,800,000
Put the magnetic strip into the epoxy machine.	$2.7X_{10}$	X_{10}	Amount of epoxy used (liter)	192
Assemble coil to the I-core.	$0.07X_{11}$	X_{11}	Number of labor (frequency/month)	6
Assemble core with I-core.	$0.14X_{12}$	X_{12}	Number of labor (frequency/month)	4
Put the inductor into the oven	$40X_{13}$	X_{13}	Number of oven operating (frequency/month)	2
The inductors undergo boundary inspection	$0.05X_{14}$	X_{14}	Waiting periods (frequency/month)	2,400
Put the inductors into a laser marking machine	$0.05X_{15}$	X_{15}	Number of laser marking machine operating (frequency/month)	1
Inspect inductors for co-planarity and then into VMI equipment.	$0.52X_{16}$	X_{16}	Number of VMI machine used (frequency/month)	1
Pack inductors	$3X_{17}$	X_{17}	Number of boxes (amount of box)	14,880

The actual time spent on this activity center per year was 1,351,293.63 min determined by substituting the volume of cost-drivers from Table 4. The total time for the wire to be winded using CNC machine in a year can be represented by X_1 equals 4,800,000 in $0.12X_1$, so that $0.12 \times 4{,}800{,}000 = 576{,}000$ min. When multiplied by capacity city cost of RM 1.99 per minute, it can be determined that the total production cost of this activity is RM 1,146,240 per year.

4 Conclusions

We will be discussing two aspects of comparison of ABC and TDABC which is accuracy and unused capacity. TDABC provides more accuracy in term of workstation activities. As shown in Table 3, there are two sub-activities for flattening while in ABC as shown in Table 1, the activities are not shown in detail. This means, TDABC capable to elaborate more detail information on the activities and sub-activities as well as practical capacity. Thus, TDABC leads to accurately assess the true cost [2]. Another aspect is unused capacity. TDABC is able to deliver unused capacity as shown in Table 5 for each activity and sub-activity. It allows the analysis of used and idle time by using time equation. TDABC is able to reduce waste as well in any activities or specifically in sub-activities by recognizing unused capacities as reported by [6]. In conclusion, the results showed that TDABC improves the costing accuracy and helps to manage the unused capacities by introducing time equation on the production cost structure. Finally, it is concluded that from this comparative study, TDABC would be beneficial to the company in a competitive environment and future further studies on product costing.

Table 5. Analysis of capacity utilization in the magnetic inductor production

Sub-activities	Practical capacity (min/year) [1]	Used time (min) [2]	Unused capacity (min) [3] = [1] - [2]	Total cost (RM/year) [2] x CCR	Loss of manufacturing costs (RM) [3] x CCR
The wires are winded using CNC machine	218,400	576,000.00	−357,600.00	1,145,801.91	−711,352.02
Pick up the coils from winding station	109,200	4,800.00	104,400.00	527.47	11,472.53
Flatten the coils by using hydraulic press machines	327,600	0.44	327,599.56	0.05	36,213.95
Pick up the coils from flattening station	109,200	4,800.00	104,400.00	527.47	11,472.53
Trim the coils by using pneumatic press machines	327,600	0.26	327,599.74	0.03	36,191.97
Bend the coils by using pneumatic press machines	436,800	0.34	436,799.66	0.04	48,191.96
Dip the coils into flux	109,200	311.04	108,888.96	71.09	24,888.91

(*continued*)

Table 5. (*continued*)

Sub-activities	Practical capacity (min/year) [1]	Used time (min) [2]	Unused capacity (min) [3] = [1] - [2]	Total cost (RM/year) [2] x CCR	Loss of manufacturing costs (RM) [3] x CCR
Then, dip the coils into solder.	109,200	21.60	109,178.40	87.84	444,016.16
Arrange core on the magnetic strip	109,200	720,000.00	−610,800.00	9,573,626.37	−8,121,626.37
Put the magnetic strip into the epoxy machine.	109,200	518.40	108,681.60	248.99	52,200.99
Assemble coil to the I-core.	655,200	0.42	655,199.58	0.35	552,402.85
Assemble core with I-core.	436,800	0.56	436,799.44	0.06	48,268.74
Put the inductor into oven	218,400	80.00	218,320.00	8.80	24,015.20
The inductors undergo boundary inspection	218,400	120.00	218,280.00	13.26	24,121.14
Put the inductors into laser marking machine	218,400	0.05	218,399.95	0.01	24,626.52
Inspect inductors for co-planarity and then into VMI equipment.	218,400	0.52	218,399.48	0.06	24,023.94
Pack inductors	218,400	44,640.00	173,760.00	9,816.88	38,211.92
Total	**4,149,600**	**1,351,293.63**	**2,798,306.37**	**10,730,730.69**	**-7,432,659.09**

Acknowledgment. This research is fully supported by RDU190156. The authors are fully acknowledged Universiti Malaysia Pahang for the approved fund which makes this important research viable and effective.

References

1. Akhavan, S., Ward, L., Bozic, K.J.: Time-driven activity-based costing more accurately reflects costs in arthroplasty surgery. Clin. Orthop. Relat. Res. **474**(1), 8–15 (2016)
2. Abu, M.Y., Jamaludin, K.R., Zakaria, M.A.: Characterisation of activity based costing on remanufacturing crankshaft. Int. J. Automot. Mech. Eng. **14**(2), 4211–4224 (2017)
3. Kamil, N.N.N.M., Abu, M.Y.: Integration of Mahalanobis-Taguchi system and activity based costing for remanufacturing decision. J. Mod. Manuf. Syst. Technol. **1**, 39–51 (2018)

4. Abu, M.Y., Nor, E.E.M., Rahman, M.S.A.: Costing improvement of remanufacturing crankshaft by integrating Mahalanobis-Taguchi system and activity based costing. IOP Conf. Ser.: Mater. Sci. Eng. **342**, 1–10 (2018)
5. Zheng, C.W., Abu, M.Y.: Application of activity based costing for palm oil plantation. J. Mod. Manuf. Syst. Technol. **2**, 1–14 (2019)
6. Helmers, R.A., Dilling, J.A., Chaffee, C.R., Larson, M.V., Narr, B.J., Haas, D.A., Kaplan, R. S.: Overall cost comparison of gastrointestinal endoscopic procedures with endoscopist- or anesthesia-supported sedation by activity-based costing techniques. Mayo Clin. Proc.: Innov. Qual. Outcomes **1**(3), 234–241 (2017)
7. Zhuang, Z.Y., Chang, S.C.: Deciding product mix based on time-driven activity-based costing by mixed-integer programming. J. Intell. Manuf. **28**(4), 959–974 (2017)
8. Zaini, S.N.A.M., Abu, M.Y.: A review on time-driven activity based costing system in various sectors. J. Mod. Manuf. Syst. Technol. **2**, 15–22 (2019)
9. Somapa, S., Cools, M., Dullaert, W.: Unlocking the potential of time-driven activity-based costing for small logistics companies. Int. J. Logist. Res. Appl. **15**(5), 303–322 (2012)
10. Pongwasit, R., Chompu-Inwai, R.: Analysis of wooden toy manufacturing costs through the application of a time-driven activity-based costing system. Mem. Muroran Inst. Tech **65**, 7–14 (2015)

Quality Improvement Model Considering Rework and Imperfect Inspection

Kuncoro Sakti Pambudi(✉), Cucuk Nur Rosyidi, and Wakhid Ahmad Jauhari

Universitas Sebelas Maret, Jl. Ir. Sutami No. 36 A, Surakarta 57126, Indonesia
teknikindustri@ft.uns.ac.id

Abstract. Quality improvement can be achieved by improving several important aspects in the production process including the inspection process. The inspection process is often considered as a trivial problem for some companies. Most companies perform inspection process in the final stages of the production. This means that the defective product has been manufactured at the time they are inspected. This causes the company should bare additional costs for reworking the defective products. To reduce the costs incurred for reworking defective products, quality improvement must be performed in the process.

Another problem that is important regarding inspection is the presence of imperfect inspection. An inspector may make mistakes in the inspection process, either it categorizing a good product as defective or a defective product as a good one. Problems related to the imperfect inspection will be detrimental to both parties, both from the company and the customers. This paper aims to develop an optimization model for quality improvement by considering the rework process and imperfect inspection. The optimization process is done based on trading off the components of the model to determine the optimal tolerance value and also the optimal investment that must be paid by the company. Optimization is not only done to obtain optimal tolerance values but also other aspects such as economic investment value, manufacturing costs, customer loss and costs due to imperfect inspection process. Two different policies will be presented to determine the impact of optimal tolerance. Numerical example is given to show the application of the proposed model.

Keywords: Tolerance design · Taguchi loss function · Imperfect inspection · Learning inspection · Rework process

1 Introduction

According to Hallgren and Olhager [1] there are several aspects of competitive strategies for manufacturing companies to win business competition in a dynamic global competition, namely quality improvement, cost reduction, and timeliness of delivery [1]. This is a challenge for manufacturing companies to produce good quality products with low production costs. Several studies have been done related to improving process quality and reducing costs to produce good quality products. One of the studies was done by Irianto [2].

M. N. Osman Zahid et al. (Eds.): iMEC-APCOMS 2019, LNME, pp. 179–185, 2020.
https://doi.org/10.1007/978-981-15-0950-6_28

Sofiana et al. [3] developed model based on the model of Irianto [2] by added a learning investment to improve the quality of the first policy [2, 3]. The investment can be in form of training, technology improvement, as well as other things related to quality improvement. However, model assumed that the inspection process is carried out when the product is finished without any inspection errors. These errors will definitely lead to costs that can be detrimental to manufacturing companies.

In this study a mathematical model is developed to determine optimal tolerance values, select better rework policies and the amount of quality investment costs by considering imperfections of the inspection process. The company has two alternatives policies in reworking defect product, (1) using the same facilities as the production line, and (2) adding new facilities only for the rework process. In both policies, will be added the inspection error model to systems. The system description of the rework process for Policies 1 and 2 is shown in Fig. 1.

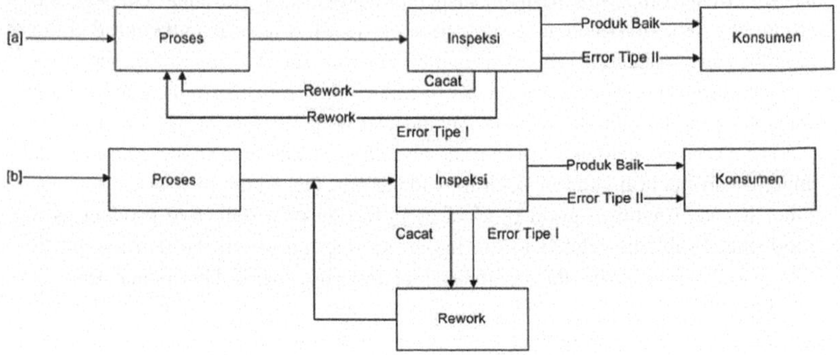

Fig. 1. [a] Policy 1 [b] Policy 2

2 Model Development

2.1 Quality Loss Function

The quality characteristics of the product are denoted by y with the average value of μ and the target value of product characteristics m. Taguchi [4] assumes that customers will be perfectly satisfied if the quality characteristics are at the target value customer dissatisfaction which can be expressed in Eq. (1) [4]:

$$L(y) = k(y - m)^2 \tag{1}$$

In Eq. (1), k is the quality loss cost coefficient which obtained from the $k = C_0/\Delta_0^2$. If x_i denotes the quality characteristic value for good product, then $m - t \leq x_i \leq m + t$. The cost of quality loss can be calculated using Eq. (2):

$$E[L(y)] = \int_{m-t}^{m+t} k(y-m)^2 . f(y) dy \tag{2}$$

In Eq. (3) the $f(y)$ denotes probability density function of y. If y normally distributed $N(m, \sigma^2)$, then the probability density function of $f(y)$ is:

$$f(y) = \frac{e^{-\frac{(y-\mu)^2}{2\sigma^2}}}{\sqrt{2\pi\sigma}} \tag{3}$$

2.2 Manufacturing Cost

All defective products assumed can be reworked using existing facilities or by using special facilities. Manufacturing cost can be calculated using Eq. (4) below:

$$MC_m = \frac{(n_{pm}.c_{pm}) + (n_{cm}.c_{cm}) + (n_{im}.c_{im})}{n_{pm}} \tag{4}$$

2.3 The Number of Components Produced, Reworked and Inspected

Probability density function $f(y)$ and $g(y)$ are assumed to be normally distributed. Probabilities of good product P are formulated in Eqs. (5) and (6):

$$P1 = \int_{\mu-t}^{\mu+t} f(y) dy \tag{5}$$

$$P2 = \int_{\mu-t}^{\mu+t} g(y) dy \tag{6}$$

Suppose that the available production operation time is T, then the total time for the process and rework must be smaller or equal to the time available T. Overall T is used for both, process and rework in Policy 1 is expressed in Eq. (7).

$$n_{p1}.t_p + n_{c1}.t_c \leq T \tag{7}$$

The number of inspected products can be calculated using the Eq. (8).

$$n_{i1} = n_{p1} + n_{c1} \tag{8}$$

The number of corrected/rework products can be estimated through Eq. (9).

$$n_{c1} = n_{p1}.\overline{P}_1 \tag{9}$$

In Eq. (9), $\overline{P}$ is the probability of the defective product. $\overline{P}$ can be calculated by $\overline{P} = P - 1$. Since the product that has been reworked is still possible to still be defective, the probability value in Eq. (9) becomes $\overline{P}/(1-\overline{P})$ so that:

$$n_{c1} = n_{p1}.\overline{P}/(1-\overline{P}) \tag{10}$$

Substituting Eq. (10) to Eq. (7) then Eq. (11) is obtained:

$$n_{p1}.t_p + n_{p1}.\frac{\overline{P}}{(1-\overline{P})}.t_c \leq T \tag{11}$$

The number of components produced through the rework facility at the first policy becomes:

$$n_{p1} \leq \frac{T}{(t_p + t_c.\overline{P}/(1-\overline{P}))} \tag{12}$$

Policy 2, almost all available time can be used for production process activities, so that:

$$n_{p2}.t_p \leq t \tag{13}$$

The rework process in large quantities must consider the time available then:

$$n_{c2}.t_c \leq T \tag{14}$$

Components that have been reworked are still possible to have defects so that the entire component in rework can be calculated by the following equation:

$$n_{c2} = n_{p2}.\frac{\overline{P}}{1-\overline{P}} \tag{15}$$

2.4 Quality Improvement Model with Quality Investment and Learning Inspection for Policy 1

The investment model was taken from the models used by previous researchers [3, 5, 6], the value of process parameters from μ_I *and* σ_I:

$$\sigma_I^2 = \sigma_L^2 + (\sigma_M^2 - \sigma_L^2)e^{(-\alpha b)}, \alpha > 0 \tag{16}$$

$$\mu_I^2 = \mu_L^2 + (\mu_0^2 - \mu_T^2)e^{(-\beta b)}, \beta > 0 \tag{17}$$

Suppose that Y is a product quality characteristic that is normally distributed, then the probability density function $f(Y, I_I)$ of the quality characteristics Y with investment value I is as follows:

$$f(Y, I_1) = \frac{1}{\sigma_I \sqrt{2\pi}} e^{\frac{(Y-\mu_I)^2}{2\sigma_I^2}} \tag{18}$$

Total cost of quality loss per unit can be calculated using Eq. (19).

$$L(Y, I_I) = \int_{m-t}^{m+t} L(Y).f(Y, I_1)dy. \tag{19}$$

Model of imperfect in the inspection process was adapted from the model of Khan et al. [7]. In this study, *m1* denotes the probability of type I error, *m2* denotes the probability of type II error, and $\overline{P}$ denotes a probability of defects before an inspection error. The probability of a defective product by considering inspection error *(γ)* becomes:

$$\gamma = \left(1 - \overline{P}\right)m_1 + \overline{P}(1 - m_2) \tag{20}$$

The learning model on inspection is adapted from the learning investment concept from the model of [2] Ganeshan, et al. [5] as shown in Eqs. (21) and (22).

$$M_1^2 = m_{1l}^2 + \left(m_{1m}^2 - m_{1l}^2\right)e^{(-\alpha d)} \tag{21}$$

$$M_2^2 = m_{2l}^2 + (m_{2m}^2 - m_{2l}^2)e^{(-\beta d)} \tag{22}$$

The imperfect inspection leads Cost of false rejection (*Cr*) and Cost of false acceptance (*Ca*) which are calculated using the following equation

$$Cr = \frac{cr(1-\gamma)m_1}{1-\gamma_e} \tag{23}$$

$$Ca = \frac{ca.\gamma.m_2}{1-\gamma_e} \tag{24}$$

So total cost for Policy 1 can be calculated using Equations

$$T_c = L(Y, I_I) + MC_I + Cr + Ca + \frac{b}{n_p} + \frac{d}{n_i} \tag{25}$$

2.5 Quality Improvement Model for Policy 2 and Selecting Rework Policy

In the second policy, the rework process is carried out using new facilities and separate from production line facilities. The addition of these facilities requires an investment of I_2 which is invested economically over a certain period. The second policy is considered better than the first policy if it meets the following requirements:

$$TC_1 + \frac{I_1}{n_{p_1}} > TC_2 + \frac{I_2}{n_{p_2}} \quad (26)$$

3 Numerical Example

A manufacturing company that producing an assembled products consisting of several components will make quality improvement in production process and rework process. There are two policies considered by company before invest their money. First policy using the same facilities for production and rework process, for second done by adding new rework facilities. This company has target process mean m 402 mm while the current process mean μ is 402.06 mm. Total time available T is 10,000 time duration. For each product need 10 time duration for production process t_p and 10 time duration for rework process t_c. The inspection process t_i need 1 time duration for each product. The production cost per unit c_p \$2, rework cost per unit c_c \$2 and Inspection cost per unit c_i \$0.3. Current standard deviation for first policy σ_1 0.60 and the second policy σ_2 is 0.528. Taguchi loss constant is k 1, variance constant α 0.00362 and mean curve constant is β 0.0150. For second policy the investment value is set at \$300. This company consider imperfect inspection on their process. Probability inspection error for error type 1 m_1 and type 2 m_2 is 2%. This error raises cost of false rejection cr \$1.5 and cost of false acceptance ca \$6.

The optimization results by using Wolfram Mathematica are shown in Table 1 bellow:

Table 1. Model result using wolfram mathematica

Rework Policy 1			Rework Policy 2		
Optimal tolerance	=	1.157 mm	Optimal tolerance	=	1.072 mm
Total cost/unit	=	\$2.81	Total cost/unit	=	\$2.93
Quality investment	=	\$201.08	Quality investment	=	\$300
The amount produced before investment	=	903 unit	The amount produced	=	987 unit
The amount produced after investment	=	968 unit			
Quality investment/unit	=	\$0.20	Quality investment/unit	=	\$0.30
Optimal mean after investment	=	402.10 mm	Optimal mean after investment	=	402.86 mm
Optimal variance after investment	=	0.46	Optimal variance after investment	=	0.52

4 Conclusion

First policy is considered to be better since it has lower total cost per unit product than the second policy. The optimal tolerance for first policy is 1.157 mm with total cost per unit $2.81 and total investment $201.08, meanwhile the optimal tolerance for second policy is 1.072 mm with total cost per unit $2.93 and total investment needed $300.

References

1. Hallgren, M., Olhager, J.: Differentiating manufacturing focus. Int. J. Prod. Res. **44**, 3863–3878 (2006)
2. Irianto, D.: Inspection and correction policies in setting economic product tolerance. Int. J. Prod. Econ. **46–47**, 587–589 (1996)
3. Sofiana, A., Rosyidi, C.N., Pujianto, E.: Product quality improvement model considering rework policies and quality investment. Industrial Engineering & Management Systems (IEMS) (2018)
4. Taguchi, G.: Introduction to Quality Engineering. White Plains, NY Unipublished 21–22 (1989)
5. Ganeshan, R., Kulkarni, S., dan Boone, T.: Production economics and process quality: a Taguchi perspective. Int. J. Prod. Econ. **71**, 343–351 (2001)
6. Abdul-Kader, W., Ganjavi, O., Solaiman, A.: An integrated model for optimisation of production and quality costs. Int. J. Prod. Res. **48**(24), 7357–7370 (2010)
7. Khan, M., Jaber, M.Y., Ahmad, A.R.: An integrated supply chain model in quality inspection and learning in production/. OMEGA Int. J. Manag. Sci. **42**, 16–24 (2014)

Develop Accessibility Design for Increase Disabilities Labor Participation in Manufactured in Indonesia with Design Thinking Approach

A. M. Hilmy Nur(✉), Roemintoyo, and Budi Siswanto

Civil Education Department of Sebelas, Maret University, Surakarta, Indonesia
hilmyalmagthani@gmail.com,
roemintoyo@yahoo.co.id, budisys@yahoo.co.id

Abstract. Indonesia is a large country with the population of disabilities among 21,84 million that still lack of participation in profession especially formal working in the manufacturing sector [1]. There is a high standard about labor requirements in much manufactory that is still stereotyped by physical conditions. Despite of the incapability of body functions, people with disabilities optimize it in another physical. The reality that founded is like a Blind person having a sense of touch and memorize is more persistent than a normal person, or a Deaf person using more visual perception and imagination in advance than a hearing person and another example. It represents their capable to work, but still difficult because on discrimination and facilities. To evolve disabilities on work labor this research is study about routes system that make it easier person with disabilities to walk around factory and more independently. To designing the routes system using qualitative method with design thinking approach and adopts the concept of "universal design" that introduced by Ronald Mace in terms of accessibility to disabilities person able to do an activity more assisted. Accessibility also can encourage the involvement of persons with disabilities to a good economic opportunity. which will be used in developing a standard of accessibility in a manufacturing work place that is friendly for disability.

Keywords: Accessibilities · Disabilities · Diversity program · Universal design

1 Introduction

Based on mandatory system data from the Ministry of Manpower, there are 440 registered companies with a worker around 237,000 people. In order, the total disability people absorbed this chance to work is only about 2,851 people or around 1.2% have a chance to place in the formal labor sector. Based on data of the National Labor Force Survey (Indonesian: SAKERNAS) in August 2017, the population of disability in productive age national amounts is about to 21,930,529 people [1]. Meanwhile, 10,810,451 people or 44.51% were jobless. While the open unemployed people with disabilities were 414,222 people or 3.69%. The Minister of Manpower has responded

M. N. Osman Zahid et al. (Eds.): iMEC-APCOMS 2019, LNME, pp. 186–192, 2020.
https://doi.org/10.1007/978-981-15-0950-6_29

to this condition by provide educational certificates, lowering age restrictions, and etc. for increasing absorption in workforce. In the regulation law, the government set a rules the obligation for company have to employ disabled person minimum of 1% for private companies and 2% for Nation Company [2]. The rules set to the companies seems not well punished that imply hold up recruitments disability in labor force. There are some problems such as discrimination on workers, or accessibility that can facilitate the disabled workers. As an effort to increase (Universal appreciate) specifically for labor sector (Table 1).

Table 1. Distribution type of disabled in Indonesia

No	Type of disabilities	Percentage
1	Physical	10,3%
2	Visual	7,9%
3	Hearing	2,7%
4	Speech	6,7%
5	Memories/concentrate	29,6%
6	Can't independently	2,8%
7	Double	40%

Economic and Social Commission for Asia and The Pacific (ESCAP) United Nations (2016) [3].

Apart from those mentioned above, people with disabilities have experienced difficulties like problems of limitations, disability, helplessness, and other assumptions that tend to get negative perceptions and leads.

1.1 Purpose

This research was carried out to develop alternative design concepts for manufacturing accessibility that focus on routes system that make it easier person with disabilities to walk around factory and more independent. Also giving facilities for better that could increase the disability workforce in order to achieve the government's target for a minimum of 2% of the workforce in the company.

2 Theoretical Review

2.1 Universal Design: Corridor and Its Parameter

According to the National Disability Authority (2014), universal design is defining the design and composition of an environment so that it can be accessed, understood and used to the greatest extent possible by all people regardless of their age, size, ability or disability [4]. The orientation of this universal design is to combine various aspects of features and usability that can finally be realized universally. The purpose of creating a universal design is to facilitate people's activities effectively and can be used by as many as possible people. For this reason, Universal design has several principles in its creation such as:

1. Equitable use
2. Flexibility in use
3. Simple and intuitive use
4. Perceptible information
5. Tolerance of error
6. Low physical effort
7. Appropriate size and space.

The necessary facilitate on the first place to construct is a route system. Checklist route parameter and size standard in this paper will be adopted from [4] Building for Everyone: A Universal Design Approach, booklet 5 (www.universaldesigns.ie). Bellows are checklist parameter that has been simplified and adjusted to the needs of research (Table 2).

Table 2. Parameter and size corridors Universal Design

No	Corridors	Surface finishes
1	Ensure recommended 1500 mm clear width for corridors in buildings	Make sure floor finishes are durable and well maintained
2	Provide passing places of 2000 mm long x 1800 mm wide in corridors less than 1800 mm wide	Optimise visual contrast between floor and wall finishes and other features, such as obstructions
3	Make sure short constrictions in width are not be less than 1200 mm	Use changes in the colour, texture and acoustic characteristics of floor finishes to delineate areas and contribute to a system of way finding
4	Recess wall-mounted items wherever possible	Avoid shiny and reflective floor finishes
5	Ensure any projections into the clear width are guarded	Consider the use of natural floor coverings to avoid the potential for aggravating allergic reactions
6	Consider using handrails for certain building types and in all cases where corridors are over 20 m long	Slip resistant when both wet and dry
7	Provide seats at no more than 20 m intervals on long corridors	

2.2 Diversity Programs

Diversity program is a form of Human Resource Management. Its definition a part of systemization processes for various diversity in the workforce, especially within the company [5]. This diversity is not only related to cultural background, but also includes diversity in terms of race, ethnicity, religion, gender, physical, and sexual orientation.

There is a more specific form of program diversity that is inclusive of those built from social phenomena involving persons with disabilities.

3 Methodology

This research was conducted in the YPAC medical rehabilitation unit (Disabled Child Development Foundation). The basis for consideration of place selection is to look at the patterns and behaviour of mobility of persons with disabilities in an inclusive area. Then this research is a development from a study entitled "Redesigning the Building Concept of YPAC Medical Rehabilitation with an Accessibility Approach to Disability in Surakarta" which was applied with readjustment to accessibility in manufacturing.

Qualitative method is used with approach of the design thinking with adopting of some literature (Gold, 1980; Riverdale & IDEO, 2011 [6]) and adjusted with research steps decide to this following work sequences:

- It starts with an inventory of data using types of primary data and secondary data. Primary data in the form of observations and interviews to and secondary data are library studies including books.
- analysis is identification the problems by comparing data with literature review
- synthesis to explore point of solution with design thinking by reference and standards.
- concepts that create ideas were then compiled into an alternative design (Master Plan).

4 Result and Discussion

4.1 Disable Mobility

From research on accessibility in the YPAC inclusion area, the mobility problem found is that new visitors with disabilities have difficulty accessing such large areas with the use of sketches. Visitors can get lost or turn around to just ensure the space they want to go to. The need for a good spatial system so that persons with disabilities are more efficient in accessing the space or unit that they want to aim to save more energy and time.

4.2 Relayout

There are needs to be a consideration of how to make or redesign the layout work place to increase production and efficiency along with ease of accessibility. Some people may have a limited walking range, especially to people with mobility issue. They may need to stop frequently, to regain strength or catch breath. Design is need to respond of differentiate stamina ability to various type disabilities. All the relayout for routes not exceed than standard from table below [5] (Building for Everyone: A Universal Design Approach, booklet 9 from www.universaldesigns.ie) (Table 3).

Table 3. Recommended maximum distances without a rest

No	Mobility difficulties	Distance (Meters)
1	People with visual difficulties	150
2	Wheelchair user	150
3	People with ambulatory difficulties without walking aid	100
4	People using walking aid, e.g. walking stick	50

After know basis criteria to plan the route then [7] we take the concept of "Class based storage" that classify material/unit by flow speed to affirm data before. The result of data will use to evaluate and then redefine new layout and placement of unit work. Here is diagram of class based storage to understanding (Fig. 1).

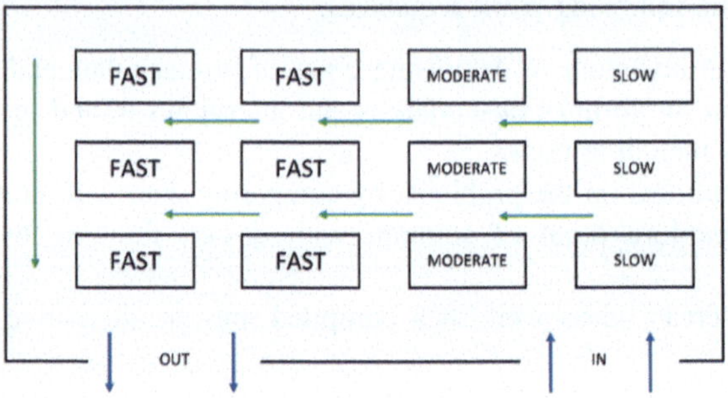

Fig. 1. Class based flow

4.3 Color Path

The system of "color path" is development from yellow block guiding line that usually functions for Blind people to guide around the road in public area and usually found on sidewalks or inclusive buildings. In the "color path" the function improved by mapping the type of color and type of texture that determined to route of specific area/unit that need to go. Here the example diagram of colour scheme when it applicate (Fig. 2).

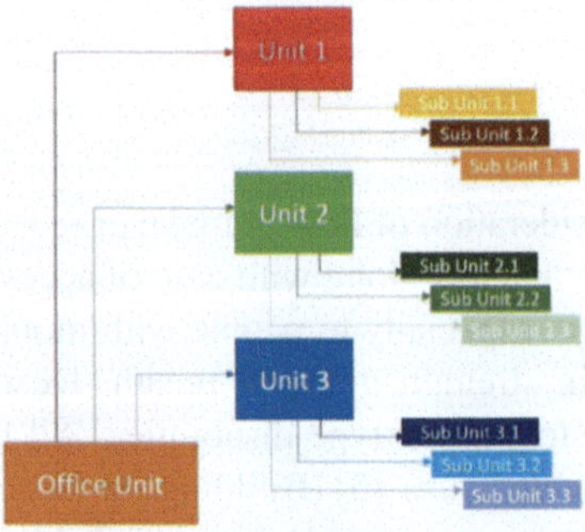

Fig. 2. Color path system

Each line is pointing to destination of unit and place. Colored and textured according to the unit and sub unit that our set it before. The "color path" theme also applied through the decoration of the color in the unit and sub unit respectively. These units should be given elevation to distinguish between the access area and the unit area. The use of "color lines" as perceptible enhancers access for employees, whether normal or specifically for the disabled, deaf, color blind (textures and accents), so that the blind are more adaptable to an area and optimal in mobility. In order to develop a better understanding to decorate and pick a color scheme we use principal of "design colour using interior psychology approach" [8] (Fig. 3).

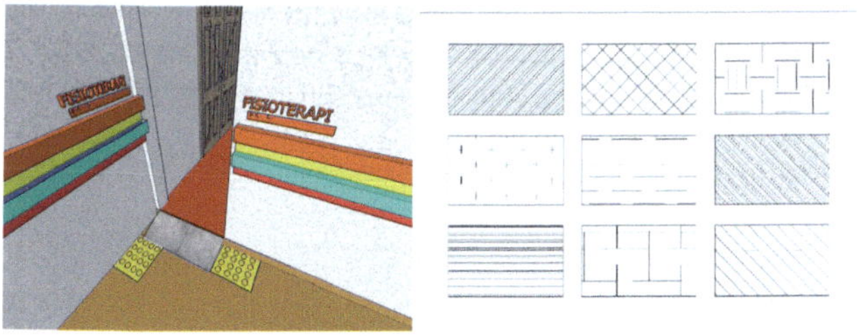

Fig. 3. Application of the path of the color pointer on the wall & stripe texture line

Letter markers is using for each room unit with the addition of braille letters and specific colors pointing. Room sign also need universally designed can be accessed by a variety of users like deaf, blind, etc. comfortably than regular. The benefit application of this "color path" is can be done at minimal cost because it only changes minimum the shape of the building by adding color and texture as a signpost, and decorating the color of the unit space. Almost all cost will only be spent on paint.

5 Conclusion

The results of this study indicate that the disability employees in companies is still lacking because the general perception of disability is still negative, and there has not been much development of inclusion infrastructure both in management or physical terms such as good accessibility. From the management side, diversity program approach can be done. Rather than worry to build new specific infrastructure for disabilities. Companies can modify infrastructure facilities simply with a design creativity approach like using "Color Path" as alternative accessibility facilities. Optimization for development of disabilities can be manufactured for the future will run properly and create inclusive culture if the level of enhance innovation getting extensive.

References

1. National Labor Force Survey (SAKERNAS), Jakarta (2017)
2. ILO: Inclusion of people with disabilities in Indonesia. ILO and Irish Aid (2013)
3. Economic and Social Commission for Asia and the Pacific (ESCAP) United Nations Building. Disability at a Glance 2015 Strengthening Employment Prospects for Persons with Disabilities in Asia and the Pacific. ESCAP United Nations Publication, Thailand (2016)
4. Building for Everyone: A Universal Design Approach. http://www.universaldesign.ie/buildingforeveryone. Accessed 30 Aug 2012
5. Effendi, A.B., Yunianto, R.: Implementasi Diversity Program Bagi Tenaga Kerja Penyandang Disabilitas Pada Pt. Wangta Agung Kota Surabaya. Indones. J. Disabil. Stud. (IJDS) **04**(02), 96–103 (2017)
6. Riverdale & IDEO: Design Thinking for Educators (2011). https://www.ideou.com
7. Muhammad, N.H., Maarif, V.: Warehouse layout planning using class-based storage-craft method in computer & office equipment distributors. J. Evolusi **6**(2) (2018) https://doi.org/10.31294/evolusi.v6i2.4425. ISSN 2338 - 8161
8. Karen, H.: Colour in interior design subsequences. Colour Des. 322–324 (2017)

Location-Allocation Model of Raw Material and Transportation Modes in Cajuput Oil Supply Chain Network (A Case in Indonesia)

Muhammad Hisjam[1(✉)], Finda Arwi Mahardika[2], Budi Widodo[3], Bobby Kurniawan[4], and Masoud Rahiminezhad Galankashi[5]

[1] Department of Industrial Engineering, Universitas Sebelas Maret, Surakarta, Indonesia
hisjam@staff.uns.ac.id
[2] Laboratory of Logistics System and Business, Faculty of Engineering, Universitas Sebelas Maret, Surakarta, Indonesia
[3] Research and Development Center, Perum Perhutani, Cepu, Indonesia
[4] Department of Industrial Engineering, University of Sultan Ageng Tirtayasa, Serang, Banten, Indonesia
[5] Department of Materials, Manufacturing and Industrial Engineering, School of Mechanical Engineering, Faculty of Engineering, Universiti Teknologi Malaysia, Johor Baru, Malaysia

Abstract. Cajuput oil is a very promising business in Indonesia due to its high people population and the demand of Cajuput oil cannot be fulfilled by the local industry. To achieve the opportunity, it is necessary to empower the local industry to operate effectively and efficiently. This study observed the cajuput oil supply chain company, to develop a location-allocation model and the transportation modes to minimize the total cost of Cajuput oil supply chain network. The mathematical model is developed using Mixed Integer Linear Programming (MILP) which considers the number and the capacity of transportation modes, the capacity of suppliers and destinations that minimize the total cost and we used IBM® ILOG®CPLEX software for supporting the computation. The supply chain network will operate at minimum cost and the demand can be fulfilled.

Keywords: Location-allocation · Transportation modes · Supply chain network · Mixed Integer Linear Programming (MILP) · Total cost minimization · Cajuput oil

1 Introduction

Melaleuca is a type of plant with a significant role in oil industry. This plant is resistant to termites [1, 2]. Two oil-producing tree species of melaleuca cajuputi and melaleuca leucadendra are widely used in the oil industry [3]. Cajuput oil is produced from the Cajuput plants (Melaleuca cajuputi), which grow naturally in the Maluku islands and northern Australia [4]. Cajuput oil product should be in acceptable level of cineole and randement [5].

M. N. Osman Zahid et al. (Eds.): iMEC-APCOMS 2019, LNME, pp. 193–198, 2020.
https://doi.org/10.1007/978-981-15-0950-6_30

One of the most important shifting paradigm of modern business management is the fact that individual businesses no longer compete exclusively, but within a supply chain network. In this competitive environment, the ultimate success of the venture will depend on management's ability to integrate complex network of companies from the business relationships [6]. The supply chain is the network of organizations that are involved, through upstream and downstream linkages, in the different processes and activities to produce value in the form of products and services in the hands of the ultimate consumer [7]. Supply Chain Management (SCM) is a set of approaches used to integrate suppliers, manufacturers, retailers, and consumers, so that goods can be produced and distributed in the right quantity, to the right location and at the right time, to minimize system- wide costs with satisfactory service [8]. A supply chain is dynamic and involves the constant flow of information, products and funds between different stages [9].

Location decisions relating to product allocation decisions. The success factors of the companies include lower costs, shorter production time, shorter lead time, less stock, larger product range, more reliable delivery time, better customer services, high quality, providing the efficient coordination between demand, supply and production and finally the trade-off between cost investment and service levels [10].

Based on query in Scopus website (www.scopus.com) on June 4, 2019 for searching document with "location allocation" and "supply chain" in article title, it is found 33 publications. With the same way by adding query "Cajuput oil", we found only one publication that is reference [11]. The reference [11] is our previous research that discusses only on location allocation problem in the Cajuput oil supply chain. In this paper we consider not only the location allocation problem, but also the transportation modes to minimize the total cost.

The existing research on location allocation of the supply chain network, Shankar research models take into account only the cost to produce the product [10]. Whereas in this study was developed by taking into account transportation costs and the amount of transportation modes used to send materials and products from suppliers to the next chain.

2 Method

Research on the location allocation models in this paper was conducted to determine the most minimal costs of the supply chain Cajuput oil in the company. The model is a mixed integer programming. Considering the existing supply chain network, we need to develop a mathematical model to minimize the total cost of the supply chain network. The primary goal of this supply chain is to deliver at low cost and acceptable quality considering the capacity of supplier, processors and trucks as transportation modess. The proposed model is a MILP (Mixed Integer Linear Programming) for multi-echelon supply chain network. This model considering the capacity and the number of owned trucks of each entity. The objective is to minimize the total cost and we used IBM® ILOG®CPLEX software for helping the computation.

3 Mathematical Model

Based on a preliminary study, Cajuput oil is considered as a prospective business. However, the production of Cajuput oil is much less than its demand in Indonesia. Supply chain network of Cajuput oil can be seen in Fig. 1. The distillery is a village owned facility located near the forest to produce distillate. The factory is refining distillate to produce good product that can used by costumer. Area that planted Cajuput is in Forest 1, Forest 2, and Forest 3. To process Cajuput leaves, the company sent Cajuput leaves to Distillery 1 and Distillery 2. Delivery of Cajuput leaves considers the distance from the location of forests to processor (distillery).

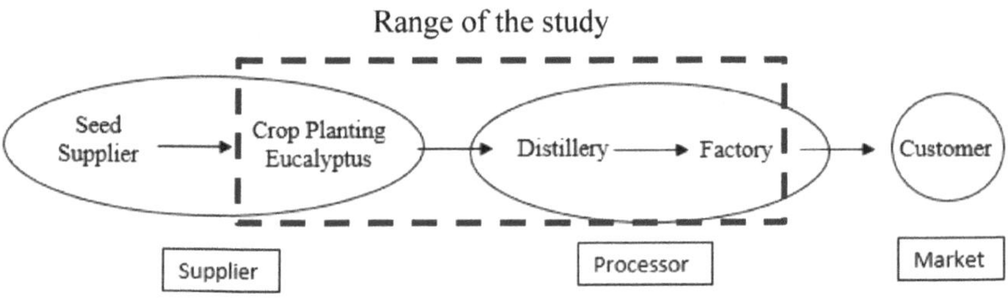

Fig. 1. Supply chain network of Cajuput oil in the case study.

In this study, Cajuput leaves collected from the crop fields and transported to distillery as raw material to produce Cajuput oil. Then, Cajuput oil is sent to Factory for further processing and packaging. To develop the model, we considered number of suppliers, Cajuput distilleries, Cajuput oil factory, materials, supplier's trucks, distillery's trucks. We also considered the capacity of each suppliers and processors, trucks for transportations, cost components of each operations.

The decision variables are whether we use the facilities or not, and the quantity of the materials in each step of operations. The objective function of the model is to minimize the total cost of supply chain network of Cajuput oil. The formula will ensure the fulfilment of demand, considering capacity of trucks, and capacity of suppliers and the destinations.

Minimize total cost (A):

$$\sum_{i=1}^{g} f_i y_i + \sum_{j=1}^{q} f_j y_j + \sum_{m=1}^{b}\sum_{h=1}^{d} FC_{mh} Y_{mh} + \sum_{n=1}^{e}\sum_{i=1}^{g} FC_{ni} Y_{ni} + \sum_{h=1}^{d}\sum_{c=1}^{r} C_{ch} X_{ch} + \sum_{i=1}^{g}\sum_{c=1}^{r} C_{ci} X_{ci} + \sum_{c=1}^{r}\sum_{h=1}^{d}\sum_{m=1}^{b}\sum_{i=1}^{g} C_{chmi} X_{chmi} + \sum_{c=1}^{r}\sum_{i=1}^{g}\sum_{n=1}^{e}\sum_{j=1}^{q} C_{cinj}(0.007X_{cinj}) \tag{1}$$

Subject to:

a. Limitation to fulfil of the factory demand.
b. Limitation of truck's capacity.
c. Limitation to deliver raw materials not exceeds supplier's capacity
d. Limitation to deliver raw materials not exceeds destination's capacity.

4 Result and Discussion

The suppliers are Forest 1, Forest 2, and Forest 3; and each capacity to process Cajuput leaves is 136,350 kgs, 9,750 kgs and 379,500 kgs, respectively. The distilleries owned by the company and its capacity to produce disyillate are Distillery 1 (84,000 kgs) and Distillery 2 (480,000 kgs). Distillery 1 is in Purwodadi and Distillery 2 is in Wonogiri. Company has only one refinery named 'Factory' located in Semarang. From the calculation using IBM ILOG CPLEX, we obtained the following results, as shown in Tables 1 and 2.

Table 1. Quantity of delivering leaves/Kgs.

Material	Supplier	Truck	Distillery	Quantity
Leaves	Forest 3	truckA	Distillery 1	0
Leaves	Forest 3	truckA	Distillery 2	0
Leaves	Forest 3	truckB	Distillery 1	136,350
Leaves	Forest 3	truckB	Distillery 2	0
Leaves	Forest 3	truckC	Distillery 1	0
Leaves	Forest 3	truckC	Distillery 2	0
Leaves	Forest 3	truckD	Distillery 1	0
Leaves	Forest 3	truckD	Distillery 2	0
Leaves	Forest 2	truckA	Distillery 1	0
Leaves	Forest 2	truckA	Distillery 2	0
Leaves	Forest 2	truckB	Distillery 1	0
Leaves	Forest 2	truckB	Distillery 2	0
Leaves	Forest 2	truckC	Distillery 1	0
Leaves	Forest 2	truckC	Distillery 2	0
Leaves	Forest 2	truckD	Distillery 1	9,750
Leaves	Forest 2	truckD	Distillery 2	0
Leaves	Forest 1	truckA	Distillery 1	0
Leaves	Forest 1	truckA	Distillery 2	0
Leaves	Forest 1	truckB	Distillery 1	75,500
Leaves	Forest 1	truckB	Distillery 2	84,000
Leaves	Forest 1	truckC	Distillery 1	120,000
Leaves	Forest 1	truckC	Distillery 2	0
Leaves	Forest 1	truckD	Distillery 1	100,000
Leaves	Forest 1	truckD	Distillery 2	0

Table 2. Quantity of delivering distillates/Kgs.

Material	Distillery	Truck	Factory	Quantity
Distillates	Distillery 1	Truck1	Factory	1,540
Distillates	Distillery 1	Truck2	Factory	1,400
Distillates	Distillery 1	Truck3	Factory	152
Distillates	Distillery 2	Truck1	Factory	0
Distillates	Distillery 2	Truck2	Factory	0
Distillates	Distillery 2	Truck3	Factory	588

Table 1 shows which truck should be used to deliver material from which supplier to which distillery, with the quantity should be loaded to ensure the total quantity must be delivered with minimum transportation cost. While Table 2 shows which truck should be used to deliver material from which distillery to the factory with the quantity should be loaded to ensure the total quantity must be delivered to minimize the transportation cost. Those selection has considered the constrains capacity of each entity and modes, and also the demand fulfillment.

The minimum total cost resulted from IBM® ILOG®CPLEX computation is IDR 855,983,600. The total costs involved are refining costs, plant costs in implementation, truck usage costs for suppliers, costs of using trucks in refineries, management costs of eucalyptus plants, costs of planting eucalyptus, the cost of delivering Cajuput leaves and the delivering costs of distillates.

When we conducted sensitivity analysis by increasing the supply of Cajuput leaves by more than 6%, it is needed to increase capacity of the distillery that means the need to build a new facilities of distillery. In that condition, other facilities are still adequate to perform their operations. It means the capacity of distilleries is the most sensitive facilities regarding to the increasing supply.

5 Conclusion

This paper proposes a model of location allocation for achieving the minimum cost of supply chain network of Cajuput oil. From the computation using ILOG®CPLEX IBM® software, it can be seen that the minimum total cost would be obtained if Company open all distilleries with production capacity are 480,000 kg and 84,000 kg each year. And all the truck owned can be used entirely at optimal capacity. From the results, we also know that the suppliers can supply all Cajuput leaves from to Distillery 1 and Distillery 2. All locations receiving material with varying amounts. Then, distillates are supply to Factory. The minimization total cost results from IBM® ILOG®CPLEX is IDR 855,983,600. The most sensitive facilities regarding to increasing supply is distillery.

Further research can be conducted by adding a market or customer zone as stakeholders. Thus, it can be seen how much cost required for supply chain networks of Cajuput oil in company to deliver product to customers. Further research can also consider for the longer operation of distilleries and factories i.e. along the year operation that will need more supply, and considerably wider area of forest.

References

1. Khomsah, N., Rimbawanto, A., Susanto, M., Baskorowati, L.: Prastyono: Budidaya dan Prospek Pengembangan Kayuputih (Cultivation and Development Prospects of Cajuput or Cajuputi Melaleuca). IPB Press, Bogor (2014)
2. Muslich, M., Rulliaty, S.: Durability of 15 local reliable wood species against dry-wood termites, subterranean termites and marine borers. For. Prod. Res. J. **29**, 67–77 (2011)
3. Widiyanto, A., Siarudin, M.: Characteristics of leaf and essential oil yield of five cajuput tree species. For. Prod. Res. J. **31**, 235–241 (2013)
4. Patty, D.J.: Products quality of cajuput oil by traditional refining in namlea (2014)
5. Perhutani, P.: Annual Report Holding Company of Perum Perhutani, Jakarta, Indonesia (2012)
6. Lambert, D.M.: Supply Chain Management Processor, Partnership, Performance. Supply Chain Management Institute (2008)
7. Christopher, M.: Logistics and Supply Chain Management, 4th edn. Prentice Hall, Upper Saddle River (2011)
8. Levi, D.S., Kaminsky, P., Levi, E.S.: Designing and Managing the Supply Chain. Irwin McGraw-Hill, New York (2000)
9. Copra, S., Meindl, P.: Supply Chain Management Strategy, Planning, and Operation, 3rd edn. Prentice Hall, Upper Saddle River (2007)
10. Shankar, B.L., Basavarajappa, S., Chen, J.C.H., Kadadevaramath, R.S.: Location and allocation decisions for multi-echelon supply chan network a multi-objective evolutionary approach. Expert Syst. Appl. **40**, 551–562 (2013)
11. Mahardika, F.A., Hisjam, M., Widodo, B., Kurniawan, B.: Location and allocation decision for supply chain network of Cajeput oil (Case in XYZ company). In: AIP Conference Proceedings, vol. 1902, no. 1, p. 020009. AIP Publishing (2017)

Job Shop Scheduling in Single Machine: An Overview

Yosi Agustina Hidayat(✉), Kiendl Valavani Setio, Harry Winata, and Nadhira Radhiyani

Department of Industrial Engineering, Faculty of Industrial Technology, Institut Teknologi Bandung, Bandung, Indonesia
yosi@mail.itb.ac.id

Abstract. Scheduling is defined as decision-making activities used in the daily activities of manufacturing or service industries. There are various types of scheduling that are related to production, personnel, vehicle, project scheduling, *etc*. In this paper, we will discuss specifically the scheduling in the context of production system, namely job shop scheduling. This paper aims to review several job shop scheduling methods on a single machine. Subsequently, this paper provides an overall overview of the development of job shop scheduling methods for single machine condition. Later on, the job shop scheduling methods for single machine condition are categorized by deterministic and stochastic job processing times.

Keywords: Job shop scheduling · Single machine · Job processing time

1 Introduction

Scheduling can be interpreted as decision-making activities used in daily activities in the manufacturing or service industries [1]. There are various types of scheduling in a company or industry. There are scheduling related to production, personnel, vehicle, project scheduling, *etc*. Each type of scheduling has experienced many developments in terms of its model and implementation.

In this paper, we will discuss more about scheduling in the context of production system, namely job shop scheduling. Scheduling in a production system consists of various scheduling on the production function [2]. This production function is related to one another and linked as a hierarchical process as shown in Fig. 1.

The purpose of this paper is to review job shop scheduling methods on a single machine. Later on, based on the job processing times, we will consider two conditions, *i.e.* deterministic and stochastic processing times. It is expected that this paper provides an overall overview and comparison of the development of job shop scheduling methods for single machine condition.

M. N. Osman Zahid et al. (Eds.): iMEC-APCOMS 2019, LNME, pp. 199–204, 2020.
https://doi.org/10.1007/978-981-15-0950-6_31

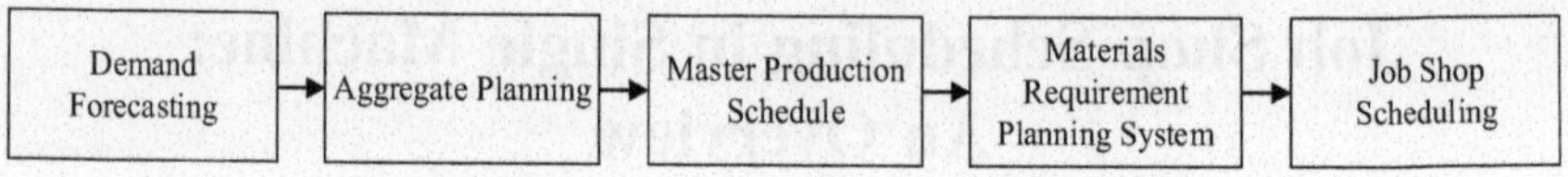

Fig. 1. Hierarchical process of production system

2 Scheduling on Production

There are several characteristics in a job shop scheduling problem. These characteristics are related to the optimum scheduling [2]. They are job arrival pattern, number and variety of machines, number of workers, flow pattern, and evaluation of alternative rules.

In a job shop scheduling problem, there are two constraints namely classic constraints and extra constraints [3]. Classic constraints consist of three things, namely precedence, capacity, and release & due date. Extra constraints or additional constraints are related to maintenance activities, resources availability, sequence dependent setup times, and other things that can occur in the implementation of the job shop scheduling model.

3 Single-Machine - Deterministic Processing Time

Machines arrangement in the manufacturing cycle are divided into several classes, *i.e.* a single machine, parallel machines, open store shop, and open shop [4]. In a typical of single machine problem, a set of independent jobs that have a single operation are available to be processed at time zero. Single machine continues to be used and never stops when jobs still available to be proceeded. The time for setting each job is independent to its position. After a job is accomplished by the machine, it means the work is also finished.

In this paper we will discuss several algorithms that can be utilized for several single machine models as performance measures, namely flowtime and lateness or tardiness.

3.1 Shortest Processing Time (SPT)

In the shortest processing time (SPT), the sequence of jobs is ordered from the smallest into the largest one. The job at the first position should have the smallest processing time [5]. The job sequence should continue in order of non-decreasing processing times, where the last job has the largest processing time. The job having least processing time is operated initially. If there are *n* jobs $\{j_1, j_2, j_3, j_4, j_5, \ldots, j_n\}$ with processing time $\{p_1, p_2, p_3, p_4, p_5, \ldots\ldots, p_n\}$ and due dates $Di = \{d_1, d_2, d_3, d_4, d_5, \ldots\ldots, d_n\}$ respectively are processed in single machine, M, then, according to SPT rule, the sequenced jobs in non- decreasing orders, respective to the processing time is shown in Eq. (1).

$$t_1 \leq t_2 \leq \ldots \leq t_{t-1} \leq t_{t-2} \tag{1}$$

SPT minimizes the flowtime on a single machine with zero release time. SPT also minimizes the average number of waiting jobs.

3.2 Weighted Shortest Processing Time (WSPT)

The problem with minimizing total flowtime using SPT method is that all the jobs are assumed to be equally important, which is not always true, especially when jobs have different priorities. Let w_i be the weight or value of job i, where the larger weight means the job is more critical to become priority, as shown in Eq. (2)

$$w_i C_i = w_1 C_1 + w_2 C_2 + w_3 C_3 + \ldots + w_n C_n \tag{2}$$

The WSPT method considers the ratio of the processing time to weight and orders the jobs in the non-decreasing ratio.

3.3 Earliest Due Date (EDD)

The due date is not considered in SPT method. Therefore, the resulted schedule with optimum flowtimes is not guaranteed. One of measurements related to the due date is tardiness, T_{max}. According to this rule, the jobs are arranged in raising order of their respective due dates; the job having earliest due date will be processed first, continued with the job having second earliest due date and so on. Hence, the jobs are sequenced according to the increasing order of their due dates (d_j), as shown in Eq. (3).

$$d_1 \leq d_2 \leq d_3 \leq \ldots\ldots \leq d_n \tag{3}$$

3.4 Hodgson's Algorithm

In an ideal condition, the EDD sequence results on timely manner accomplishment. However, if some jobs are delayed (due to the machine breakdown, broken tools, *etc.*), one of disadvantages of using EDD is that many jobs may be somewhat tardy. In a job shop scheduling, it is expected to have as many timely manner jobs as possible, or in other words, it is to minimize number of tardy jobs. Hodgson's algorithm is used in this situation. Here are the steps of Hodgson's algorithm.

- Step 1. Compute the tardiness for each job in the EDD sequence, set $N_T = 0$ and let k be the first position containing a tardy job. If there is no tardy job go to Step 4.
- Step 2. Find the job with the largest processing time in position 1 to k.
- Step 3. Remove j (job with the largest processing time among the first k jobs) from the sequence, set $N_T = N_{T+1}$, and repeat Step 1.
- Step 4. Place removed N_T jobs in any order at the end of the sequence. This sequence minimizes the number of tardy jobs.

4 Single-Machine - Stochastic Processing Time

In this section, we will discuss about single machine models with stochastic processing times with non-preemptive and preemptive arbitrary distributions as discussed in [1].

4.1 Arbitrary Distribution Without Preemptions

For stochastic problems, finding the optimal solution is equivalent with deterministic scheduling problem. Usually the solution of deterministic problems can be obtained by replacing all random variables with their means. In the same way, the optimal schedule for the stochastic problem can be obtained by replace the means with the expectation value. The models that will be used for arbitrary distribution without preemptions are Weighted Shortest Expected Processing Time (WSEPT) and Weighted Discounted Shortest Expected Processing Time (WDSEPT).

Weighted Shortest Expected Processing Time (WSEPT). It is easy to find the optimal permutation schedule for the stochastic counterpart when the processing time of job j is X_j, from an arbitrary distribution F_j, and the objective is $E(\sum w_j C_j)$. The stochastic version of the Weighted Shortest Processing Time (WSPT) is called Weighted Shortest Expected Processing Time (WSEPT). The only difference between WSPT and WSEPT is that p_j in WSPT need to be replaced by $E(X_j)$. The WSEPT rule minimize the expected sum of the weighted completion times in the class of non-preemptive static list policies as well as in the class of non-preemptive dynamic policies.

Weighted Discounted Shortest Expected Processing Time (WDSEPT). The equivalences between the single machine stochastic models and their deterministic counterparts go even further. If in the stochastic counterpart the jobs are subject to precedence constraints that take the form of chains, then Weighted Discounted Shortest Expected Processing Time (WDSPT) can be used for minimizing the expected sum of the weighted completion times. The stochastic version of the WDSPT is called Weighted Discounted Shortest Expected Processing Time (WDSEPT). The concept is same with WSPT and WSEPT where p_j is replaced by $E(X_j)$, so the order ratio is change from $\sum w_j(1 - e^{-rC_j})$ is shown in Eq. (4).

$$\frac{w_j E(e^{-r_j})}{1 - E(e^{-r_j})} \tag{4}$$

The WDSEPT rule minimizes the expected weighted sum of the discounted completion times in the class of non-preemptive static list policies as well as in the class of non-preemptive dynamic policies. The graphical explanation can be seen in Fig. 2.

Figure 2 shows that the WDSEPT rule fulfills all conditions starting from the stochastic, equal weight and discounted properties.

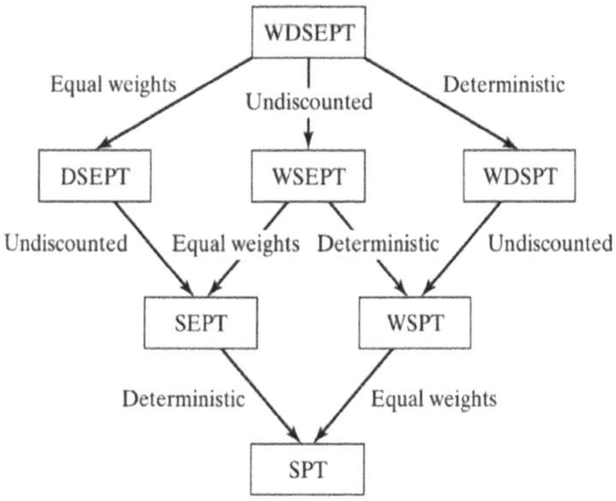

Fig. 2. WDSEPT rule

4.2 Arbitrary Distribution with Preemptions: The Gittins Index

For stochastic scheduling with preemptive conditions, it is permitted to preempt the machine at discrete times 0, 1, 2, etc. If job j is completed at the integer completion time C_j a reward $w_j\beta^{C_j}$ is received, where β is a discount factor between 0 and 1. The value of β is typically close to 1. It is of interest to determine the policy that maximizes the total expected reward. The model that can be used to maximize the total expected reward is Gittins Index that is shown in Eq. (5).

$$G_j(x_j) = \max_{\tau > 0} \frac{\sum_{s=0}^{\tau-1} \beta^{s+1} w_j P(X_j = x_j + 1 + s)}{\sum_{s=0}^{\tau-1} \beta^{s+1} P(X_j \geq x_j + 1 + s)} \tag{5}$$

5 Current Development

Currently, there have been many developments from job shop scheduling. One example of developing job shop scheduling is to consider dynamic environment. Dynamic environments in scheduling such as the number of random jobs and the number of operable machines make the scheduling method not optimum [6]. A reinforcement learning approach with a Q-factor algorithm is used to improve the performance of the dynamic scheduling method. The scheduling method is based on the variable neighborhood search (VNS) method [6]. The results of this study indicate that the approach taken gives better results compared to other general methods.

Another example of the development of job shop scheduling is to make a model that is suitable for manufacturing companies that have many factory locations. This is known as distributed job shop scheduling problem (DJSP). In this case, the factories owned are spread in different geographical locations and each factory has several machines that will be used to process jobs. One approach developed to solve this problem is to use the Modified Ant Colony Optimization algorithm as done by [7].

In addition to the two new development approaches above, there are also developments in existing methods. One of the examples of the development was carried out by [8] who developed the Multi-Objective Heuristic Kalman Algorithm (MOHKA) so that they could complete the Multi-Objective Flexible Job Shop Scheduling Problem (MOFJSSP) well. This development is based on the existing method, namely Heuristic Kalman Algorithm (HKA).

6 Conclusion

This paper reviews some job shop scheduling methods for deterministic processing time using Shortest Processing Time (SPT), Weighted Shortest Processing Time (WSPT), Earliest Due Date (EDD), and Hodgson's algorithm. We also review some job shop scheduling methods for stochastic processing time *i.e.* depend on with or without preemptions. Currently there have been many developments from job shop scheduling so that the existing job shop scheduling model can be more in-line with the real situations. One example of developing job shop scheduling is to consider dynamic environment.

References

1. Pinedo, M.: Scheduling: Theory, Algorithms, and Systems, 3rd edn. Prentice Hall, New York (2008)
2. Nahmias, F., Olsen, T.: Production and Operation Analysis, 7th edn. Waveland Press, Illinois (2015)
3. Nezhad, M., Abdullah, S.: Job shop scheduling: classification, constraints, and objective function. Int. J. Comput. Inf. Technol. **11**(4), 429–434 (2017)
4. Sipper, D., Bulfin, R.: Production: Planning, Control, and Integration. McGraw-Hill, New York (1997)
5. Tyagi, N.: Single machine scheduling model with total tardiness problem. Indian J. Sci. Technol. **9**(37), 4–14 (2016)
6. Shahrabi, J., Adibi, M.A., Mahootchi, M.: A reinforcement learning approach to parameter estimation in dynamic job shop scheduling. Comput. Ind. Eng. **110**, 75–82 (2017)
7. Chaoch, I., Driss, O.B., Ghedira, K.: A modified ant colony optimization algorithm for the distributed job shop scheduling problem. Proc. Comput. Sci. **112**, 296–305 (2017)
8. Robert, O., Hankun, Z., Shifeng, L., Borrut, B.: Improved heuristic Kalman algorithm for solving multi-objective flexible job shop scheduling problem. Proc. Manuf. **17**, 895–902 (2018)

Potential Failure Modes of Cement Production Process: A Case Study

Elita Amrina(✉) and Mutty Oktaviani

Department of Industrial Engineering,
Andalas University, Padang 25163, Indonesia
elita@eng.unand.ac.id

Abstract. Cement is one of the strategic commodities that support the infrastructure development and the economic growth of a country. This requires the cement companies to continuously improve the production process to become more effective and efficient. This paper aims to identify the failure modes of waste occurred during the cement production process using the Failure Mode and Effect Analysis (FMEA) method. The action plans are then developed based on the prioritized failure modes. The failure modes are determined on the three main machines of cement production process consist of raw mill, kiln, and cement mill. The results indicated the highest failure mode in the cement production process is the low quality of raw materials. It is followed by the high vibration of cement mill, the kiln is stop, and voltage is not constant. The failure modes are largely due to the quality of raw material and the production machinery. Therefore, it is necessary to improve the quality control of raw material and the maintenance program. The results are expected to help the cement companies to improve the efficiency and effectiveness of the production process.

Keywords: Cement production · Failure modes · Waste

1 Introduction

The cement industry is one of the oldest manufacturing industries in the world. The demand for cement has increased rapidly over the last decades and has become the second substance after the water [1]. It forced the cement companies to increase the production amount which increases the production activities. The cement companies must be able to carry out the efficient and effective production process. The increased efficiency of the production line is widely acknowledged to reduce the downtime and satisfy the high market demand for products [2].

Based on the preliminary study, it was found that there are some wastes occurred during the cement production process such as overproduction, transportation, delays, and inventory. Waste will produce the negative effects on the company such as increasing the production cost, create the work in process between workstations and increase the level of redundancy operations [2]. These will cause the company unable to meet the customer demand [3]. Waste is caused by failures that occur in a process.

M. N. Osman Zahid et al. (Eds.): iMEC-APCOMS 2019, LNME, pp. 205–210, 2020.
https://doi.org/10.1007/978-981-15-0950-6_32

Therefore, it is needed to identify the failure modes of wastes and develop the action plans to improve the efficiency and effectiveness of the production process.

2 Methodology

This research is conducted in PT X, a cement manufacturing company located in Padang, Indonesia. The company is established in 1910 and certified by ISO 9001, ISO 14001, Occupational Safety and Health Management System, and Security Management System. Currently, the company has a total production capacity of 10.4 million tons per year. The cement production process has three main machines consist of raw mill, kiln, and cement mill. In this research, the Failure Mode and Effect Analysis (FMEA) method is applied to assess the failure modes during the cement production process.

The methodology started with the identification of failure modes in the cement production process of PT X. The causes and effects are then determined for each of the failure modes. Eleven managers and staffs from the Department of Production of PT X are consulted through a questionnaire to determine the failure modes on the three main machines of cement production process. The causes and the effects of each of the failure modes are then determined.

The relative risk of failure and its effects are determined by three factors consisting of severity, occurrence, and detection [4]. The severity is the consequence of the failure should it occur. The occurrence is the probability of frequency of the failure occurring. The detection is the probability of the failure being detected before the impact of the effect is realized. The same experts are consulted to assign the severity, occurrence, and detection for each of the failure modes in the cement production process of PT X. A scale ranging from 1 to 10 is used adopted from [4]. The Risk Priority Number (RPN) is then calculated for each of the failure modes by multiplying the ranking of the three factors (severity, occurrence, detection) has assigned. The RPN will be used to rank the need for corrective actions to eliminate or reduce the potential failure modes.

3 Results and Discussions

3.1 Identification of Failure Modes

A total of 32 failure modes are identified from the three main machines of cement production process in PT X. The causes and effects of the failure modes are then determined through discussions with the industry experts. The results are shown in Table 1.

3.2 Evaluating the Risks of Failure Modes

In this stage, the severity (LPZ), the occurrence (LPW), and the detection (LPO) of the failure modes of cement production process in PT X are assigned. The ranking from the eleven experts are then averaged and the results are shown in Table 2.

Table 1. The failure modes, effects, and causes.

Failure modes	Effects	Causes
1. Voltage is not constant	Process is stop	Problem in GI
2. Drive not ready	Process is stop	Interlocking of tools
3. Hopper of material is empty or in critical condition	Decreasing of feeding rate	Raw material is difficult to get
4. Storage is empty	Process is stop	Problem of raw material transport
5. Feeding is stop	Feeding is stop	Block in chute
6. Maximum vibration of mill	Feeding is stop	Strange material in mill
7. Low quality of material	Decreasing of feeding rate	Material is wet
8. Belt conveyor is damaged	Maintenance	Belt is sway
9. BCR works is not maximal	Decreasing of feeding rate	Low material on pile
10. Reject bin is full	Mill process is stop	Delay of cleaning activity
11. Raw mix stock is minimum	Decreasing of feeding rate	Raw mill process is stop
12. Fine coal stock is minimum	Decreasing of feeding rate	Coal mill process is stop
13. Voltage is not constant	Process is stop	Problem in GI
14. Kiln is stop	Decreasing of clinker production	Cooler maintenance
15. Feeding number is fluctuative	Material is overcook	Low quality of material
16. GCT is not well running	Hot gas and fine material passing the chimney	GCT Pump is Not Running
17. Coal mill is hang	Process is stop	Vibration
18. Pressure inlet BHF is increasing	Kiln is stop, raw mill is stop	Pump is not working
19. Abnormalities of machine sound	Checking by staff, process stop	Milling process is hampered
20. Damper mill fan is not open	Coal trapped inside the mill	Problem in instruments
21. RAL motion detector	Process is stop	Detecting motion
22. Bus fault	Process is stop	Problem in instruments
23. Maximum vibration of mill	Mill is stop	Roller and Table collide each other
24. Belt conveyor is damaged	Feeding is stop	Belt is Flaked and Folded
25. Pull rope/pull cord is not ready	Feeding is stop	Hit By Material
26. Clinker is out of stock	Decreasing of feeding rate, decreasing of production	Kiln is Stop
27. Bucket elevator silo level high alarm	Material is not transported	Interlocking
28. Block in air slide	Cement accumulated	Canvas is shot
29. Hopper of gypsum is empty or in critical condition	Decreasing of feeding rate	Gypsum is wet
30. Reject bin is full	Mill is stop	Delay of cleaning activity
31. Feeding is not smooth	Material is in critical condition	Feeder is trip
32. Hydraulic station is not ready	Mill is stop	Problem in oil circulation

Table 2. The severity, occurrence, and detection of failure modes.

Failure modes	LPZ	LPW	LPO
1. Voltage is not constant	6.33	2.33	6.83
2. Drive not ready	2.50	4.17	4.50
3. Hopper of material is empty or in critical condition	4.17	6.00	1.67
4. Storage is empty	4.00	2.33	1.50
5. Feeding is stop	2.83	7.83	2.17
6. Maximum vibration of mill	3.33	7.83	5.00
7. Low quality of material	4.67	6.50	4.33
8. Belt conveyor is damaged	3.00	3.17	4.67
9. BCR works is not maximal	2.17	3.00	3.50
10. Reject bin is full	2.33	2.00	1.83
11. Raw mix stock is minimum	5.83	4.17	1.67
12. Fine coal stock is minimum	4.67	2.50	2.00
13. Voltage is not constant	6.17	1.67	8.00
14. Kiln is stop	5.67	3.67	5.50
15. Feeding number is fluctuative	4.50	4.33	2.33
16. GCT is not well running	2.83	2.83	5.33
17. Coal mill is hang	4.00	3.67	5.33
18. Pressure inlet BHF is increasing	3.00	2.83	4.50
19. Abnormalities of machine sound	3.50	3.17	5.17
20. Damper mill fan is not open	3.83	3.33	5.17
21. RAL motion detector	2.83	2.83	3.17
22. Bus fault	3.33	4.50	6.67
23. Maximum vibration of mill	4.00	2.83	5.83
24. Belt conveyor is damaged	3.67	3.17	4.67
25. Pull rope/pull cord is not ready	4.00	2.83	3.17
26. Clinker is out of stock	3.50	1.50	1.50
27. Bucket elevator silo level high alarm	2.00	2.00	3.67
28. Block in air slide	3.33	1.33	3.33
29. Hopper of gypsum is empty or in critical condition	4.33	2.83	1.83
30. Reject bin is full	2.17	2.17	1.83
31. Feeding is not smooth	2.33	2.83	1.83
32. Hydraulic station is not ready	3.17	1.67	3.50

It can be seen that the highest severity ranking is voltage is not constant with a value of 6.33. The most occurrence failure is feeding is stop with a value of 7.83. In term of detection, voltage is not constant identified as the most detection failure with a value of 8.00. The failure of non-constant voltage occurs suddenly and causes the production process to become disturbed and even stop.

3.3 Calculating the Risk Priority Number

The Risk Priority Number (RPN) is calculated for each of the failure modes of cement production process in PT X. The results are presented in Fig. 1.

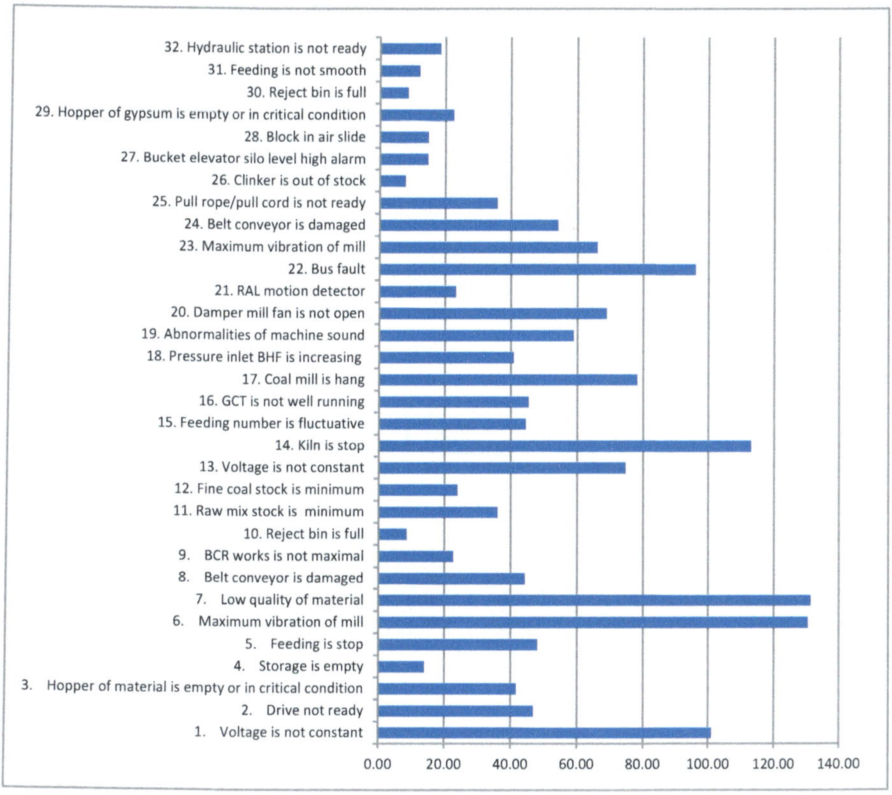

Fig. 1. The RPN of failure modes.

It is obtained that the highest failure mode is low quality of material with an RPN of 131.44, followed by maximum vibration of mill (130.56), kiln is stop (112.93), and voltage is not constant (100.98). The low quality of material is mostly caused by the wet material. It will affect the cement production process, especially in the milling machine. It also disturbs the material movement on the belt conveyor. In order to overcome this failure, the company usually uses the rotary dryer to minimize the water content in the wet material. This will increase the production time and the production cost. Therefore, it needs high attention from the production manager to conduct a comprehensive inspection of the material supplied. Besides, the supplier selection and evaluation need to be improved to provide a good quality of the material. The quality inspection is important to ensure the conformity of materials with requirements [5].

The failure of maximum vibration in the milling process is might be due to the physical quality of raw material. It is suggested to check the raw material before entering the production stage. With regard to the failure of kiln stop, it is recommended to improve the maintenance program. The performance of machine and equipment is critical to maintain the production process [6]. Action needed to minimize the failure of non-constant voltage is by using a stabilizer for the engine so that if a failure occurs, it will not affect directly to the engines which can stop the process. Although the non-constant voltage is rare, it can cause a serious effect on the production process.

4 Conclusions

This research has applied the Failure Mode and Effect Analysis (FMEA) method to identify the failure modes of waste occurred in the cement production process. A total of 32 failure modes are identified from the three main machines of cement production process in PT X. The causes and effects of the failure modes are also determined. The failure modes are then evaluated on the severity, the occurrence, and the detection. Finally, the Risk Priority Number (RPN) is calculated for each of the failure modes of cement production process in PT X. It is obtained the highest failure modes are low quality of material, maximum vibration of mill, kiln is stop, and voltage is not constant. Then, action plans are developed to improve the quality control of raw material and the maintenance program. This study hoped can help the cement companies to improve their production process to be more efficient and effective.

Acknowledgments. The authors would like to thanks to Andalas University, Padang, Indonesia and Directorate General of Higher Education, Ministry of Research, Technology and Higher Education, Republic of Indonesia.

References

1. Karim, M.R., Zain, M.F.M., Jamil, M., Lai, F.C.: Significance of waste materials in sustainable concrete and sustainable development. In: Proceedings of International Conference on Biotechnology and Environment Management (IPCBEE), pp. 43–47. IACSIT Press, Singapore (2011)
2. Khalil, R.A., Stockton, D.J., Tourki, T., Mukhongo, L.M.: Implementation of lean in continuous process-based industries. Int. J. Sci. Eng. Res. **4**(10), 723–735 (2013)
3. Saifuddoha, A.M., Habib, M.A., Sumi, S.Y., Jennurine, M., and Islam, M.S.: Minimization of waste by applying value stream mapping in the supply chain of cement industry. IOSR J. Bus. Manag. (IOSR-JBM) **9**(3), 79–84 (2013)
4. McDermott, R.E., Mikulak, R.J., Beauregard, M.R.: The Basics of FMEA, 2nd edn. CRC Press/Taylor & Francis, Boca Raton/Routledge (2009)
5. Amrina, E., Andryan, R.: Assessing wastes in rubber production using lean manufacturing: a case study. In: Proceedings of International Conference on Industrial Engineering and Applications (ICIEA), pp. 328–332. IEEE (2019)
6. Kurniati, N., Yeh, R., Lin, J.: Quality inspection and maintenance: the framework of interaction. Proc. Manuf. **4**, 244–251 (2005)

Six Sigma Implementation to Reduce Rejection Rate in Textile Mills

Angelia[1], Wildan Trusaji[2](✉), Wisnu Aribowo[1], and Dradjad Irianto[1]

[1] Industrial Engineering Department, Institut Teknologi Bandung, Bandung, Indonesia

[2] Engineering Management Department, Institut Teknologi Bandung, Bandung, Indonesia

wildan@mail.ti.itb.ac.id

Abstract. NGX Inc. is one of the biggest producers of uniform fabric in Indonesia. It faces difficulty to fulfill export demand for XYZ123 fabric because of its 14% of total production is classified as a low-grade quality. To reduce the number of low-grade fabric, Six Sigma approach consisting define, measure, analyze, improve, and control is used. The focus of the study is defined on the greige cloth of XYZ123 up to weaving stage, and its production capacity, stability, and capability of the existing process are measured. Then, the Pareto chart is used to determine types of defects need to be studied, and Fishbone Diagram, Delphi Method, and Failure Mode and Effect Analysis (FMEA) are incorporated for further analysis. 7 solutions are proposed, but only 4 solutions are accepted by management. The result of the implemented solutions is production efficiency increase from 78.9% to 98.7% and process capability improvement from 2.66 sigma to 3.35 sigma.

Keywords: Six Sigma · Textile mills · Fabric quality

1 Introduction

NGX Inc. is an Indonesian textile manufacturer that was founded in 1982. Since from the inception, the company has rapidly grown. Currently, it serves the domestic distribution channel (DDC) in Indonesia and the international distribution channel (IDC) in Japan. Since the profit from IDC is twice form DDC, the company prioritize the IDC demand. In addition to that, the export demand has tighter quality standard and has a strict connecting date in Singapore.

NGX Inc. produces 16 types of export demand, and one of them is XYZ123. Even though the demand is newly submitted, the quantity of the demand is relatively large. However, after 2 months of the production, the capacity of XYZ123 production is still below the planned effective capacity.

Based on the analysis and the interview with the stakeholder the unmet planned effective capacity is caused by the very high rate of C grade fabric. Since the international standard is tight, the buyer wouldn't accept the C grade fabric. In other the hand, the XYZ123 is very customized fabric, so it will very hard to sell it to the

M. N. Osman Zahid et al. (Eds.): iMEC-APCOMS 2019, LNME, pp. 211–216, 2020.
https://doi.org/10.1007/978-981-15-0950-6_33

domestic market. Therefore, the C grade fabric is a defect fabric, and the company wants to reduce it.

One of the popular methodology to reduce defect product and increase the production capacity is Six Sigma [1]. There is a lot of success story of the six sigma implementation to improve the quality of the product either in SME or world-class manufacturer [2, 3]. Furthermore, Adikorley, Rothenberg, Guillory [4] implemented Six Sigma in the textile industry, and Muruganantham and Muraleedharan [5] create a review on Six Sigma implementation in the textile industry.

This paper explains the implementation of Six Sigma in NGX Inc. to reduce the defect fabric. The paper is organized in four parts namely introduction, implementation, and analysis and conclusion.

2 Implementation

The Six Sigma methodology usually consisting of five-phase that is define, measure, analyze, improve, and control (DMAIC) [6]. The goal in each phase is described in Table 1 [7]. DMAIC procedure is adopted to solve the defect problem. Although it seems the DMAIC procedure seems like the sequential process, iteration may happen during the implementation.

Table 1. The goal of DMAIC phase [7]

Phase	Goals
Define	To define problem and customer expectation
Measure	To measure defect rate and process operation
Analyze	To analyze data and to identify defect cause factors
Improve	Process improvement to eliminate defect cause factors
Control	Process control to assure the defect will not occur

2.1 Define Phase

Based on the huge production volume and large defect proportion, the weaving of XYZ123 fabric (greige) is defined as the research object. The production processes of XYZ123 fabric are warping, sizing, leasing, reaching, and weaving. Defect types of the XYZ123 fabric are broken warp, thick layer, broken weft, thin layer, loose/tight warp, piecing defects in combing, double weft, and dirty stain. There is six critical quality factor, warping machine, sizing machine, weaving machine, yarn inspection, and environment.

2.2 Measure Phase

The effective production capacity of XYZ123 is 182,9 yard/machine/day and the actual production capacity is 144,465 yard/machine/day. The number of defects is mapped in the control chart with demerit systems in Fig. 1. The existing quality level of SCD 409570 is 2,66 sigma. The calculation is based on Eqs. 1 and 2.

$$Defective\ Units\ per\ Million\ (DPM) = \frac{Number\ of\ Defective\ Units \times 1.000.000}{Number\ of\ units\ produced} \tag{1}$$

$$Sigma\ Value = NORMSINV\frac{1.000.000 - DPM}{1.000.000} + 1,5 \tag{2}$$

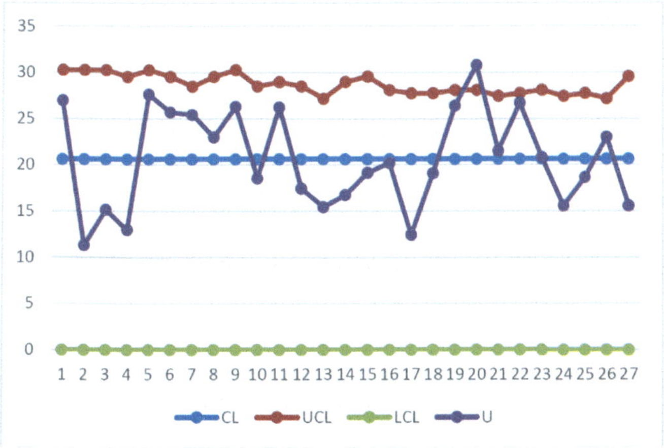

Fig. 1. Number of defects control chart

2.3 Analyze Phase

Production efficiency based on the ratio of actual output and effective production capacity is 78,9%. Based on the control chart, it can be seen that the existing process is stable, but, the quality level is under three sigma. Thus, the Pareto analysis from identified defects is done to find dominants defects. 80% of total defects comes from the broken warp, thick layer, broken weft, and double wrap. The Pareto analysis is illustrated in Fig. 2.

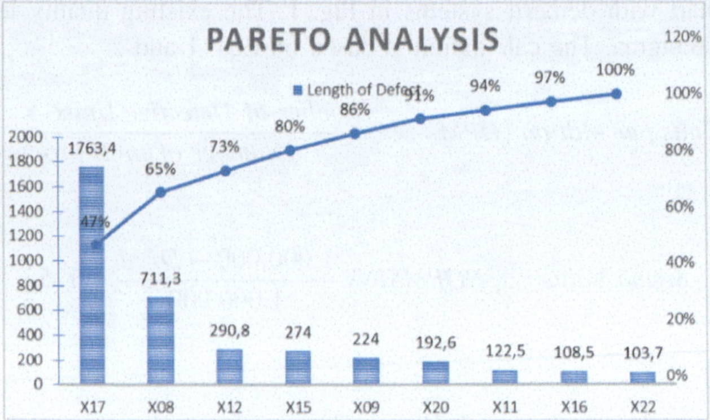

Fig. 2. Pareto analysis

Critical factors of four defects are identified by the cause and effect diagram and the Delphi method. The Delphi method iteration is illustrated in Fig. 3. All those identified factors are prioritized by Failure Mode and Effect Analysis. Based on that, 5 factors with the highest Risk Priority Number are selected to be solved/improved. Those five factors are warping machine spare part, weaving operator accuracy, yarn inspection method, warping operator accuracy, and starch recipe.

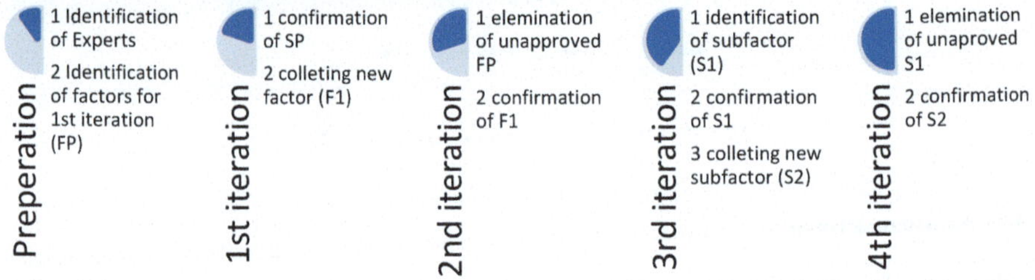

Fig. 3. Iteration of the Delphi method

Table 2. Excerpt of FMEA Result

Defect type	Cause factor	Cause sub-factor	Critical factor	S	O	D	RPN
Broken warp (X17)	Broken warp thread in the sizing process	The presence of a big starch crust	Preparation of starch solution setting	5	3	5	75
Broken warp (X17)	Fragile yarn	Yarn from the supplier is not conforming with the specification	Yarn inspection method	5	3	5	75

(continued)

Table 2. (*continued*)

Defect type	Cause factor	Cause sub-factor	Critical factor	S	O	D	RPN
Broken pick (X12)	Insensitivity of weaving machine sensor	Feeler 1 and/or feeler 2 is not working	Maintenance	3	6	4	72
Broken pick (X12)	Insensitivity of weaving machine sensor	Feeler 1 and/or feeler 2 difficult to distinguish between yarn or dust	Initial brightness sensor setting	3	6	4	72
Broken pick (X12)	The weft thread is left behind when revocation is applied	Operator handling is poor	Competence of weaving operator	6	3	3	54
Double warp (X15)	Handling mistake in sizing machine	Sizing machine operator not thoroughly check	Contentious of operator	3	3	5	45

2.4 Improve Phase

Four solution/improvement alternatives are proposed and accepted by management. Those solutions are (1) alteration of warping machine from automatic dropdown system to gravity and photoelectric sensor system. The function of the systems is to stop the warping operation when warp yarn breaks. (2) Adding one more operator to manually supervise the warping process. (3) create a sampling plan based on MIL-STD-414 [8]. There are 3 inspection variables: breaking strength, elongation, and turn per meter. (4) redesign of the starch solutions viscosity from 21–30 s to 8–9 s to accommodate the thin raw material of XYZ123.

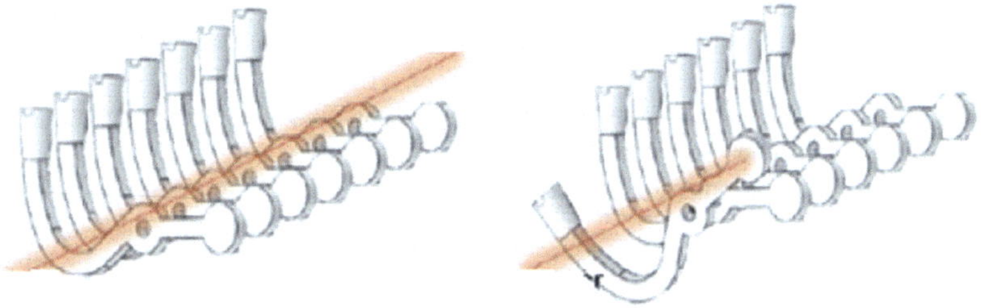

Fig. 4. Gravity and photoelectric system

2.5 Control Phase

To ensure all four solutions are well implemented, supporting document such as standard operating procedure of warping machine, work instruction for sampling plan, and starch recipe in machine sizing process are arranged.

3 Analysis and Conclusion

The XYZ123 fabric is defined as the project focus, and 8 defects and 6 critical factors are identified. It is known that the actual production capacity is 78% from its effective production capacity, and the quality level is in 2,66 sigma. Based on Pareto analysis four defects contribute to 80% of total defects. From those four defects, all critical factors are identified by the cause and effect diagram and the Delphi method. 5 critical factors with the highest RPN in FMEA are selected to be improved/solved, and 4 solutions are suggested and approved. Those solutions are (1) alteration of warping machine to gravity and photoelectric sensor system, (2) adding one more operator to manually supervise the warping process. (3) create a sampling plan based on MIL-STD-414. (4) Redesign of the starch solutions viscosity to accommodate the thin raw material of XYZ123.

Two from four approved solutions was implemented, those are adding one more operator and redesign of the starch solution viscosity. The adding one operator reduced the broken warp defect in the warping process, and redesign of the starch solution reduced the double warp in the sizing process. Based on that reduction, the production efficiency is increased from 78,9% to 98,7%, and the quality level is improved to 3,35 sigma.

References

1. Groover, M.P.: Work Systems and the Methods, Measurement, and Management of Work. Pearson Education, Inc., New Jersey (2007)
2. Wijaya, A.J., Trusaji, W., Akbar, M., Ma'ruf, A., Irianto, D.: Improving quality of shoe soles product using six sigma. In: IOP Conference Series: Material Science Engineering, vol. 319, p. 012049 (2018)
3. Vivekananthamoorthy, N., Sankar, S.: Lean Six Sigma, Six Sigma Projects, and Personal Experiences, Abdurrahman Coskun, IntechOpen (2011). https://www.intechopen.com/books/six-sigma-projects-and-personal-experiences/lean-six-sigma
4. Adikorley, R.D., Rothenberg, L., Guillory, A.: Lean six sigma applications in the textile industry: a case study. Int. J. Lean Six Sigma **8**(2), 210–224 (2017)
5. Muruganantham, V.R., Muraleedharan, P.: A review on six sigma implementation in textile industries. Int. J. Recent Sci. Res. **8**(9), 20052–20055 (2017)
6. Montgomery, D.C.: Introduction to Statistical Quality Control, vol. 6. Wiley, Jefferson City (2009)
7. Pande, P.S., Neuman, R.P., Cavanagh, R.R.: The Six Sigma Way Team Fieldbook: An Implementation Guide for Process Improvement Teams. McGraw-Hill, New York (2002)
8. MIL-STD-414. Military Standard: Sampling Procedures and Tables for Inspection by Variables for Percent Defective. U.S. Government Printing Office, Washington, D.C. (1957)

Application of Technometric to Improve Productivity in Indonesian Small Medium Industries (SMI)

Augustina Asih Rumanti[1,2(✉)], Iwan Inrawan Wiradmadja[2], Praditya Ajidarma[2], and Melita Hidayat[3]

[1] School of Industrial and System Engineering, Telkom University, Bandung, Indonesia
rumanti.augustina@gmail.com
[2] Industrial Engineering Program Study, Bandung Institute of Technology, Bandung, Indonesia
[3] Industrial Engineering, Atma Jaya Catholic University of Indonesia, Jakarta, Indonesia

Abstract. Small Medium Industry (SMI) is a group of small company that are heterogeneous in size and attitude and contribute significantly towards Indonesia's gross labor production. SMIs are expected to keep producing in a good quantity and quality to satisfy consumer and to maintain its productivity. The object of this research is Ira Silver, an SMI located in Yogyakarta, Indonesia. This study aims to measure current state of technology level within the SMI and find a correlation between such technology components and productivity. There are three concluding points made in this study. First, technometric method shows that the score of humanware and orgaware components are considerably low in terms of state of the art. Second, the result shown by SmartPLS software shows that technology components have a positive correlation to productivity, which means that SMI's productivity can be leveraged through the four technology components: technoware, humanware, infoware, and orgaware. However, only humanware component is proven to have a significant correlation to the SMI's productivity. Lastly, through this result, productivity could be enhanced by fixing humanware component at Ira Silver, so that state of the arts core in humanware component and productivity at Ira Silver can be increased.

Keywords: Humanware · Productivity · Small Medium Industry (SMI)

1 Introduction

Industrial landscape in Indonesian rural regions are filled with Small and Medium Enterprises (SMIs) due to the government's inability to tackle unemployment and poverty issues. SMIs have important role in both economic and industrial development of a country [1]. According to the Indonesian Central Bureau of Statistics and the Ministry of Cooperatives and Small and Medium Enterprises, in 2003, there were 42.4 million registered SMIs which absorbed 79 million workers. Further growth of

M. N. Osman Zahid et al. (Eds.): iMEC-APCOMS 2019, LNME, pp. 217–223, 2020.
https://doi.org/10.1007/978-981-15-0950-6_34

Indonesian SMIs is expected to decrease the unemployment rate in the country. SMI growth in Indonesia is expected to increase productivity, both in terms of quality and quantity, such that customer satisfaction and the company's profitability are optimized. The quality of output is inseparable from the quality of resource being used, for instance, the creativity, innovation and technology. Globalization demands SMI to maintain its technological advancement in order to keep their competitiveness in the market.

Technometric method is an approach to analyze the degree of implementation of technological components in a certain firm. The method is often used as a basis for decision making and formulation tools for policymaking or technological content-based analysis. There are four components included: hardware, humanware, infoware, and orgaware. Through the calculation of these components, the level of technology currently implemented by a company could be assessed and improved [2]. Previous studies on technology transfer, planning, and forecasting are also studied [3, 4].

Yogyakarta is renowned for the culture, tradition, art, business, and its signature series of handicraft. Ira Silver is a Small and Medium-sized Industry (SMI) which produces a variety of silver-based handicrafts in Kotagede district. The SMI is chosen based as the object of the study based on several factors. First, the SMIs in Yogyakarta is in imminent need of business counseling and advocacy. Second, most SMIs outside of Jakarta are inclined to collaborate through research with academic entity. Based on previous description, the objectives of this research are as follows:

1. To identify State of The Art (SOTA) values and each component's contribution with regard to the technological component of the SMI
2. To identify the impact of technological components to SMI's productivity
3. To propose technological-based recommendations to improve productivity

Furthermore, the scope of this research is defined within the following scope. Primary observation is conducted between 17 and 22 March 2014, and the data is collected on September 2014. The products being observed is limited to silver-based handicrafts.

2 Methodology

Prior observation, preliminary study and topic generation were conducted. The acquired data is then analyzed with a technometric approach to measure the joint contribution of all 4 technological components within input-to-output transformation. According to Economic and Social Commission for Asia and Pacific (ESCAP) of United Nations [5], technology is divided into four components. Despite the classification, technological transformation cannot be conducted if even one of the components are neglected. The 4 components are as follows:

1. Technoware (T) is object-embodied technology or physical facilities, which comprises equipment, machines, tools, machines, automobiles, factories, and physical infrastructure which human uses to operate the transformation.

2. Humanware (H) is person-embodied technology or human abilities, which comprises knowledge, skills, wisdom, creativity, achievements, and experience of an individual or group of people in utilizing the available resource.
3. Infoware (I) is document-embodied technology or document fact which includes process, procedure, technique, method, theory, specification, design, observation, user manual, and other fact-based information which is communicated through publication, documentation, or blueprints.
4. Orgaware (O) is institution-embodied technology or organizational frameworks, which is necessary to facilitate physical infrastructures, human's ability, and information, including managerial practice, hierarchy, and regulation within an organization to achieve positive impacts.

Technometric model is used to analyze the level of technological content in terms of contribution of the four components in an input-to-output transformation process. Joint contribution is further defined as the technological contribution, and is measured in terms of Technology Contribution Coefficient (TCC), as shown on Table 1.

Table 1. TCC's level for technology.

TCC value	Classification
$0.1 \leq$ TCC ≤ 0.3	Traditional
$0.3 \leq$ TCC ≤ 0.7	Semi-modern
$0.7 \leq$ TCC ≤ 1	Modern

3 Results and Analysis

Upon collecting the data, calculation is conducted. Historical data is observed and used for the input of technometric measurement. The technometric measurement process starts by estimating the contribution of technological components' advancement degree. Equations 1–4 shows the calculation of State of the Art Assessment (SOTA), and Eqs. 5–8 shows the calculation of contribution of each component.

$$ST_I = \frac{1}{10} \times \left[\frac{\sum_k ik}{k_i}\right] \text{ for } k = 1, 2, \ldots, k_i;\ ST_I = 0.2125 \tag{1}$$

$$SH_I = \frac{1}{10} \times \left[\frac{\sum_i hi}{i_h}\right] \text{ for } k = 1, 2, \ldots, i_h;\ SH_I = 0.45 \tag{2}$$

$$SI_I = \frac{1}{10} \times \left[\frac{\sum_m fm}{m_f}\right] \text{ for } k = 1, 2, \ldots, m_f;\ SI_I = 0.33 \tag{3}$$

$$SO_I = \frac{1}{10} \times \left[\frac{\sum_n on}{n_o}\right] \text{ for } k = 1, 2, \ldots, n_o;\ SO_I = 0.5 \tag{4}$$

$$\text{Technoware}_{\text{Smelting}} : T_i = \frac{1}{9} \times [LT_i + ST_i \times (UT_i - LT_i)] = 0.158 \quad (5)$$

$$\text{Humanware}_{\text{Worker}} : H_j = \frac{1}{9} \times [LH_i + SH_i \times (UH_i - LH_i)] = 0.322 \quad (6)$$

$$\text{Infoware} : I = \frac{1}{9} \times [LI_i + SI_i \times (UI_i - LI_i)] = 0.296 \quad (7)$$

$$\text{Orgaware} : O = \frac{1}{9} \times [LO_i + SO_i \times (UO_i - LO_i)] = 0.667 \quad (8)$$

Table 2 shows the calculation result of each technological component. The intensity of contribution is shown in Table 3. Further, the calculation proceeds until Technology Contribution Coefficient (TCC) is known, as indicated in Table 4.

Table 2. Estimated sophistication degree and calculation result of technological component.

Component			LL	UL	Contribution score	Description
Technoware	1	Smelting	1	3	0.158	Manual facilities
	2	Casting	1	3	0.158	Manual facilities
	3	Forging	2	4	0.269	Powered facilities
	4	Tooling	1	3	0.158	Manual facilities
	5	Cutting	1	3	0.158	Manual facilities
	6	Burning	2	4	0.269	Powered facilities
	7	Washing	1	3	0.158	Manual facilities
	8	Scratching	2	4	0.269	Powered facilities
Humanware	1	Owner	6	8	0.767	Improving
	2	Worker	2	4	0.322	Setting-up
Infoware			2	4	0.296	Describing facts
Orgaware			5	7	0.667	Stabilizing

Table 3. The Result of technometric data processing.

Technometric component	SOTA	Contribution	Intensity of contribution
Technoware	0.2125	0.199	1.067
Humanware	0.4500	0.544	0.827
Infoware	0.3300	0.296	1.117
Orgaware	0.5000	0.667	0.750

Table 4. Recapitulation of TCC computation.

Manifested variables	Code	AVE	Composite reability	T-statistic	Significance
Latent variables: Technological component (A)					
Technoware	(A.1)	0.454	0.797	0.656	Insignificant
Humanware	(A.2)	0.410	0.770	4.064	Significant
Infoware	(A.3)	0.505	0.847	0.245	Insignificant
Orgaware	(A.4)	0.324	0.681	0.637	Insignificant
Latent variables: Productivity (B)					
Investment	(B.1)	0.265	0.192	7.978	Significant
Labor capital ratio	(B.2)	0.610	0.154	0.545	Insignificant
Research and Development	(B.3)	0.515	0.001	1.424	Insignificant
Capacity usage	(B.4)	0.297	0.313	3.262	Significant
Government regulation	(B.5)	0.960	0.987	5.909	Significant
Age of tool and factory	(B.6)	0.594	0.801	2.936	Significant
Energy cost	(B.7)	0.559	0.206	3.001	Significant
% of significant labor	(B.8)	0.972	0.990	21.375	Significant
Work ethics	(B.9)	0.505	0.058	9.100	Significant
Fear of unemployment	(B.10)	0.530	0.770	9.130	Significant
Influence of labor union	(B.11)	0.556	0.789	2.145	Insignificant
Management	(B.12)	0.434	0.244	8.091	Significant
Technological component	**A**	**0.383**	**0.716**	**37.242**	**Significant**
Productivity	**B**	**0.362**	**0.799**		

After the score of Technology Contribution Coefficient (TCC) is acquired, this study maps each score of Technoware, Humanware, Infoware, and Orgaware (THIO) in a diagram (Fig. 1). The technological component technoware, humanware, infoware, and orgaware are scored at 0.2125, 0.45, 0.33, and 0.5, respectively.

Based on the figure, technoware component has the lowest value compared to the other technological components. Yet, this does not necessarily conclude that the technoware component is not being utilized optimally by the firm. Upon further comparison with regard to the components' contribution in terms of THIO, the opposite is true: the technoware technological component has been utilized perfectly by the SMI, such that the score exceeds the potential threshold of improvement in terms of technological component. The maximum attainable technoware score could be derived based on the component's contribution, which equals 0.199. In Fig. 1, it is concluded that the SMI Ira Silver focuses mostly the on humanware and orgaware technological components. This is mainly due to the fact that those two aspects score lower in terms of state of the art, compared to its component's contribution value.

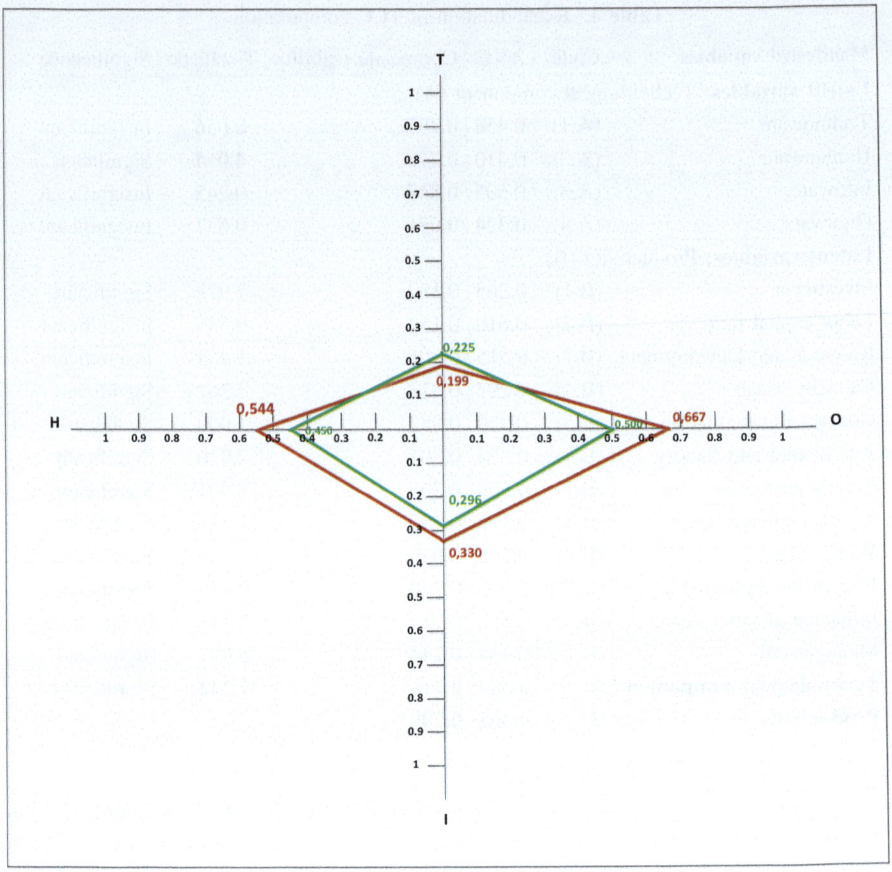

Fig. 1. The comparison of THIO diagram based on SOTA and component contribution.

4 Conclusion

Based on the technometric calculation conducted on Ira Silver, there are two technological components with considerably low scores, which could be improved further: humanware and orgaware components. On the other hand, technoware element in Ira Silver has been optimized such that SOTA value on the component is higher compared to its component's contribution value. Based on similar deduction, infoware technological component has also been fully utilized by Ira Silver.

Through statistical significance test on SmartPLS software, it is concluded that technological components positively correlate with SMI's productivity. Thus, Ira Silver could increase its productivity by improving the technological components which scores are considered low. Based on technometric and SmartPLS software calculation, it is concluded that improving humanware component in Ira Silver will automatically boost its productivity. Four potential recommendations are proposed, such as adequate training for the labors, promoting discussion between owner and workers regarding product development, conducting further research to develop silver products, and promoting discussion between artisan SMIs in Yogyakarta.

References

1. Tambunan, T.: Promoting small and medium enterprises with a clustering approach: a policy experience from Indonesia. J. Small Bus. Manag. **43**(2), 138–139 (2005)
2. Ellitan, L.: Peran sumber daya dalam meningkatkan pengaruh teknologi terhadap produktivitas. Jurnal Manajemen & Kewirausahaan **5**(2) (2003)
3. Wiratmadja, I.I., et al.: Tingkat kesiapan humanware dalam proses alih teknologi studi kasus: Tenaga Insinyur Perusahaan Jasa. Jurnal Teknik dan Manajemen Industri Forum Ilmiah Bagi Ilmuwan dan Praktisi Industri **24**(1) (2004)
4. Wiratmadja, I.I., et al.: Diktat Kuliah Program Magister Teknik & Manajemen Industri ITB: Perencanaan Teknologi dan Peramalan Teknologi. Institut Teknologi Bandung, Bandung (2006)
5. Purwaningsih, R., et al.: Penilaian teknologi dengan metode teknometrk di PT. Indo Acidatama Chemical Industry Solo. Jurnal Transistor **5**(1), 13 (2005)
6. Ellitan, L.: Peran sumber daya dalam meningkatkan pengaruh teknologi terhadap produktivitas. Jurnal Manajemen & Kewirausahaan **5**(2) (2003)
7. Fauzan, A.: Penilaian tingkat teknologi dok pembinaan UPT BTPI Muara Angke Jakarta. Tugas Akhir Sarjana Fakultas Perikanan dan Ilmu Kelautan, Institut Pertanian Bogor, Bogor (2009)
8. Ghozali, I.: Structural equation modeling metode alternatif dengan partial least square (PLS). Badan Penerbit Universitas Diponegoro, Semarang (2006)
9. Khalil, T.: Management of Technology: The Key to Competitiveness and Wealth Creation. The McGraw-Hill Book Co., Singapore (2000)
10. Tandianputera, Y.: Analisis pengaruh produktivitas terhadap kinerja karyawan studi kasus: CV. Kajeye Food, Malang. Tugas Akhir Sarjana Fakultas Teknik, Universitas Katolik Atma Jaya, Jakarta (2013)

Ant Colony Optimization-Based Multiple-AGV Route-and-Velocity Planning for Warehouse Operations

Anugrah K. Pamosoaji[1(✉)] and Sarifah Putri Raflesia[2(✉)]

[1] Universitas Atma Jaya Yogyakarta, D. I. Yogyakarta 55281, Indonesia
kusumo_pamosoaji@mail.uajy.ac.id
[2] Universitas Sriwijaya, Ogan Ilir, Indralaya, Sumatera Selatan 30662, Indonesia
sarifah@unsri.ac.id

Abstract. In this paper, a route-and-velocity planning algorithm for a group of automated guided vehicles (AGVs) in a simple warehouse is investigated. It decides waypoint-to-waypoint routes and the velocity decision between any pair of waypoints is presented. The main objective of this paper is generating free-collision routes for each vehicle while minimizing the minimum traveling time of each vehicle. The maximum velocity limitation of each vehicle's motion is considered. The warehouse used is of the type of simple warehouse is modeled as a matrix of nine waypoints. A modification of typical Ant Colony Optimization (ACO) is used as the search algorithm. In the proposed ACO algorithm, two tasks are accomplished: waypoints routes and the velocities between waypoint pairs. The selection of waypoint uses the paradigm of Artificial Potential Field (APF) such that collision among vehicles can be avoided. The simulation results will be presented and evaluated. Simulation result shows that the proposed algorithm applied for simple warehouse performs convergent results in a number of iterations. The resulted minimum traveling time of the slowest vehicle is convergent and the resulted routes are collision-free.

Keywords: Ant Colony Optimization · Warehouse · Vehicle routing problem · Route and velocity planning

1 Introduction

A scenario of vehicle routing problem (VRP) applied on multiple automated guided vehicles (AGV) that is equipped with velocity planning feature is presented. The planner provides a chain of paths of all AGVs together with velocity applied on the paths. This study is motivated by the needs of a warehouse operating a group of AGVs to prevent any collision in the AGVs' operation. However, the existence of share points among paths in typical warehouses that potentially cause collisions [1, 5, 6]. Also, since the resulted plan will be tracked by the autonomous AGV, the vehicle's velocity must be determined as well [4].

Various solutions, especially of the metaheuristic class, were reported [1, 7, 8]. According to [9, 10], metaheuristic methods can be used to produce suboptimal solution with acceptable computation time. Ant Colony Optimization (ACO) was first coined by

M. N. Osman Zahid et al. (Eds.): iMEC-APCOMS 2019, LNME, pp. 224–229, 2020.
https://doi.org/10.1007/978-981-15-0950-6_35

Dorigo and Gambardella in [2] aimed to solve travelling salesman problem (TSP). Recently, various applications of this method have been reported. However, mostly the VRP problems are approached by using TSP scenario, which eliminates the possibility that a particular waypoint in a graph must be visited at most by one vehicle [3].

In our study, a variant of VRP algorithms that accommodates the possibility of multiple accesses for some points by different vehicles is introduced. In order to guarantee that the routes of all vehicles are collision-free, then a modified Ant Colony Optimization (ACO) algorithm is proposed. The main feature of the algorithm is the addition of velocity applied by each vehicle on the routes. Also, for each waypoint-to-waypointedge, we provide more than 1 velocity alternatives. The pheromone updates are then dropped to each edge.

The organization of this paper is described as follows. Section 2 describes the problem description. Section 3 presents the method used to solve the problem. Section 4 reveals the simulation results and some discussions. Finally, Sect. 5 concludes the discussions.

2 Problem Description

Consider a graph $\mathbf{G}(\mathbf{P}, \mathbf{E})$ consists of a set of waypoint $\mathbf{P} = \{p_i\}$, where $i = 1, \ldots, N_\text{v}$ and a set of waypoint-to-waypoint connectors $\mathbf{E} = \{e_{i,j}\}$, where i and j are the label for the start waypoint and the destination waypoint, respectively. Let us define an adjacency matrix $\mathbf{A} = \{a_{i,j}\}$ where $a_{i,j} = 1$ if there exists a direct route from the i-th waypoint to the j-th waypoint, and $a_{i,j} = 0$ otherwise. Also, let us define $f(e_{i,j}) = \{0, 1, 2, \ldots, N_\text{t}\}$ as an edge-occupation function for the i-to-j edge $e_{i,j}$, that indicates the number of vehicles pass through the edge in the i-to-j direction at a time instance t. Additionally, it is possible that $f(e_{i,j}) \neq f(e_{j,i})$.

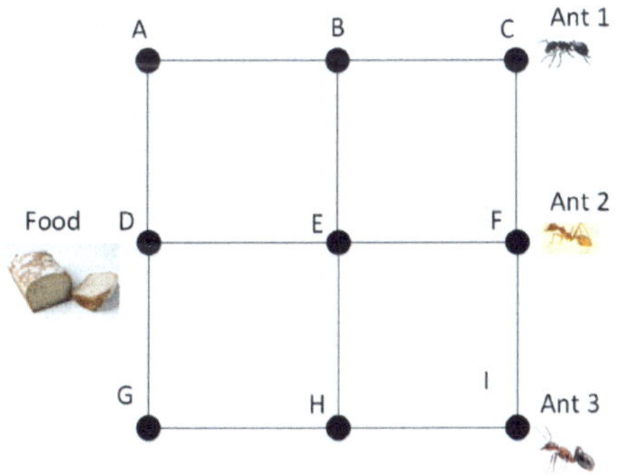

Fig. 1. The layout of a simple warehouse.

Suppose that there exists N_t vehicles whose the i-th vehicle's initial position is in $p_{i,0} \in \mathbf{P}$, where $i = \{1, 2, \ldots, N_t\}$ and all of vehicles are initially separated. Suppose that all the vehicles have to reach a common target $p^t \in \mathbf{P}$. The objective is that finding a path for each vehicle such that the traveling time of the slowest vehicle to reach the goal is minimized, or formally,

$$\min J = \{\max(T_1, T_2, \ldots, T_N)\} \tag{1}$$

subject to

(1) $0 \le v_i \le v_{\max}$, for all $t \ge 0$,
(2) For all finite time $0 \le t \le T$, the maximum number of vehicles allowable to occupy a waypoint is one.
(3) For all finite time $0 \le t \le T$, if $f(e_{i,j}, t) > 0$, then $f(e_{j,i}, t) = 0$, and vice versa,
(4) $p_i(0) = p_{i,0}$ for all $i = \{1, 2, \ldots, N_t\}$,
(5) $p_i(T_i) = p^t$ **for all** $i = \{1, 2, \ldots, N_t\}$.

where T_i is the travelling time of the i-th vehicle. The solution is modeled as $S = \{\mathbf{P}_s, \mathbf{V}_s\}$, where $\mathbf{P}_s = \{\mathbf{p}_{s,1}, \mathbf{p}_{s,2}, \ldots, \mathbf{p}_{s,N_t}\}$ and $\mathbf{V}_s = \{\mathbf{v}_{s,1}, \mathbf{v}_{s,2,\ldots}, \mathbf{v}_{s,N_t}\}$ are the set of routes $\mathbf{p}_{s,i}$ of the i-th vehicle and the set of the i-th vehicle's velocity applied on the route $\mathbf{p}_{s,i}$, respectively. The i-th route $\mathbf{p}_{s,i}$ is defined as $\mathbf{p}_{s,i} = \{p_{s,i,1}, p_{s,i,2}, \ldots, p_{s,i,|\mathbf{p}_{s,i}|}\}$, where $p_{s,i,j}$ is defined as the j-th waypoint visited by the i-th vehicle. Furthermore, $\mathbf{v}_{s,i}$ is defined as $\mathbf{v}_{s,i} = \{v_{s,i,1}, v_{s,i,2}, \ldots, v_{s,i,|\mathbf{v}_{s,i}|}\}$, where $v_{s,i,j}$ is the velocity applied on the j-th to the (j + 1)-th points in the i-th route. Note that the operator $|.|$ is defined as the size of the argument (which is, a set).

In this study, we consider a simple warehouse layout as shown in Fig. 1. Suppose that the warehouse consists of 9 waypoints **P**. Some of the waypoints are connected such that the paths perform boundaries of four squares areas. Suppose that the distance of any connected waypoint pairs is the same. Three vehicles are provided to pick a load in waypoint D. The 1st, 2nd, and 3rd vehicles start from waypoints C, F, and I, respectively.

3 Methods

The algorithm is described as follows.

for each iteration h
  for each ant i,
    $\mathbf{p}_{s,i}^{h} \leftarrow \{\}$; $\mathbf{p}_{s,i}^{h} \leftarrow p_{\text{curr},i}$. //Initial assignment
  end
  while at least one ant does not reach the goal waypoint.
    for each ant i
      $\mathbf{p}_{s,i}^{h} \leftarrow selectNextPoint\ (\text{neighbor}(p_{c,i}))$. //The nearest-to-goal neighbor.
      $\mathbf{v}_{s,i}^{h} \leftarrow determineVelocity\left(e_{m,n}\right),\ m = p_{c,i}\ ,\ m = last(\mathbf{p}_{s,i}^{h})$.
      $t_{s,i}^{h} \leftarrow calcTravelingTime\left(\text{last}(\mathbf{p}_{s,i}^{h}),\ last(\mathbf{v}_{s,i}^{h})\right)$
    end
    $freeCollisionVerification(\mathbf{p}_{s,i}^{h}, t_{s,i}^{h})$ for all ants.
  end
  calculateMinimumSlowestVehicleTravelingTime()
  $updatePheromone(\mathbf{p}_{s,i}^{h} \leftarrow)$
end

The procedure *determineVelocity* $\left(e_{m,n}\right)$ generates velocity for the i-th ant from waypoint m to waypoint n. Here, we define N_c velocity centroids where each centroid represents a velocity alternative. The applied velocity is that with high probability. The probability of choosing the velocity centroid is calculated by the *updatePheromone* (.) procedure. In the *updatePheromone* (.), pheromones are injected to each centroid of the edges passed by each ant. This is the main feature of the proposed method: instead of injecting the pheromone in the edges, the pheromone in this paper is injected into the velocity centroid of the edges.

4 Results and Discussions

A simulation was performed for 1000 iterations and the result can be revealed in Table 1 and Fig. 2. We set the maximum allowable velocity for each vehicle is 5 m/s. The distance of any pair of connected points is set to be 20 m. All vehicles have to reach target waypoint D.

As shown in Table 1, the same routes were resulted in the 2nd, 3rd, and 4th trials and the remaining trials gave different ones. The "Velocity" column can be interpreted as the velocity when the vehicle moves from the current waypoint to the next waypoint. For instance, in Trials no. 1, vehicle 1 moves from C to F with velocity 4.9642 m/s; from F to E with velocity 4.9958 m/s, and so forth. Overall, all trials give the minimum slowest vehicle's traveling time is 12.03498 s in average.

To verify the validity of the traveling time results, assume that each vehicle' velocity is 5 m/s. Since the largest traveling distance is 60 m, then the most possible fastest traveling time is 12 s. Therefore, the resulted traveling time in Table 1 can be concluded to be near-optimal since it is very close to 12 s.

Free collision feature of the results can be checked by investigating the arrival time to some shared points. A solution is free of collision if and only if there is no route with the same arrival time to any single shared waypoint. For instance, in Trials no. 3, vehicles 2 and 3 have share points, i.e., H, G, and D. It can be investigated that vehicle 3 reach H, G, and D at t = 4.0029 s, t = 8.0191 s, and t = 12.0.355 s, respectively, and vehicle 2 reach the same points at t = 0 s, t = 4.0032 s, and t = 8.0064 s, respectively.

Table 1. Simulation results

Trials no.	Vehicle no.	Routes	Velocity (m/s)	Traveling time (s)
1	1	C – F – E – D	4.9642–4.9958–4.9957	12.0357 (15.4909)
	2	H – E – D	4.9988–4.9988	
	3	I – H – G – D	4.9895–4.9987–4.9962	
2	1	C – F – E – D	4.9900–4.9946–4.9902	12.0260 (15.5533)
	2	H – G – D	4.9960–4.9960	
	3	I – H – G – D	4.9828–4.9924–4.9924	
3	1	C – F – E – D	4.9982–4.9921–4.9906	12.0355 (15.5533)
	2	H – G – D	4.9924–4.9371	
	3	I – H – G – D	4.9964–4.9797–4.9797	
4	1	C – F – E – D	4.9864–4.9752–4.9967	12.0336 (15.1633)
	2	H – G – D	4.9950–4.9903	
	3	I – H – G – D	4.9836–4.9978–4.9959	
5	1	C – B – E – D	4.9987–4.9619–4.9899	12.0441 (15.5689)
	2	H – G – D	4.9916–4.9916	
	3	I – F – E – D	4.9477–4.9989–4.9989	

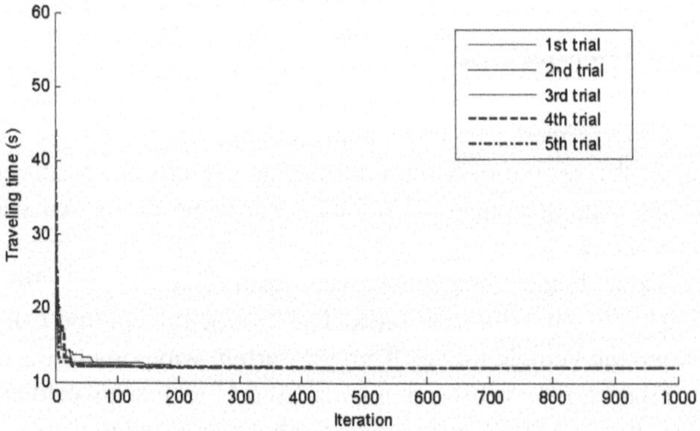

Fig. 2. Search progression for 5 trials and 1000 iterations.

5 Conclusions

A scenario of vehicle routing problem (VRP) with velocity planning applied on multiple automated guided vehicles (AGV) for warehouse operations is presented. A modified Ant Colony Optimization (ACO) designed for generating velocities for all AGVs' routes is proposed. The algorithm is simulated in a simple warehouse with 9 points. The simulation results of the proposed algorithm show that the traveling time of the slowest vehicle can be minimized into a convergent value. Also, the proposed method is proven to produce collision-free trajectories. Future works will be focused on the addition of some more realistic aspects, such that vehicle-to-vehicle distance, the acceleration, and the tracking controller design for each vehicle.

References

1. Chen, F., Wang, H., Xie, Y., Qi, C.: An ACO-based online routing method for multiple order pickers with congestion consideration in warehouse. J. Intell. Manuf. **27**(2), 389–408 (2016)
2. Dorigo, M., Gambardella, L.: Ant colony: a cooperative learning approach to the travelling salesman problem. IEEE Trans. Evol. Comput. **1**(1), 53–66 (1997)
3. Dorigo, M., Stutzle, T.: Ant colony optimization: overview and recent advances. In: Gendreau, M., Potvin, J.Y. (eds.) Handbook of Metaheuristics. International Series in Operations Research and Management Science, vol. 146. Springer, Boston (2010)
4. Pamosoaji, A.K.: Trajectory tracking control strategy using co-reference for rear-steered vehicle. In: 3rd International Conference on Control and Robotics Engineering, pp. 74–78. IEEE, Nagoya (2018)
5. Pamosoaji, A.K., Hong, K.-S.: Group-based particle swarm optimization for multiple-vehicles trajectory planning. In: 15th International Conference on Control, Automation, and Systems, pp. 862–867. IEEE, Busan (2015)
6. Pamosoaji, A.K., Piao, M., Hong, K.-S.: PSO-based minimum-time motion planning for multiple-vehicle systems considering acceleration and velocity limitations. Int. J. Control Autom. Syst. (2019). https://doi.org/10.1007/s12555-018-0176-9
7. Schyns, M.: An ant colony system for responsive dynamic vehicle routing. Eur. J. Oper. Res. **245**, 704–718 (2015)
8. Yuan, Y., Wang, D.: Path selection model and algorithm for emergency logistics management. Comput. Ind. Eng. **56**(3), 1081–1094 (2009)
9. Wang, X., Choi, T.-M., Liu, H., Yue, X.: Novel ant colony optimization methods for simplifying solution construction in vehicle routing problems. IEEE Trans. Intell. Transp. Syst. **17**(11), 3132–3141 (2016)
10. Xiang, W., Lee, H.P.: Ant colony intelligence in multi-agent dynamic manufacturing scheduling. Eng. Appl. Artif. Intell. **21**, 73–85 (2008)

Dual-Channel Warehouse Raw Material Inventory Model for Probabilistic Demand

Docki Saraswati(✉) and Hana Tyasari

Department of Industrial Engineering, Universitas Trisakti,
Jakarta 11440, Indonesia
docki_saraswati@trisakti.ac.id,
hana.tyasari@gmail.com

Abstract. In manufacturing industries, raw material inventories play a big role. This paper discusses the merging of orders of two types of raw materials for two products in two different warehouses. Both types of raw materials are used to manufacture components for products; automotive and educational teaching aids. The demand for both products are probabilistic, while the two warehouses for raw materials have capacity constraints. The merger of ordering the same raw materials for the two warehouses is called dual-channel warehouse. This study shows that combined orders of raw material with dual-channel warehouse can save 29.61% compared to the two warehouses with individual orders.

Keywords: Dual-channel warehouse · Probabilistic demand · Capacity constraint

1 Introduction

The dual-channel warehouse structure is one of the multi-channel inventory management structures that connect two warehouse areas. Centralized inventory strategy in a dual-channel system is a strategy in which manufacturing companies concentrate inventory management to meet the demand that has two demand lines [1]. The advantages of a centralized and integrated ordering structure include reducing facility costs and increasing flexibility as well as service levels [2]. Several studies for multi-channel supply has been done before. Xie *et al.* [1] in a study designed three inventory strategies from the perspective of the manufacturing industry with two supply channels. Research with models of inventory has been carried out by Radhi and Zhang [3] with the consideration of cross-channel consumer behaviors on product returns policy. Additionally, Alawneh and Zhang [2] proposed dual-channel inventory and warehouse management with probabilistic demand and constraint.

This research proposes the inventory design and calculation that relates to the procurement of raw material to meet the production needs of two product categories. The approach used considers the policy of centralized inventory control and the limited area of raw materials warehouse. Determination of the amount and time of ordering optimal raw materials is needed to meet probabilistic demand and based on the minimum total inventory cost.

M. N. Osman Zahid et al. (Eds.): iMEC-APCOMS 2019, LNME, pp. 230–237, 2020.
https://doi.org/10.1007/978-981-15-0950-6_36

2 Methodology

2.1 Research Methodology

This research was conducted at a plastic injection manufacturing company that manufactures components for automotive and educational teaching aids. These components use raw materials in the form of plastic granules and coloring pigments. The make-to-order system is implemented. The current control of inventories in the company is carried out separately for each production type, and the raw materials are stored in two different warehouses. Each product type has its own warehouse, but the capacity of the warehouse is limited.

Data Collection. The data collected in this study are; bill of material, product demand, material specifications, purchase costs, storage area, order costs, stock-out of raw materials, and lead time.

Inventory Model. The raw material inventory model used is the Probabilistic Economic Order Quantity (EOQ) model with a continuous review system for probabilistic demand conditions and constrained for warehouse areas. The determination of inventory model parameters is carried out with the following steps: (1) Determine the ordering lot size and the reorder points with backorder; (2) Calculate the probability and the number of expectations of inventory shortage; (3) Calculate the area requirements for raw materials; (4) Calculate the order quantity and reorder point, with backorder and constraints for warehouse area; (5) Finally, determine the expected total inventory cost.

Sensitivity Analysis. The parameters used in sensitivity analysis are the demand for finished products. Sensitivity analysis was conducted to determine the effect of changes in parameters on optimal ordering lot size and total inventory costs.

Inventory Model Simulation. A daily perpetual inventory simulation for the stages of raw materials is carried out according to the time horizon used using the demand routine inventory simulation model. Optimal order lot size, reorder point, safety stock, and lead time for raw materials are used as inputs in the simulation.

2.2 Inventory Model Approach

This study conducted a dual-channel warehouse inventory model approach with the company as a buyer related to the procurement of raw materials from the supplier. The channel is used to produce automotive components and educational teaching aids. The upstream of the model is a centralized warehouse area, while downstream is the production floor for the two-channel production category. The procurement of raw materials is designed to be carried out from suppliers to centralized storage areas, before finally being distributed to each warehouse according to the production category and used as needed. Decision variables used to minimize the total inventory cost in this study are the same as the ones used in previous studies done by Radhi [3] and Alawneh [2], which are the order quantity and reorder point.

3 Inventory Model

3.1 Notations and Assumptions

The mathematical model used in this paper refers to a research conducted by Alawneh and Zhang [2]. The notations used in this research are given as follows:

i = raw material index
j = stage index
k = finished good/product index
d_k = demand quantity of k
D_{ij} = demand of material i in stage j
h_{ij} = holding cost per unit material i in stage j
b_{ij} = stock-out cost per unit material i in stage j
A_{ij} = order cost per unit material i in stage j
μ_{ij} = mean of material i demand in stage j
σ_{ij}= standard deviation of material i demand in stage j
x_{ij} = demand during lead time (DDLT) of material i in stage j
L_{ij} = ordering lead time of material i in stage j
Q_{ij} = order quantity of material i in stage j
r_{ij} = *reorder point* of material i in stage j
z_{ij} = safety factor of material i in stage j
$\int f(x_{ij})dx_{ij}$ = stock out probability of material i in stage j
θ = *lagrange* multiplier for the warehouse space constraint
$z_{1-\alpha}$=cumulative probability distribution of demand at point $1 - \alpha$
γ_{ij} = space requirement per unit material i in stage j
S = available space
TC = total inventory cost

The assumptions used in this research are given as follows [2]:

a. The demand rate per unit time is a random variable and the raw materials lead times are constant.
b. The inventory policy used is a continuous review (Q, r).
c. Backorder conditions occur when a request cannot be met from the existing provisions with a prescribed penalty fee.
d. Each stage has its own reorder point value obtained from the sum of DDLT expectation values and safety stock.

3.2 Inventory Model Design

There are two stages of ordering in the application of a dual-channel warehouse inventory system; ordering in stage 1 and stage 2 [2]. The model is shown in Fig. 1. The description of the stages in the model is as follows: (1) Stage 1 is a raw material warehouse area, where stage 1A is the warehouse of raw material used for automotive components, while stage 1B is to store raw material used for educational teaching aids; (2) If the quantity of raw materials in stage 1A and 1B have achieved the reorder point, then the replenishment is done from stage 2; (3) Stage 2 is the area for receiving raw materials. Raw materials imported from suppliers are stored in stage 2 before being sent to be stored in stage 1; (4) The quantity of raw material requirements in stage 2 is an aggregation of demand for raw materials at stage 1A and 1B.

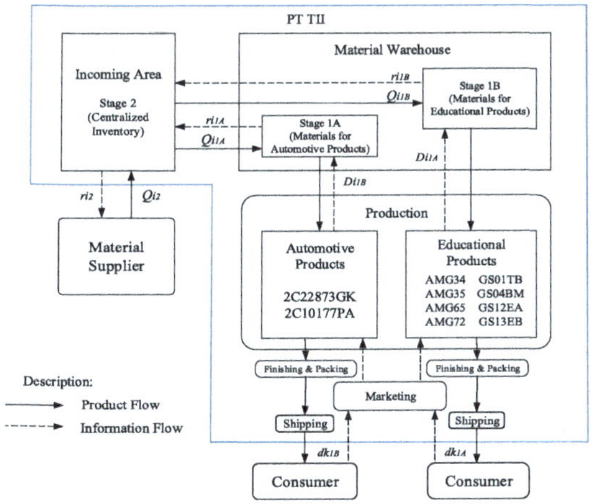

Fig. 1. Inventory system model design

3.3 Mathematical Models

The objective of this study is to minimize the expectations of total inventory costs, which consist of ordering, holding, and stockout costs. The expectation of total inventory costs can be seen in Eq. (1), where θ is the *Lagrange* multiplier for the warehouse space constraint. The optimal solution can be obtained using Kuhn-Tucker first order conditions.

$$E\{TC(Q_{ij}, r_{ij}) = \sum \frac{A_{ij}D_{ij}}{Q_{ij}} + \sum h_{ij}\left[\left(\frac{Q_{ij}}{2}\right) + (r_{ij} - \mu_{xij})\right] + \sum \frac{b_{ij}D_{ij}}{Q_{ij}}\left[\int_{rij}^{\infty} (x_{ij} - r_{ij})f(x_{ij})dx_{ij}\right]$$
$$+\theta\left[\sum (\gamma_{ij}(Q_{ij} + \sigma_{xij}z_{ij}) - S - \mu_Y - z_{1-\alpha}\right] \tag{1}$$

where $\mu_{xij} = \mu_{ij} \times L_{ij},$ $\boldsymbol{\sigma}_{xij}=\sqrt{\sigma_{ij}^2 \times L_{ij}}$, and $\mu_Y = \sum_i \sum_j \gamma_{ij}\mu_{ij}$

4 Results and Discussion

4.1 Material Demand

Demand for automotive and educational teaching aid are normally distributed. These products use the same raw materials; Polypropylene Polytam PF1000 and PEM 2020 Black. Raw material requirements are based on product's bill of material. Raw material requirements at each stage are shown in Table 1.

Table 1. Material demand

No.	Period	PP polytam PF1000 material (i = 1)			Unit	PEM 2020 black material (i = 2)			Unit
		$D_{1,1A}$	$D_{1,1B}$	D_{12}		$D_{2,1A}$	$D_{2,1B}$	D_{22}	
1	18-Jan	13	4	17	Sacks	8	1	9	Containers
2	18-Feb	13	8	21		4	5	9	
3	18-Mar	11	7	18		8	4	12	
4	18-Apr	10	10	20		5	5	10	
5	18-May	11	9	20		5	5	10	
6	18-Jun	7	7	14		3	3	6	
7	18-Jul	14	6	20		9	4	13	
8	18-Aug	15	7	22		6	3	9	
9	18-Sep	9	9	18		5	5	10	
10	18-Oct	9	7	16		6	4	10	
11	18-Nov	11	8	19		8	4	12	
12	18-Dec	15	4	19		12	3	15	
Total		**138**	**86**	**224**		**79**	**46**	**125**	

4.2 Inventory Model

The decision variables and total inventory costs is carried out for a 12-months time horizon. The decision variables are order lot size and reorder point of raw material i at each stage j as; stage 1A ($Q_{i,1A}$, $r_{i,1A}$), stage 1B ($Q_{i,1B}$, $r_{i,1B}$), and stage 2 (Q_{i2}, r_{i2}). The inventory model with space constraints is only carried out for PP Polytam PF1000 raw material at stage 2. The number of ordering lots to suppliers uses an inventory model with constraints for PP Polytam PF1000 and PEM 2020 Black; 27 units and 74 units per order, consecutively. Outputs of the inventory model are shown in Table 2.

Table 2. Outputs of constrained inventory model

i	j	Q_{ij}	ss_{ij}	r_{ij}	Unit	M_{ij}
1	1A	4	0	0	Sacks	35
1	1B	3	0	0	Sacks	27
1	2	27	1	5	Sacks	9
2	1A	8	0	0	Containers	10
2	1B	6	0	0	Containers	8
2	2	74	1	2	Containers	2

Based on the calculation, total inventory cost for two types of raw materials can be reduced 29.61% from the actual cost. The centralized inventory model provides efficiency in terms of ordering and procurement activities. Moreover, the calculation of safety stock helps to avoid stock-out conditions occurring in the company. The integrated structure of the dual-channel warehouse also provides hierarchical inventory structure, which makes stock monitoring easier to conduct.

4.3 Sensitivity Analysis

Sensitivity analysis is performed by increasing and decreasing the demand for 10% out of the actual quantity. The results of the varying parameter on the raw materials order sizes are shown in Table 3. Optimum order size in stage 1A did not change when the demand for products was increased by 10%, but there was a change when the demand was reduced by 10%, while the size of orders at stage 1B did not change. Overall, the size of the ordering lot tends to increase with increasing demand for finished goods.

Table 3. Order quantities for different demand quantity conditions

Parameter changes	Order size (Qij)						Optimum order size (Qij*)					
	Before rounding						After rounding					
	i = 1			i = 2			i = 1			i = 2		
	j = 1A	j = 1B	j = 2	j = 1A	j = 1B	j = 2	j = 1A	j = 1B	j = 2	j = 1A	j = 1B	j = 2
−10%	3.28	2.59	27	7.17	5.39	68.85	3	3	27	7	5	69
Actual	3.47	2.74	27	7.68	5.86	74.07	4	3	27	8	6	74
+10%	3.62	2.85	27	7.77	5.98	75.25	4	3	27	8	6	75
Unit	**Sacks**			**Containers**			**Sacks**			**Containers**		

4.4 Inventory Model Simulation

Inventory simulation is carried out with the following assumptions: (1) Demand for raw material i at stage j is known at the beginning; (2) If the demand for raw material i on the stage j exceeds the existing supply, the downstream will wait until the request can be fulfilled; (3) If raw materials are not used for production, it will be returned to stage 1 and can be used for the next day. As a result (Table 4), there are no stock-out units in stages 1A and 1B, it indicates that the company can fully meet the requirements for raw materials from the production floor. The stock-out unit in stage 2 did not cause stock-outs at stages 1A and 1B due to safety stock and reorder points.

Table 4. Number of stock-out units based on simulation results

Parameter	*i*	*j*	Product demand quantity			Units
			−10%	Actual	+10%	
Number of stock-out units	1	1A	0	0	0	Sacks
	1	1B	0	0	0	Sacks
	1	2	0	1	0	Sacks
	2	1A	0	0	0	Containers
	2	1B	0	0	0	Containers
	2	2	0	0	0	Containers

4.5 Service Level

Service level is determined based on the proposed centralized inventory model as well as the current company's decentralized inventory policies. Based on the results (Table 5), it is known that the centralized inventory policy provides a higher service level for each product type with a value >90%, compared to the decentralized policy.

Table 5. Service level calculation results

i	*j*	Centralized inventory policy		Decentralized inventory policy	
		Stock-out probability	Service level	Stock-out probability	Service level
1	1A	0.0649	94%	0.11	89%
1	1B	0.0782	92%	0.18	82%
2	1A	0.0343	97%	0.19	81%
2	1B	0.0442	96%	0.33	67%

5 Conclusions

Centralized raw material inventory policy with the consideration of safety stocks helps to avoid stock-out conditions in fulfilling probabilistic demand. Future research could consider the variable costs, variable lead times, cross-channel policies, with different inventory strategies regarding raw material on dual-channel warehouse.

References

1. Xie, F., Nie, Q., Yu, H.: Inventory strategy of dual-channel supply chain from manufacturer's perspective. Int. J. Manag. Fuzzy Syst. **4**(2), 29–34 (2018)
2. Alawneh, F., Zhang, G.: Dual-channel warehouse and inventory management with stochastic demand. Transp. Res. Part E **112**, 84–106 (2018)
3. Radhi, M., Zhang, G.: Optimal cross-channel return policy in dual-channel retailing systems. Int. J. Prod. Econ. **210**, 184–198 (2019)

Eye Segment Movements as Indicators of Mental Workload in Air Traffic Control Tasks

Vivi Triyanti(✉), Hastian Abdul Azis(✉), Hardianto Iridiastadi, and Yassierli

Faculty of Industrial Technology, Institut Teknologi Bandung, Jalan Ganesha 10, Bandung, Indonesia
vivi.triyanti@student.ti.itb.ac.id, hastianabdulazis@gmail.com

Abstract. Air Traffic Control job endures working processes that result in mental workloads, mainly due to the increase in air traffic. This study was aimed at evaluating mental workload among ATC controller by examining changes in facial landmarks (especially the eyes segments) that could be indicative of high workload, compared to moderate workload. Through observation of ATC's tasks in the simulator, the recording of facial movement during the experiment were analyzed by facial landmark detection software. The facial movement was average, maximum, and a number of change direction moves, both in x and y axis. Among the four analyzed points (left and right eye-brows and left and right eye outer-tips), the eyebrows segments could differentiate moderate vs. high mental workload. The average x axis movements in the in medium load is more than 2.5 units, higher than value in high load, between 1–2.5 units. Y axis movement tend to have bigger movement, above 30 units for medium load and below 25 units for high load. The outcome of this study showed that the differences in facial movements can be employed as potential indicators of workload among ATC operators.

Keywords: Facial expression · Mental workload · Air Traffic Control

1 Introduction

According to the International Civil Aviation Organization (ICAO), Air Traffic Control (ATC) is one of the jobs that require continuous monitoring of fatigue [1]. ATC has the duty to regulate the smoothness of traffic so that there is no conflict and accident of the aircraft, with reference to certain procedures. Taking into account the limitations of time, Air Traffic Controller (ATCo) must continuously regulate the flow of air traffic and simultaneously provides information and guidance to pilots and other parties [2]. The obligation to maintain performance (zero tolerance) under the complexity of the task and limited work time often causes high mental workload on ATCo [3].

For high-risk jobs that require alertness at all times such as ATC, the ability to monitor workload is fairly critical. One approach that has been widely adopted is based on behavior responses such as ocular responses [4]. The advantages of these

M. N. Osman Zahid et al. (Eds.): iMEC-APCOMS 2019, LNME, pp. 238–244, 2020.
https://doi.org/10.1007/978-981-15-0950-6_37

approaches are that they do not need particular responses from the operator (other than the body's natural responses) and can be carried out continuously. For the ocular-based approach, however, the necessity to always calibrate the device and maintain the position of the eye causes difficulties in dynamic work situations. A different real-time monitoring approach that has the potency to be developed is based on facial recognition because the setup and adjustments needed are not as complex as the previous approaches.

A number of studies have noted differences in ocular movements as a function of differences in mental workload [5, 6], including saccade and fixation. Saccade is a quick, simultaneous movement of both eyes between two or more phases of fixation in the same direction [5, 6]. Throughout the working period, an ATCo has to continuously see the air traffic flow on the screen. While observing traffic conditions, it is likely that not only the eyes would move, but also the face. Studies have also mentioned that face might give certain expression in a certain situation, such as during fatigue [7] or during certain emotions [8]. Related to ATC task, most likely the direction of head movement would be in accordance with the direction of eye movement. Therefore, if there is a change in the direction of eye movement, there is a possibility that there would be a change in the direction of movement of the head, according to the direction of the eye.

This research was a preliminary study that aimed at looking for indications of changes in the location of facial points when doing ATC tasks comprising of high vs. moderate mental workload. The lower workload was not studied because it tended to cause fatigue associated with sleepiness [9]. Sleepiness can trigger substantially different facial expressions [7, 9]. This research focused on facial changes due to only workload factors.

2 Methods

A series of experiments was conducted with the purpose of examining the relationship between workload, time on task, and facial expression, using an ATC simulator.

2.1 The Experiment

Experiments in the laboratory were intended to see whether there were differences in facial expressions associated with workloads and time on task (TOT). In this study, mental workload was associated with the level of work difficulty (task load), especially in terms of difficulty in managing traffic [2, 3]. In this simulation, the moderate load task was set as traffic with 72 aircraft per hour, using 2 runways. Meanwhile, high load task was set as traffic with 84 aircraft per hour with 1 runway. Low load task was not used, because this study focused more on the identification of facial cues associated with moderate or high mental loads. If the mental workload was low, the expression that occurs might be more related to boredom or sleepiness, not the task load itself.

Six male participants took part in this experiment; their mean age was 20.42 years with a standard deviation of 0.67 years. Each participant produced more than 2000 data in each condition. The amount of data is sufficient based on the statistical power 0.81 used in the experiment. All participants did not have any ATC experience before. All participants did not use glasses. Before the experiment, the participants had learned and tried the task of ATC simulation software until they could run the simulator with correct procedures. For each participant, the experiment consisted of 10 min of a practice session and 2 experimental sessions, which were 3 h nonstop tasks. After the first session, there was at least 1-day break period before a participant run the second session. During the experiment, a video camera recorded facial movements and appearances. Some data related to fatigue, sleepiness, and participant performance were also examined throughout the study period. But the data were not used in this paper.

2.2 Data Processing Procedure

Facial record of each experiment was divided into a series of 2-minute videos. For this research, only the last 2 min videos for each hour were used. Therefore, facial data used in this paper began at two minutes before the 60th, 120th, and 180th minutes, in moderate and high loads. Using the Open CV software, the numbers of facial landmarks were detected. Two axis coordinates were calculated from the top left corner. In this study, only 4 points around the eye were used to detect changes in position, which were a point on the right eyebrow, left eyebrow, right eye outer tip, and left eye outer tip (Fig. 1).

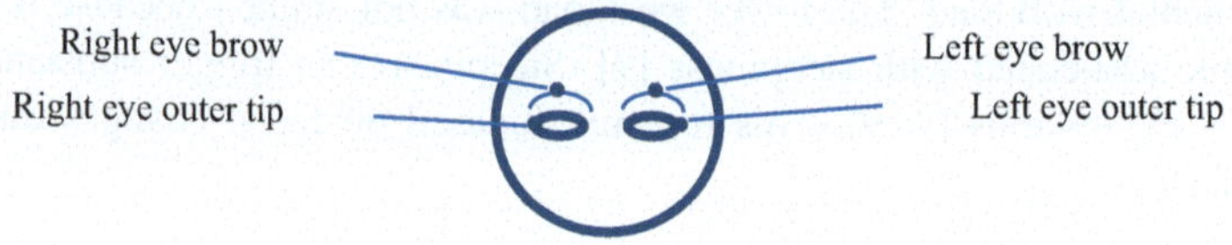

Fig. 1. Eye segments points that were analyzed

For each point, changes in location were calculated from time to time. Because software Open CV captured images every 1/10 of a second, the unit associated with the speed of change in this study is 1/10 s. Changes were calculated for the x and y axis separately to detect the vertical (down and down) and horizontal (left and right) movement patterns. The calculations were emphasized on the magnitude of the change per unit time, the maximum change, and the number of changes in direction compared to the whole movement. Some terminology used were:

1. X-axis average move: sum of eye "change direction" movement (left-to-right or right-to-left)/number of all eye x-axis movements (in pixel per 1/10 s)
2. Y-axis average move: sum of eye "change direction" movement (up-to-down or down-to-up)/number all eye y-axis movements (in pixel per 1/10 s)
3. X-axis maximum move: maximum of eye "change direction" movement (left-to-right or right-to-left)/all eye x-axis movements (in pixel per 1/10 s)
4. Y-axis maximum move: maximum of "change direction" movement (up-to-down or down-to-up)/all eye y-axis movements (in pixel per 1/10 s)
5. Normalized number of x-axis move: count of eye "change direction" movement (left-to-right plus right-to-left)/all eye x-axis movements

6. Normalized number of y-axis move: count of eye "change direction" movement (up-to-down and down-to-up)/all eye y-axis movements

Data processing was then carried out using 2-way ANOVA. Related factors were task demand (moderate vs. high), time on task or TOT (60', 120', 180'), and the interactions.

3 Result

ANOVA results showed that between the two factors (workload and TOT), only workload factor distinguished most of the dependent variables tested. The results of ANOVA related to workload factors are presented in Tables 1 and 2.

Table 1 ANOVA results for eyebrows and eye-tip points

Parameter	Right eye brow			Left eye brow			Right eye tip			Left eye tip		
	WL	TOT	Int.	WL	TOT	Int.	WL	TOT	Int.	WL	TOT	Int.
Avg. value of right-to-left head movement (x-axis)	**0.02**	0.69	0.80	0.79	0.39	0.19	0.09	0.39	0.45	0.13	0.40	0.47
Avg. value of left-to-right head movement (x-axis)	**0.02**	0.80	0.53	0.98	0.44	0.19	0.09	0.46	0.48	0.10	0.46	0.51
Avg. value of down-to-up head movement (y-axis)	0.09	0.61	0.83	0.52	0.45	0.33	0.13	0.76	0.73	0.11	0.65	0.91
Avg. value of up-to-down head movement (y-axis)	**0.04**	0.78	0.69	0.60	0.48	0.30	0.60	0.48	0.30	0.07	0.59	0.99
max. right-to-left head movement value (x-axis)	0.06	0.05	0.82	**0.04**	0.35	0.34	0.09	0.21	0.45	0.09	0.25	0.39
max. left-to-right head movement value (x-axis)	0.11	0.20	0.14	**0.03**	0.68	0.68	0.16	0.49	0.72	0.16	0.66	0.54
max. bottom-to-up head movement value (y-axis)	**0.04**	0.31	0.44	0.63	0.54	0.24	0.06	0.20	0.30	0.11	0.35	0.47
max. up-to-bottom head movement value (y-axis)	**0.04**	0.28	0.40	0.78	0.43	0.48	0.16	0.16	0.89	0.33	0.22	0.54
Norm. Number of change direction (in x-axis)	0.17	0.78	**0.03**	0.92	0.47	0.29	0.86	0.47	0.21	0.96	0.47	0.23
Norm. Number of change direction (in y-axis)	0.09	0.53	0.15	0.48	0.58	0.70	**0.04**	0.63	0.86	0.17	0.51	0.87

Note: WL = workload, TOT = Time on Task, Int. = interaction between WL and TOT

Based on a 2-way ANOVA, it was concluded that the eyebrows were more sensitive than eye-tips in detecting changes in facial movements, associated with differences in workload. This indicates that the movement in the eyebrows can differ workload conditions. By using a 0.05 level of significance, eyebrows changed more facial patterns, especially related to the average and maximum magnitude of the movement value when changes in direction occurred (the unit is pixels per 1/10 s). Among the four analyzed points (left and right eye-brows and left and right eye outer-tips), the eyebrows segments could differentiate moderate vs. high mental workload. Table 2 shows that the average x

Table 2. Means of each level for significant variables related to workload

Parameter	Right eye brow			Left eye brow		
	F	Means		F	Means	
		Med.	Med.		Med.	Med.
Avg. value of right-to-left head movement (x-axis)	**0.02**	−2.56	−1.85	0.79	−2.56	−2.45
Avg. value of left-to-right head movement (x-axis)	**0.02**	2.71	1.94	0.98	2.62	2.64
Avg. value of down-to-up head movement (y-axis)	0.09	−2.66	−2.27	0.52	−2.37	−3.16
Avg. value of up-to-down head movement (y-axis)	**0.04**	2.75	2.22	0.60	2.45	3.03
max. right-to-left head movement value (x-axis)	0.06	−35.78	−18.06	**0.04**	−2.56	−49.78
max. left-to-right head movement value (x-axis)	0.11	43.22	24.94	**0.03**	2.62	49.94
max. bottom-to-up head movement value (y-axis)	**0.04**	−34.00	−20.44	0.63	−27.22	−33.72
max. up-to-bottom head movement value (y-axis)	**0.04**	35.39	20.44	0.78	27.11	31.33
Norm. Number of change direction (in x-axis)	0.17	0.18	0.24	0.92	0.23	0.22
Norm. Number of change direction (in y-axis)	0.09	0.14	0.31	0.48	0.20	0.24

axis movements in the in medium load is more than 2.5 units, higher than value in high load that are between 1–2.5 units. Y axis movement tend to have bigger movement, above 30 units for medium load and below 25 units for high load. Bigger movement of Y axis mainly because the participant tens to move closer to screen occasionally.

When one of the data shown in graphical form, it can be seen that the average value of the x-axis head movement on high workload was smaller than the moderate workload. Meanwhile, TOT did not show a consistent change in movement patterns (Fig. 2).

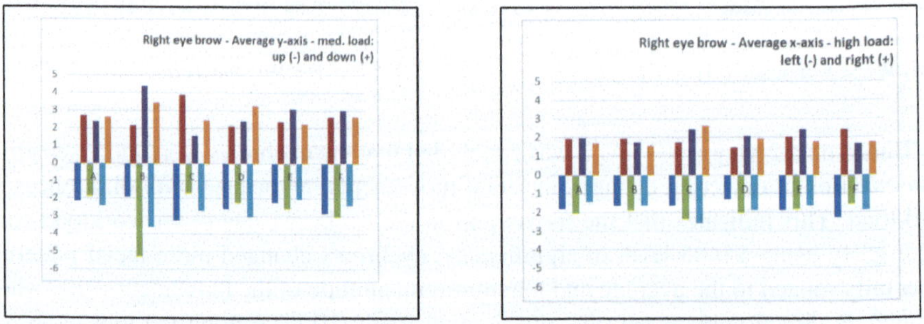

(a) Moderate load (b) High load

Fig. 2. Average movement change (pixels per unit time) of right eyebrow in x-axis direction

4 Analysis

Of the 3 factors analyzed (workload, TOT, and their interactions), only the workload factor that consistently distinguished the facial variables value. The eyebrow segment played a major role in detecting this difference. On the right eyebrow, the workload distinguished facial variables related to the average magnitude of changes in direction per unit time (towards the x and y axis) and the maximum value changes in direction per unit time (toward the y axis). Meanwhile, the change in workload also affected the variable number-of-direction-changes (x axis) on the left eyebrow (Table 1). The eye-tip segment was not sensitive to identify differences in factor's levels. Only normalized number of changes (y axis) which can be concluded to be related to the right eye-tip.

When seen in Table 1, it is interesting to see that the three of four observed points turned out to be complementary in indicating facial changes in all dependent variables, especially those related to changes in workload. Each variable was only significantly related to 1 point of certain facials. Therefore, it was suggested that all points should be used together to detect a change in workload. So that the analysis of the movement of facial points is not only done partially.

With reference to Table 2, almost all variables, high workload caused lower value than the moderate workload. This is consistent with changes in ocular response. Previous studies concluded that saccade duration increased at high workloads [4, 5]. In this study, a smaller value states the value of change per unit of time was smaller. This meant that the duration of time needed to reach a certain point was greater.

The limitations of this research were mainly on the limited number of participants, making it difficult to generalize the results of this study. In addition, in future studies, other facial points that may be more sensitive will be examined. In addition, some parameters for other facial changes will also be examined, especially those related to the duration and speed of movement, as well as the distance between points on the face.

This research is expected to be the first step in developing a system for detecting mental workload experienced by workers, facial recognition-based, especially in the work of Air Traffic Control. Because of the non-intrusive nature of the approach, this approach has the potential to be developed into a real-time monitoring device that can be used on a real system.

5 Conclusion

Based on a series of data processing and analysis, it was concluded that facial movement has the potency to be indicators of high mental workload, compared to moderate mental workload. Since there is a possibility of subjects to nod, eyebrows segment was proven more sensitive in detecting the facial cues differences, compares to eye segment.

Acknowledgement. We thank Direktorat Pendidikan Tinggi Indonesia for funding this research. We also thank our colleagues from Institut Teknologi Bandung and Atma Jaya Catholic University of Indonesia who provided insight and expertise that greatly assisted the research.

References

1. International Civil Aviation Organization (ICAO): Fatigue Management Guide for Air Traffic Service Providers, 1st edn. International Civil Aviation Organization (2016)
2. Gawron, V.J., Kaminski, M.A., Serber, M.L., Payton, Hadjimichael, M., Jarrott, W.M., Estes, S.L., Neal, T.A.: Human performance and fatigue research for controllers revised. MTR Mitre Technical Report, Federal Aviation Administration (2011)
3. Tobaruela, G., Schuster, W., Maajumdar, A., Ochieng, W.Y., Martinez, L., Hendrickx, P.: A method to estimate air traffic controller mental workload based on traffic clearances. J. Air Transp. Manag. **39**, 59–71 (2014)
4. Di Stasi, L.L., Canas, J.J., Antoli, A.: Main sequence: an index for detecting mental workload variation in complex tasks. Appl. Ergon. **42**(6), 807–813 (2011)
5. Marchitto, M., Benedetto, S., Baccino, T., Canas, J.J.: Air traffic control: ocular metrics reflect cognitive complexity. Int. J. Ind. Ergon. **54**, 120–130 (2016)
6. Benedetto, S., Carbone, A., Drai-Zerbib, V., Pedrotti, M., Baccino, T.: Effects of luminance and illuminance on visual fatigue and arousal during digital reading. Comput. Hum. Behav. **41**, 112–119 (2014)
7. Triyanti, V., Iridiastadi, H.: Challenges in detecting drowsiness based on driver's behavior. In: IOP Conference Series: Material Science and Engineering, vol. 277, no. 012042, pp. 1–8. IOP Publishing (2017)
8. Triyanti, V., Yassierli, Iridiastadi, H.: Basic emotion recognition using automated facial expression analysis software. Jurnal Optimasi Sistem Industri **18**(1), 55–64 (2019)
9. Jabon, M.E., Bailenson, J.N., Pontikakis, E., Takayama, L., Nass, C.: Facial-expression analysis for predicting unsafe driving behavior face and gesture recognition. In: Pervasive computing, October–December 2011

Mixed Model Assembly Line Balancing for Human-Robot Shared Tasks

Susanto Yaphiar(✉), Cahyadi Nugraha, and Anas Ma'ruf

Bandung Institute of Technology, Bandung 40132, Indonesia
susanto.yaphiar@gmail.com

Abstract. This paper presents a new mathematical model for mixed model assembly line balancing that includes human-robot collaboration. Assembly line with human-robot collaboration means that each workstation may consist of human only as an operator, a robot only, or a human and robot which work simultaneously. The proposed mathematical model will be solved with a mixed-integer linear programming model to minimizes the total relevant cost by assigning tasks to workstations and determine which kind of resources (human, robot, or human-robot collaboration) will be placed on workstations to produce various kinds of products. The type of assembly line is based on a simple assembly line: straight line, deterministic time, and mixed product variant model.

Keywords: Mixed model assembly line balancing · Human-robot collaboration · Shared tasks · Mathematical model

1 Introduction

Nowadays, a lot of industry needs to adjust to environmental changes so they could maintain their existence. This is due to the increasing competitiveness between companies so that the method is needed to make the company can compete and maintain its existence. The challenge faced by the industry is the increasing level of product diversity as well as product customization by consumers that need to be met with low production time [1]. The challenge can be overcome by interacting with humans and robots working together. Robots can be utilized to do specific and repetitive work automatically, and humans take action in determining decisions or interventions in a work station [1]. So the problem that needs a change in the production process every certain period can be overcome by the presence of humans who can process changes and provide jobs for robots to do specific tasks.

Collaboration between humans and robots is possible because technology has developed with the existence of sensors so that robots can work with humans without harming humans [2]. In contrast to robots that were previously placed in a special area and there shouldn't be a human around it because robot can cause accidents for workers. Human-robot collaboration (HRC) is a new advancement in human-machine interaction. HRC is one of the characteristic of technologies in the fourth industrial revolution era, which is popularly known as Industry 4.0. The aim of HRC is to get the best performance from the production system, by simultaneously utilizing human

M. N. Osman Zahid et al. (Eds.): iMEC-APCOMS 2019, LNME, pp. 245–252, 2020.
https://doi.org/10.1007/978-981-15-0950-6_38

advantages, such as flexibility, dexterity, and high-level intelligence, as well as the advantages of robots, such as accuracy, strength, and consistency [2].

Even though the technology is still in its infancy, some automotive companies and universities have already developed applications with HRC [3]. One of the potential uses of HRC that has begun to be implemented is in assembly production systems [4]. One important problem for implementing HRC in assembly systems is the decision to allocate tasks between humans and robots.

This paper discusses the allocation of work in one work station, and generally assume that the manual work station already exists and will be upgraded to HRC. Work allocation at a work station on one production line also considers the resources to be used related to production and investment costs. Work allocation is intended so that the production line works in a balanced manner. However, there are currently no researches about the allocation of human-robot shared tasks in an assembly line consisting of several work stations. Research related to the allocation of collaborations between humans and robots was found only in a work station [5]. Indeed, research on assembly line balancing has been very much developed [6], but none has been related to human-robot shared tasks. This paper proposes a new mathematical model to optimize the assembly line that implements human-robot shared tasks for a mixed model of the product due to current demand characteristic based on the mathematical formulation of balancing mixed-model two-sided assembly line balancing [7].

2 Balancing of Mixed-Model Assembly Lines with HRC

2.1 Problem Definition

A mixed-model assembly line balancing for human-robot collaboration (HRC) shared tasks is designed to carry out a set of product models ($\forall m \in M$) with similar production characteristics in any model sequence and model mix. Each model has its own set of task precedence relationships, and they can be combined into a single precedence diagram. The tasks ($\forall i \in I$) in the combined precedence diagram for all models are performed on a set station ($\forall j \in J$). A task i on a model m is performed in a certain time (t_{ism}) with one particular resource s (between only humans H, only robots R, or human-robot collaboration (HRC)). The product models are produced over a pre-specified planning horizon, PH. The demand, over the planning horizon, for model m is Dm. The required cycle time (τ) of the line for this problem is calculated by Eq. (1), and the overall proportion of the number of units of model m (q_m) is computed by Eq. (2) based on Simaria and Vilarinho (2007) in [7].

$$\tau = \frac{PH}{\sum_{m \in M} . D_m} \tag{1}$$

$$q_m = \frac{D_m}{\sum_{m \in M} . D_m} \quad \forall m \in M \tag{2}$$

The assembly line in this paper has characteristics such as straight line, single-sided, and workstation only using on resources alternative (H/R/HRC).

2.2 Problem Assumptions

The mathematical model considered in this paperwork under the following assumptions:

- Product models with similar production characteristics are produced on the same assembly line.
- Precedence diagrams of different models are known. The combined precedence diagram concept of Macaskill (1972) in [7] is employed.
- Task times are deterministic and independent of the assigned station.
- Common tasks among different models exist. A task completion time may differ from one model to another, and also it may be equal to zero.
- The travel times of operators are ignored.
- No work-in-process inventory is allowed. All the parameters are deterministic.
- There is only one type of robot considered in the design of the assembly line.
- Setup time is negligible or implicitly included in the processing time.

2.3 Notation

The notations used in the mathematical formulations are given as follows.

Indices:

i, h, p	task
j, q	workstation
s	human-robot task-sharing resource-type: 1 = human only, 2 = robot only, 3 = human-and-robot simultaneously
m	product model

Parameters:

I	Set of tasks; $I = \{1, 2, \ldots, n_T\}$
J	set of workstations; $J = \{1, 2, \ldots, n_W\}$
P	set of pairs (h, i) where task h is a direct predecessor of task i
Q	set of pairs (i_1, i_2) where task i_1 and i_2 have no direct-nor-indirect precedence relationship
t_{ism}	process time of task i if performed by resource-type s for model m
c_{I1}	investment cost of a human operator (recruitment, training, etc.)
c_{I2}	investment cost of a robot
c_{O1}	operational cost per unit time of a human operator
c_{O2}	operational cost per unit time of a robot
b_{2i}	quality and or ergonomic-related benefit (in currency), throughout the planning horizon, for using a robot in task i
b_{3i}	quality and or ergonomic-related benefit (in currency), throughout the planning horizon, for using human-robot collaboration in task i
τ	cycle time (takt time) requirement for the assembly line

ψ a very large positive number
W_{jm} subset of all task that can be assign to station j of model m
$||W_{jm}||$ number of task in subset W_{jm}
M set of product models; $M = \{1,2,...,n_m\}$
$||M||$ number of model
D total number of demand from all models

Decision variables:
x_{ijs} 1, if task i is assigned to workstation j and performed by resource-type s; 0, otherwise
f_{im} finish time of task i for model m
z_{ip} sequencing position for the pairs of tasks $(i, p) \in Q$ if i and p are assigned in the same workstation: 1, if task i is performed before task p; 0, if task p is performed before task i
v_{jm} 1, if station j is utilized for model m; 0, otherwise
U_j utilization of station j in the line: 1, if station j is utilized; 0, otherwise

Indicator variables:
α Number of human operators needed in the line
ρ number of robots needed in the line.

2.4 Mathematical Formulation

Ozcan et al. [7] developed a mathematical model for the Mixed Model Two-sided Assembly Line Balancing (MTALB-I). The primary objective of the model is minimizing the number of mated-stations and the number of stations. In this study, the mathematical model formulation MTALB-I problem is modified for MALB-I problem with consideration resources (human, robot, or HRC shared tasks) for given cycle time. The proposed mathematical model formulation for this problem is as follows:

Objective function

$$\text{Minimize } TC = c_{I1}.\alpha + c_{I2}.\rho + (c_{O1}.\alpha + c_{O2}.\rho).D.\tau - \sum_{i=1}^{n_T}\sum_{j=1}^{n_W} b_{2i}x_{ij2} - \sum_{i=1}^{n_T}\sum_{j=1}^{n_W} b_{3i}x_{ij3} \tag{3}$$

where:

$$\alpha = \sum_{j=1}^{n_W}\left[\min\left\{1, \left(\sum_{i=1}^{n_T}(x_{ij1} + x_{ij3})\right)\right\}\right] \tag{4}$$

$$\rho = \sum_{j=1}^{n_W}\left[\min\left\{1, \left(\sum_{i=1}^{n_T}(x_{ij2} + x_{ij3})\right)\right\}\right] \tag{5}$$

Subject to:

$$\sum_{j=1}^{n_W}\left(\sum_{s\in\{1,2,3\}} x_{ijs}\right) = 1 \qquad \forall i \in I \tag{6}$$

$$\sum_{s\in\{1,2,3\}} x_{ijs} \le \sum_{q=1}^{j}\left(\sum_{s\in\{1,2,3\}} x_{hqs}\right) \qquad \forall j \in J,\ \forall h,\ i \in P \tag{7}$$

$$f_{im} \le \tau \qquad \forall i \in I,\ \forall m \in M \tag{8}$$

$$f_{im} \ge \sum_{j=1}^{n_W}\left(\sum_{s\in\{1,2,3\}} t_{ism}x_{ijs}\right) \qquad \forall i \in I,\ \forall m \in M \tag{9}$$

$$f_{im} - f_{hm} + \psi\left(1 - \sum_{s\in\{1,2,3\}} x_{ijs}\right) + \psi\left(1 - \sum_{s\in\{1,2,3\}} x_{hjs}\right) \ge \sum_{s\in\{1,2,3\}} t_{ism}x_{ijs}$$
$$\forall j \in J,\ \forall h,\ i \in P \tag{10}$$

$$f_{im} - f_{pm} + \psi\left(1 - \sum_{s\in\{1,2,3\}} x_{ijs}\right) + \psi\left(1 - \sum_{s\in\{1,2,3\}} x_{pjs}\right) + \psi z_{ip} \ge \sum_{s\in\{1,2,3\}} t_{ism}x_{ijs} \tag{11}$$

$$f_{pm} - f_{im} + \psi\left(1 - \sum_{s\in\{1,2,3\}} x_{ijs}\right) + \psi\left(1 - \sum_{s\in\{1,2,3\}} x_{pjs}\right) + \psi(1 - z_{ip}) \ge \sum_{s\in\{1,2,3\}} t_{psm}x_{pjs}$$
$$\forall j \in J,\ \forall i,p \in Q \tag{12}$$

$$\sum_{i\in I} x_{ijs} - |W_{jm}||v_{jm} \le 0 \qquad \forall j \in J,\ \forall m \in M \tag{13}$$

$$\sum_{m\in M} v_{jm} - ||M||U_j \le 0 \qquad \forall j \in J \tag{14}$$

$$U_j \le U_{j-1}k \qquad j = 2,\ldots,n_w \tag{15}$$

$$where\ U_j \le \min\left\{1, \sum_{i=1}^{n_T}\sum_{1,2,3} x_{ijs}\right\} \qquad \forall j \in J \tag{16}$$

$$x_{ijs} \in \{0,1\} \qquad \forall i \in I,\ j \in J,\ s = 1,2,3 \tag{17}$$

$$z_{ip} \in \{0,1\} \qquad \forall i \in I,\ p \in P \tag{18}$$

$$U_j \in \{0,1\} \qquad \forall j \in J \tag{19}$$

$$v_{jm} \in \{0,1\} \qquad \forall j \in J,\ m \in M \tag{20}$$

The objective function (3) minimizes the total cost of investment costs and operational costs. Operational costs are calculated by multiplying the number of resources with the cost of resources in units of time multiplied by the number of units produced with the cycle time. Investment costs include the cost of purchasing machinery for use during the planning period as well as the costs of recruiting and training workers.

Operational costs include energy costs for working machinery (such as electricity) and labor costs. Total costs also include savings from using robots or collaboration between humans and robots. Equations (4) and (5) show that each station can do several jobs but can only use one type of the same resource that will be used if several tasks are allocated in one work station using the same resources. Equation (4) limits that a work station consists of only one human worker even though it consists of several jobs, and it is possible that humans do some of the work. In Eq. (4) also consider work involving collaboration between humans and robots. This is because collaboration between humans and robots also include humans, so that is also considered in this equation. The same thing applies to Eq. (5) which limits the number of robots at work stations to one.

Constraint (6) is the assignment constraint, which ensures that each task is assigned to exactly one station as well as using one kind of resource. Constraint (7) is the precedence constraint, which ensures that all precedence relations among tasks are satisfied. Constraint (8) and (9) are the cycle time constraint, which ensures that each of the finish time of tasks for each model ($\forall m \in M$) doesn't exceed the cycle time.

Constraints (10)–(12) control the sequence-dependent finishing time of tasks for each model m ($\forall m \in M$). For every pair of tasks i and h, if task h is an immediate predecessor of task i, and if they are assigned to the same station j, then the constraint (10) becomes active. If two tasks do not have any precedence relations, and if they are assigned to the same station (j), then the constraints (11)–(12) become active. Constraints (13) and (14) are station constraints which ensure that the number of stations is same for all product models. Constraints (15)–(16) are used so that task assignments at work stations are carried out on consecutive indexes. Constraints (17)–(20) are the integrality constraints.

3 Numerical Experiment and Discussion

Numerical testing is carried out to consider and validate the resulting model using optimization software. Tests are carried out using data that makes a model for selecting robotic resources using very little production time and low investment and operating costs. The same thing is done so that the model decides only human resources or human-robot collaboration for all work stations. After using data that makes the model choose one of the resources, another data is used so that the model will choose resources based on cycle times and costs of each resource. Based on the test results, the model will meet the cycle time and then select the resources that produce the lowest total cost. Here is one example of data used in the experiment. Figure 1 shows the precedence diagram in the experiment.

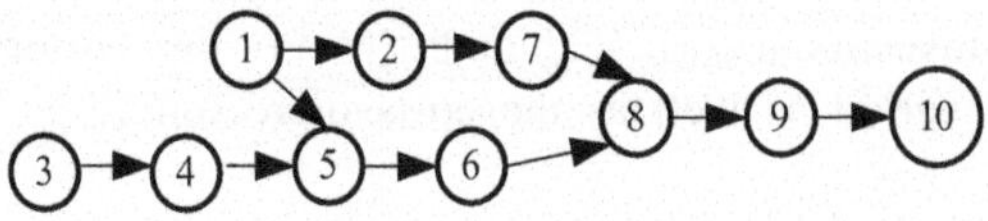

Fig. 1. Precedence diagram of example [8]

The experiment was conducted to make 2000 total products (D) consisting of 2 product models ($\|M\|$) with 10 task (n_t) and cycle times of 20 s (τ). Investment costs for humans (C_{I1}) and robots (C_{I2}) are Rp10 million and Rp 1 billion respectively while operational costs for humans (C_{O1}) and robots (C_{O2}) are Rp 4/s and Rp 1/s respectively. Benefit throughout the planning horizon for using a robot (b_{2i}) and human-robot collaboration (b_{3i}) in every task i are Rp 10 million and Rp 20 million respectively. Time data can be seen in Table 1. Mathematical models produce an arrangement of 4 work stations with a total cost of Rp. 2,930,600,000.

Table 1. Time data (in seconds) for each task of each product model with each resource

Task-		1	2	3	4	5	6	7	8	9	10
Human	Model-1	30	30	30	0	8	9	30	40	7	3
	Model-2	25	40	50	50	4	5	40	20	9	2
Robot	Model-1	40	30	30	0	30	40	7	8	30	30
	Model-2	30	40	50	50	40	50	4	2	50	20
Human-robot collaboration	Model-1	8	10	3	0	30	40	70	40	30	30
	Model-2	12	6	5	5	40	50	40	20	40	20

4 Conclusion

In this paper, we presented a new mathematical model for solving the mixed-model assembly line balancing problem that includes human-robot collaboration to minimizing total cost. The results of the numerical experiments show the program produces good results with short computational times (1 s). Development of mathematical models based on the proposed mathematical model considering another aspect of collaborative robot characteristics should be interest in further studies.

References

1. Wang, X.V., Kemény, Z., Váncza, J., Wang, L.: Human–robot collaborative assembly in cyber-physical production: classification framework and implementation. CIRP Ann. Manuf. Technol. **66**(1), 5–8 (2017)
2. Tsarouchi, P., Makris, S., Matthaiakis, A.S., Chryssolouris, G.: On a human–robot collaboration in an assembly cell. Int. J. Comput. Integr. Manuf. **29**(8), 916–931 (2016)
3. Vysocky, A., Novak, P.: Human-robot collaboration in industry. MM Sci. J. **1**, 903–906 (2016)
4. Michalos, G., Makris, S., Papakostas, N., Mourtzis, D., Chryssolouris, G.: Automotive assembly technologies review: challenges and outlook for a flexible and adaptive approach. CIRP J. Manuf. Sci. Technol. **2**(1), 81–91 (2010)
5. Takata, S., Hirano, T.: Human and robot allocation method for hybrid assembly systems. CIRP Ann. Manuf. Technol. **60**(1), 9–12 (2011)

6. Sivasankaran, P., Shahabudeen, P.: Literature review of assembly line balancing problems. Int. J. Adv. Manuf. Technol. **73**(9–12), 1665–1694 (2014)
7. Ozcan, U., Toklu, B.: Balancing of mixed-model two-sided assembly lines. Comput. Ind. Eng. **57**(1), 217–227 (2009)
8. Scholl, A.: Balancing and sequencing of assembly lines, 2nd edn. Physice-Verlag, Heidelberg, New York (1999)

Project Manufacture Scheduling Using Resource Constrained Multi Project Optimization Model (A Case Study in Machine Manufacturing Company Solo)

Seamus Tadeo Marpaung, Cucuk Nur Rosyidi(✉), and Wakhid Ahmad Jauhari

Industrial Engineering Department, Universitas Sebelas Maret, Jalan Ir. Sutami 36 A, Surakarta, Indonesia
seamusmarpaung@gmail.com, cucuk@uns.ac.id, wakhidjauhari@gmail.com

Abstract. The machine manufacturing company specialized in building customized (engineered-to-order) machines ordered for certain purposes. The company faces a problem in which it can not be able to finish the received orders within the estimated duration. In the last 5 years, the percentage of production delay in manufacturing the engineered-to-order machines is 80%. The inability to meet the production due date is caused by the company does not have the right scheduling mechanism to schedule every activity in production. Due to the similarity of characteristics, the production of customized machines can be considered as a project. This paper aims to find an efficient yet effective way to schedule project manufacture and allocate the resources for such projects. To serve this purpose, the resource constrained multi project optimization model was applied in a real case study. The objective function of the model is to minimize the completion time of the projects when run simultaneously. The resource constrained multi project optimization model was solved using LINGO 15.0 and shows that the projects can be completed in 133 days. This result will make the company to save IDR 17,200,000 compared to the existing project schedule.

Keywords: Project manufacture · Scheduling · Optimization model · Case study

1 Introduction

Project manufacturing aims to produce and assemble certain unique product which must be done in certain available time. Although the activities are conducted in manufacturing environment, the activities follow the definition of the project of being temporary and unique [1]. ETO or project based manufacturer is usually called "custom" manufacturer. This term refers to manufacturers that produce products that are

M. N. Osman Zahid et al. (Eds.): iMEC-APCOMS 2019, LNME, pp. 253–259, 2020.
https://doi.org/10.1007/978-981-15-0950-6_39

unique and often complex since the production must be started from the design phase. Researcher listed some challenges faced by custom manufacturers, some of these are the facts that customer is given the full authority to determine the design, change of priorities among customers make the schedule to stop and resume several times during the timespan, and the high number of components that are needed for the assembly of a single sub-product [2].

This research was taken in PT. ATMI, a machine manufacturing company in Solo Indonesia. The problem that occurs in the machine manufacturing company is the delay in the completion of the project. Based on company data, from 2013–2017 in average they received 12 projects per year with the average lateness percentage of 80%. The inability to complete the project happened because the company does not have any appropriate scheduling methods with their production characteristics yet. In general, a machine manufacturing project has three stages, namely design engineering stage, procurement stage, and manufacturing stage. Most of the project time duration is used mainly on the first two stages, leaving the last stage very little time to execute. The company received multiple orders simultaneously. Those simultaneous orders give other difficulties as the company does not have any method to optimally allocate resources in order to meet all the due dates.

According to previous study, engineered-to-order production systems and projects have many similarities, namely both are temporary, have certain due dates, and have price that have been agreed upon from the start [3]. According to its main activities, a project can be divided into several types, namely construction projects, manufacturing industry projects, and research and development projects [4]. The activities carried out by PT. ATMI Solo is the type of manufacturing industry project, because the machine production process starts from the design engineering stage (Suharto, 1995). Because of the similarity of characteristics between machine production activities at PT. ATMI Solo with the project, the scheduling method used in this study will use the project management approach.

The research on resource constrained project scheduling was initiated [5–7]. Bowman used the binary decision variable to determine the schedule. Manne [7] formulated an integer variable to schedule the beginning and ending activities, while Wagner [6] develop a binary optimization model to allocate certain resources to a certain work activities in a certain time sequence. Pritsker, Watters, and Wolfe formulated boundaries and variables from previous studies to develop resource constrained multi-project optimization model with binary programming approach [8]. In this paper, we apply the model of Pritsker et al. [8] to solve the problem faced by PT. ATMI. There are several objectives to be achieved, but this study will focus on minimizing the total throughput time for all projects. The purpose of the optimization model in this research is to determine the schedule of each activity in the three machine manufacturing projects carried out by PT. ATMI to minimize the duration of the projects when run simultaneousy.

This research aims to determine the optimal project schedule and resource allocation using a resource constrained multi project scheduling optimization model. The generated schedule is expected to minimize the project cost and duration. Later, the optimal resource allocation will help the company to consider whether they will take specific project. The result from the model give the staff valuable information about their ability to finish the project within due date. If the result show that the company will not be able to finish the project within the given date, then the company will have two choices. The first is to turn down the offer and the other option is to add resource in order to meet the project duration.

2 Methodology

The first step in this research is to conduct a field study by conducting interviews with PPIC and marketing staff of PT. ATMI Solo to find out the machine production process. In addition, this field study is used to identify several constraints that must be considered in the scheduling process. To support the field study, literature reviews were carried out to determine the appropriate model that can be used to solve the problems. The next step is data collection which performs by collecting the needed data in the form of bill of material from the three machines that are used as objects of case studies, production sheets, types and the amount of resources needed in each project, and the company's budget to carry out the activities. Interviews are conducted to obtain the information on business processes and machine production flow, mainly about the activities involve in the project, the duration of each activity and the precedence relation among activities within project. The information are provided by the PPIC, Marketing, and Engineering division staffs. The estimated activity duration given is the most likely duration. The last step is to translate the data into the optimization model and run the model using the LINGO 15.0 software and analyze the results.

3 Results and Discussion

Based on the problem setting and data, the mathematical model for the resource constrained multi project scheduling can be described as follows [8]:

The Objective Function

The objective function of the model is to minimize the time starting from the last activity.

$$Min \sum_{t=1}^{T} t \times Xijt \tag{1}$$

Constraints

1. Every activity in each ETO engine manufacturing project must be completed.

$$\sum_{t=1}^{T} Xijt = 1 \quad (2)$$

2. This constraint is used for activities that cannot be started until one or several other activities are completed. For example, activity m must precede activity n, then $t_{im} + d_{in} \leq t_{in}$.

$$\sum_{t=l_{im}}^{u_{im}} tX_{imt} + d_{in} \leq \sum_{t=l_{in}}^{u_{in}} tX_{int} \quad (3)$$

3. This constraint states that activities in a project cannot be started if the resources needed to carry out activities are not available. The r_{ijk} represents the number of type k resources needed to carry out j activities on the project i.

$$\sum_{i=1}^{I} \sum_{j=1}^{J} \sum_{k=1}^{K} r_{ijk} x_{ijt} \leq R_{kt} \quad (4)$$

In this research, we use three machine manufacturing projects, namely Printability Tester Machine, Punch Hosti Machine, and Mold Hosti Machine. The precedence relation of activities in each project and time duration of such activities are shown in Fig. 1, while Table 1 shows the number of resources to run the projects.

The multi project scheduling model was run using LINGO 15.0 software on a computer with an Intel Core i3-4005U processor (1.7 GHz, 3 MB L3 cache) and 2 GB RAM. After running the model, the optimum global results are obtained after 14120630 iterations in 4 h, 45 min, and 46 s with an objective value of 133 days. Table 2 shows the estimated costs resulting from the optimization model using resource constrained multi-project scheduling and actual production that has been done by the company. If the company schedules the projects using the multi-project scheduling model, they can save IDR. 17,200,000. The delay in production causes the company to incur additional costs, because the company has to pay the personnel wages and more operating costs than estimated.

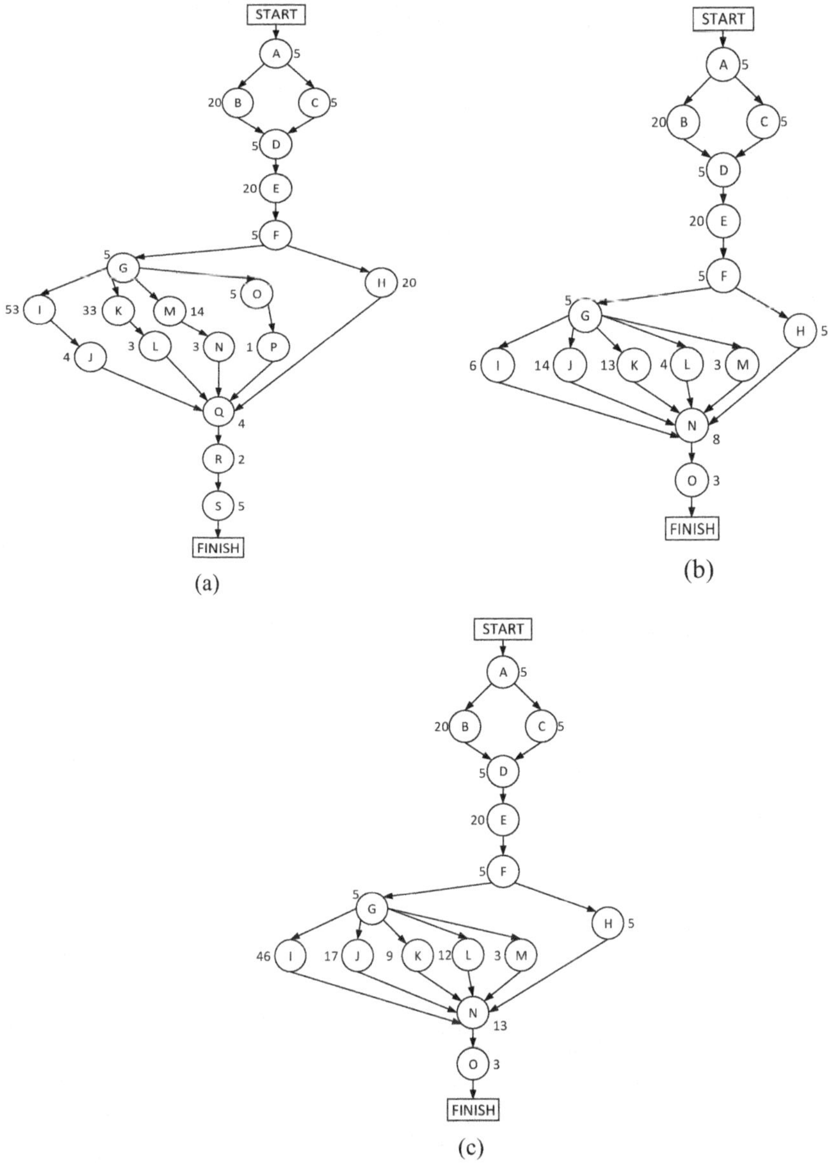

Fig. 1 Activity network diagram of (a) Printability Tester Machine, (b) Punch Hosti Machine, and (c) Mold Hosti Machine

Table 1. The number of resources to run the project

Code	Resource	Amount	Code	Resource	Amount
M1	Welding machine	8	O3	Cutting operator	6
M2	Bending machine	5	O4	Drilling operator	3
M3	Cutting machine	6	O5	Turning operator	6
M4	Drilling machine	3	O6	Hardening operator	3
M5	Turning machine	6	O7	Milling operator	6
M6	Hardening machine	3	O8	Punching operator	9
M7	Milling machine	6	O9	Grinding operator	8
M8	Punching machine	9	O10	Welding operator	8
M9	Grinding machine	8	O11	Engineering staff	7
M10	Welding machine	8	O12	PPIC staff	7
O1	Welding operator	8	S1	Supervisor	10
O2	Bending operator	5			

Table 2. Project cost and duration comparison

Project	Cost (IDR)		Duration (days)	
	Actual	Optimization	Actual	Optimization
Printability tester	74,976,732	82,326,732	133	152
Punch Hosti	35,088,003	39,688,003		105
Mold Hosti	58,774,983	64,024,983		148

4 Conclusion

The optimization resulted a schedule with time duration of 133 days. This schedule was better than the actual scheduling by the company, where each project was completed in 152 days, 105 days and 148 days. With optimum scheduling, the company saved costs of Rp 17,200,000. The objective of the research is to find an appropriate method for scheduling several ETO engine manufacturing projects that are run simultaneously. The resource constrained multi-project scheduling optimization model method is suitable for achieving this goal. The model also results in the allocation of resources so that the company also has a schedule of resources usage for each activity in each project. The next advantage is that companies can use the model optimization results to consider whether they will take the project offered or not by calculating the duration of the project in accordance with the resources owned by the company. Companies can consider increasing the resources they have so they can take on certain projects. Then, companies can anticipate problems such as machine damage and problems with personnel from afar because they know when certain resources are needed.

References

1. El-Mehalawi, M.: Scheduling and controls of project manufacturing. In: Proceedings of 8th Annual PMI College of Scheduling Conference, San Fransisco, US, pp. 1–7 (May 2011)
2. Fox, S., Jokinen, T., Lindfors, N., Ylen, J.P.: Formulation of robust strategies for project manufacturing business. Int. J. Manag. Proj. Bus. **2**(2), 217–237 (2008)
3. Taghizadeh, H., Zeinalzadeh, A.: Application of project management approaches for planning in engineer-to-order production. Middle-East J. Sci. Res. **10**(5), 581–591 (2011)
4. Dimyati, H., Nurjaman, K.: Project Management, 1st edn. (in Bahasa). Pustaka Setia, Bandung (2014)
5. Bowman, E.H.: The schedule-sequencing problem. Oper. Res. **7**, 621–624 (1959)
6. Wagner, H.: An integer linear programming model for machine scheduling. Naval Res. Logistics Q. **6**, 131–140 (1959)
7. Manne, A.S.: On the job-shop scheduling problem. Oper. Res. **8**, 219–223 (1960)
8. Pritsker, A.A.B., Watters, L.J., Wolfe, P.M.: Multiproject scheduling with limited resources: a zero-one programming approach. Manag. Sci. **16**, 93–108 (1969)

Scheduling an Aircraft Maintenance Shop with Dedicated Technician and Dedicated Machine Constraints

Wisnu Aribowo(✉), Oktifian Windhi Prastomo, and Abdul Hakim Halim

Department of Industrial Engineering, Institut Teknologi Bandung, Bandung, Indonesia
wisnu@ti.itb.ac.id

Abstract. Scheduling problems in manufacturing industries are usually associated with production system, where the typical kind of resource is machine. However, the problems also arise in non-production setting, for example in an aircraft MRO (maintenance, repair, and overhaul) shop. In the shop, the primary resource is technician. There are three stages: stripping, repairing, and testing, with spare part procurement delay between stripping and repairing stages. Once a repair job is allocated to a technician, it will be processed by the same technician in subsequent stages, hence dedicated technician constraint. In the testing stage, jobs require not only technician but also dedicated testing machine. This paper models the scheduling problem as a mixed integer linear programming problem with makespan as the objective function. Numerical experiments are conducted to explore the model behaviors as well as to demonstrate the capability of the model.

Keywords: Aircraft maintenance · Flexible flow shop scheduling · Dedicated resource

1 Introduction

Scheduling deals with the allocation of resources to tasks over given time periods with the goal of optimizing one or more objectives [1]. In manufacturing industries, scheduling is often associated with production system, in which the typical kind of resource is machine. However, the scheduling problems also arise in non-production settings, for example in an aircraft MRO (maintenance, repair, and overhaul) shop. Aircraft maintenance is an important element to optimize because it is estimated that the aircraft maintenance takes as much as 10–15% of the aircraft operating cost [2]. In aircraft maintenance, the workforce is considered the highest priority resource because maintenance tasks are labour intensive [3].

The typical aircraft maintenance job in an MRO shop involves three main consecutive stages: stripping, repairing, and testing. In the stripping stage technicians observe the aircraft condition, locate the problem, and determine the broken components to be repaired. The replacement components or spare parts may need to be

M. N. Osman Zahid et al. (Eds.): iMEC-APCOMS 2019, LNME, pp. 260–265, 2020.
https://doi.org/10.1007/978-981-15-0950-6_40

ordered from suppliers and the procurement lead time constitutes the waiting time. In that case, the repairing process in the next stage can only be started after the components arrived. The last stage is the testing process, where the repair result is checked and verified. While technicians are the main resources in the stripping and repairing stages, the testing stage also need specific testing machine in addition to technicians.

The three-stage aircraft maintenance scheduling problem, which involves heterogeneous technicians and parallel dedicated machines in the last stage, is classified as flexible flow shop scheduling problem. In the literature the flexible flow shop scheduling problem has been addressed by quite many research, whether the resources are parallel (undedicated) [4–6] or dedicated [7, 8]. The scheduling problem addressed in this paper poses a few differences from the typical flexible flow shop scheduling problem in the literature. The aircraft maintenance scheduling problem involves two kinds of resource: heterogeneous technicians in all stages and dedicated testing machines in the last stage. In the last stage the two kinds of resource have to be considered simultaneously. The allocation of technicians to jobs is also dedicated, but the definition is somewhat different from that in 'dedicated machine'. It does not mean that the technician for the jobs is allocated prior to scheduling, but it means that once a technician is allocated to a job, the job has to be handled by the same technician in all subsequent stages. The reason is that the maintenance jobs often require personal attention. The technician who does the stripping to a job is considered as the one who knows the specific problem better than other technicians. Therefore, the person has to do the repairing and testing for the job as well.

The rest of the paper is organized as follows. Section 2 explains the development of the mathematical model, both the non-dedicated technician case and the more general dedicated technician case. Section 3 describes and discusses the result of the model execution for a numerical example. Finally, the paper is concluded in Sect. 4 along with the plan for possible extensions of the research in the future.

2 Mathematical Formulation

2.1 Problem Definition

In a planning period, there are n repairing jobs to be repaired by m available heterogeneous technicians. Each job is handled by exactly one technician, and each technician can work on only one job at a time. The processing time of each job in each stage depends on the technician because experienced or skillful technician may work faster than others for the job. There are two cases – the non-dedicated technician case and the dedicated technician case. In the dedicated technician case, some jobs have to be handled by the same technician in all stages. The processing times and the waiting times are assumed deterministic and known in advance. The shop wants to minimize makespan, which is the maximum job completion time. Figure 1 illustrates the problem description of the dedicated technician case. A job stays in its line once it enters the stripping stage. In other words it is handled by the same technician from start to finish.

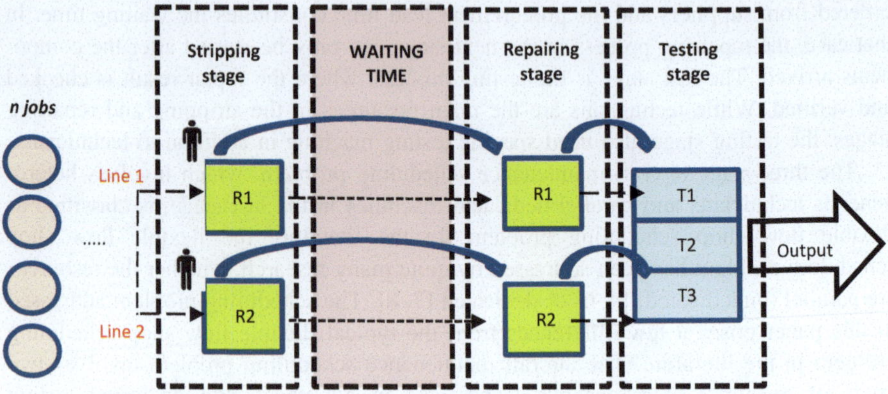

Fig. 1. MRO scheduling problem with dedicated technician constraint.

2.2 Mathematical Model – Non-dedicated Technician

The notations used in the mathematical model are as follows:

n : number of jobs
m : number of available technicians
p_{ijk} : processing time of job i in stage k by technician j
q_i : waiting time of job i
a_{it} : assignment of testing machine ($a_{it} = 1$ if job i is assigned to testing machine t; $a_{it} = 0$ otherwise)
C_{ik} : completion time of job i in stage k
$C^{\max}$: makespan
X_{ijk} : binary decision variable ($X_{ijk} = 1$ if job i is allocated to technician j in stage k; $X_{ijk} = 0$ otherwise)
Y_{iljk} : binary decision variable ($Y_{iljk} = 1$ if job i is processed before job l in stage k by technician j; $Y_{iljk} = 0$ otherwise)
T_{il} : binary decision variable ($T_{il} = 1$ if job i is processed before job l in stage 3; $T_{il} = 0$ otherwise).

The mathematical model is formulated as mixed integer linear programming (MILP) problem as follows:

$$\text{Minimize}\, Z = C^{\max} \tag{1}$$

$$\sum\nolimits_{j=1}^{m} X_{ijk} = 1 \quad \forall i, k \tag{2}$$

$$\sum\nolimits_{j=1}^{m} p_{ij1} X_{ij1} \leq C_{i1} \quad \forall i \tag{3a}$$

$$\sum\nolimits_{j=1}^{m} p_{ij2} X_{ij2} + C_{i1} + q_i \leq C_{i2} \quad \forall i \tag{3b}$$

$$\sum_{j=1}^{m} p_{ij3}X_{ij3} + C_{i2} \le C_{i3} \quad \forall i \tag{3c}$$

$$C_{ik} + p_{ljk}X_{ljk} \le C_{lk} + M\left(1 - Y_{iljk}\right) \quad \forall i,l,j,k(i \ne l) \tag{4a}$$

$$Y_{iljk} + Y_{lijk} \le 1 \quad \forall i,l,j,k(i \ne l) \tag{4b}$$

$$Y_{iljk} + Y_{lijk} \ge X_{ijk} + X_{ljk} - 1 \quad \forall i,l,j,k(i \ne l) \tag{4c}$$

$$C_{ik} + p_{lj[k+1]} \le C_{l[k+1]} + M\left(2 - X_{ijk} - X_{lj[k+1]}\right) \quad \forall i,l,j,k(k<3) \tag{5}$$

$$C_{i3} + \sum_{j=1}^{m} p_{lj3}X_{lj3} \le C_{l3} + M(1 - T_{ilt}) \quad \forall i,l,j,t\begin{pmatrix} i \ne l; \\ a_{it} = 1; a_{lt} = 1 \end{pmatrix} \tag{6a}$$

$$T_{il} + T_{li} = 1 \quad \forall i,l,j,t\begin{pmatrix} i \ne l; \\ a_{it} = 1; a_{lt} = 1 \end{pmatrix} \tag{6b}$$

$$C_{i3} \le C^{\max} \quad \forall i \tag{7}$$

The objective function to minimize is the makespan. Equation (2) ensures that each job is assigned to only one technician. Equations (3a, 3b, 3c) define the relationship between processing times, waiting times (in stage 2), and completion times of jobs. The disjunctive constraint in Eqs. (4a, 4b, 4c) ensures that jobs processed in a stage by a technician do not overlap with one another. Meanwhile, Eq. (5) sets similar constraint but for jobs in two different stages assigned to a technician. Equations (6a, 6b) guarantee that a testing machine can be used by only one job at a time. Finally, Eq. (7) sets the makespan as the maximum completion time of jobs.

2.3 Mathematical Model – Dedicated Technician

In the dedicated technician case, some jobs may need to be handled by a dedicated technician from the stripping stage until the test stage, while other jobs may not. The additional constraint is added to the mathematical model. The result is a more general model, from which the non-dedicated technician model can be derived. The additional constraint is as follows:

$$X_{ij1} = X_{ij2} \text{ and } X_{ij2} = X_{ij3} \quad \forall i,j(d_i = 1) \tag{8}$$

For a job with dedicated technician, Eq. (8) forces the job to be handled by the same technician in all stages. The parameter d_i is a binary parameter, which states that job i is to be handled by dedicated technician if the value is 1. If the value of d_i is 0 for all jobs then the model is essentially equivalent with the non-dedicated technician model.

3 Numerical Examples and Discussions

The models are executed to find the optimal solutions to a numerical example case. An MRO shop has to plan for five repair jobs to be handled by two technicians and three testing machines. Table 1 lists the processing times for all stages and the waiting time of the jobs.

Table 1. Processing times and waiting time of jobs

Job	Stripping stage		Waiting time	Repairing stage		Testing stage		
	Tech 1	Tech 2		Tech 1	Tech 2	TM 1	TM 2	TM 3
J1	5	5	10	4	4	3		
J2	3	3	8	3	4		1	
J3	3	3	5	3	4			2
J4	4	4	8	4	5	1		
J5	3	3	8	4	5		1	

Figures 2 and 3 show the optimal solutions found by the non-dedicated model and the dedicated model, respectively. For the dedicated model, the first three jobs (J1–J3) are to be handled by dedicated technician, while the rest two jobs (J4–J5) are not. The solutions can be found in a relatively short time by the default MILP solver in the IBM ILOG CPLEX Optimization Studio 12.7.1.0. The makespans in the two solutions are 25 and 26, respectively. The dedicated model case takes shorter time to find the optimal solution. Indeed, by the restriction the solution space becomes smaller. In both cases, all constraints are respected.

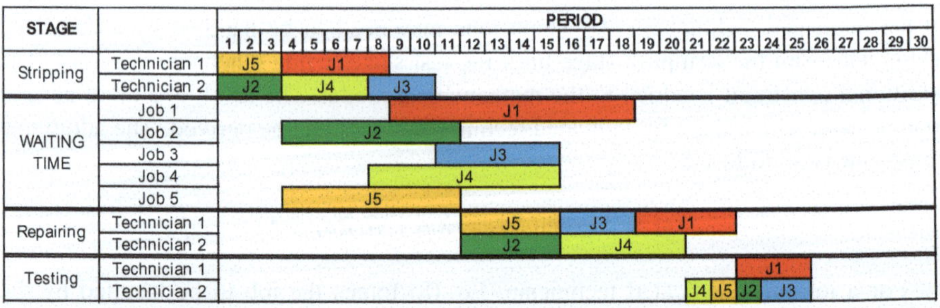

Fig. 2. Gantt chart of the optimal solution found by the non-dedicated model.

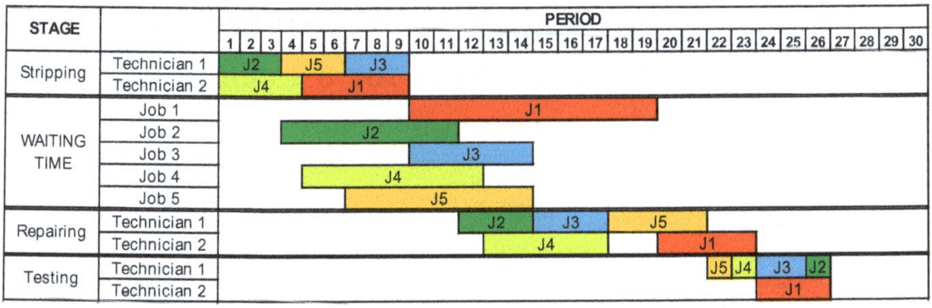

Fig. 3. Gantt chart of the optimal solution found by the dedicated model.

4 Concluding Remarks

This paper has outlined the initial development of a mixed integer linear programming optimization model of a maintenance, repair, overhaul (MRO) shop which has three stages (stripping, repairing, and testing stages) in a flow shop arrangement. Between the stripping and repairing stages the jobs spend some time to wait for the purchased components to arrive. The problem was motivated from a real problem in an aircraft industry. However, the model should be general enough to be applied in many other MRO shops which share similar characteristics.

The next step in this line of work is to analyze the performance of the models in various problem sizes and job characteristics while vying for simplifying the models. Future extension to this work is directed toward finding an efficient algorithm to yield good solutions in a short time. Such algorithm would be beneficial in solving large real problems where optimal solutions are hard to find.

References

1. Pinedo, M.L.: Scheduling – Theory, Algorithms, and Systems, 5th edn. Springer, Heidelberg (2016)
2. McFadden, M., Worrells, D.S.: Global outsourcing of aircraft maintenance. J. Aviat. Technol. Eng. **1**(2), 63–73 (2012)
3. Safaei, N., Banjevic, D., Jardine, A.K.S.: Workforce-constrained maintenance scheduling for military aircraft fleet: a case study. Ann. Oper. Res. **186**, 295–316 (2011)
4. Halim, A.H., Jenny: Penjadwalan pada flowshop dengan kelompok mesin heterogen untuk meminimasi makespan. Jurnal TMI (2003)
5. Naderi, B., Gohari, S., Yazdani, M.: Hybrid flexible flowshop problems: models and solution. Appl. Math. Model. **38**, 5767–5780 (2014)
6. Tyagi, N., Tripathi, R.P., Chandramouli, A.B.: Flexible flowshop scheduling model with four stages. Indian J. Sci. Technol. **9**(42), 1–11 (2016)
7. Bedhief, A.O., Dridi, N.: Minimizing makespan in a three-stage hybrid flow shop with dedicated machines. Int. J. Ind. Eng. Comput. **10**(2), 161–176 (2019)
8. Besbes, W., Loukil, T., Teghem, J.: A two-stage flow shop with parallel dedicated machines. In: 8th International Conference of Modeling and Simulation, Hammamet (2010)

Value Stream Mapping – A Tool to Detect and Reduce Waste for a Lean Manufacturing System

Noraini Mohd Razali[1(✉)] and Mohd Nizam Ab Rahman[2]

[1] Faculty of Manufacturing Engineering, Universiti Malaysia Pahang, 26000 Pekan, Pahang, Malaysia
norainimbr@ump.edu.my

[2] Faculty of Engineering and Built Environment, Universiti Kebangsaan Malaysia, 43600 Bangi, Selangor, Malaysia

Abstract. Waste is a non-value added activity that may exist everywhere in the production line. This waste is an activity that consumes resources but does not directly contribute to product or service and does not add value to the customer. This paper presents the classification of waste and the steps to implement value stream mapping (VSM) which is one of the powerful lean practice to detect and analyze waste activities in the manufacturing operation.

Keywords: Value stream mapping · Lean manufacturing · Waste

1 Introduction

The goal of lean manufacturing is to reduce waste in human effort, inventory, time to market and manufacturing space to become highly responsive to customer demand while producing quality products in the most efficient and economical manner. This approach emphasis on the elimination of waste. Waste takes many forms and can be found at any time and in any place in an organization. It may be found hidden in policies, procedures, process and product designs, and in operations [1]. This waste is an activity that consumes resources but does not directly contribute to product or service and does not add value to the customer [2]. Lean manufacturing uses tools like value stream mapping (VSM), one-piece flow, visual control, Kaizen, cellular manufacturing, inventory management, Poka-yoke, standardized work, and workplace organization to reduce manufacturing waste [3].

VSM which popularized by Rother and Shook [4] is one of the tools in lean that is often used by the company as a starting point in implementing lean transformation. VSM is not only used as a tool for employees to understand process flow in production but also provides useful information on information flow as well as material flow from supplier to the customer (i.e. for the entire supply chain). Usually, before the lean transformation can be realized, VSM is used as a tool for identifying waste, analyzing waste by finding the source of the waste and identifying the activities that need to be done to eliminate or reduce waste using other lean tools.

M. N. Osman Zahid et al. (Eds.): iMEC-APCOMS 2019, LNME, pp. 266–271, 2020.
https://doi.org/10.1007/978-981-15-0950-6_41

Although VSM looks as easy as a tool for mapping the process from start to finish, manufacturing personnel are still unable to implement VSM successfully due to lack of understanding of the concepts and a correct procedure of implementing VSM. They are incapable to construct the VSM in detail and accurately because they do not understand what data should be taken and analyzed and what is the next action after the current VSM has been performed. Hence the main purpose of this paper is to demonstrate a simple step of using VSM as a means of identifying waste. This paper also describes the types of waste found in production.

The next sections address the following topics: Sect. 2 presents the literature review. The VSM implementation steps are included in Sect. 3; Sect. 4 provides the conclusions and future research directions.

2 Literature Review

2.1 Lean Manufacturing Principle

Lean manufacturing (LM) is an assembly line methodology developed originally for Toyota and the manufacturing of automobiles. Also known as the Toyota Production System (TPS) and Just-in-Time (JIT) production, developed by the chief engineer of Toyota Taiichi Ohno after World War II between 1948 and 1975 [5]. LM was first introduced in the West Country by Womack, Jones and Roos in 1990 in their best-selling book, "The machine that Changed the World" [6]. The main objective of LM is to produce finished goods according to the customer requirements with very less waste or without waste [7]. Womack et al. [8] defined the five principles of Lean manufacturing, which is Define value, Map the value stream, Create flow, Establish flow, and Pursue for perfection.

VSM is a lean tool that is used to understand where wastes exist in the process. According to Rother & Shook, the value stream definition is all added and non-value added activities needed to produce products throughout the process from raw materials to customers, and design flow from concept to launch. They also stated that VSM is used to define and analyze the current state product value stream and design future state which focus on reducing waste, improving lead time and improving workflow. Monden [2] categorizes activities into three operations namely value adding, non-value adding and necessary but non-value adding. Value adding activity is an activity that will transform or improve the product for the customer. While non-value added activities are defined as activities that consume resources (man, material, equipment) but are not directly contributing to the product or service.

2.2 Waste Definition and Classification

Waste is any non-value added activities in which Ohno's classified them into seven categories that are: overproduction, waiting, transportation, inappropriate processing, unnecessary inventory, unnecessary motions, and defects [5].

- ***Inventory*** – material not currently being processed, including raw materials, work-in-process, and finished product inventory.

- ***Motion*** – operator walking around the equipment to get where they are needed, or to get changeover parts or tools. It is also related to ergonomics and is seen in all instances of bending, stretching, walking, lifting, and reaching.
- ***Overprocessing*** – doing more to the material than the customer requires. Any operation or process that does not add value to the company can be considered a production waste and it can potentially increase the incidence of defects in the products.
- ***Overproduction*** – making more than the customer needs, or making it sooner than needed. Production of more items than required by the customer.
- ***Defects*** – parts or material that do not meet required specifications.
- ***Transportation*** – movement of material either from one process step to the next step or into or out of inventory.
- ***Waiting*** – time that operator spend waiting for material or for the equipment to be ready to use. Time wasted waiting for people, materials or equipment.
- ***Talent*** – this is an additional waste that has more recently been pointed out by [10]. The waste of human potential can lead to missed improvement opportunities, considering that lean philosophy advocates that every individual is a thinker and can contribute with positive outcomes.

2.3 Application of VSM in Manufacturing Operation

Singh et al. [9] applied VSM to identify waste such as WIP, lead time, and manpower by bridging the gap between the current state and future state of the production industry. Andreas et al. [10] in their paper discussed the benefits of VSM such as a reduction in lead time, improved productivity, reduction in cycle time and reduction in inventory. However, other lean tools should also be subsequently implemented in order to support a more effective reduction of waste. Alaya et al. [11] stated that VSM is a good starting tool for transforming any enterprise into lean. They incite researchers and practitioners to start to enhance their approach towards lean by integrating the standard VSM approach with additional lean tools to enable people not only to see but also to communicate and to manage their processes. Prashar [12] employed the Lean-Kaizen approach using VSM for process improvement through redesigning an assembly line in a manufacturing steering system of automobiles. In recent years, many publications have extensively documented the implementation of VSM as a key method to implementing lean concepts. A case study has been widely used in the field of operations management as a method for evaluating the applicability of the VSM tools to improve company performance. More recent case studies can be referred to [13, 14]. This paper will only discuss the step required to implement VSM without demonstrating any case study in the manufacturing operation.

3 VSM Implementation Steps

The steps to implement VSM in this paper is based on Rother & Shook [4] and King and King [15]. The basic steps of VSM are presented in Fig. 1 and the three main steps are described below.

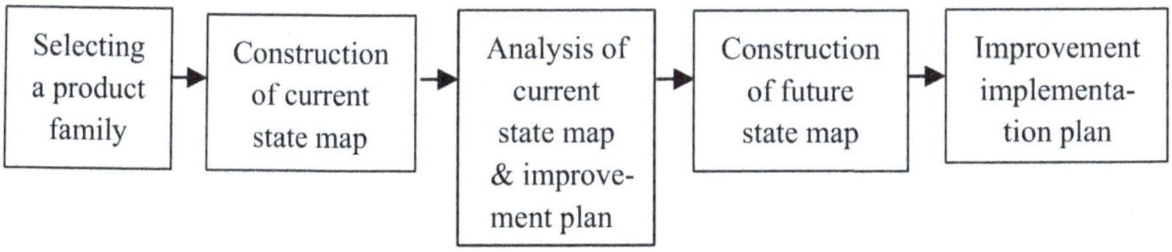

Fig. 1. Basic steps of VSM

Selecting a Product Family

At this stage, it is important to focus on one product family only. Although there are many products produced on the shop floor, however, the customers care about their specific product, and not all the other products produced by the company. VSM means walking and drawing the processing steps (material and information) for one product family from door to door in the plant. A product family is a group of products that pass through similar processing steps and over common equipment in the downstream process.

Construction of Current State Map

Having selected a product family the next step is to draw a current state map for the sub-frame as depicted in Fig. 2. The map is constructed from "door-to-door" production flow inside the firm. The process activity map sets out the sequence of flow by recording all the individual steps that take place to produce a product or service. When constructing a current state map, it always begins at the shipping end and works upstream. It must begin with the processes that are linked most directly to the customer, which should set the pace for other processes further upstream. Usually, the stopwatch is used to collect data on cycle time for the process. An average cycle time must be obtained from a set of at least ten data. For a manual process which uses the operator in performing task or process, the standard time must be used that considers performance rate and allowance rate. The whole value stream must be mapped by only one same person, even if several people are involved. This is because if different people map different process, then no one will understand the whole process.

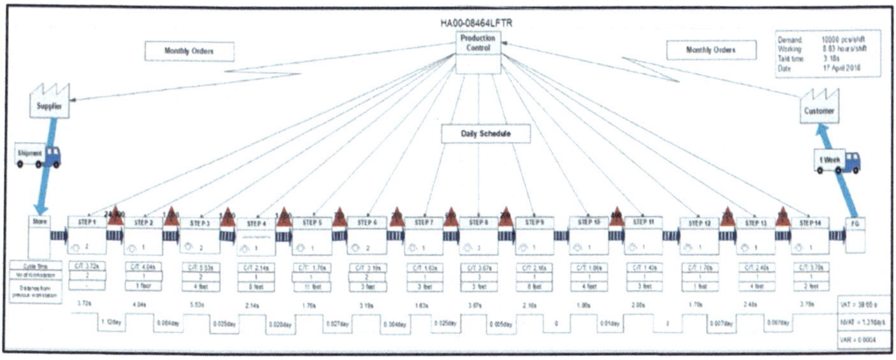

Fig. 2. Example of the current state VSM

In general, the current state map consists of three main components;

- Material flow – the flow of material as it progresses from raw materials, through each major process step, to finished goods moving toward the customer. It shows only the major processing system, with data boxes that illustrate the performance of each process. All inventories along the flow are also shown.
- Information flow – the flow of all major types of information that govern what is to be made and when it is to be made. This starts with orders from the customers and ends with schedules and control signals to the production floor.
- Timeline – it shows the value added (VA) time and non-value added (NVA) time. It is a line at the bottom of the VSM in the form of a square wave. This is a key indicator of waste in the process

Construction of Future State Map
Future state map is constructed to show all the improvement plans that minimize the non-value added time. In order to develop the future state map, the current state map is analyzed using the following guidelines developed by Rother and Shook [4]:

- Produce to the TAKT time based on the demand and the available working time of processes closest to the customer.
  TAKT time = Available production time per shift/Target output per shift
- Develop continuous flow wherever possible (move a single product through every step of the process instead of grouping work items into batches)
- Use the "supermarkets" concept to control production where continuous flow is not possible upstream.
- Distribute the production of different products evenly over time at the pacemaker process (level the production mix – "heijunka").
- Create an "initial pull" by releasing and withdrawing small, consistent increments of work at the pacemaker process (level the production volume).
- Develop the ability to make "every part every day" in the processes upstream of the pacemaker process.

4 Conclusion and Future Works

As a conclusion, any organization wishes to achieve lean manufacturing level is essential to firstly identify non-value added activities and eliminate those activities using appropriate lean tools. This paper has highlighted the importance of value stream mapping as one of the lean tools to detect and reduce waste. This paper also had briefly explained the principles of lean manufacturing, the definition and classification of waste, and a simple yet clear step to implement VSM effectively. In the future, it is necessary to include the real application of VSM through case studies in the production line. The case study should indicate step by step of VSM application to detect waste, and analyze waste thus suggesting improvement activity to reduce or eliminate waste.

Acknowledgment. This research is fully supported by RDU170390 (Universiti Malaysia Pahang).

References

1. Seth, S., Gupta, V.: Application of value stream mapping for lean operations and cycle time reduction: an Indian case study. Prod. Plann. Control **6**(1), 44–59 (2005)
2. Monden, Y.: Toyota Production System: An Integrated Approach to Just-In-Time, 2nd edn. Industrial Engineering and Management Press, Norcross (1993)
3. Russell, R.S., Taylor, B.W.: Operations Management, 2nd edn. Prentice Hall, Upper Saddle River (1999)
4. Rother, M., Shook, J.: Learning to See: Value Stream Mapping to Create Value and Eliminate Muda. Lean Enterprise Institute, Boston (1999)
5. Wilson, L.: How to Implement Lean Manufacturing, 2nd edn. McGraw Hill Education, New York (2015)
6. Dennis, P.: Lean Production Simplified. A Plain Language Guide to the World's Most Powerful Production System, 2nd edn. Productivity Press, New York (2015)
7. Liker, J.K.: The Toyota Way: 14 Management Principles from the World's Greatest Manufacturer. McGraw Hill, New York (2003)
8. Womack, J., Jones, D., Roos, D.: The Machine That Changed the World. Rawson Associates, New York (1990)
9. Singh, B., Garg, S.K., Sharma, S.K.: Value stream mapping: literature review and implications for Indian industry. Int. J. Adv. Manuf. Technol. **53**, 799–809 (2011)
10. Andreadis, E., Garza-Reyes, J.A., Kumar, V.: Towards a conceptual framework for value stream mapping (VSM) implementation: an investigation of managerial factors. Int. J. Prod. Res. **55**(23), 7073–7095 (2017)
11. Alaya, L.B.F.: VSM a powerful diagnostic and planning tool for a successful Lean implementation: a Tunisian case study of an auto parts manufacturing firm. Prod. Plann. Control **27**(7–8), 563–578 (2016)
12. Prashar, A.: Redesigning an assembly line through Lean-Kaizen: an Indian case. TQM J. **26** (5), 475–498 (2014)
13. Seth, D., Seth, N., Dhariwal, P.: Application of value stream mapping (VSM) for lean and cycle time reduction in complex production environments: a case study. Prod. Plann. Control **28**(5), 398–419 (2017)
14. Ali-Asghar Dadashnejad, A., Changiz Valmohammadi, C.: Investigating the effect of value stream mapping on operational losses: a case study. J. Eng. Des. Technol. **16**(3), 478–500 (2018)
15. King, P.L., King, J.S.: Value Stream Mapping for the Process Industries. CRC Press, Taylor & Francis Group, Boca Raton (2015)

Make or Buy Decision with Price and Quality Dependent Demand

Cucuk Nur Rosyidi(✉)

Industrial Engineering Department, Universitas Sebelas Maret,
Jalan Ir. Sutami 36 A, Surakarta, Indonesia
cucuk@uns.ac.id

Abstract. The aim of this research is to developed an optimization model to determine an optimal process parameter to help a decision maker in making decision concerning make or buy considering price and quality dependent demand. In this paper, two models are developed for make regime and buy regime. The price and quality level will determine the demand of the product. The quality level was treated as parameters in many research, but in this paper the quality level is treated as model component and expressed in term of process parameter. The objective function of the model is to maximize total profit of both manufacturer and supplier. The cost components consist of material purchasing cost, manufacturing cost, and quality loss. The results of optimization shows that both model gives the same value of optimal process parameter in manufacturer side.

Keywords: Make or buy decision · Quality loss · Manufacturing cost · Price and quality dependent demand

1 Introduction

A manufacturing company has two alternatives in providing the needed components for final product assembly. First, the company produced all the needed components using its own production facilities which commonly known as in-house production or make. Second, the company outsources all the needed components from its suppliers which known as buy. Both alternatives will have consequences to the company. By make alternative, the company has an ability to fully control the quality of the components and increase the utility of equipment. The buy alternative will make the company more focus to its core business and reduce the needs of investments in term of equipment and operational cost.

Product quality has been known as one of the effective strategies for a company to be competitive in the market. The quality of a product should be maintained and improved by the company to satisfy the customers. The company must perform quality improvement as a part of continuous improvement and the company should develop some programs for the improvement. The programs need some investment funds that must be provided by the company. The funds are used to train the operators in order to increase their skills and capabilities in certain manufacturing processes. Each operator has his own learning rate and in average this learning rate will determine the needed

M. N. Osman Zahid et al. (Eds.): iMEC-APCOMS 2019, LNME, pp. 272–277, 2020.
https://doi.org/10.1007/978-981-15-0950-6_42

investment in the quality improvement training programs. The higher the learning rate, the quicker the operator achieves the standard skill set by the company.

According to Taguchi concept, then there are two sources of quality problems. First is bias which resulted from the difference between mean and target value of a certain quality characteristic. The second source of the problem is variance which measures the data spread of the quality characteristic. Taguchi expressed the quality loss in a quadratic form to measure the loss of customer in using a product. This kind of loss includes in external failure quality cost, while the internal quality loss deals with the out of specification product. The out of specification product will determine the quality level of the product. The quality level along with the product price will determine the number of product demand. In this paper, we develop a make or buy optimization model considering investment cost and price and quality dependent demand.

2 Literature Review

Make or buy decision is a strategic decision that must be made by a company to determine which components that must be made in-house using its own production facilities and which ones that must be bought from suppliers. Conventionally, the sourcing decision is simply be done by comparing internal production costs with the prices from external suppliers and choosing the least costly alternative [1]. Further, they noted that the decision may be more complex in which sourcing decisions can be influenced by fears of supplier hold-up, concerns about leakage of proprietary information, the need to ensure timely and reliable supply of high-quality inputs, and prospective gains from cultivating long term alliances with suppliers. According to [2], the make or buy decision implies two types of sourcing, namely hybrid and plural sourcing. In hybrid sourcing, the company purchases the entire volume from a single mode which exhibits mixed governance. While plural sourcing involves both make and buy in the same time. The decision involves several considerations such as production capacity, process capability, and technological capability.

Make or buy has attracted many researchers. For example, [3] developed an optimization model to determine the optimal quality investment in single supplier and manufacturer to reduce defective rate. The model considered outbound inspection in supplier site while the manufacturer performed inbound for the incoming materials or components and outbound before shipping the final product to customers. A research by [4] developed an optimization model in the case of plural sourcing to determine the optimal suppliers and its corresponding quantity of component orders. The model considered quality improvement in term of learning investment to maximize the return of such investment. In more recent research, [5] developed two optimization models which solved sequentially to determine the optimal suppliers, order quantity from selected suppliers and learning investments. The model considered not only learning but also forgetting in the model to maximize the return on investments.

In a more recent research, [6] analyzed the impact of two conditions, namely exogenous and endogenous carbon tax, to the make or buy decisions. In exogenous carbon tax, the government has no regulation on the carbon tax rate, while in endogenous case the government will impose a regulation about carbon tax rate. In another context, [7] developed a model to determine the optimal process mean, number of shipments, order quantity, and maximum shortage to maximize the profit of supply chain. In this research, we used the system scheme of make or buy analysis model in [6] as the basis to develop a more comprehensive make or buy decision model. We consider several aspect in this research including inspection error, quality investment, and product demand. The decision variables of the model are the variance of final product as the indicator of quality level and investment level as the representation of product quality improvement.

3 Proposed Model

In this research, we consider a supply chain consists of single supplier and single manufacturer. The manufacturer adopts hybrid sourcing in which the needed components or materials will be provided by manufacturer itself or bought from suppliers. Those alternatives were known as make regime and buy regime respectively [6]. The demand of final product will depend on the price and quality level of the product. In many research, the quality level was treated as a parameter, hence the value was given. In this research, the quality level is defined as a function of variance as one of important process parameters. The objective of the model is to maximize the total profit of manufacturer in make regime and maximize the total profit of both manufacturer and supplier in buy regime.

The demand function follows the research of [8] Christy et al. as expressed in Eq. (1), where d, p_m, and q_m denote base demand, product price, and quality level of the product respectively. While α and β denote the parameters of the demand function.

$$D(p_m, q_m) = d - \alpha p_m + \beta q_m \tag{1}$$

The quality level depends on the process parameter of quality characteristics of the product as shown in Eq. (2). In the equation, μ_m and σ_m denote the mean and standard deviation of certain quality characteristic of the product, while LSL and USL denote lower and upper specification limits of the characteristic respectively which defined by $\mu_m \pm 3\sigma_m$.

$$q_m = \int_{LSL}^{USL} \frac{e^{-\frac{(x-\mu_m)^2}{2\sigma_m^2}}}{\sigma_m \sqrt{2\pi}} \tag{2}$$

3.1 Make Regime

In the Make Regime, the manufacturer make all the needed components or materials using its own production facilities. The total cost consists of manufacturing cost, quality lost, and material production cost. The manufacturing cost of the product will depend on the product tolerance. The tighter the tolerance, the higher the manufacturing cost and reversely the looser the tolerance, the lower the manufacturing cost. In this paper, we use the inverse function as in Eq. (3). In the equation, a and b denote the manufacturing cost parameters while t denotes the product tolerance. By the definition of process capability index (Cp), standard deviation of the product can be expressed in term of tolerance as in Eq. (4).

$$M = \frac{a}{bt} \tag{3}$$

$$\sigma_m = 3tCp \tag{4}$$

We use Taguchi loss function to quantify the quality loss of the product as expressed in Eq. (5).

$$TLF = k\left(\frac{t}{3Cp}\right)^2 \tag{5}$$

The number of material used to manufacture the product depends on the quality level of the product as expressed in Eq. (6). Total material cost used is the product of unit material cost (C_m) with Eq. (6).

$$Q = \frac{D(p_m, q_m)}{q_m} = \frac{d - \alpha p_m + \beta q_m}{\int_{LSL}^{USL} \frac{e^{\frac{(x-\mu_m)^2}{2\sigma_m^2}}}{\sigma_m \sqrt{2\pi}}} \tag{6}$$

Equation (7) expresses total profit function in the Make Regime.

$$\pi_{mr} = p_m D(p_m, q_m) - Q\left(\frac{a}{3bCp\sigma_m} + k\sigma_m^2 + C_m\right) \tag{7}$$

3.2 Buy Regime

In Buy Regime, the manufacturer buy all the needed materials or components to its supplier. The objective function of the model is to maximize the profits of both supplier and manufacturer. The total profit of the supplier is the function of the material ordered by manufacturer and the manufacturing process of the material. The supplier's profit expression is shown in Eq. (8). In the equation, p_s denotes the material price, C_{sm} denotes the processing material cost, while q_s denotes the quality level of material processing in supplier side. We assume in this paper that the price of material purchased by the manufacturer depends on the quality level of certain material quality characteristic. Hence, the price of the material follows a function as expressed in Eq. (9).

$$\pi_{sp} = p_s Q - C_{sm} \frac{Q}{q_s} \tag{8}$$

$$p_s = p_b + \gamma q_s \tag{9}$$

The total profit of the manufacturer is the subtraction of revenue from total cost consists of manufacturing cost, quality loss, and material purchasing cost. Equation (10) expresses the profit function of the manufacturer. The total profit of the supply chain is then the summation of Eqs. (7) and (8).

$$\pi_{mr} = pD(p,q) - Q\left(\frac{a}{bt} + k\left(\frac{t}{3Cp}\right)^2 + p_s\right) + p_s Q - C_{sm}\frac{Q}{q_s} \tag{10}$$

4 Numerical Example

The parameters used in the numerical example are given in Table 1. We used Wolfram Mathematica 7 for Students to solve the model. We used maximize function of the software to find the solution. Table 2 shows the optimization results. The optimal value of the variance was the same for both make and buy regime. The make regime results a profit of $34,298.2, while for the buy regime results a lower profit of $29,556.4. From the results, the make regime gives better profit than buy regime even the cost of material production in make regime was set about 30% higher than material price from the supplier. The decision making whether the material should be make or buy will depend on other considerations such as investment cost to install the production facilities, the availability and quality requirements of the material, and other relevant considerations.

Table 1. Table captions should be placed above the tables.

Parameter	Value	Parameter	Value
p	35	μ_m	403
d	1000	μ_s	405
α	0.4	Cp	1
β	0.2	C_m	10
γ	0.5	C_{sm}	5
a	0.1	p_b	7.5
b	3	k	30

Table 2. Optimization results

Make regime		Buy regime	
Variable	Optimal value	Variable	Optimal value
σ_m	0.0570	σ_m	0.0570
q_m	0.9973	σ_s	0.0185
		q_m	0.9973
		q_s	0.9973
		p_s	7.0014

5 Conclusion

In this paper, we developed a make or buy decision model under hybrid sourcing in which the manufacturer must make all the needed materials or components using its production facilities or outsourced to its supplier. The demand depends on price and quality level of the final product. The objective function of the model was to maximize the total profit of the manufacturer for make regime and maximize the total profit of both manufacturer and supplier for buy regime. Product and material variances represented two decision variables of the model. For further research, firstly we still have to check the optimality of the model solution. Second, how to include the quality cost of raw material production in the make regime. Third, how to include the learning investment and perform sensitivity analysis for the whole model.

References

1. Arya, A., Mittendorf, B., Sappington, D.E.M.: The make-or-buy decision in the presence of strategic outsourcing to a common supplier. Manag. Sci. **54**(10), 1747–1758 (2008)
2. Serrano, R.M., Ramirez, M.R.G., Gasco, J.L.G.: Should we make or buy? An update and review. Eur. Res. Manag. Bus. Econ. **24**, 137–148 (2018)
3. Hsieh, C.-C., Liu, Y.-T.: Quality investment and inspection policy in a supplier-manufacturer supply chain. Eur. J. Oper. Res. **202**, 717–729 (2010)
4. Rosyidi, C.N., Pratama, M.A.: Two-stage optimization model for process/supplier selection, component allocation, and quality improvement. Cogent Eng. **5**(1), 1–16 (2018)
5. Pratama, M.A., Rosyidi, C.N., Pujiyanto, E.: Two stages optimization model on make or buy analysis and quality improvement considering learning and forgetting curve. J. Ind. Eng. Manag. **11**(4), 794–813 (2018)
6. Meng, X., Yao, Z., Zhao, Y.: Make or buy? It is the question: a study in the presence of carbon tax. Int. J. Prod. Econ. **195**, 328–337 (2018)
7. Chuang, C.-J., Wu, C.-W.: Determining optimal process mean and quality improvement in a profit-maximization supply chain model. Qual. Technol. Quant. Manag. **16**(2), 154–169 (2019)
8. Christy, A.Y., Fauzi, B.N., Kurdi, N.A., Jauhari, W.A., Saputro, D.R.S.: A closed-loop supply chain under retail price and quality dependent demand with remanufacturing and refurbishing. IOP Conf. Ser.: J. Phys.: Conf. Ser. **855**, 1–7 (2017)

Participatory Ergonomics Intervention for Exploring Risk Factors Lead to Work-Related Musculoskeletal Disorders Among Automotive Production Workers

Fazilah Abdul Aziz(✉), Zakri Ghazalli, and Nik Mohd Zuki Nik Mohamed

Faculty of Mechanical and Manufacturing Engineering, University Malaysia Pahang, 26600 Pekan, Pahang, Malaysia
fazilahaa@ump.edu.my

Abstract. Risk factors related to work activity and ergonomics can make it more challenging to maintain this balance and raise the probability that some individuals may develop musculoskeletal disorders (MSD). This study was designed to identify the ergonomics risk factors that increase work-related musculoskeletal disorders (WMSD) among automotive production workers. The participatory ergonomics (PE) was employed to assess the risk factors related to WMSD by involving production workers and management team. The study was initiated by reviewing previously established studies regarding critical body region pains risk factors. Expert interviews were then conducted to share knowledge from senior management staff members to identify the potential risk factors. The predicted risks factors were assessed with a mixed group of senior workers from three automotive manufacturers through a survey questionnaire. In all, twenty-six dominant risk factors related to WMSD, specifically in the context of automotive production plant operations were found, and these factors can be considered as points for targeting ergonomics intervention efforts.

Keywords: Ergonomics risk factors · Work-related musculoskeletal disorders · Participatory ergonomics · Production workers · Automotive

1 Introduction

The automotive industry is one of the catalyst of the country's development through a high rate of work [1]. In line with this, automotive workers are assets to the nation as they play an important role in ensuring a sustainable growth and market expansion for the industry. However, automotive workers are often exposed to various ergonomics risks due to the nature of their working environment where they are required to deal with machines as well as heavy tools and materials. Moreover, automotive workers are more likely to be involved in repetitive work tasks. Argubi-Wollesen et al. [2] reported that 10% of working processes in the automotive sector involve pushing and pulling

M. N. Osman Zahid et al. (Eds.): iMEC-APCOMS 2019, LNME, pp. 278–283, 2020.
https://doi.org/10.1007/978-981-15-0950-6_43

with more than 40% of them require the manipulation of heavy objects. Thus, the occurrence of work-related musculoskeletal disorders (WMSD) is high in the automotive industry.

There is a consensus among scholars on the multifactorial nature of WMSDs [3]. The WMSD development is influenced by physical demands of certain tasks, work organizational hazards and psychosocial contexts [4, 5]. Besides, personal factors, such as individual perceptions and other related characteristics are equally important in dealing with risk management [6]. The proactive ergonomics should focus on the counteractive action to overcome WMSDs through early detection and reducing risk factors at work, and identifying relevant risk factors in the future [7].

Therefore, this study aims to identify the dominance of risk factors related to WMSD among production workers using PE approach in an automotive component manufacturer. The findings will contribute to a body of knowledge needed to assist the occupational safety and health (OSH) managers by producing effective WMSD prevention strategies in the workplace for the worker's well-being, productivity, and organizational sustainability.

2 Materials and Method

2.1 Investigate the Related WMSD Risk Factor

Literature analysis was used to recognize the risk factors associated with MSD. To identify the relevant literature, established studies related to back, neck, shoulder and arm pain risk factors were identified. The risk factors identified from the literature analysis were reviewed based on the tacit knowledge provided by the industry experts. Senior management team members were selected for interview analysis based on their expertise, experience, and convenience.

The group of subject matter experts consists of a manager and above level, including general manager and managing director. Managers were from the production, engineering, and safety health and environment departments. The interview analysis comprised a question and answer session, verbal problem solving, and observation analysis associated with workplace ergonomics risks factors. Then, the potential risk factors were grouped according to the ergonomics domains, namely individual, organizational, physical and psychosocial domains.

2.2 Survey WMSD Risk Factor Analysis

The questionnaire survey was distributed to a mixed group of production workers working in three automotive manufacturers. A total of 110 automotive production workers were approached based on their expertise, experience, and accessibility as respondents. Male contributed 99% of respondents and more than 40% age between 31 and 35 years. Majority of respondents (29%) had been working for more than 15 years.

The concluded list of risk factors from tacit knowledge analysis was presented to the respondents in the form of questionnaires. The respondents were required to respond on whether or not the factors are critical consideration in evaluating risk factors related to WMSD in the automotive production plant using a 5-point Likert scale. The concluded list of risk factors from tacit knowledge analysis was presented to the respondents in the form of questionnaires. The respondents were required to respond on whether or not the factors are critical consideration in evaluating risk factors related to WMSD in the automotive production plant using a 5-point Likert scale.

The mean value of each factor was determined by multiplying the percentage of respondents with the values of 1, 2, 3, 4 and 5. These values represent the level of agreement: "strongly disagree", "disagree", "not sure", "agree" and "strongly agree", respectively, adding to the resulting factors. The cut-off value (refer Eq. 1) was used and these factors were identified based on the relevant criteria, for which the mean values are greater than or equal to cut-off value.

$$\text{Cut-off value} = \frac{\sum mean\ value}{number\ of\ factor} \quad (1)$$

The selected value appeared to be the natural cut-off point as it was found to be the average mean rating value for all factors included in the survey instrument.

3 Results

The results of the literature analysis were reviewed based on the potential risk factors for body pain identified to be the upper back, lower back, shoulders, neck, and arms. The group of SME experts has identified the potential risk factors related to WMSD (see Tables 1, 2, and 3). Each of the risk factors is placed in the appropriate ergonomics domain, including individual, organizational, physical-job task, physical-workplace and equipment, and psychosocial ergonomics.

Table 1. Dominant individual ergonomics-related risk factors

No.	Sub-risk factors	Cronbach's alpha, α	Mean	Cut off value
IF1	Negligence of workers	0.873	***4.03***	3.61
IF2	Improper use of PPE		***3.92***	
IF3	Level of education		***3.90***	
IF4	Working experience		***3.68***	
IF5	Age		***3.64***	
IF6	Body size		3.59	
IF7	Body weight		3.45	
IF8	Sedentary lifestyle of worker		3.39	
IF9	Employment duration		2.88	

Table 2. Dominant organizational and physical-job task ergonomics-related risk factors

No.	Sub-risk factors	Cronbach's alpha, α	Mean	Cut off value
	Organizational related risk factors	0.822		3.75
OF1	High work load		***3.94***	
OF2	Frequent workdays		***3.85***	
OF3	Exposure to physical demands		***3.80***	
OF4	Worker lack of rest		***3.76***	
OF5	Tight production schedule		***3.75***	
OF6	Long working hour		3.74	
OF7	Irregular working schedule		3.61	
OF8	Shift work		3.56	
	Physical-job task related risk factors	0.890		3.99
PhyJF1	Frequent work lifting		***4.21***	
PhyJF2	Carrying and lifting heavy loads		***4.19***	
PhyJF3	Poor working practice		***4.10***	
PhyJF4	Heavy physical work		***4.08***	
PhyJF5	Forced exertion in job task		***4.05***	
PhyJF6	Poor working posture		***4.03***	
PhyJF7	Static loading of the spine		3.98	
PhyJF8	Prolong static posture		3.96	
PhyJF9	Prolong standing posture		3.88	
PhyJF10	Overexertion		3.72	
PhyJF11	Repetitive of tasks		3.70	

The reliability analysis of 45 risk factors related to WMSD is shown in Tables 1, 2, and 3. The Cronbach's α for the five factors are above 0.7 with the ranged between 0.821 and 0.890. Thus, all studied risk factors are deemed reliable [8]. The survey results for risk factor related to WMSD in each ergonomic domain also presented in Tables 1, 2 and 3. The value of 3.61 appeared as the natural cut off point as it was found to be the average of mean rating values of all the individual ergonomics-related risk factors. Thus, the five most critical risk factors have presented in Table 1.

Similarly, the respondents rated the factors considered in organizational ergonomics-related risk factors affecting WMSD, and the cut-off value is 3.75 for mean ratings. Hence, the five most important risk factors have shown in Table 2.

The physical ergonomics risk factors were divided into job tasks, and workplace and equipment. The results for job task related risk factors established, and the cut-off value is 3.99 for mean ratings. Consequently, the six most serious risk factors have exposed in Table 3. The results for workplace and equipment related risk factors are summarized and the cut-off value is 3.96 for mean ratings. So, the five most important risk factors have shown in Table 2.

Table 3 displays the dominant psychosocial ergonomics-related risk factors affect WMSD. The value of 3.61 appeared to be the natural cut-off point. Therefore, the five most critical risk factors for psychosocial ergonomics have shown in Table 3. Based on

Table 3. Dominant Physical-workplace and equipment and psychosocial ergonomics related risk factors

No.	Sub-risk factors	Cronbach's alpha, α	Mean	Cut off value
	Physical-workplace and equipment related risk factors	0.821		3.96
PhyWF1	Poor ventilation in working environment		***4.08***	
PhyWF2	Poor workspace		***4.01***	
PhyWF3	Poor temperature in working environment		***3.99***	
PhyWF4	Hand tools and vibration		***3.98***	
PhyWF5	Noise in working environment		***3.96***	
PhyWF6	Poor workstation		3.95	
PhyWF7	Poor work surface height		3.93	
PhyWF8	Heavyweight equipment or tools		3.77	
	Psychosocial related risk factors	0.863		3.61
PsyF1	Fatigue		***4.03***	
PsyF2	Work stress		***3.92***	
PsyF3	Emotional stress		***3.90***	
PsyF4	Frustration with work and not work-related		***3.68***	
PsyF5	Low job support		***3.64***	
PsyF6	Work fast		3.59	
PsyF7	Job task rotation		3.45	
PsyF8	Long working time (overtime)		3.40	
PsyF9	Work intensely		2.89	

the reliability and descriptive analysis, 26 sub-factors that act as the key to assess the risk factors related to WMSD specifically in the context of the automotive production assembly plant were identified.

4 Discussions and Conclusions

The most critical individual ergonomics-related risk factor is the negligence of workers. Negligence is one of the main causes of body pain, accident, and deadly injuries among industry workers [9]. The results of the study reveal that the high workload, is perceived as the most critical organizational ergonomics-related risk factors affecting WMSD. These findings supported the verdict of Weale et al., [10] workers have a high workload experienced MSD pain in a variety of body regions. Furthermore, frequent work lifting, perceived as the most critical physical-job task related to risk factors. The result of this study supports the finding of others, which found the increments of low back pain intensity are more common among workers performing frequent work lifting and lifting heavy loads [11].

The most severe physical-workplace and equipment related risk factors is poor ventilation in working environments. This finding supports the latest studies which addressed the quality of the environment of a workplace like temperature, ventilation, and noise have influenced the productivity and efficiency of the workers [12]. The finding of the study shows that the most serious psychosocial risk factors is fatigue. Fatigue could lead to work-related disorders and declining in productivity and efficiency [13].

In conclusion, the dominance of risk factors related to WMSD has discovered. Hence, OSH practitioners and engineers can cooperate to minimize or eliminate that risk factors while ergonomics workplace design stage. Finally, WMSD risk can be avoided because of the safe, healthy, and well-being culture created in the organization.

Acknowledgement. This research has been supported by University Malaysia Pahang (PGRS 170325).

References

1. Rashid, N., Jabar, J., Yahya, S., Samer, S.: State of the art of sustainable development: an empirical evidence from firm's resource and capabilities of Malaysian automotive industry. Procedia Soc. Behav. Sci. **195**, 463–472 (2015)
2. Argubi-Wollesen, A., Wollesen, B., Leitner, M., Mattes, K.: Human body mechanics of pushing and pulling: analyzing the factors of task-related strain on the musculoskeletal system. Saf. Health Work **8**, 11–18 (2017)
3. Mossa, G., Boenzi, F., Digiesi, S., Mummolo, G., Romano, V.A.: Productivity and ergonomic risk in human based production systems: a job-rotation scheduling model. Int. J. Prod. Econ. **171**, 471–477 (2016)
4. Oakman, J., Rothmore, P., Tappin, D.: Intervention development to reduce musculoskeletal disorders: is the process on target? Appl. Ergon. **56**, 179–186 (2016)
5. Occhipinti, E., Colombini, D.: A toolkit for the analysis of biomechanical overload and prevention of WMSDs: criteria, procedures and tool selection in a step-by-step approach. Int. J. Ind. Ergon. **52**, 18–28 (2016)
6. Aziz, R.A., Rebi, M.A.T., Rani, A., Rohani, J.M.: Work-related musculoskeletal disorders among assembly workers in Malaysia. J. Occup. Saf. Heal. **11**, 33–38 (2014)
7. Chander, D.S., Cavatorta, M.P.: Multi-directional one-handed strength assessments using AnyBody Modeling Systems. Appl. Ergon. **67**, 225–236 (2018)
8. Pallant, J.: SPSS Survival Manual: a step by step guide to data analysis using SPSS. In: SPSS Survival Manual. National Library of Australia (2007)
9. Biswas, G., Bhattacharya, A., Bhattacharya, R.: Occupational health status of construction workers: a review. Int. J. Med. Sci. Public Health **6**, 1 (2017)
10. Weale, V.P., Wells, Y., Oakman, J.: Self-reported musculoskeletal disorder pain: the role of job hazards and work-life interaction. Am. J. Ind. Med. **61**, 130–139 (2018)
11. Andersen, L.L., Fallentin, N., Ajslev, J.Z.N., Jakobsen, M.D., Sundstrup, E.: Association between occupational lifting and day-to-day change in low-back pain intensity based on company records and text messages. Scand. J. Work Environ. Health **43**, 68–74 (2017)
12. Katz, J.D.: Control of the environment in the operating room. Anesth. Analg. **125**, 1214–1218 (2017)
13. Wee, H.M., Fu, K., Chen, Z., Zhang, Y.: Optimal production inventory decision with learning and fatigue behavioral effect in labor intensive manufacturing. Sci. Iran. (2018)

A Closed-Loop Supply Chain Model for Manufacturer-Collector-Retailer with Rework, Waste Disposal, Carbon Cap and Trade Regulation

Niimas Ayu Frensilia Putri Adam(✉), Wakhid Ahmad Jauhari, and Cucuk Nur Rosyidi

Department of Industrial Engineering, Sebelas Maret University, Surakarta, Indonesia
nfrensilia@gmail.com

Abstract. The importance of an industrial system that is both economically and environmentally sustainable has gained considerable attention in this last decade. In this paper, we address a closed-loop supply chain in which the manufacturer produces the brand-new product and the retailer sells it through the market. A collector will collect the used product from the consumer and sends it to the manufacturer to be remanufactured or refurbished. The used items which have not passed the acceptable quality level of the collector will be considered as a waste and need to be disposed. The defective item produced through the manufacturing system will be reworked. In order to restrict carbon emissions generated by the production phases in this system, trade regulation and carbon cap are used. The proposed model simultaneously optimizes the quality level of the product that should be produced by the manufacturer and the pricing decision of the retailer to maximize the total joint profit of the supply chain. A numerical example is taken to illustrate the proposed model.

Keywords: Remanufacturing · Manufacturing · Closed-loop supply chain · Refurbishing · Carbon emissions

1 Introduction

In recent decades, many companies have implemented closed-loop supply chains (CLSC) system due to the increasing awareness of environmental sustainability. Approximately 2.01 billion metric tons of municipal solid waste (MSW) are being produced each year worldwide. It is estimated that by 2050, this number will increase significantly to 3.40 billion metric tons. It is estimated that only 5.5% of today's waste is composted and 13.5% is recycled and [1].

Recently, there has been a significant growth in research and applications on managing supply chain in reverse chains. Maiti and Giri [2] developed a closed loop supply chain where reverse supply involve third parties as a recycle dealer or collector. Maiti and Giri [2] assumed that all of the used items are remanufacturable. In fact, not all used products can obtained the same quality as new products so that other recovery

M. N. Osman Zahid et al. (Eds.): iMEC-APCOMS 2019, LNME, pp. 284–290, 2020.
https://doi.org/10.1007/978-981-15-0950-6_44

processes are needed. Later, Christy [3] developed a research based on Maiti and Giri [2] where there are two type of recovery processes, remanufacturing and refurbishing with a demand that depends on the retail price and quality of the product. One common assumption in all of these models is that the manufacturing process is perfect where no items are defective during the manufacturing process. In fact, in any production system, defective products cannot be avoided due to many reasons [4]. The closed-loop supply chain model developed by Moshtagh and Taleizadeh [5] considers the existence of imperfect production that produces defective products during the production process. The defective product is then carried out by reworking process to improve the quality of the product so the quality will be the same as the product that does not have defect. Some used item that has not pass the acceptable quality will be considered as a waste and need to be disposed with waste disposal process [6]. In reducing and curb the carbon emissions, trade regulation and carbon cap as a market-based approach have been widely used by several researchers. In the regulation, government agencies allocate a number of carbon emissions to a company (carbon cap), and companies can buy or sell shortages or residual carbon emissions held in carbon trading markets such as the European Union Emissions Trading System (EU ETS) [7]. Kundu [8] examines the impact of emission carbon policies on decisions on manufacturing, remanufacturing, and collecting used items.

Motivated by the description above, in this study, we developed a CLSC model that uses the Maiti and Giri [2] model as a main reference. There is remanufacturing and refurbishing, and taking into consideration the process of reworking, waste disposal, and the cost of carbon emissions on the model. The models involve three parties, manufacturer, retailer, and collector. The demand function is linearly dependent on the retail price, and product quality. We aim to obtain maximum profit in the CLSC system.

2 Notations and Assumptions

The notations that will be used to develop the proposed model are shown in Table 1.

The following assumptions for the proposed models are presented as follows:

(1) The CLSC involve a manufacturer, a retailer, and a collector.
(2) The demand is linearly dependent on the retailer selling price and the quality of the product. ($D = d - \alpha P_R + \beta Q$)
(3) There are used products that collectors cannot collect.
(4) The recoverable products that are returned to manufacturer will be processed by remanufacturing and the rest are processed with refurbishing, the cost for each process is different ($C_M > C_{REM} > C_{REF}$).
(5) The products produced by the remanufacturing process have the same quality as new products ($Q > Q_r$). Products produced by the refurbishing process have better quality than used products, but are lower than the new products and will be sold to the secondary market ($Q_2 > Q_f$).
(6) The proportion of product defects (f) is constant.
(7) All defective products can be repaired with the reworking process and all products reworked have good quality.

Table 1. The notations and parameters in this model

Manufacturer's parameters			
ρ	Proportion of remanufacturable items	f	Proportion of defective items
P_M	Unit wholesale price ($/unit)	C_q	Quality improvement cost ($/unit)
C_S	Cost of raw materials ($/unit)	P_{2nd}	Wholesale price of secondary market ($/unit)
C_M	Cost of manufacturing end product from raw materials ($/unit)	C_{insp}	Cost of inspection process for manufactured end product ($/unit)
C_{REM}	Remanufacturing cost ($/unit)	C_{RW}	Reworking cost ($/unit)
C_{REF}	Cost of refurbishing end product from return products ($/unit)	C_p	Unit trading price of carbon emission permit ($/grCO$_2$)
C_{IM}	Cost of inspection process for recoverable items ($/unit)	$E_m(s)$	Total emission of the supply chain (grCO$_2$/unit)
Q_f	Product quality of refurbishable products,	C	Carbon emission cap (grCO$_2$)
Q_r	Product quality of remanufacturable products	e_1	Carbon emissions per unit in the manufacturing stage (grCO$_2$/unit)
Q_2	Product quality to be sold to secondary market	e_2	Carbon emissions per unit in the remanufacturing stage (grCO$_2$/unit)
e_4	Carbon emissions per unit in the manufacturing stage when (grCO$_2$/unit)	e_3	Carbon emissions per unit in the refurbishing stage (grCO$_2$/unit)
Collector's parameters			
τ	Return rate of used items	C_{IT}	Cost of inspection process ($/unit)
λ	Proportion of unrecoverable items	C_{WD}	Cost of waste disposal ($/unit)
C_T	Recycling cost of used products ($/unit)	P_T	Recoverable items selling price ($/unit)
Other parameters			
D	Demand rate of the product (unit)	α	Sensitivity factor of selling price
d	Basic market demand (unit)	β	Sensitivity factor of quality
Decision variable			
P_R	Unit selling price ($/unit)	Q	Product quality maintained by the manufacturer, $0 < Q < 1$

3 Model Formulation and Analysis

3.1 Problem Description

As shown in Fig. 1, the proposed model concerns a supply chain model for closed-loop three-party supply chain systems (retailer, manufacturer, and collector) for single products. The diagram is modified model of Maiti and Giri [2] and Christy [3]. The collector collect τD product from consuments and λτD products must go through waste disposal process because it does not meet the acceptable quality level of the manufacturer. From $(1-\lambda)\tau D$ products that are sent to the manufacturer, only $(1-\lambda)\rho\tau D$ products are remanufacturable, the rest $(1-\lambda)(1-\rho)\tau D$ products will be refurbished and sold to the secondary market. To satisfy the demand D, the manufacturer must manufacture $(1-(1-\lambda)\rho\tau)D$ products. In the manufacturing process, there are $f(1-(1-\lambda)\rho\tau)D$ defective products and need to go through the reworking process.

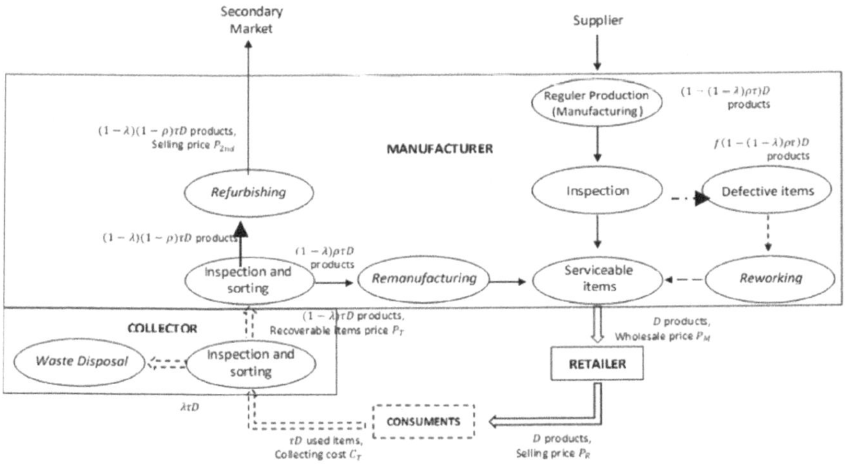

Fig. 1. Graphical presentation of the proposed CLSC system.

3.2 Mathematical Model for the Manufacturer

Total cost of the manufacturer includes manufacturing cost, remanufacturing cost, refurbishing cost, collecting recoverable items cost, reworking cost, and carbon emission cost. The total amount of carbon emission can be formulated as follow.

$$E_m = e_1(1-(1-\lambda)\rho\tau)D + e_2(1-\lambda)\rho\tau D + e_3 f(1-(1-\lambda)\rho\tau)D + e_4(1-\lambda)(1-\rho)\tau \quad (1)$$

The total profit of the manufacturer is obtained by subtracting total cost with total revenue as expressed below.

$$\begin{aligned}\Pi_M(P_R, Q) &= DP_M - (1-(1-\lambda)\rho\tau)D\left(C_M + C_S + C_{insp} + C_q Q^2\right)\\ &- (1-\lambda)\rho\tau D\left(C_{REM} + C_q\left(Q^2 - Q_r^2\right)\right) - (1-\lambda)(1-\rho)\tau D\left(C_{REF} + C_q\left(Q_2^2 - Q_f^2\right) - P_{2nd}\right)\\ &- (1-\lambda)\tau D(P_T + C_{IM}) - f(1-(1-\lambda)\rho\tau)DC_{rw} - C_p[E_m - C]\end{aligned}$$

3.3 Mathematical Model for the Retailer

Retailer profit (Π_R) is obtained by subtracting total cost (TC_R) from total revenue (TR_R). The total cost of the retailer TC_R consists of purchasing cost of the products from the manufacturer.

$$\Pi_R(P_R, Q) = DP_R - DP_M \tag{3}$$

3.4 Mathematical Model for the Collector

Total cost of the collector (TC_C) consists of collecting used items cost (TC_{RC}) and waste disposal cost (TC_{WD}). The collecting cost (TC_{RC}) can be obtained from purchasing cost of the used items and inspection and sorting cost. Thus, the collector profit (Π_T) is formulated by subtracting total revenue (TR_C) with total cost (TC_C).

$$\Pi_T(P_R, Q) = TR_C - TC_C = (1-\lambda)\tau DP_T - \tau D(C_T + C_{IT}) - \lambda\tau DC_{wd} \tag{4}$$

Finally, the joint total profit of the closed-loop supply chain system ($\Pi(P_R, Q)$) can be calculated by summing up the total profit of the manufacturer, the retailer, and the collector.

$$\Pi(P_R, Q) = \Pi_M + \Pi_R + \Pi_T \tag{5}$$

$$\begin{aligned}\Pi(P_R, Q) &= DP_R + (1-\lambda)\tau DP_T - (1-(1-\lambda)\rho\tau)D\big(C_M + C_S + C_{insp} + C_qQ^2\big)\\ &- (1-\lambda)\rho\tau D\big(C_{REM} + C_q(Q^2 - Q_r^2)\big) - (1-\lambda)(1-\rho)\tau D\left(C_{REF} + C_q\left(Q_f^2 - Q_r^2\right) - P_{2nd}\right)\\ &- (1-\lambda)\tau D(P_T + C_{IM}) - f(1-(1-\lambda)\rho\tau)DC_{rw} - C_p[E_m - C] - \tau D(C_T + C_{IT}) - \lambda\tau DC_{wd}\end{aligned} \tag{6}$$

3.5 Solution Procedure

The optimal values of P_R and Q to maximize $\Pi(P_R, Q)$ are started from finding the partial derivatives of $\Pi(P_R, Q)$ with respect to P_R, while variable quality of the product (Q) is obtained through iteration with the help of the algorithm given that $0 < Q < 1$. The optimal function of P_R* is formulated and the proposed algorithm of iterative procedure to find the optimal decision is listed below, respectively.

$$\begin{aligned}P_R* &= -\frac{1}{2\beta}(-d - Q\gamma + \beta(-f(1-(1-\lambda)\rho\tau)C_{\mathrm{RW}} - (1-(1-\lambda)\rho\tau)(C_{\mathrm{insp}} + C_M\\ &+ Q^2C_q + C_S) + \tau(C_{\mathrm{IT}} + C_T) + \lambda\tau C_{\mathrm{WD}} + C_P(-(1-(1-\lambda)\rho\tau)e_1 - (1-\lambda)\rho\tau e_2\\ &- f(1-(1-\lambda)\rho\tau)e_3 - (1-\lambda)(1-\rho)\tau e_4) - (1-\lambda)\tau P_T - (1-\lambda)\tau(C_{\mathrm{IM}} + P_T)\\ &- (1-\lambda)(1-\rho)\tau(C_{\mathrm{REF}} - P_{\mathrm{2nd}} + C_q(Q_2^2 - Q_f^2)) - (1-\lambda)\rho\tau(C_{\mathrm{REM}} + C_q(Q^2 - Q_r^2))))\end{aligned} \tag{7}$$

Step 1 Set the initial value of $Q = 0.01$ and $\Pi\left(P^*_{R_{Q-1}}, Q-1\right) = \infty$
Step 2 Calculate the value of P_R* using Eq. (7)
Step 3 Calculate the value of $\Pi(P_R*, Q)$ using Eq. (6).

Step 4 If $\Pi\left(P^*_{R_Q}, Q\right) \geq \Pi\left(P^*_{R_{q-0.01}}, Q - 0.01\right)$, repeat step 2–3 by setting $Q = Q + 0.01$. If not go to step 5.

Step 5 Calculate $\Pi\left(P^*_{R_Q}, Q\right) \geq \Pi\left(P^*_{R_{q-0.01}}\right), Q^* - 0.01)$, then the value $\left(Q^*, P^*_R\right)$ is the optimal solution.

4 Numerical Example

This section illustrates the performance of the developed model derived in the previous section. The parameters values are mainly adapted from the works of Maiti and Giri [2] and Christy et al. [3]. The parameters that are used, C_m = 60, C_S = 70, C_{insp} = 15, C_q = 6, C_{rem} = 30, C_{ref} = 20, C_{rw} = 50, C_{im} = 10, C_t = 15, C_{it} = 5, C_{wd} = 8, P_{2nd} = 500, P_m = 800, and P_T = 30. The basic demand d = 120 with α = 0.07, β = 0.70. We use e_1 = 40, e_2 = 37, e_3 = 23, e_4 = 10 and use C_p = 2.5 and C = 10,000. The rest of the variable are Q_r = 0.2, Q_f = 0.1, Q_2 = 0.5Q, f = 0.04, ρ = 0.7, λ = 0.13, τ = 0.75. The obtained result of optimal solution are presented in Table 2.

Table 2. Result of optimal solution

Optimal decisions	Optimum values
Q	0.83
P_R	988.277 \$/unit
Π_M	\$ 5.6281×10^4
Π_R	\$ 9.6777×10^3
Π_T	\$ 3.2136×10^3
Π	\$ 6.9172×10^4

Carbon emission cap (C) and unit trading price of carbon emission permit (C_p) is adopted from Bai et al. [7].

Based on the results it is known that the manufacturer receives the greatest profit. This is because the manufacturer plays the biggest role in the system. Meanwhile, third parties receive the lowest profits due to a very small role in the system. We observe that carbon cap and trade regulation is an effective way to reduce carbon emission because it has a quite significant impact to the firm's overall cost. Thus, the company will try to lower the emission.

5 Conclusions

In this paper we construct a closed loop supply chain model for manufacturer-retailer-collector by considering rework processes, waste disposal activity, and carbon emission costs where the level of demand depends on the quality of the product and the selling price. This paper applies two recovery processes, remanufacturing and refurbishing. This paper also applies carbon cap and trade regulation which are effective methods to reduce carbon emissions.

For future research, the model can be extended by allowing the inclusion of stochastic environments, such as stochastic demand and stochastic returns.

Acknowledgement. This research was partially funded by Research and Community Service (P2M) PNBP funds 2019 budgeting year with contract number 516/UNS27.21/PM/2019.

References

1. WorldBank. What a Waste: Updated Look into the Future of Solid Waste Management (2018)
2. Maiti, T., Giri, B.C.: A closed loop supply chain under retail price and product quality dependent demand. J. Manuf. Syst. **37**, 624–637 (2015)
3. Christy, A.Y., Fauzi, B.N., Kurdi, N., Jauhari, W.A., Saputro, D.A.: CLSC under retail price and quality dependent demand with remanufacturing and refurbishing. J. Phys. Conf. Ser. **855**(1), 012009 (2017)
4. Giri, B.C., Sharma, S.: Optimizing a closed-loop supply chain with manufacturing defects and quality dependent return rate. J. Manuf. Syst. **35**, 92–111 (2015)
5. Moshtagh, M.S., Taleizadeh, A.A.: Stochastic integrated manufacturing and remanufacturing model with shortage, rework, and quality based return rate in a closed loop supply chain. J. Cleaner Prod. **141** (2016)
6. Hasanov, P., Jaber, H.H., Zanoni, S., Zavanella, L.: Close-loop supply chain system with energy, transportation, and waste disposal cost. Intl. J. Sustain. Eng. **6**(4), 1–7 (2013)
7. Bai, Q., Gong, Y., Jin, M., Xu, X.: Effects of carbon emission reduction on supply chain coordination with vendor-managed deteriorating product inventory. Int. J. Prod. Econ. **2018**, 83–99 (2018)
8. Kundu, S., Chakrabarti, T.: Impact of carbon emission policies on manufacturing, remanufacturing, and collection of used item decisions with price dependent return rate. OPSEARCH **5**(2), 532–555 (2018)

Optimization of Woven Fabric Production Process on Picanol Omniplus Air Jet Machine Using Taguchi Multi-response and Grey Relational Analysis Methods

Yunus Nazar[1,2](✉), Eko Pujiyanto[1,2](✉), and Cucuk Nur Rosyidi[1,2](✉)

[1] Master Program of Industrial Engineering, Universitas Sebelas Maret, JL. Ir. Sutami 36A, Surakarta 57126, Indonesia
yunusnazar@kemenperin.go.id, ekopujiyanto@ft.uns.ac.id, cucuk@uns.ac.id
[2] AK-Tekstil Solo, JL. Ki Hajar Dewantara, Surakarta 57126, Indonesia

Abstract. The purposes of this research are to determine the optimal process parameters of woven fabric production process and the rank of factors that influence the responses of the fabric, namely tear strength and air permeability. In this research, Taguchi method was used in the optimization of woven fabric production by involving four factors, namely weft yarn type, weft density, air pressure, and warp tension each with the three levels. Orthogonal Array L9 is used for experiments with two responses, namely the fabric tear strength and air permeability. Grey relational analysis was used to determine the optimal process parameters. The results of the experiments showed that the fourth experiment has the highest rank with a value of 0.887 with TC 30 of yarn types, 62 of weft density, 4 bar of air pressure and 2 kN of warp tension. We also found that weft yarn type become the most influence factor of the responses. For the next stage, a variance analysis must be carried out to find the significance of the factors. Subsequent research makes it possible to add other responses, namely the tensile strength of the fabric, the pilling performance, and the drape test.

Keywords: Air permeability · Fabric tear strength · Grey relational analysis · Multi-response · Taguchi

1 Introduction

Textiles are products that will always be needed in human life. Therefore, the textile industry is required to meet the needs in accordance with population growth. The textile industry is one of the Indonesian government's priorities in the era of industrial revolution 4.0. This industry sector priority selection is based on the highest demand sector in the world. In addition, based on statistical data obtained from Central Bureau of Statistics and the Ministry of Industry, from 2011 to 2015, the textile industry became the fifth largest non-oil and gas processing industry contributing to Indonesia's Gross Domestic Product (GDP) with an average of 1.32%.

M. N. Osman Zahid et al. (Eds.): iMEC-APCOMS 2019, LNME, pp. 291–297, 2020.
https://doi.org/10.1007/978-981-15-0950-6_45

The textile industry in general includes yarn making, fabric making (weaving, knitting, and non wovens) and fabric improvement (dyeing and printing). Meanwhile, the process from fabric to "apparel" is referred to as garment or textile products. In the industrial system, there are 6 factors that influence the sustainability of the production process, namely human, material, machine, methods, money, and information technology. Recently, the competition in textile industry is getting higher in terms of productivity and business. In terms of productivity, the industry has to increase its effectiveness and efficiency by eliminating non-value-added activities. Shanshan et al. [1] explained that quality is an important factor for companies to succeed in the market. Costa [2] pointed the importance of maintaining quality and ensuring the quality characteristics in term of target values and variability must be kept at minimal.

Woven fabrics can be classified according to the type of woven used, trade name, weight of fabric, color-giving method, and end-use. Fabrics have flexibility which is one of the important characteristics of woven fabric [3]. Ga [4] stated several factors influence the characteristics of woven fabrics, namely woven structures, warp threads, and weft yarns. Turhan and Eren [5] stated that the main purpose of woven fabrics is to develop new fabrics that have the most appropriate properties for end-use applications to achieve an increase in the level of fabric quality. Masaeli et al. [6] conducted experiments by combining the types of weft yarn type, woven fabrics, weft density, air pressure, and warp closure time with two responses namely the angle of fabrics folding and fabrics pilling. Sarpkaya et al. [7] conducted experiments by combining the factors of woven fabrics, the intensity of dyeing and the number of the fabrics lifting with two responses namely the tensile strength in fabrics and the fabrics elongation. Meanwhile, Umair et al. [8] conducted experiments by combining the angular factors of the degree of yarn, woven fabrics and types of yarn with seven responses namely crimped grey and after washed fabric, weight of fabric, thickness of fabric, stiffness of fabric, flexibility of fabric, and air permeability.

In this research, a combination experiment of factors of the weft yarn type, weft density, air pressure, and warp tension with a response of the fabric tear strength and air permeability was conducted using the method of multi-response Taguchi and grey relational analysis.

2 Literature Review

Masaeli et al. [6] used L18 orthogonal array to obtain optimal settings of factors and levels, with two responses using the grey relational analysis method. The results showed that the fabric pilling response was mostly influenced by woven fabric and weft density, while fabric fold response was mostly affected only by fabric woven factor.

Sarpkaya et al. [7] used L9 orthogonal array to obtain optimal settings of factors and levels with 4 responses, two of them were correlated. Those responses were the tensile strength in fabrics (direction of warp and weft), stretch of fabric (direction of warp and weft) which then processed using grey relational analysis and converted into one response. The results showed that the optimal setting was twill 2/1 fabric cover, 1% dyeing intensity and without lifting the fabric combination

Umair et al. [8] proposed a simultaneous optimization approach of the properties of woven fabrics using factor analysis. The research also indicated that the properties of woven fabrics were influenced by the structure of yarn and/or woven fabrics. The results of the research showed that the woven fabric with a low angle of twisting yarn on the twill weave had the most optimal results.

3 Taguchi Method

Taguchi explained that quality improvement is a continuous effort to reduce product variance around the target value. To fulfill this, Taguchi proposed experiments using Orthogonal Arrays (OA). The Taguchi method serves to determine the number of experiments to be carried out and determine the optimal value for each desired response. Hsu [9] stated that Taguchi method was one of the most widely used approaches to solve parameter design problems.

Anthony [10] stated Taguchi parameter design is widely accepted methodology among researchers and practitioners to control quality, improve product quality and process performance at a low cost. In this problem, optimization of some responses only received limited attention among researchers and practitioners. In complex experimental processes, we often have to optimize multiple responses simultaneously rather than optimizing one response at a time. In a multi-response problem, the goal is to determine the optimal settings for factors and process variables that will simultaneously optimize multiple responses.

4 Grey Relational Analysis

Sarpkaya et al. [5] explained that grey relational analysis allowed to optimize of more than one responses characteristic. Grey relational analysis steps as follows:

1. Data Normalization

$$\text{The Larger - The Better}\, x_i(k) = \frac{x_i^0(k) - min\, x_i^0(k)}{\max\ x_i^0(k) - \min x_i^0(k)} \tag{1}$$

2. Calculate the grey relational coefficient of normalization with equations:

$$\xi_i(k) = \frac{\Delta_{min} + \xi\,\Delta_{max}}{\Delta_{oi}(k) + \xi\,\Delta_{max}} \tag{2}$$

where, Δ_{oi} is the deviation sequence of the reference sequence and the comparability sequence and

$$\Delta_{oi} = x_0(k) - x_i(k), \tag{3}$$

where $x_0(k)$ denotes the reference sequence and $x_i(k)$ termed as comparability sequence. Δ_{min} and Δ_{max} are the minimum and maximum value of the absolute

(Δ_{oi}) of all comparing sequences. ξ is distinguishing or identification coefficient and the range is between 0 to 1. Usually, the value of ξ is taken as 0.5.

3. Calculate the grey relational grade with equation

$$\gamma_i = \frac{1}{n}\sum_{k=1}^{n} \xi_i(k) \tag{4}$$

where, γ_i is the required grey relational grade for i th experiment and n = number response characteristics. The grey relational grade represents the level of correlation between the reference sequence and the comparability sequence and the overall representative of all the quality characteristics.

5 Research Method

In this research, we use Picanol Omniplus air jet machine because the machine has a very high speed with a sumo motor, precision fabric production and easy maintenance. Firstly in this research, a literature study was conducted to find some literature related to experiments on woven fabrics including the factors, level and responses. Taguchi method is used in the optimization of the weaving fabric experiment against two responses, namely tear strength and air permeability. The responses, factors, levels and orthogonal array that will be used in the experiment must be determined first. In this research, we set two machine parameters at constant level, namely machine speed and shed closing time. Grey relational analysis method is used to determine the optimal setting of factors and levels in the experiment.

6 Result and Discussion

The results of processing the data using Orthogonal Array (OA) L9 for four factors, namely weft yarn type, weft density, air pressure, and warp tension using two responses, namely tear strength and air permeability are shown in Table 1.

The grey relational analysis passes through the following steps:

Step 1

Normalization was done for both responses using Eq. (1). The results of normalization shown in Table 2.

Step 2

The grey relation coefficient was calculated by first calculating the deviation sequence as shown in Table 3.

Step 3

In step 3, grey relational grade calculation was done based on deviation sequence and grey relational coefficient. The grey relational grade is shown in Table 3.

Step 4

The optimal parameter setting was found by averaging correspondent GRG according to each factor and level. Table 4 shows the results. From the table, we found

Table 1. Experimental result OA (L9)

Run no	Process parameters				Experimental results	
	WYT	WD	AP	WT	TS	AP
1	Cotton	62	3	1.5	842.4	20
2	Cotton	65	4	1.75	864.6	20
3	Cotton	68	5	2	863.6	19
4	TC	62	4	2	2416	26
5	TC	65	5	1.5	2366	21
6	TC	68	3	1.75	1990	17
7	TR	62	5	1.75	2542	25
8	TR	65	3	2	2486	20
9	TR	68	4	1.5	2722	16

Table 2. Normalization data

Run no	Tear strength (Larger-the-better)	Air permeability (Larger-the-better)
1	0.000	0.400
2	0.012	0.400
3	0.011	0.300
4	0.837	1.000
5	0.811	0.500
6	0.611	0.100
7	0.904	0.900
8	0.874	0.400
9	1.000	0.000

Table 3. Result GRC and GRG

Run no	Deviation sequence		GRC		GRG	Rank
	TS	AP	TS	AP		
1	1.000	0.600	0.333	0.455	0.394	8
2	0.988	0.600	0.336	0.455	0.395	7
3	0.989	0.700	0.336	0.417	0.376	9
4	0.163	0.000	0.754	1.000	0.877	1
5	0.189	0.500	0.725	0.500	0.613	5
6	0.389	0.900	0.562	0.357	0.460	6
7	0.096	0.100	0.839	0.833	0.836	2
8	0.126	0.600	0.799	0.455	0.627	4
9	0.000	1.000	1.000	0.333	0.667	3

level 3 of weft yarn type, level 1 of weft density, level 2 of air pressure and level 3 of warp tension to be the optimal setting for woven fabrics production process on Picanol Omniplus Air jet machine.

From Table 4, we can also observe that weft yarn type becomes the most influential factor in the woven fabric production process. Yarn has several specific characteristics, namely yarn count, twist, and characteristics of materials that will affect the results of the response in production fabric. Since yarn becomes the most influential factors in the experiment, then it will be interesting to study the effect of those specific characteristics of the yarn to some responses of the fabric.

Table 4. Mean GRG

Factors	Mean GRG			Max-Min	Rank
	Level 1	Level 2	Level 3		
Weft yarn type	0.39	0.65	0.71	0.32	1
Weft density	0.70	0.54	0.5	0.20	2
Air pressure	0.49	0.65	0.61	0.15	3
Warp tension	0.56	0.56	0.63	0.07	4

7 Conclusion

In this research, the Taguchi method was used in the optimization of the experiment of woven fabrics production by involving four factors, namely weft yarn type, weft density, air pressure, and warp tension with each of three levels. Orthogonal Array L9 was used in the experiments with two responses, namely fabric tear strength and air permeability. Grey relational analysis was used to determine the optimal production setting. The result of a grey relational grade was the fourth experiment obtained the highest optimal rank with a value of 0.887 with a combination of yarn type TC 30, weft density 62, air pressure 4 bar, and 2 kN warp tension. Further research it is possible to add other responses such as tensile strength, pilling performance and the drape test. Analysis of variance needs to be performed to find the significant of the factors. Subsequent research makes it possible to add other responses, namely the tensile strength of the fabric, the pilling performance, and the drape test.

References

1. Shanshan, L., He, Z., Vining, G.: Simultaneous optimization of quality and reliability characteristics through designed experiment. Qual. Eng. **29**(3), 344–357 (2017)
2. Costa, N.R.P.: Simultaneous optimization of mean and standard deviation. Qual. Eng. **22**(3), 140–149 (2010)
3. Adanur, S.: Handbook of Weaving. CRC Press Taylor and Francis, Boca Raton (2001)
4. Ga, N.: Using taguchi methodology to optimize woven fabrics air permeability. J. Text. Sci. Eng. **7**(6), 325–328 (2017)

5. Turhan, Y., Eren, R.: The effect of loom settings on weaveability limits on air-jet weaving machines. Text. Res. J. **82**(2), 72–182 (2011)
6. Masaeli, R., Hasani, H., Shanbeh, M.: Optimizing the physical properties of elastic-woven fabrics using Grey-Taguchi method. J. Text. Inst. **106**(8), 814–822 (2014)
7. Sarpkaya, C., Ozgur, E., Sabir, E.C.: The optimization of woven fabric tensile strength with taguchi method based on grey relational analysis. Tekstil ve Konfeksiyon **25**(4), 293–299 (2015)
8. Umair, M., Shaker, K., Ahmad, N., Hussain, M., Jabbar, M., dan Nawab, Y.: Simultaneous optimization of woven fabric properties using principal component analysis. J. Nat. Fibers **14**(6), 846–857 (2017)
9. Hsu, C.-M.: Improving the lighting performance of a 3535 packaged hi-power LED using genetic programming, quality loss functions and particle swarm optimization. Appl. Soft Comput. **12**(9), 2933–2947 (2012)
10. Anthony, J.: Multi-response optimization in industrial experiments using Taguchi's quality loss function and principal component analysis. Qual. Reliab. Eng. Int. **16**(1), 3–8 (2000)

Re-designing an Assembly Lines at an Automotive Manufacturing Company

Leonard Leymena(✉), Cucuk Nur Rosyidi, and Wakhid Ahmad Jauhari

Sebelas Maret University, Surakarta, Indonesia
leonardleymena@student.uns.ac.id

Abstract. One of the major industries in manufacturing is the automotive industry. the production process of motorcycle is separated into three sections: manufacturing, assembly and distribution activities. One of which is the section of the Assembly Engine that assembles engine components into the main engine on a motorcycle. This research focuses on Assembly Line, which covers 41 stages of the machining process. To meet the set production targets, companies must increase their productivity by reducing the bottleneck. The purpose of this paper is to improve the balance of assembly lines by assign the tasks, machines, and operators to each work station. To fulfill this objective, an optimization model developed with the objective function is used to minimize Mean Absolute Deviation (MAD) as a measure of the time difference between work stations. The search for optimal solutions with models can produce task, machine, and operator allocations to each work station with MAD of 17.5. These results were obtained after 4001 iterations with runtime for 6721 min and 14 s with Crystal Ball software.

Keywords: Assembly line balancing problem · Combinatorial optimization model · Manufacturing industry

1 Introduction

The assembly line consists of a set of workstations in which a specific task in a predefined sequence is processed. The problem of line balancing starts from the unbalanced workload at each work station. Line balancing needs to be done to create the a balance of the production line so that the production process will run smoothly. Then the application of line balancing in a production system will increase the line efficiency of the company's production system. The company engaged in the automotive industry that produces motorbikes. The company has several sections, one of which is Assembly Engine Section that assembles the engine components into the main engine of a motorcycle. The problem of assembly lines is known as the Assembly Line Balancing Problem (ALBP). ALBP deals with the allocation of tasks to each work station by taking into account the sequence of processes so as to minimize/maximize a certain objective functions. ALBP is NP-hard class as combinatorial optimization [1]. Chen et al. [2] conducted a study of ALBP with differences in the level of worker ability and workload smoothing on the sewing line in a garment industry. With the

M. N. Osman Zahid et al. (Eds.): iMEC-APCOMS 2019, LNME, pp. 298–304, 2020.
https://doi.org/10.1007/978-981-15-0950-6_46

research resulted shorter cycle times and higher worker utilities. In this study, the problem of assembly lines in the Assembly Engine will be conducted using an optimization model developed by Chen et al. [2].

2 Literature Review

2.1 Assembly Line Balancing Problem (ALBP)

The assembly line balancing (ALB) problems can be classified based on the number of models produced in the line, the nature of task times (deterministic or probabilistic) and the nature of flow (straight-type or U-type). In the same assembly line, one or more models of a product may be assembled. If only a single model is assembled in the line, then the production system is defined as a single-model assembly system; otherwise, it is called a multi-model assembly system. The processing times of the tasks may be either deterministic or probabilistic. If the tasks are performed using all sophisticated tools and fixtures by highly skilled labors, then the processing times of the tasks may be approximated to deterministic quantity, because the variability in the processing times may be less under such situation. This is because of the facilitating nature of tools and availability of operators with required skills [3]. But, normally, in assembly-type o erations, the processing times will vary, which can be characterized in the form of some probability distribution. The arrangement of the workstations of the assembly line may be in a straight-line layout or in a U-shape layout [4, 5].

2.2 Determination the Number of Work Stations

The number of work stations is needed as a constraint in optimization. The number of work stations can be calculated by the following equation:

$$W_{\min} = \frac{\sum_{J=1}^{W} ti}{TT} \tag{1}$$

Where,
$W_{\min}$ = Minimum number of work stations
t_i = Standard processing time for activities i
TT = Takt Time

Takt time can be explained as the time needed to produce one unit of product based on the speed of customer demand. Takt Time can be calculated by the following equation:

$$TaktTime = \frac{Production\ Time\ /\ period}{t\ Period} \tag{2}$$

2.3 Productivity and Efficiency

Productivity deals with actual working hours in a given period compared to production capacity based on machines or the number of workers available. The formula used to determine the productivity is as follow:

$$Productivity = \frac{Number\ of\ actual\ working\ hours}{Total\ work\ amount} \tag{3}$$

Efficiency is calculated by comparing actual production time in a certain period against the predetermined production time. The formula used to calculate the efficiency is shown in Eq. (4):

$$Efficiency = \frac{Standard\ Production\ Time}{Actual\ Production\ Time} \tag{4}$$

3 Model Adjustment

The following assumptions are used in this:

- Processing time, Precedence relation-ship, and machine types and their respective number are known.

Table 1. List of notations

i	Index of task, where i = 1, 2,. .., N
j	Index of workstations j = 1,2, …, W
k	Index of machine type k = 1,2, …, M
u	Index of labor type u = 1,2, …, P
N	Total number of tasks
W	Total number of Workstations
M	Total number of machine
P	Total number of labor
PN	Maximum number of operators in a workstation
t_i	Standard processing time of task i
CT	Cycle time (the average time between two successive products coming out of the end of the line)
I_j	Subset of tasks assigned to workstation j
U_j	Subset of labor type assigned to workstation j
K_j	Subset of machine types assigned to workstation j
H_j	Number of operators assigned to workstation j
X_{ij}	1: if task i is assigned to station j; 0: otherwise
Y_{jk}	1: if machine type k is allocated to workstation j; 0: otherwise
L_{uj}	1: if labor type u is assigned to workstation j; 0: otherwise
R_{uk}	Labor efficiency of labor type u for machine type k
T_j	Workload of workstation j (i.e., the total operating time of labors assigned to workstation j)
$\bar{T}$	Average workload of work stations of the sewing line

- The assembly line consists of a series of workstations and a number of non-permanent work stations.
- The processing time includes setup time.
- Only one operator can be assigned to each work station (Table 1).

In this study, the model of Chen et al. [2] was used to solve the assembly line balancing problem at the Assembly Engine section of the company. By paying attention to the condition of the assembly line system at the Assembly Engine section, it is necessary to modify the model by reducing the constraints. Equation (3)–(11) constitutes the model to solve the problem.

$$\text{Minimize MAD } = \frac{1}{W}\sum\nolimits_{j=1}^{W} |T_j - \bar{\text{T}}| \tag{5}$$

$$\sum\nolimits_{j=1}^{W} X_{ij} = 1 \tag{6}$$

$$1 \leq \sum\nolimits_{k=1}^{M} Y_{jk} \leq M \tag{7}$$

$$\sum\nolimits_{j} jX_{aj} \leq \sum\nolimits_{j} jX_{bj} \text{ if task a precedes task b} \tag{8}$$

$$\sum\nolimits_{k \in K_j} \sum\nolimits_{u \in U_j} \sum\nolimits_{i \in I_j} t_i.X_{ij}R_{uk} \leq CT \tag{9}$$

$$X_{ij} \in \{0, 1\} \tag{10}$$

$$Y_{jk} \in \{0, 1\} \tag{11}$$

$$L_{uj} \in \{0, 1\} \tag{12}$$

$$Tj = \frac{\sum_{k=K_j} \sum_{u=U_j} \sum_{i \in I_j} t_i R_{uk}}{H_j} \tag{13}$$

$$\bar{T} = \frac{1}{W}\sum\nolimits_{j=1}^{W} T_j \tag{14}$$

4 System Description

4.1 Primary and Secondary Data Collection

Primary data are obtained from direct observation at the Assembly Engine section using a stopwatch. Whereas secondary data obtained from the company are standard time of 40 s for each activity and historical sales data from February to July 2018 as presented in Table 2. There are 41 activities in the engine assembly line, each of which has 1 operator and 15 activities that require a machine.

Table 2. Historical sales data 2018

February	March	April	May	June	July
10000	11621	13350	15325	15800	17000

In the process of the company applies 6 working days with 15 h per day (07.00–16.00 and 18.00–24.00).

4.2 Operator Efficiency

Each operator has a different level of efficiency from each other. Here is an example of calculating Operator Efficiency:

$$Operator\ Efficiency = actual\ time\ /\ standard\ time \tag{15}$$

The following is a Table 3 that explains the efficiency of each operator at the work station:

Table 3. Operator efficiency for every activity

Station Code	Operator	Efficiency	Station Code	Operator	Efficiency
S11	Operator 1	1.15	M08	Operator 21	1.55
S12	Operator 2	1.09	M09	Operator 22	1.34
S13	Operator 3	0.90	S05	Operator 23	1.33
S14	Operator 4	0.84	M10	Operator 24	1.32
Q2	Operator 5	1.13	M11	Operator 25	1.01
S15	Operator 6	1.03	M12	Operator 26	1.55
M01	Operator 7	1.12	S06	Operator 27	0.63
S09	Operator 8	0.92	M13	Operator 28	0.64
S10	Operator 9	1.33	M14	Operator 29	1.41
M02	Operator 10	1.00	S07	Operator 30	1.27
S01	Operator 11	0.87	M15	Operator 31	1.30
M03	Operator 12	2.02	M16	Operator 32	1.55
S02	Operator 13	1.19	S08	Operator 33	0.60
S03	Operator 14	1.13	M17	Operator 34	1.21
Q1	Operator 15	1.01	M18	Operator 35	1.27
M04	Operator 16	1.00	M19	Operator 36	1.13
M05	Operator 17	1.72	M20	Operator 37	1.12
M06	Operator 18	1.26	M21	Operator 38	1.08
S04	Operator 19	1.31	M22	Operator 39	1.00
M07	Operator 20	1.32	Q3	Operator 40	1.56
			M23	Operator 41	1.27

4.3 The Number of Work Stations

The processing time available in 1 month is 1404000 s and the number of production is 17000 units/month. Takt Time calculations are as follows:

$$Takt\,Time = 1404000 / 17000 = 82.6 \tag{16}$$

The total processing time is 1939.5, so the number of work stations is calculated as follows:

$$Wmin = 1939.5 / 82.6 = 23.5 \approx 24\,work\,stations \tag{17}$$

5 Search for Optimal Solutions

Figure 1 illustrates the optimal value with an objective function to minimize MAD by using oracle Crystal Ball software. The optimal results were obtained after 4001 iterations which required 112 h 01 min and 14 s with optimum solution during the iteration was 17.5. criteria for stopping the simulation by setting a maximum limit of 4001 iterations and the convergent value limit does not change for 1000 consecutive iterations (Table 4).

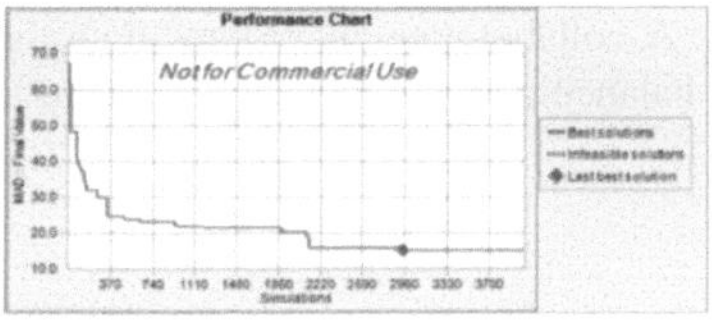

Fig. 1. Convergence of MAD

Table 4. Assignment of tasks and machines

No	Station code	Task	Machine	No	Station code	Task	Machine
1	Station code 1	Task 13, Task 15	Machine 2	13	Station code 13	Task 10, Task 23	Machine 5
2	Station code 2	Task 9, Task 35	Machine 10	14	Station code 14	Task 6, Task 16	Machine 15
3	Station code 3	Task 1, Task 11	Machine 1, Machine 11	15	Station code 15	Task 21, Task 33	Machine 8
4	Station code 4	Task 30, Task 36	Machine 7	16	Station code 16	Task 22	
5	Station code 5	Task 24		17	Station code 17	Task 20, Task 28	
6	Station code 6	Task 31		18	Station code 18	Task 26	
7	Station code 7	Task 3, Task 14	Machine 13, Machine 3	19	Station code 19	Task 17, Task 19	Machine 4
8	Station code 8	Task 2, Task 4	Machine 12, Machine 14	20	Station code 20	Task 18, Task 27	Machine 6
9	Station code 9	Task 5, Task 34		21	Station code 21	Task 29	
10	Station code 10	Task 32, Task 8	Machine 9	22	Station code 22	Task 37, Task 38	
11	Station code 11	Task 7, Task 25		23	Station code 23	Task 39, Task 40	
12	Station code 12	Task 12		24	Station code 24	Task 41	

6 Conclusion

In this study, a model was applied to balance the assembly line in the company. To fulfill this objective, an optimization model was developed by Chen, et al. (2012) [2] with an objective function, minimizing Mean Absolute Deviation (MAD) as a measure of the time difference between work stations. Searching for an optimal solution with a model can produce assignments, machines, and operators to each work station with MAD of 17.5. Comparison of initial conditions and proposals is shown by Line Efficiency 53.49% to 64.97% and Smoothness index 35.4% to 28.6%

Acknowledgement. This research was fully funded by Research and Community Service (P2 M) PNBP funds 2019 budgeting year with contract number 516/UNS27_21/PM/2019.

References

1. Karp, R.M.: Reducibility among Combinatorial Problems. In: Complexity of Computer Computations, pp. 85–103 (1972)
2. Chen, J.C., Chen, C.C., Su, L.H., Wu, H.B., Sun, C.J.: Assembly line balancing in garment industry. Expert Syst. Appl. **39**, 10073–10081 (2012)
3. Sivasankaran, P., Shahabudeen, P.: Literature review of assembly line balancing problems. Int. J. Adv. Manuf. Technol. **73**(912), 1665–1694 (2014)
4. Aase, G.R., Olson, J.R., Schniederjans, M.J.: U-shaped assembly line layouts and their impact on labor productivity: an experimental study. Eur. J. Oper. Res. **156**, 698–711 (2004)
5. Agrawal, S., Tiwari, M.K.: A collaborative ant colony algorithm to stochastic mixed-model U-shaped disassembly line balancing and sequencing problem. Int. J. Prod. Res. **46**(6), 1405–1429 (2008)

A Three-Echelon Inventory Model for Deteriorated and Imperfect Items with Energy Usage and Carbon Emissions

Aldy Fajrianto(✉), Wakhid Ahmad Jauhari, and Cucuk Nur Rosyidi

Industrial Engineering Department,
Sebelas Marct University, Surakarta, Indonesia
Fajriantoaldy@gmail.com

Abstract. This research develops an integrated inventory model for deteriorated products consisting of a supplier, a 3PL and buyer. Carbon emissions are generated by the production process, storage, disposal of deteriorated products and transportation activities. The production process is imperfect, thus producing a proportion of defective items that needs to be reworked by the supplier before being sent to 3PL. The objective function of the model is to minimize joint total cost of supply chain by simultaneously determining the buyer consuming time, and the frequency of delivery. An iterative procedure is proposed to find the decision variables. A numerical example is given as illustrations of the model. The result from numerical example shows that the optimal buyer consuming time (Tb) and 3PL's delivery frequency (n) are 0.0598 years and 3, respectively with the joint total cost is $ 548,158. 97.

Keywords: Deteriorated products · Carbon emissions · Energy usage · Imperfect production · Three-echelon

1 Introduction

The infinite shelf-life of products while in storage is one of the basic implicit assumptions of most inventory model. In many situations, such as spoilage of fruits and vegetables or decay of radioactive substances explicit consideration of deterioration is required. Deterioration can be defined as change, damage, decay, spoilage, obsolescence and loss of utility or loss of marginal value that results in decreasing usefulness [1]. Reference [1] developed a model of deteriorating items with integrated production inventory model under imperfect production process and inflation. The objective of Ref. [1] model is to minimize the total inventory cost in an integrated production-inventory by adjusting production period and retailer's out of stock period.

The integrated inventory management system can be used to provide economic advantages for parties involved in the supply chain. The coordination can be achieved by synchronizing the vendor's production decision and buyer's ordering decision [2]. Reference [3] developed a model of deteriorating items in three-echelon supply chain considering carbon emission cost. The aim of the Ref. [3] model is to minimize the inventory cost in an integrated inventory system involving supplier, thirs party logistic and a buyer by adjusting number of delivery per cycle and buyer consuming time.

M. N. Osman Zahid et al. (Eds.): iMEC-APCOMS 2019, LNME, pp. 305–312, 2020.
https://doi.org/10.1007/978-981-15-0950-6_47

With the increasing of carbon emissions and the diminishing of critical resources such as food, water, and minerals environmental concerns are becoming a serious threat. As a result, many organizations are now integrating environmental concerns into their supply chain practices [4]. since the energy prices are rising and volatile the evaluation of the energy consumption in manufacturing processes has become more important [5]. Reference [5] developed an inventory model incorporating greenhouse gases emissions, energy usage, imperfect process under different coordination decision. The model Ref. [5] aims to minimize the inventory cost by determining the number of shipment, vendor production rate and ordering lot size. From the description above, it can be seen that there is no research that developed a model of three-echelon inventory system for a supplier, third party logistic and buyer considering deteriorated products, emissions and energy usage. In this research, we extend and combine the research of Refs. [1, 3] and [5] by developing a three-echelon inventory model considering deteriorated products, energy usage, and carbon emissions.

2 Assumptions and Notations

This paper has the assumptions, which are:

(1) The demand is fixed, deterministic and shortages are not allowed
(2) The rate of deterioration that occurs in suppliers, 3PL and buyers is constant.
(3) For every 3PL order, supplier make one production set-up, and delivering the item to 3PL (single-delivery single setup) once, and has n deliveries per production cycle
(4) The replenishment of the 3PL and buyers are instantaneous, and the 3PL do the first delivery directly after getting supplies from supplier
(5) Carbon emissions occur from transportation, warehousing, production processes, and disposal of a deteriorated unit
(6) All defective products are detected at the end of production cycle, and there are costs of reworking the defective products.

The following notations are used in this paper:

Decision variables

n	Frequency of deliveries each cycle
Tb	Consuming time for buyer (year)

Parameter

P	Production rate (unit/year)
Tp	Supplier production period (year)
Qp	Supplier production unit (unit)
$Is(t)$	Inventory level for supplier (unit)
Ss	Set up cost for supplier ($/cycle)
hs	Holding cost for supplier ($/unit/year)
dcs	Supplier deteriorating cost ($/unit)
des	Carbon emission cost from deteriorated unit for supplier ($/unit)

fs	fixed supplier transportation cost ($)
vs	Fuel cost for supplier ($/liter)
c1s	Fuel consumption for vehicle with empty condition for supplier (liter/km)
c2s	Supplier additional fuel consumption from transporting each item (liter/km/ton)
e1s	Carbon emission cost from supplier's vehicle ($/km)
e2s	Supplier's additional emission cost from transporting each item (liter/km/ton)
d1	Supplier to 3PL distance (km)
TCS	Supplier total cost ($/year)
C5	Rework cost ($/unit)
a	Production process emission function parameter (ton-h2/unit3)
b	Production process emission function parameter (ton-h/unit2)
c	Production process emission function parameter (ton/unit)
cen	Supplier energy cost ($/kWh)
Ep	Greenhouse gas emissions from the production process (ton/unit)
es	Production emission cost ($/ton)
W0p	"idle" power for the production process when the machine in ready position (kW)
k	Constant variable of the power used (kWh/unit)
IC	Inspection cost ($/unit)
μ	Defect percentage
Q2	3PL Delivery quantity (unit)
Ip(t)	Inventory level at time t 3PL (unit)
Td	Interval for each Delivery (year)
dTd	Deteriorating quantity at 3PL inventory during Td interval (unit)
Op	Ordering cost for 3PL($/order)
hp	Holding cost for 3PL ($/unit/year)
dcp	Deterioration cost for 3PL ($/unit)
dep	Carbon emission cost from deteriorated unit for 3PL ($/unit)
fp	Fixed 3PL transportation cost ($)
p	Fuel cost for supplier 3PL ($/liter)
c1p	Fuel consumption for vehicle with empty condition for supplier (liter/km)
c2p	3PL additional fuel consumption from transporting each item (liter/km/ton)
e1p	3PL carbon emission cost from vehicle ($/km)
e2p	3PL additional emission cost from transporting an item (liter/km/ton)
d2	3PL to buyer distance (km)
TCP	3PL total cost per unit time ($/year)
d	Buyer demand rate (unit/year)
dTb	Deteriorating quantity at buyer's inventory during Tb interval (unit)
Ib(t)	Inventory level for buyer (unit)
Ob	Ordering cost for buyer ($/cycle)
hb	Holding cost for buyer ($/unit/year)
dcb	Deteriorating cost for buyer ($/unit)
deb	Carbon emission cost from deteriorated unit for buyer ($/unit)
TCB	Buyer total cost per unit time
TC	The integrated total cost of the buyer, 3PL, and supplier

T	Time interval for one cycle (year)
θ	Rate of deterioration (0 < θ < 1)
we	Carbon emission cost from warehousing activity ($/unit/year)

3 Model Development

In this mathematical model, we develop an integrated inventory system involving a individual buyer, a single 3PL and a single supplier with the objective to determine optimal buyer consuming time, and the frequency of delivery in order to minimize the joint total cost. Figure 1 illustrates the inventory level of the the buyer, 3PL, and supplier with the effect of item deterioration. The development of each cost is described below.

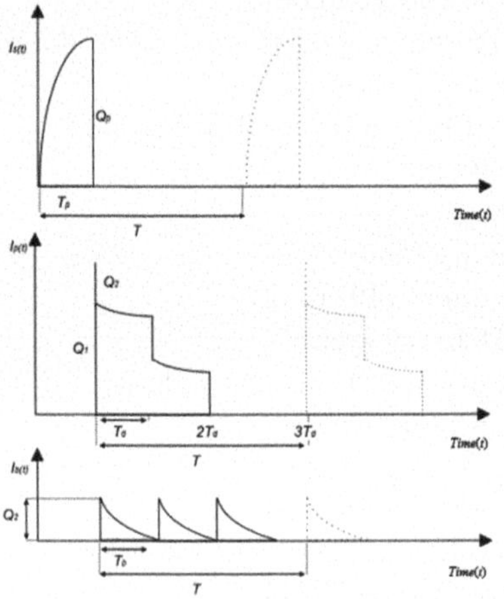

Fig. 1. Inventory level of supplier, 3PL, and buyer

3.1 Buyer Cost

Buyer total cost per unit time (*TCB*) consists of the following cost ordering cost for buyer (*COB*), holding cost for buyer (*CHB*), and deteriorating cost for buyer (*CDB*). Where,

$$COB = Ob/Tb \tag{1}$$

buyer holding cost and carbon emission from warehousing activities is given by

$$CHB = (hb + we)\frac{1}{Tb}\frac{d\left(e^{Tb\theta} - Tb\theta - 1\right)}{\theta^2} \tag{2}$$

Buyer deteriorating quantity can be derived from the delivery lot size of 3PL minus the buyer demand during the period Tb. Hence,

$$CDB = (\mathrm{dcb} + \mathrm{deb})\left(\frac{\frac{d}{\theta}\left(e^{\theta \mathrm{Tb}} - 1\right)}{\mathrm{Tb}} - d\mathrm{Tb}\right) \tag{3}$$

3.2 3PL Cost

3PL total cost per unit time (*TCP*) consists of ordering cost for 3PL (*COP*), transportation cost for 3PL (*CTP*), holding cost for 3PL (*CHP*) and deterioration cost for 3PL(*CDP*).

With thc assumption of single delivery single set up, the 3PL only make one order per cycle and incurs an ordering cost

$$COP = \frac{Op}{nTb} \tag{4}$$

The transportation cost for 3PL considering variable and fix transportation costs and emission cost is given by

$$CTP = \frac{1}{Tb}\left(fp + \left(2d2\,vp\,c1p + d2\,vp\,c2p\frac{d}{\theta}\left(e^{\theta Tb} - 1\right)\right) + \left(2\,d2\,e1p + d2\,e2p\frac{d}{\theta}\left(e^{\theta Tb} - 1\right)\right)\right) \tag{5}$$

Then 3PL deteriorating cost and holding cost per unit time can be found using following expressions, respectively

$$CHP = \frac{(hp + we)}{nTb}\left(\frac{d}{\theta}\left(e^{\theta Tbn} - 1\right) - n\left(\frac{d}{\theta}\left(e^{\theta Tb} - 1\right)\right)\right) \tag{6}$$

$$CDP = \frac{(dcp + dep)}{nTb}\left(\frac{d}{\theta}\left(e^{\theta Tbn} - 1\right) - n\left(\frac{d}{\theta}\left(e^{\theta Tb} - 1\right)\right)\right) \tag{7}$$

3.3 Supplier Cost

Supplier total cost per unit time (*TCS*) consist of supplier setup cost (*CSS*), holding cost (*CHS*), deteriorating cost (*CDS*), production energy cost (*CPS*) transportation cost (*CTS*) emission cost from production (*CES*) and rework and inspection cost (*CRIS*). In which,

$$CSS = \frac{Ss}{T} \tag{8}$$

Supplier holding cost and carbon emission from warehousing activities is given by

$$CHS = \frac{(hs + we)}{T}\frac{P\left(\theta Tp + e^{-\theta Tp} - 1\right)}{\theta^2} \tag{9}$$

With the equations *Qp* equal to *Q1* we have

$$Tp = -\frac{ln\left(\frac{d+P-de^{\theta Tbn}}{P}\right)}{\theta} \tag{10}$$

Supplier deteriorating quantity can be derived from the delivery lot size of supplier minus the buyer demand during the period *T*, hence

$$CDS = \frac{(dcs + des)}{T}\left(PTp - \frac{P\left(1 - e^{-\theta Tp}\right)}{\theta}\right) \tag{11}$$

Production energy usage is calculated from the condition when machine is operating and when the production machine is idle. Therefore,

$$CPS = (W0p + kP)Tp\, Cen\frac{1}{T} \tag{12}$$

The transportation cost for supplier is identical with 3PL as shown in Eq. (7), which is

$$CTs = \frac{1}{T}(fs + (2\,d1\,vs\,c1s + d1\,vs\,c2s\,Qp) + (2d1\,e1s + d1\,e2s\,Qp)) \tag{13}$$

The supplier production emission depends on production rate of the supplier therefore

$$Ep = c - bP + aP^2 \tag{14}$$

$$CES = es(P\,Tp\,Ep)\frac{1}{T} \tag{15}$$

Then rework and inspection cost per unit time can be found using the following equation

$$CIRS = (C5\,\mu + IC)Qp\frac{1}{T} \tag{16}$$

Then the integrated total cost (*TC*) can be found by summing up all costs incurred by supplier, 3PL and buyer.

3.4 Solution Procedure

To find the optimal solutions of the model, first we use the taylor expansion theory and neglecting the $\theta^2 Tb^2$ and higher terms, then we find the derivative of *TC(tb,n)* with respect to *Tb*. Next, we develop an algorithm to find the optimal value of *n* and *Tb*. The algorithm to solve the above problem is as follows

(1) Set n = 1, initial value of Tb and $TC\left(Tb^*_{n-1,}n-1\right)=\infty$
(2) Compute *Tb* by substituting value of *Tb* and n into equation derivative of *TC* equal to zero
(3) Repeat step 2 until no change occurs in the value of *Tb*
(4) Set *Tb* = *Tb** and compute $TC(Tb^*, n)$ using the integrated total cost equation
(5) If $TC(Tb^*_n, n) \leq TC(Tb^*_{n-1}, n-1)$ repeat step 3 to 5 with *n* = *n*+1, otherwise go to step 6
(6) Compute $TC(Tb^*_n, n) = TC(Tb^*_{n-1}, n-8)$, then *n** and *Tb** are the optimal solution and *TC(Tb*,n*)* are the optimal total cost.

4 Numerical Example

Here, we provide a numerical example to show the application of proposed model. The following are numerical values taken from Refs. [1, 3] and [5]. The parameters used in this model can be seen at Table 1.

Table 1 Numerical example parameter

Parameter	Value	Parameter	Value	Parameter	Value	Parameter	Value
d	= 8000 unit/year	*vs*	= 0.75 $/liter	*C5*	= 10 $/unit	*e1p*	= 0.04 $/km
θ	= 0,1	*c1s*	= 0.3 liter/km	*Op*	= 600 $	*e2p*	= 0.00000231 $/unit km
hs	= 5 $/unit	*c2s*	= 0.005 liter/km ton	*hp*	= 1.5 $/year	*Ob*	= 300 $/order
we	= 0.445 $/unit	*e1s*	= 0.048 $/km	*dcp*	= 100 $/unit	*hb*	= 3 $/unit year
dcs	= 30 $/unit	*e2s*	= 0.0000034 $/unit km	*dep*	= 0.074 $/unit	*dcb*	= 200$/unit
des	= 0.074 $/unit	*a*	= 0.0000003	*fp*	= 100 $/delivery	*deb*	= 0.074 $/unit
WOp	= 100 kWh	*b*	= 0.00012	*d2*	= 25 km	*IC*	= 0.1 $/unit
Cen	= 0.158 $/kWh	*c*	= 1.4	*c1p*	= 0.25 liter/km	*k*	= 8 kWh/unit
fs	= 200 $/delivery	*μ*	= 0.01	*vp*	= 0.75 $/liter	*Ss*	= 2000 $/ setup
d1	= 500 km	*es*	= 0.18 $/ton	*c2p*	= 0.0036 liter/km ton	*P*	= 28000 unit/year

The optimal solution is given as follows: the number of 3PL delivery per cycle is 3, the buyer consuming time is 0.0598 year and the total cost is 548158. 97 $/year.

5 Conclusion

In inventory management, there are products that could deteriorate and lose its value while stored in warehouse. In this study, we formulate a mathematical inventory model for three-echelon system consisting of supplier, 3PL, buyer with four mains carbon emission sources which are warehousing, transportation, production emission, and disposing deteriorated product. This research contributes the current literature by incorporating production emission cost and energy cost. Minimum total cost is found by simultaneously optimizing buyer consuming time, and the frequency of delivery. For future research, Incorporating the case of inflation issues may also provide insight.

Acknowledgement. This research was partially funded by Research and Community Service (P2 M) PNBP funds 2019 budgeting year with contract number 516/UNS27.21/PM/2019

References

1. Lo, S.T., Wee, H.M., Huang, W.C.: An integrated production-inventory model with imperfect production processes and Weibull distribution deterioration under inflation. Int. J. Prod. Econ. **106**(1), 248–260 (2007)
2. Jauhari, W.A.: Investigating carbon emissions in a production-inventory model under imperfect production, inspection errors and service-level constraint. Int. J. Logist. Syst. Manag. **34**(1), 29–55 (2019)
3. Daryanto, Y., Wee, H.M., Astanti, R.D.: Three-echelon supply chain model considering carbon emission and item deterioration. Transp. Res. Part E Logist. Transp. Rev. **122**, 368–383 (2019)
4. Jauhari, W.A., Laksono, P.W., Saga, R.S., Dwicahyani, A.R.: A collaborative inventory model for vendor-buyer system with stochastic demand, defective items and carbon emission cost. Int. J. Logist. Syst. Manag. Indersci. Enterp. Ltd. **29**(2), 241–269 (2018)
5. Marchi, B., Zanoni, S., Zavanella, L.E., Jaber, M.Y.: Supply chain models with greenhouse gases emissions, energy usage, imperfect process under different coordination decisions. Int. J. Prod. Econ. **211**, 145–153 (2019)

An Assignment Model to Support the Assembly Line Activities by Considering the Operator's Unique Classification – The Computational Results

Rudy Prijo Utomo[1], Mohammad Mi'radj Isnaini[2(✉)], and Anas Ma'ruf[2]

[1] Indonesian Aerospace, Bandung, West Java 40174, Indonesia
[2] Industrial Engineering Study Program, Bandung Institute of Technology, Bandung, West Java 40132, Indonesia
iis@ti.itb.ac.id

Abstract. An assembly line model needs to be designed in order to be able to manage the assignment of operators and balance the workloads at once to meet the targeted cycle time. Each station has operations to do which need certain operators' skills and classification. The proposed model is an analytical model of Mixed Integer Linear Programming (MILP) which uses the makespan as the objective. The inputs of this model are the historical demand, operation data, operator's data and the corresponding precedence diagrams. The outputs are the operating placement on the workstation, operator assignment on exact operations, the start time of each operation, the last operation start time, and the last operation process time. The proposed model was able to reduce the makespan and reduce the number of operators. Furthermore, the computational results show that the proposed model was able to produce an assembly line which less sensitive by the given main parameter such as the demand (takt time) by comparing the line efficiency, smoothness index and its corresponding makespan in each parameter changes.

Keywords: Assembly operation · Distinct classification · Line balancing · Cycle-time · MILP

1 Introduction

In the assembly process, the job shop production system has irregular sequencing lines and required tools which are used together or alternately for various operations in the process. The delay in the final assembly process might be caused by several reasons. Among them is to balance the workloads and assignment due to the limitation number of available operators [1]. Scheduled operations often must be delayed due to the lackness of available operators that can be assigned by the required classification. The operation scheduling and operator assignments much likely have limited attention to the availability of operators which corresponds to certain classification.

M. N. Osman Zahid et al. (Eds.): iMEC-APCOMS 2019, LNME, pp. 313–319, 2020.
https://doi.org/10.1007/978-981-15-0950-6_48

Several former studies have discussed cross-assembly design issues and operator assignments which considers operator's skills or classification. Moon et al. [2] design an integrated cross assembly with the assignment of operators who have various skills (multi-functional workers). Sungur and Yavuz [3] propose a cross-assembly model and operator assignment for operations carried out by operators that have hierarchical levels. By considering a practical constraints, Battini et al. [4] design cross-assembly and assignment of operators by considering the type of equipment which corresponds to specific needs of operator's skill. Giglio et al. [5] design cross-assembly and assignment of operators by developing models from [2], namely by adding operator assignment characteristics that were previously limit to only serialized process to become parallel and simulants. Moreover, Martignago et al. [6] propose a cross assembly model and operator assignment for operations carried out by operators who have single-skill and are cross-trained.

This paper explains a model development for cross assembly and assignment of operators who can at the same time balance workloads by considering classification and multi operators. This paper elaborates the development by showing two main following sections. The first section explains the adjustment on previous model. The second section shows the elaboration of the hypothetical and real data on the developed model. Eventually, the last section shows the conclusion of all the elaboration.

2 Mathematical Model Development

2.1 Problem Description

The reference model is determined by matching the characteristics between the models to be selected and the condition of the existing assembly line at a single aircraft manufacturer. From the preliminary observations, it is known that the characteristics of the assembly activities have more than 300 operations. The assembly line currently uses the concept of fixed layout assembly with 4 stages (station). The series of operations in the assembly process follows the precedence diagram. Each operation is carried out simultaneously by operators with 1 to 4 operators and specific classification. Rules for grouping operations on workstations are based on operating time, number, and operator classification.

2.2 The Reference Model and Its Development

The mathematical model in [6] has the closest suitability and presents a mathematical model to solve line balancing problems by arranging operators in complex assembly lines, namely the final assembly of large and bulk products, such as cars, trucks, buses, and other large products. This model is a development of the problems of Simple Assembly Line Balancing (SALB). SALB assumes all workstations are considered to have the same level of difficulty so that each operation can be carried out by any operator [7]. Every operation requires 1 specific type of operator skill. The operation will be assigned to operators who have skills according to the type of operation. The objective function is to minimize the total cost of workers.

Table 1. The objective function and its constraints.

No	Equation	Remarks
(1)	$Min\,C_{max} = S_{LT} + p_{LT}$	The makespan minimization (C_{max}) on assembly line
(2)	$\sum_{m=1}^{N_W} x_{i,m} = n_i$	Bounds to limit the number of operators in each operation. This boundary will be used to each classification
(3)	$u_{i,j} + u_{j,i} \leq 1$	Bounds to allow only one sequence in working on the pair of operations when using the same operator
(4)	$u_{i,i} = 0$	Bounds to assign each operator to different operations
(5)	$u_{i,j} + u_{j,i} \geq x_{i,m} + x_{j,m} - 1$	Bounds to follow the precedence for each assigned operator. If two operations (*i, j*) are carried out by the same operator, then the value of the decision variable for $u_{(i,\,j)}$ or $u_{(j,\,i)}$ must be 1. This expression indicates the precedence for the operator in multi-operations
(6)	$S_j - S_i \geq p_i - M\left(1 - u_{i,j}\right)$	Bounds to allow only one operation at a time for each operator
(7)	$\sum_{s \in K} y_{i,s} = 1$	Bounds to place each operation only on one station
(8)	$\sum_{s \in K} s \cdot (y_{i,s} - y_{j,s}) \leq 0$	Bounds to meet the given precedence for each operation in each station
(9)	$\sum_{s \in K} j_{m,s} \leq 1$	Bounds to assign operator to only particular station
(10)	$\sum_{m=1}^{N_W} x_{i,m} = n_i \cdot y_{i,s}$	Bounds to set the needed number of operator in each operation in each station
(11)	$\sum_{s \in K} \left(j_{m,s} - j_{m+1,s}\right) \geq 0$	Bounds to ensure that the operator (*m* + *1*) can only be assigned to the station if the m-operator has been assigned previously, this applies to each operator classification
(12)	$S_i \geq \left(-1 + \sum_{s \in K} \left(s \cdot y_{i,s}\right)\right) \cdot TaktTime$	Set a lower limit when starting from an operation. This boundary ensures that if the operation is placed at station 1 then the start of the operation is $\geq$ 0, if the operation is placed at station 2 then the starting the operation is $\geq$ takt time
(13)	$\sum_{i \in A} x_{i,m} \leq M \cdot \sum_{s \in K} j_{m,s}$	Bounds to calculate the total active operators in all stations
(14)	$S_i \leq S_j - p_i$	States when the operation will start
(15)	$S_{LT} = S_i$	States when the last operation will start
(16)	$p_{LT} = p_i$	States the last operation process time
(17)	$C_{max} = S_{LT} + p_{LT}$	States the maximum time needed to work on all operations which exist in the assembly line

The modification in the operation set is done by substituting the set of operations based on the needs of the type of operator that exists. In [6], the process considers 3 types of single-skill and 1 type cross-trained skill. This condition will be modified to not consider any cross-trained workers, but all are single-skill workers. The proposed model considers 4 types classification of single-skill workers.

The proposed model modifies the objective function to match the production target. The production target can be achieved by adjusting the makespan to be more time-wisely [8]. To achieve this goal, the objective function in the model will be changed to minimize the makespan. Additional decision variables to show the start of and the duration of the last operation are needed as the basis for makespan calculation. Therefore, the form of *S_LT* (at the start of the last operation) and *p_LT* (last operation process time) are added as the basis for makespan calculation. The cost related variables from previous model are no more considered since the study will not consider any cost factors.

Parameter elimination is carried out on α, *ESi*, and *LSi* because the three parameters are not relevant to the conditions. α is related to the ratio of cross-trained operator costs to single-skill operators. Similarly, the parameters *ESi* (earliest start) and *LSi* (latest start) are eliminated because of from the given condition stated each operation can be done as soon as possible, if the required material and operators are available. Operator requirements for each operation in [6] specifically intended for one single-skill operator or cross-trained. To match the actual conditions, the boundary of the reference model related to the type of operator needs to be changed to meet the operator requirements of each classification. Moreover, further boundaries related to the earliest start and the latest start in the reference model will also be eliminated.

3 Case Study and Analysis

3.1 Model Verification and Validation on Hypothetical and Real Assembly Data

The verification process is carried out by running the proposed model using a hypothetical data which derived from a small data set presented in [5]. Figure 1 shows the precedence diagram of an assembly line with 6 operations and its parameters. Available operators are multi-functional operators that can be assigned to any operation.

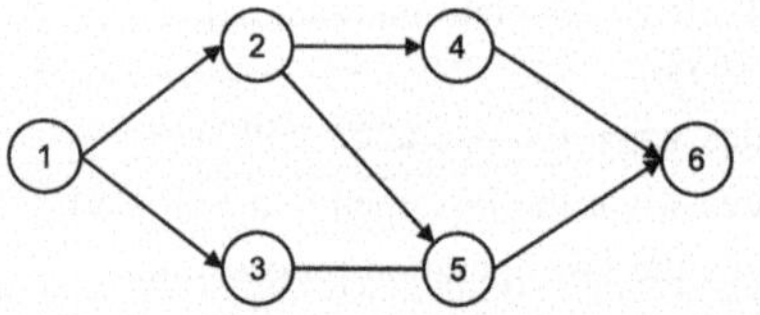

Total Operation	6 ops
Number of Operator	5 people
Number of Station	3 stations
Takt Time	5 hours
Total Time	15 hours

Fig. 1. A hypothetical precedence chart and its parameter.

The proposal model validation process will also be tested with a data set that represents the actual conditions. The data set used in the validation process is a third stage data set from the X-type aircraft assembly line that represents the actual condition. The third stage consist of 85 operations. Each operation can be carried out by 1 or several operators at once and according to the classification needed. Four types of available operators are considered, namely: 1. Mechanical Installer, 2. Mechanical Tester, 3. Electrical Installer, and 4. Electrical Tester. The parameters are shown in Table 1.

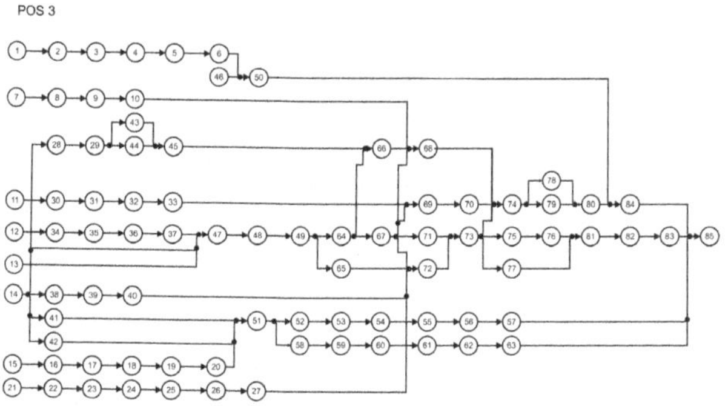

Total Operation	85
Number of Station	7
Takt Time (hours)	390
Total Time (hours)	2730

Fig. 2. The real precedence chart in the third station and its parameter.

The proposed model in Sect. 2.2 is implemented CPLEX version 12.8.0.0 environment to solve the given data in Figs. 1 and 2. The proposed model is considered to be valid because the objective function and boundaries have been met by visually confirming the solution, including an assignment to each operation to the specific number and required classification. Each operation was placed on one workstation. A preemptive operation was placed at an earlier workstation. Each operator was assigned at most to one workstation. Operators assigned to a workstation can complete several operations at once. The maximum time needed to process all operations (makespan) was not less than the amount from the start of the last operation and the last operation process. The maximum time needed to work on all operations (makespan) was no more than the total time. When starting each operation, it did not violate constraints precedence.

3.2 Model Performance on the Real Assembly Data

Further computation was taken to compute to the entire 4 stations in making the X-type aircraft. The computational results showed the total (optimum) makespan of 1131 h. The average makespan of each station was still below the takt time (see Table 2). The changes ±25% of demand did not influence the makespan. However, if the demand rise up to 75%, then it showed a significant difference in the line efficiency and smoothness index. The makespan was also had a slight change by the given raise on demand. From this evaluation, the proposed model still can produce significant result if the demand (in this case the takt time) varies around ±25%.

Table 2. Sensitivity results caused by demand and takt time changes.

Changes			Result			
+/- (%)	Aircraft demand (unit)	Takt time (hours)	Station (unit)	Line efficiency (%)	Smoothness index	Makespan (hours)
+75	7	223	8	71,47	211,09	1275
+50	6	260	8	56,15	383,53	1168
+25	5	312	4	96,5	28,51	1131
0	4	390	4	96,5	28,51	1131
−25	3	520	4	96,5	28,51	1131

4 Conclusion and Future Works

There are two main conclusions obtained from this study, first, the proposed model is successful in reducing the makespan to meet the given takt time while also reducing the number of operator requirements which has distinct classification. Secondly, the proposed model has an enough sensitivity to accommodate the changes in demand (demand). Several suggestions are given as follows to incorporate the proposed model: 1. Periodically evaluate performance and the assembly line conditions to anticipate the changes in demand; 3. Improve the discipline and understanding of operators and the aircraft supervisors regarding the critical tasks and related changes in working methods in the assembly process; 4. Re-evaluate the standard operating time data; 5. Increase an excellent cooperation between related stations or departments to achieve production targets.

Further research will be needed to revise and improve the precedence diagram which is adjusted to the condition of the new station to consider a two level platform. The model still need an enhancement to accommodate three objective functions at once, namely optimization of makespan, smoothness index, and efficiency. Furthermore, an allowance of operation sets is also needed to accommodate an assignment for multi-shift working time for each operator.

References

1. Rekiek, B., Delchambre, A.: Assembly Line Design—The Balancing of Mixed-model Hybrid Assembly Lines with Genetic Algorithms, 1st edn. Springer, London (2006)
2. Moon, I., Logendran, R., Lee, J.: Integrated assembly line balancing with resource restrictions. Int. J. Prod. Res. **47**(19), 5525–5541 (2009)
3. Sungur, B., Yavuz, Y.: Assembly line balancing with hierarchical worker assignment. J. Manuf. Syst. **37**(1), 290–298 (2015)
4. Battini, D., Faccio, M., Gamberi, M., Persona, A., Pilati, F.: Assembly line balancing with equipment and parallel worker: an heuristic algorithm.. In: XVIII Summer School "Francesco Turco" – Industrial Mechanical Plants, pp. 61–66. Italy (2015)

5. Giglio, D., Paolucci, M., Roshani, A., Tonelli, F.: Multi-manned assembly line balancing problem with skilled workers: a new mathematical formulation. IFAC-PapersOnLine **50**(1), 1211–1216 (2017)
6. Martignago, M., Battaïa, O., Battini, D.: Workforce management in manual assembly lines of large products: a case study. IFAC-PapersOnLine **50**(1), 6906–6911 (2017)
7. Boysen, N., Fliedner, M., Scholl, A.: A classification of assembly line balancing problems. Eur. J. Oper. Res. **183**(2), 674–693 (2007)
8. Gafarov, E., Dolgui, A.: Two-dedicated-machine scheduling problem with precedence relations to minimize makespan. Optim. Lett. **8**(4), 1443–1451 (2014)

Capacity Planning Model for Make-To-Order Companies Considering Lateness Penalty Cost Based on Critical Resources

Wisnu Aribowo(✉), Muhammad Afandi Hudzaifah, and Abdul Hakim Halim

Department of Industrial Engineering, Institut Teknologi Bandung, Bandung, Indonesia
wisnu@ti.itb.ac.id

Abstract. This paper addresses the problem of capacity planning in a Make-To-Order aircraft manufacturer company, in which order rejection is not an option. An optimization model is developed with the objective of minimizing total cost that is comprised of processing cost and lateness penalty cost. A numerical example is given to demonstrate the capability of the model. This paper also proposes a procedure for implementing the model that considers only critical resources in order to reduce the problem size such that optimal solution can be found in reasonable time.

Keywords: MRP-II · Capacity planning · Lateness penalty cost · Critical resources

1 Introduction

In the Manufacturing Resource Planning (MRP-II) framework Master Production Schedule (MPS) must be verified by Rough Cut Capacity Planning (RCCP) before it is released to production [1, 2]. The goal of the verification is to ensure that the plan can be realized considering limited capacity in the factory. If the capacity is not enough for realizing the plan, or in other words the MPS is infeasible, then it has to be revised and checked again iteratively. Upon the feasibility verification, the MPS is then released to and further detailed in the Materials Requirement Planning (MRP) stage.

There are several options that can be used in the capacity planning, such as working overtime, adding worker and machines, inventory, outsourcing, and order rejection. For some companies, however, order rejection is not an option. Aircraft manufacturers typically use Make-To-Order strategy and once the orders are accepted and contracts are signed they cannot be rejected. However, the manufacturers may opt to deliver late with penalty cost implication.

There have been numerous research studies in capacity planning in the literature [3–6]. However, most of them cannot be used as-is for the specific problem in the industrial case that we faced. Firstly, in our real case all orders have been accepted and must be delivered. Next, the capacity planning problem encompasses the part making

M. N. Osman Zahid et al. (Eds.): iMEC-APCOMS 2019, LNME, pp. 320–325, 2020.
https://doi.org/10.1007/978-981-15-0950-6_49

and assembly stage, necessitating the need to put assembly takt times and in-between auxiliary processes into consideration.

Research in the literature on capacity planning often results in models that are not suitable for large data because of the long time it takes to find solution. One possible strategy to tackle the problem is, based on the Theory of Constraints, by focusing on the critical resources in the capacity planning [5]. The aim of this paper is to develop a mathematical model that considers the conditions described above. To be specific, the novelty of this paper lies in the consideration of the penalty cost, the consideration of assembly production stage, and the strategy to use only critical resources. The model in this paper is developed based on a previous research [4] of capacity planning in Make-To-Order environment, which is closest to our problem description.

The rest of the paper is organized as follows. Section 2 explains the development of the mathematical model. Section 3 describes and discusses the result of the model execution for a numerical example. The section also proposes an implementation procedure that considers critical resources only. Finally, the paper is concluded in Sect. 4 along with the plan for possible extensions of the research in the future.

2 Mathematical Formulation

An aircraft manufacturer company receives production orders of a few types of aircraft. The orders should be delivered before their respective due-dates because lateness in the delivery incurs penalty cost to the manufacturer. The manufacturing system consists of part-making stage and assembly stage. The part-making stage involves a number of resources (work centers), which are available to be used in three sources: regular time, overtime, and outsource. The capacity in each source is limited. Contracts with customers state that outsourced resources may be used only after the resources in regular time and overtime of the same period are fully utilized. Meanwhile, in the assembly stage each type of aircraft has its own dedicated assembly line and takt time. Between the part-making stage and the assembly stage there is a fixed time for auxiliary processes such as surface treatment and primer painting.

The plan to fulfill the orders is made in the form of Master Production Schedule (MPS), which states when each product will be manufactured. Before released to production, the MPS needs to be verified to check whether there is enough capacity to realize the plan. In the case that the capacity is not enough, some orders will have to be delivered late. The total cost consists of the processing cost and the lateness penalty cost. The manufacturer wants to determine the optimal allocation of resources and sources to production orders such that the total cost is minimized.

The notations used in the mathematical model are as follows:

SETS

I: aircraft types, $\{i \in I\}$

S_i: ordered items of aircraft i, $\{s \in S_i\}$

M: resources, $\{m \in M\}$

R: sources, $\{r \in R\}$, where the last source $|R|$ is outsource

T: time periods, $\{t \in T\}$

PARAMETERS

p_{im}: processing time of aircraft i at resource m
d_{is}: due date of item s of aircraft i at resource m
k_{mrt}: capacity of resource m of source r in period t
c_{mr}: processing cost per time unit of resource m of source r
l_{is}: lateness penalty cost per time period of item s of aircraft i
a_i: assembly takt time of aircraft i
b_i: auxiliary processing time of aircraft i between part-making stage and assembly stage

VARIABLES

X_{ismrt}: time allocated to process item s of aircraft i at resource m of source r in period t
Y_{ismrt}: binary variable; $Y_{ismrt} = 1$ if item s of aircraft i is processed at resource m of source r in period t; $Y_{ismrt} = 0$ otherwise
C^K_{ism}: completion time of item s of aircraft i at resource m in part-making stage
C^A_{is}: completion time of item s of aircraft i in assembly stage

The mathematical model is formulated as a mixed integer linear programming (MILP) problem as follows:

MIN $TC =$

$$\sum_{i\in I}\sum_{s\in S_i}\sum_{m\in M}\sum_{r\in R}\sum_{t\in T} X_{ismrt}\cdot c_{mr} + \sum_{i\in I}\sum_{s\in S_i}\sum_{m\in M}\max\left(C^A_{is} - d_{is}, 0\right)\cdot l_{is} \tag{1}$$

$$\sum\nolimits_{i\in I}\sum\nolimits_{s\in S_i} X_{ismrt} \le k_{mrt} \qquad \forall m, r, t \tag{2}$$

$$\sum\nolimits_{r\in R}\sum\nolimits_{t\in T} X_{ismrt} = p_{im} \qquad \forall i, s, m \tag{3}$$

$$\sum\nolimits_{i\in I}\sum\nolimits_{s\in S_i} X_{ism|R|t} = \max\left(\sum\nolimits_{i\in I}\sum\nolimits_{s\in S_i}\sum\nolimits_{r\in R} X_{ismrt} - \sum\nolimits_{r\in R\setminus\{|R|\}} k_{mrt}, 0\right) \qquad \forall m, t \tag{4}$$

$$X_{ismrt} \ge Y_{ismrt} \qquad \forall i, s, m, r, t \tag{5a}$$

$$X_{ismrt} \le p_{im} Y_{ismrt} \qquad \forall i, s, m, r, t \tag{5b}$$

$$C^K_{ism} \ge Y_{ismrt} t \qquad \forall i, s, m, r, t \tag{6}$$

$$C^A_{is} \ge C^K_{ism} + b_i + a_i \qquad \forall i, s, m \tag{7}$$

$$C^A_{i[s+1]} \ge C^A_{is} + a_i \qquad \forall i, s \in S_i\setminus\{|S_i|\} \tag{8}$$

$$Y_{ismrt} \in \{0, 1\} \qquad \forall i, s, m, r, t \tag{9}$$

The objective function in Eq. (1) to minimize is the total cost, which is comprised of processing cost and tardiness (positive lateness) penalty. Equation (2) ensures that the allocation does not exceed capacity. Equation (3) ensures that the required processing time has been allocated fully. Equation (4) is derived from the requirement that outsourced resources can only be used after resources inside factory – in regular time and overtime – have no more capacity left in the period. The Eqs. (5a) and (5b) link the allocation variables and the binary Y variables: if allocated time is not zero then Y would equal 1, otherwise Y would be 0. Equation (6) calculates the completion time of each item in each resource in part-making stage. Equation (7) states that the completion time in the assembly stage must be later than the completion time in part-making stage plus waiting time and takt time. Equation (8) ensures that the items are assembled one by one in succession. Eqs. (1) and (4), which contain non-linear terms, are linearized beforehand in the implementation in order to keep the whole model linear.

3 Numerical Examples and Discussions

We will use a small numerical example to show the capability of the model. There are two types of aircraft ordered (I1 and I2), each has two items (S1 and S2). Table 1 lists the processing time at two machines (M1 and M2), processing cost per hour, and capacity per month. R1, R2, and R3 in the table refer to regular time, overtime, and outsource, respectively. Table 2 lists the due date and penalty cost per month. The takt times are 3 and 2 periods for aircraft type 1 and type 2, respectively. The auxiliary processing times are also 3 and 2 periods for type 1 and type 2, respectively.

Table 1. Processing and capacity data

Machine	Processing time (hour)		Processing cost per hour (USD)			Capacity per month (hour)		
	I1	I2	R1	R2	R3	R1	R2	R3
M1	879.99	501.97	29.57	29.57	28.09	182	109	200
M2	455.71	394.37	12.33	12.33	11.71	130	55	200

Table 2. Due date and penalty cost data

Type	Due date (month)		Penalty cost per month (thousand USD)	
	S1	S2	S1	S2
I1	7	9	457.16	723
I2	6	8	756.6	313.36

The optimization model is implemented in LINGO 12.0 software and the global optimal solution can be found for the above problem. Figures 1 and 2 show the allocation of capacity at resource M1 and M2, respectively. In each source (regular, overtime, and outsource), the allocation does not exceed the capacity limit. The outsource capacity are used only when the regular and overtime sources are fully used in the period. Completion times and tardiness of each job is listed in Table 3. The total cost in the optimal solution is USD 3,184.

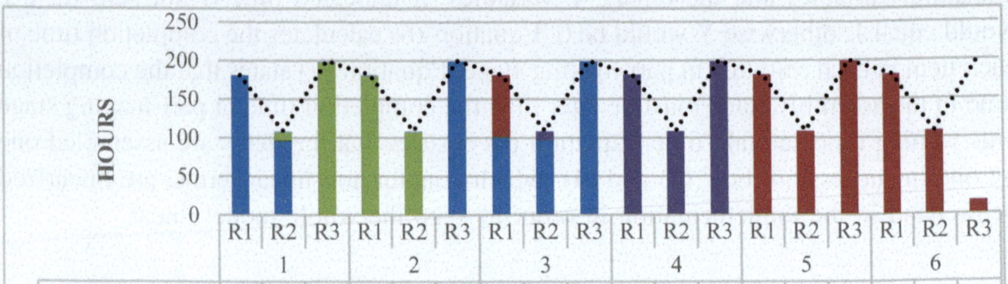

	R1	R2	R3	R1	R2	R3	R1	R2	R3	R1	R2	R3	R1	R2	R3	R1	R2	R3
		1			2			3			4			5			6	
X_I2S2	0	0	0	0	0	0	0	11	0	182	109	200	0	0	0	0	0	0
X_I2S1	0	11	200	182	109	0	0	0	0	0	0	0	0	0	0	0	0	0
X_I1S2	0	0	0	0	0	0	80	0	0	0	0	0	182	109	200	182	109	18
X_I1S1	182	98	0	0	0	200	102	98	200	0	0	0	0	0	0	0	0	0
Capacity	182	109	200	182	109	200	182	109	200	182	109	200	182	109	200	182	109	200

Fig. 1. Capacity allocation at resource M1

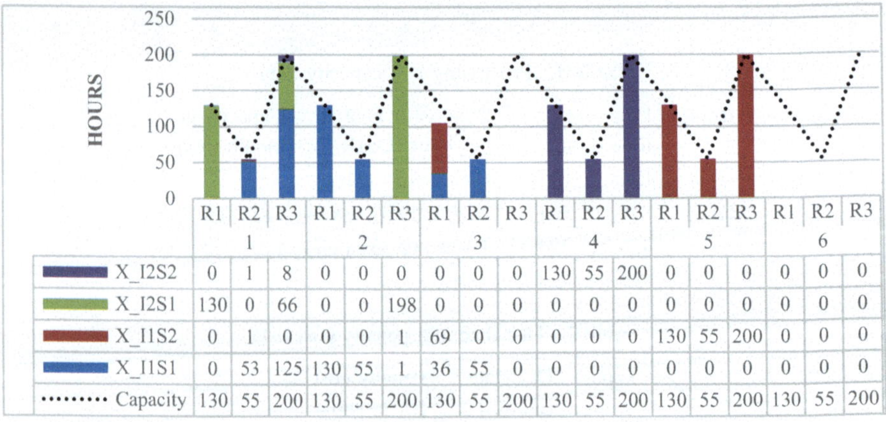

	R1	R2	R3	R1	R2	R3	R1	R2	R3	R1	R2	R3	R1	R2	R3	R1	R2	R3
		1			2			3			4			5			6	
X_I2S2	0	1	8	0	0	0	0	0	0	130	55	200	0	0	0	0	0	0
X_I2S1	130	0	66	0	0	198	0	0	0	0	0	0	0	0	0	0	0	0
X_I1S2	0	1	0	0	0	1	69	0	0	0	0	0	130	55	200	0	0	0
X_I1S1	0	53	125	130	55	1	36	55	0	0	0	0	0	0	0	0	0	0
Capacity	130	55	200	130	55	200	130	55	200	130	55	200	130	55	200	130	55	200

Fig. 2. Capacity allocation at resource M2

Table 3. Completion time and tardiness

Job (Type_item)	Due date	Completion period	Tardiness
I1_S1	7	9	2
I1_S2	9	12	3
I2_S1	6	6	0
I2_S2	8	8	0

The optimization model as described in Sect. 3 can find optimal solutions for small problems in relatively short time. However, real capacity planning problems may involve many orders and many resources so that it will take a long time to find the optimal solutions. The strategy that can be used to overcome that, based on the Theory of Constraints, is to consider only the critical resources and optimize the allocation in the critical resources. Here is our implementation procedure for large problems:

a. Perform RCCP using the resource profile (or other suitable) method for regular work time. The method will show the estimated utilization of each resource in the shop.
b. Identify the critical resources, i.e. resources whose allocation is more than the regular capacity.
c. Execute the optimization model with only those critical resources.
d. Verify that the optimal solution does not make any other resources infeasible.

4 Concluding Remarks

This paper has described the initial development of an optimization model of capacity planning with multiple sources (regular working time, overtime, and outsource) and lateness penalty option. In the model, outsource resources are allowed only after the resources in regular and overtime are fully used. After linearization, the model can be implemented as a mixed integer linear programming (MILP) problem. However, the calculation time of the optimization model may be prohibitive for large problems because of limited computation resources. In that case, based on the Theory of Constraints idea to focus on the critical resources, we have also proposed an implementation procedure that makes it possible to implement the model to larger problems.

As the next step, we plan to devise heuristic or metaheuristic algorithms to find good solutions in reasonable time even for large problems. We also plan to include other considerations to the model, such as inventory and demand uncertainty.

References

1. Sheikh, K.: Manufacturing Resource Planning (MRP II) with Introduction to ERP. McGraw-Hill, New York (2002)
2. Ullah, S., Guan, Z., Lei, Y., Zhang, F., dan Chong, H.: Intelligent integrated MRP, MPS, and detail scheduling. In: Proceeding of the 2016 International Conference on Industrial Engineering and Operation Management (2016)
3. Özdamar, L., Yazgaç, T.: Capacity drive due date settings in make-to-order production systems. Int. J. Prod. Econ. **49**, 29–44 (1997)
4. Chen, C., Mestry, S., Damodaran, P., Wang, C.: The capacity planning problem in make-to-order enterprise. Math. Comput. Model. **50**, 1461–1473 (2009)
5. Phruksaphanrat, B., Ohsato, A., Yenradee, P.: Aggregate production planning with fuzzy demand and variable system capacity based on theory of constraints measures. Int. J. Ind. Eng. **18**(5), 219–231 (2011)
6. Wang, C., Liu, X.: Integrated production planning and control: A multi-objective optimization model. J. Ind. Eng. Manag. **6**(4), 815–830 (2013)

An Optimization Model for Coal Procurement Networks with Coal Blending Facilities

Muhammad Imaduddin and Sukoyo(✉)

Department of Industrial Engineering, Institut Teknologi Bandung, Bandung, Indonesia
sukoyo@itb.ac.id

Abstract. This paper discusses the development of LRP (location routing problem) model for the problem of coal procurement networks with hubs which are called Coal Blending Facilities (CBF). This study aims to determine the location of suppliers and CBFs as well as the quantities of supply coal supply allocation and vessel distribution routes. The problem could be formulated as a mixed integer linear programming (MILP) to cover multiple supply allocation under capacitated facilities and fixed cost. Since the MILP of LRP is in the NP-hard category, we proposed a decomposition method to solve a large-scale problem of the LRP. This method is firstly used to select the nodes involved by determining the location of opened hub (CBF), and then applying the MILP for each selected CBF to solve allocation and vessel routes. A numerical example is presented to show effectiveness of the proposed method.

Keywords: Optimization · Location Routing Problem · Coal procurement · Coal Blending · Distribution routes

1 Introduction and Problem Description

In line with the conservation view, a coal blending method is introduced to utilize not only high-quality coal but also low-quality coal. The coal blending is a method to obtain the desired coal quality by blending high-quality and low-quality coals at coal blending facilities (CBFs). By applying the CBFs, the selection of suppliers of coal sources can be more flexible. In power plant, coal procurement costs which consist of coal prices, CBF's operation, and distribution costs should be minimized. [1] and [2] have developed the optimization of coal transportation and coal blending costs for power plants but they did not consider the LRP. [3, 4] and [5] have conducted research on many-to-many LRP under un-capacitated facility without CBFs.

The coal supply chain system of power plant consists of three layers, namely suppliers, CBFs as hubs, and power plants as shown in Fig. 1. Those three layers will be connected by two echelons vessels that will start and finish the trip at CBF (hub) point. The vessels of echelon-I are to connect suppliers to hubs, and echelon-II are to connect customers to hubs. This research will develop a LRP model which considers multiple suppliers, capacitated facility, and hubs of product blending process.

M. N. Osman Zahid et al. (Eds.): iMEC-APCOMS 2019, LNME, pp. 326–334, 2020.
https://doi.org/10.1007/978-981-15-0950-6_50

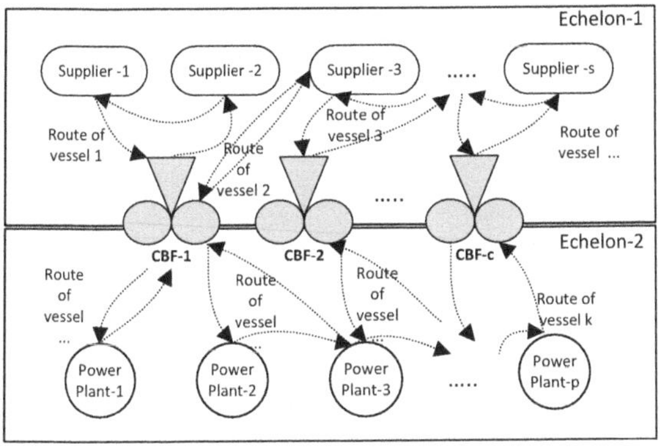

Fig. 1. Supply chain and coal distribution system of power plant

2 Model Formulation

2.1 LRP Model

Based on the problem system (Fig. 1), this problem is within the scope of many-to-many location routing problems and could be formulated in a mixed integer linear programming (MILP). We need to decide the location of hubs and suppliers, allocation of demands from power plants, allocation and amount of supply from each supplier, number of vessels needed and distribution routes.

The decision variables of the model are consisting of:

y_c = 1 if hub $c \in C$ is established, 0 others.
z_{cp}^{k} = 1 if hub $c \in C$ supplies to power plant $p \in P$, 0 others.
u_c = quantities of coal demand in hub $c \in C$.
z_{cs}^{k} = 1 if hub $c \in C$ is supplied by supplier $s \in S$ using vessel $k \in K$, 0 others.
v_{cs}^{k} = quantities of coal supplied in hub $c \in C$ from supplier $s \in S$ by vessel $k \in K$.
x_{cs}^{k} = 1 if arc c-s is traversed by vessel $k \mid c \in C, s \in S, k \in K$; 0 others.
x_{hi}^{k} = 1 if arc h-i is traversed by vessel $k \mid h, i \in C$, $k \in K$; 0 others.
x_{sc}^{k} = 1 if arc s-c is traversed by vessel $k \mid s \in S, c \in C, k \in K$; 0 others.
x_{cp}^{k} 1 if arc c-p is traversed by vessel k | c ∈ C, p ∈ P, k ∈ K; 0 others.
x_{mn}^{k} = 1 if arc m-n is traversed by vessel k | m, n ∈ P, k ∈ K; 0 others.
x_{pc}^{k} = 1 if arc p-c is traversed by vessel k | p ∈ P, c ∈ C, k ∈ K; 0 others.
$JK_c^{E_1}$ and $JK_c^{E_2}$ = number of vessels used on hub c ∈ C in echelon I and II

The objective function is to minimize total cost of coal procurement which consists of coal purchasing from suppliers, coal distribution from suppliers to hubs, cost of using CBFs (hubs), and mixed-coal distribution from hubs to power plants, as follows:

$$\begin{aligned} \textit{Minimum} \quad & \sum_{k=1}^{K}\sum_{c=1}^{C}\sum_{s=1}^{S} v_{cs}^{k} P_s + \sum_{k=1}^{K}\sum_{c=1}^{C}\sum_{s=1}^{S} C_{cs} V_1\, x_{cs}^{k} + \sum_{k=1}^{K}\sum_{h=1}^{S}\sum_{i=1}^{S} C_{cs} V_1\, x_{hi}^{k} + \sum_{k=1}^{K}\sum_{s=1}^{S}\sum_{c=1}^{C} C_{cs} V_1\, x_{sc}^{k} \\ & + \sum_{k=1}^{K}\sum_{c=1}^{C}\sum_{s=1}^{S} F_1\, x_{cs}^{k} + \sum_{c=1}^{C} 0 w_c y_c + \sum_{k=1}^{K}\sum_{c=1}^{C}\sum_{s=1}^{S} 0 p_c v_c^{k} + \sum_{k=1}^{K}\sum_{c=1}^{C}\sum_{p=1}^{P} C_{pc} v_2 x_{cp}^{k} \\ & + \sum_{k=1}^{K}\sum_{m=1}^{P}\sum_{n=1}^{P} C_{mn} v_2 x_{mn}^{k} + \sum_{k=1}^{K}\sum_{p=1}^{P}\sum_{s=1}^{C} C_{pc} V_2\, x_{pc}^{k} + \sum_{k=1}^{K}\sum_{c=1}^{C}\sum_{p=1}^{P} F_2\, x_{cp}^{k} \end{aligned} \tag{1}$$

Subject to:

$$\sum_{k\in K}\sum_{c\in C} z_{cp}^{k} = 1, \forall p \in P \tag{2}$$

$$\sum_{k\in K}\sum_{p\in P} h_p z_{cp}^{k} = u_c,\ \forall c \in C \tag{3}$$

$$\sum_{k\in K}\sum_{s\in S} v_{cs}^{k} = u_c,\ \forall c \in C \tag{4}$$

$$\sum_{k\in K}\sum_{s\in S} CV_s v_{cs}^{k} \geq \sum_{k\in K}\sum_{p\in P} h_p z_{cp}^{k} CV_p, \forall c \in C \tag{5}$$

$$\sum_{k\in K}\sum_{s\in S} TM_s v_{cs}^{k} \leq \sum_{k\in K}\sum_{p\in P} TM_p h_p z_{cp}^{k}, \forall c \in C \tag{6}$$

$$\sum_{k\in K}\sum_{s\in S} TS_s v_{cs}^{k} \leq \sum_{k\in K}\sum_{p\in P} TS_p h_p z_{cp}^{k}, \forall c \in C \tag{7}$$

$$\sum_{k\in K}\sum_{s\in S} AS_s v_{cs}^{k} \leq \sum_{k\in K}\sum_{p\in P} AS_p h_p z_{cp}^{k}, \forall c \in C \tag{8}$$

$$\sum_{k\in K}\sum_{s\in S} v_{cs}^{k} \leq A_s, \forall s \in S \tag{9}$$

$$\sum_{k\in K}\sum_{s\in S} v_{cs}^{k} \leq B_c, \forall c \in C \tag{10}$$

$$\sum_{k\in K} z_{cs}^{k} \leq y_c, \forall c \in C; \forall s \in S \tag{11}$$

$$v_{cs}^{k}/B_c \leq z_{cs}^{k}, \forall k \in K; \forall c \in C; \forall s \in S \tag{12}$$

$$\sum_{k\in K} z_{cp}^{k} \leq y_c, \forall c \in C; \forall p \in P \tag{13}$$

$$z_{ah}^{k} + z_{bi}^{k} \leq 1, \forall k \in K; \forall a, b \in C, a \neq b; \forall h, i \in S \tag{14}$$

$$z_{am}^k + z_{bn}^k \leq 1, \forall k \in K; \forall a, b \in C, a \neq b; \forall m, n \in P \quad (15)$$

$$\sum_{s \in S} x_{cs}^k \leq \sum_{s \in S} x_{sc}^k, \forall k \in K; \forall c \in C \quad (16)$$

$$x_{cs}^k \leq z_{cs}^k, \forall k \in K; \forall c \in C; \forall s \in S \quad (17)$$

$$x_{sc}^k \leq z_{cs}^k, \forall k \in K; \forall c \in C; \forall s \in S \quad (18)$$

$$\sum_{s \in S} x_{cs}^k \leq 1, \forall k \in K; \forall c \in C \quad (19)$$

$$\sum_{p \in P} x_{cp}^k \leq \sum_{p \in P} x_{pp}^k, \forall k \in K; \forall c \in C \quad (20)$$

$$x_{cp}^k \leq z_{cp}^k, \forall k \in K; \forall c \in C; \forall p \in P \quad (21)$$

$$x_{pc}^k \leq z_{cp}^k, \forall k \in K; \forall c \in C; \forall p \in P \quad (22)$$

$$\sum_{p \in P} x_{cp}^k \leq 1, \forall k \in K; \forall c \in C \quad (23)$$

$$\sum_{k \in k} \sum_{s \in S} x_{cs}^k = JK_c^{E_1}, \forall c \in C; E_1 \quad (24)$$

$$\sum_{k \in k} \sum_{p \in P} x_{cp}^k = JK_c^{E_2}, \forall c \in C; E_2 \quad (25)$$

$$x_{hi}^k \leq \sum_{c \in C} z_{ch}^k, c; \forall k \in K; \forall h, i \in S, h \neq i \quad (26)$$

$$x_{hi}^k \leq \sum_{c \in C} z_{ci}^k, c; \forall k \in K; \forall h, i \in S, h \neq i \quad (27)$$

$$\sum_{i \in S, h \neq i} x_{ih}^k + \sum_{c \in C} x_{ch}^k = \sum_{c \in C} z_{ch}^k, c; \forall k \in K; \forall h \in S \quad (28)$$

$$\sum_{i \in S, h \neq i} x_{hi}^k + \sum_{c \in C} x_{hc}^k = \sum_{c \in C} z_{ch}^k, c; \forall k \in K; \forall h \in S \quad (29)$$

$$x_{mn}^k \leq \sum_{c \in C} z_{cm}^k, c; \forall k \in K; \forall m, n \in P, m \neq n \quad (30)$$

$$x_{mn}^k \leq \sum_{c \in C} z_{cn}^k, c; \forall k \in K; \forall m, n \in P, m \neq n \quad (31)$$

$$\sum_{k\in K}\sum_{n\in P,m\neq n} x_{nm}^{k} + \sum_{k\in K}\sum_{c\in C} x_{cm}^{k} = 1; \forall m \in P \tag{32}$$

$$\sum_{k\in K}\sum_{n\in P,m\neq n} x_{mn}^{k} + \sum_{k\in K}\sum_{c\in C} x_{mc}^{k} = 1; \forall m \in P \tag{33}$$

$$x_{hi}^{k} + x_{ha}^{k} + z_{ah}^{k} + \sum_{b\in C,a\neq b} z_{bi}^{k} \leq 2; \forall k \in K; \forall a \in C; \forall h,i \in S \tag{34}$$

$$x_{mn}^{k} + x_{ma}^{k} + z_{am}^{k} + \sum_{b\in C,a\neq b} z_{bn}^{k} \leq 2; \forall k \in K; \forall a \in C; \forall m,n \in P \tag{35}$$

$$\sum_{c\in C} x_{ch}^{k} + \sum_{i\in S,h\neq i} x_{ih}^{k} - \sum_{i\in S,h\neq i} x_{hi}^{k} - \sum_{s\in C} x_{hc}^{k} = 0; \forall k \in K; \forall h,i \in S \tag{36}$$

$$\sum_{c\in C} x_{cm}^{k} + \sum_{n\in S,m\neq n} x_{nm}^{k} - \sum_{i\in S,h\neq i} x_{mn}^{k} - \sum_{c\in C} x_{mc}^{k} = 0; \forall k \in K; \forall m,n \in P \tag{37}$$

$$\sum_{s\in S} x_{cs}^{k} + \sum_{h\in S}\sum_{i\in S} x_{hi}^{k} + \sum_{s\in S} x_{sc}^{k} \geq \sum_{s\in S} z_{cs}^{k}; \forall k \in K; \forall c \in C \tag{38}$$

$$\sum_{p\in P} x_{cp}^{k} + \sum_{m\in P}\sum_{n\in P} x_{mn}^{k} + \sum_{p\in P} x_{pc}^{k} \geq \sum_{p\in P} z_{cp}^{k}; \forall k \in K; \forall c \in C \tag{39}$$

$$x_{hi}^{k} + x_{ih}^{k} \leq 1; \forall k \in K; \forall i \in S \tag{40}$$

$$\sum_{k\in K} x_{mn}^{k} + \sum_{k\in K} x_{nm}^{k} \leq 1; \forall m,n \in P \tag{41}$$

$$x_{hi}^{k} = 0; \forall k \in K; \forall h,i \in S, h = i \tag{42}$$

$$x_{mn}^{k} = 0; \forall k \in K; \forall m,n \in P, m = n \tag{43}$$

$$W_{c}^{kE_1} = 0; \forall k \in K; \forall c \in C; E_1 \tag{44}$$

$$W_{s}^{k} \geq W_{c}^{kE_1} + V_{cs}^{k} - Q_1(1 - x_{cs}^{k}); \forall k \in K; \forall s \in S; \forall c \in C; E_1 \tag{45}$$

$$W_{s}^{k} \leq Q_1 z_{cs}^{k}; \forall k \in K; \forall s \in S; \forall c \in C \tag{46}$$

$$W_{i}^{k} \geq W_{h}^{k} + V_{ci}^{k} - Q_1(1 - x_{hi}^{k}); \forall k \in K; \forall c \in C; \forall h,i \in S; h \neq i \tag{47}$$

$$W_{c}^{kE_2} = 0; \forall k \in K; \forall c \in C; E_2 \tag{48}$$

$$W_{p}^{k} \geq W_{c}^{kE_2} + H_p z_{cp}^{k} - Q_2(1 - x_{pc}^{k}); \forall k \in K; \forall p \in P; \forall c \in C; E_2 \tag{49}$$

$$W_p^k \leq Q_2 z_{cp}^k; \forall k \in K; \forall c \in C; \forall p \in P \tag{50}$$

$$W_m^k \geq W_n^k + h_m z_{cm}^k - Q_{21}(1 - x_{mn}^k); \forall k \in K; \forall c \in C; \forall m, n \in P; m \neq n \tag{51}$$

$$\sum_{s \in S} v_{cs}^k \leq Q_1; \forall k \in K; \forall c \in C \tag{52}$$

$$\sum_{c \in C} \sum_{p \in P} H_p z_{cp}^k \leq Q_2; \forall k \in K \tag{53}$$

$$JKE_1 = \sum_{c \in C} \sum_{s \in S} c_{cs} x_{cs}^k + \sum_{h \in S} \sum_{i \in S} c_{hi} x_{hi}^k + \sum_{c \in C} \sum_{s \in S} c_{cs} x_{sc}^k \leq 2520; \forall k \in K; E_1 \tag{54}$$

$$JKE_2 = \sum_{c \in C} \sum_{p \in P} c_{cp} x_{cp}^k + \sum_{m \in P} \sum_{n \in P} c_{mn} x_{mn}^k + \sum_{c \in C} \sum_{p \in P} c_{pc} x_{pc}^k \leq 2520; \forall k \in K; E_2 \tag{55}$$

Constraints for decision variables.
$y_c \in \{0,1\}$; $z_{cs}^k \in \{0,1\}$; $z_{cp}^k \in \{0,1\}$; $x_{cs}^k \in \{0,1\}$; $x_{sc}^k \in \{0,1\}$;
$x_{hi}^k \in \{0,1\}$; $x_{cp}^k \in \{0,1\}$; $x_{pc}^k \in \{0,1\}$; $x_{mn}^k \in \{0,1\}$; $v_{cs}^k \geq 0$

Constraint (2) ensures that each power plant is only served by one hub. Coal allocation from supplier to hub must be in accordance with the quantities of coal demand from customer to hub, constraints (3) and (4). Constraints (5)–(8) guarantee that coal in each hub has to meet the required quality from customer, including calorific value, total moisture, total sulphur, and ash content. Constraint on supplier capacity is (9) and hub capacity is (10). Constraints (11)–(13) ensures that each opened hub is supplied by suppliers, and it supplies mixed coal to customers. Each vessel in each echelon only traverse on single trip and only visits one hub, constraints (14) and (15). Vessels departing from a hub must return to the same hub (constraints (16)–(19) for echelon I and (20)–(23) for echelon II). Constraints (24)–(25) is the number of vessels to be provided at each hub to serve echelon I and II. Constraints (26)–(29) ensures that each supplier who will supply coal to hubs must be visited by vessels as many as number of hubs to be served. Constraints (30)–(33) ensure that each customer is only served once and visited by one vessel. Constraints (34)–(35) is used to avoid the formation of illegal routes in echelon I and II, where the route does not start and finish at the same node. Constraints (36)–(37) are flow conservation constraints in echelon I and II. Constraints (38)–(39) restrict the formation of distribution lane and route. Each lane that is traversed by the same vessel must be connected to each other and form exactly one route, constraints (38) and (39). A vessel that traverses through a lane between nodes, cannot return to the previous node, constraints (40) and (41). Each formed lane must connect two different nodes, constraints (42) and (43). Constraints (44)–(47) is to specify the product quantities on vessels in echelon I, and (48)–(51) on echelon II. The vessels must be empty when leaving a hub, and fully or partially loaded when returning to the hub. Product quantities on each vessel must not exceed the capacity of vessel, constraints (52) and (53). Constraints (54)–(55) ensures that length of each route at echelon I and II does not exceed the maximum route length (2520 km).

2.2 Network Design Model

The LRP model is included in the NP-hard category. To overcome it, we firstly decompose the LRP to be a network design model to determine the opened-hubs and the selected-suppliers. The network design model is built by eliminating vessel index of the LRP model. Furthermore, operational problems decisions are solved using the LRP model based only on the opened-hubs and the selected suppliers.

2.3 Solution Method

We propose a heuristic method to solve un-directly the LRP model through the network design model. This propose heuristic as follows:

- Solve the problem LRP with the network design model to get the opened-hubs and the selected-suppliers;
- Simplify the LRP model only using the opened-hubs and the selected-suppliers to solve allocation demand from plant to the opened-hubs and the opened-hubs to the selected-supplies, number of vessels and vessel routes at echelon-I and II.

Solution quality between the proposed-heuristic and the LRP are estimated with three hypothetical data scenarios. We use LINGO version 18 to build and to solve the LRP and network design models into. The results in Table 1 showed the gap of the objective function values is less than 1% with the optimal solution.

Table 1. Estimation solution gap

Scenarios		4S 3C 4P	5S 3C 3P	3S 3C 4P
Total quantities of demands (ton)		82,000	82,000	77,000
Total quantities of potential supplies (ton)		120,000	130,000	120,000
Selected suppliers	LRP	S1, S2, S3 & S4	S1, S2, S3 & S5	S1, S2 & S3
	Heuristic	S2, S3 & S4	S1, S2, S3 & S5	S2 & S3
Opened hubs	LRP	C2 & C3	C2 & C3	C2 & C3
	Heuristic	C1 & C3	C1 & C3	C2 & C3
Number of vessels used	LRP	6 vessels	6 vessels	6 vessels
	Heuristic	6 vessels	7 vessels	6 vessels
Data processing (seconds)	LRP	4,130.02	3,490.40	2,153.14
	Heuristic	11.84	10.21	4.49
Value of objective function	LRP	USD 6,726,040	USD 5,478,243	USD 6,455,157
	Heuristic	USD 6,754,784	USD 5,492,014	USD 6,482,076
Gap of the objective function	Nominal	USD 28,744	USD 13,771	USD 26,919
	Percentage	0.43%	0.25%	0.42%

Table 2. Solution of numerical example

No	Decision Variables	Solution				
1	Opened hub (CBF)	C1, C2 & C3				
2	Allocation of demands to opened-hubs	P6 (64,632 ton)				C1: 64,632 ton
		P3 (32.316 ton), P4 (42,657 ton), P7 (61,401 ton)				C2: 136,374 Ton
		P1 (64.632 ton), P2 (10,180 ton), P5 (32,316 ton)				C3 :107,128 Ton
4	Selected suppliers	9 suppliers : S8, S10, S11, S12, S13, S14, S15, S16 & S19				
5	Allocation and quantities of demands from each opened hub to selected suppliers	C1	S10	14,632 ton	S13	50,000 ton
		C2	S8	4,592.56 ton	S12	8,414 ton
			S10	45,170.44 ton	S13	50,000 ton
			S11	824 ton	S15	27,373 ton
		C3	S11	42,734.23 ton	S16	16,438 ton
			S12	11,312 ton	S19	3,767.77 ton
			S14	32,876 ton		
6	Number of vessels	Echelon 1	8 vessels		Echelon 2	6 vessels
7	Coal distribution routes	E1	K1	C1→S10→C1	K4	C2→S8→S10→C2
			K2	C1→S13→C1	K5	C2→S12→S11→S15→C2
			K3	C2→S13→C2	K6	C3→S19→S11→12→C3
			K7	C3→S16→C3	K8	C3→S14→S11→C3
		E2	K1	C1→P6→C1	K4	C2→P3→C2
			K2	C2→P6→C2	K5	C3→P5→P2→C3
			K3	C2→P4→C2	K6	C3→P1→C3
Objective Function Values		**USD 20,296,506**				

3 Numerical Example

The model is tested in a case study of the PLN coal procurement network for 7 power plants to be built on Java Island year 2019–2020. There are 19 potential suppliers from Sumatera and Kalimantan as recorded in Indonesian Coal Book 2016/2017, where they have a value of HGI is 45-6, production capacity more than of 400,000 tons/year for at least 5 years. The are 4 candidates of CBFs at East Kalimantan, South Kalimantan, Banten, and Lampung with capacity of 164.383 ton/three days. By using the proposed method, we have a LPR solution as in Table 2.

4 Conclusion

This research has developed a LRP model of two-echelons many-to-many location-routing problem for coal procurement network by applying coal blending method. A heuristic procedure is proposed to solve the LRP to get network design and operational decision to provide vessel distribution routes as well as coal demand allocation. The heuristic method has a solution gap value less than 1% with the optimal solution. This research can be extended to heterogeneous vehicle types, multiple trips problems.

References

1. Arigoni, A., Newman, A., Turner, C., Kaptur, C.: Optimizing global thermal coal shipment. Omega **72**, 118–127 (2017)
2. Liu, C.M., Sherali, H.D.: A coal shipping and blending problem for an electric utility company. Int. J. Manag. Sci. **28**, 433–444 (2000)
3. Rieck, J., Ehrenberg, C., Zimmermann, J.: Many-to-many location-routing with inter-hub transport and multi-commodity pickup-and-delivery. Eur. J. Oper. Res. **236**, 863–878 (2014)
4. Çetiner, S., Sepil, C., Süral, H.: Hubbing and routing in postal delivery systems. Ann. Oper. Res. **181**, 109–124 (2010)
5. De Camargo, R.S., de Miranda, G., Løkketangen, A.: A new formulation and an exact approach for the many-to-many hub location-routing problem. Appl. Math. Model. **37**, 7465–7480 (2013)
6. Rodrígues-Martin, I., Salazar-Gonzalez, J.-J., Yaman, H.: A branch-and-cut algorithm for the hub location and routing problem. Comput. Oper. Res. **50**, 161–174 (2014)

Competing Risk Models in Reliability Systems, an Exponential Distribution Model with Gamma Prior Distribution, a Bayesian Analysis Approach

Ismed Iskandar[1(✉)], Muchamad Oktaviandri[1,3], Rachmawati Wangsaputra[2], and Zamzuri Hamedon[1]

[1] Faculty of Mechanical and Manufacturing Engineering, Universiti Malaysia Pahang, 26600 Pekan, Pahang, Malaysia
ismed@ump.edu.my
[2] Bandung Institute of Technology, Bandung, Indonesia
[3] Fakultas Teknologi Industri, Universitas Bung Hatta, Padang 25143, Indonesia

Abstract. This paper is a second paper on the use of Exponential distribution in competing risk problems. The difference is this model is developed using Gamma distribution as its prior distribution. For the cases where the failure data together with their causes of failure are simply quantitatively inadequate, time consuming and expensive to perform the life tests, especially in engineering areas, Bayesian analysis approach is used. This model is limited for independent causes of failure. In this paper our effort is to introduce the basic notions that constitute an exponential competing risks model in reliability using Bayesian analysis approach and presenting their analytic methods. Once the model has been developed through the system likelihood function and individual posterior distributions then the parameter of estimates are derived. The results are the estimations of the failure rate of individual risk, the MTTF of individual and system risks, and the reliability estimations of the individual and of the system of the model.

Keywords: Reliability · Competing risks · Exponential distribution · Bayesian

1 Introduction

Reliability theory concerns the ability of a component or a system, either a life creature or an equipment, to be functioning during its expected length of life. The estimation of the parameters of the assumed failure models may be based on data collected over life tests or obtained from past engineering experience or from handbook data, if available. Analysis of data, either historical or from life tests, may use any of the classical parameter estimation methods such as maximum likelihood, matching moments, regression, etc. However, most of recent analysis so far has been limited to single risk-mode models. A system or component only considered as in function or fail only without considering the causes of its failure. But in the real systems, the failure may be a phenomenon within a single item such as the tread wear, puncture, or defective side

M. N. Osman Zahid et al. (Eds.): iMEC-APCOMS 2019, LNME, pp. 335–341, 2020.
https://doi.org/10.1007/978-981-15-0950-6_51

walls or from other causes of an automobile tire. This phenomenon is commonly called as competing risks. The competing risks theory describes how many causes of failure act together to affect the performance of a system and only one of them will cause the failure.

Another problem in this area: the failure data together with their causes of failure are simply quantitatively inadequate, time consuming and expensive to perform the life tests, especially in engineering areas. The competing risks phenomenon together with the scarcity of the data will lead us to the imprecise estimations of the parameters of the models interests. A new approach is needed by using the Bayesian analyses to analyse the investigated competing risk models [1, 2]. The Bayesian estimation analyses allow us to combine past knowledge or experience with less data available (which reflect the competing risks phenomenon) in the form of an appropriate distribution to make inferences of the parameter of interest. Although Bayesian estimation has been used widely in reliability analysis, the application so far has been limited to single risk-mode models.

2 Literature Review

The theory of competing risks has been in use in bio-medicine for a long time. It began in 1760 with Daniel Bernoulli when he tried to determine mathematically what would happen to a population mortality structure at different ages if smallpox were eliminated from that population. Most of the currently accepted techniques were developed during the 19th century when the problem had been of great interest and importance to actuaries for over 100 years.

Some scientists have worked on competing risks analysis using classical approaches. In 2016, the competing risk data in the presence of complete information of failure cause has been analyzed by considering the occurrence of missing causes as well as interval censored failure time [3]. The inference for a simple step-stress model with progressively censored competing risks data from Weibull distribution has been studied using the simple step-stress model under progressive Type-II censoring [4]. Other researcher used of the Marshall–Olkin bivariate Weibull distribution for dependent causes of failure to estimate the unknown model parameters with a parameter transformation [5]. A new Weibull inter-arrival shock process in reliability and maintenance modeling for competing risk processes in four different hard failure models was proposed and analytic forms of the reliability functions were derived [6].

Numerous scientists have contributed and provided the philosophical basis for Bayesian analysis in reliability systems and use it in many areas of application and research. They are not to be mentioned here because so many of them doing research in this area and because their Bayesian method of analysis were for single risk-mode model only.

There were a limited number of scientists who studied and analysed the competing risks problems by using a Bayesian analyses approach in order to estimate the parameters of interest. Their models were mostly for applications in a specific condition but not for the basic understanding of general model applications. One of these researchers used data from randomized clinical trial for patients with node negative breast cancer, illustrated the approach to handle lost to follow up observation. Nelson-Aalen estimator and Bayesian approach are considered to handle the lost to follow up in the data set [7]. Bayesian nonparametric mixed-effects joint model for longitudinal-competing risks data analysis in presence of multiple data features was proposed by another researcher [8].

For competing risks models where the causes of failures are being considered, the life test data scarcity for each cause is inevitable. In many situations, especially in engineering areas, these life test data are time consuming to obtain and expensive to perform the life tests. For these cases, Bayesian analyses are more beneficial than classical one. Author has published three papers that deal with this matter using Weibull, Gamma, Exponential, as posterior functions and Uniform distribution as prior distribution for the posteriors [9–11]. Another paper was for attribute type of life test data with Multinomial distribution models as the posterior and Beta distribution function as its prior [12]. Author also recognized that there were some more potential competing risks models need to be developed for the future research.

3 Exponential Distribution and Its Likelihood Function

The exponential distribution is the most widely used reliability analysis. This distribution is very suitable for representing the lengths of life of many cases and is available in a simple statistical form. The characteristic of this distribution is a constant hazard rate. The exponential distribution is the lower rank of the Weibull distributions.

This section describes the likelihood function of this distribution and follows with the description of the posterior function. A Gamma prior is used in our analysis. It is followed by the estimation of the failure rate and hazard function. The net, crude, and partial crude probabilities are also included. In this paper, we will use the terms "individual" for every function that refers to an individual risk.

The probability distribution function (p.d.f.) of the length of life t_i due to a risk C_i is given by $f(t_i|\lambda_i) = \lambda_i e^{-\lambda_i t_i}$ for $i = 1, \ldots, k$, $\lambda_i > 0$, and $t_i \geq 0$. The parameter λ_i is known as the failure rate.

Let us consider a life test of m items in which n items have failed and s items have survived. Suppose n_i items have failed and s_i have survived from cause C_i where $n = \sum_{i=1}^{k} n_i$, $s = \sum_{i=1}^{k} s_i$, $m_i = n_i + s_i$, and $m = \sum_{i=1}^{k} m_i$. The failed items due to each cause are in the time order of $t_{i1}, t_{i2}, \ldots, t_{in_i}$ and the surviving items have been operating for $t_{i(n_i+1)}, t_{i(n_i+2)}, \ldots, t_{im_i}$, respectively. The likelihood function of the system is given by

$$L(\lambda_1, \lambda_2, \ldots, \lambda_k) \propto \prod_{i-1}^{k} \left[\prod_{i=1}^{n_i} f(t_j)\right] \left[\prod_{i=n+1}^{m} (1 - F(t_i))\right]$$

$$\propto \prod_{i=1}^{k} \lambda_i^{n_i}{}^{e-\lambda_i\left(\sum_{j=1}^{n_i} t_{ij} + \sum_{j=n_1+1}^{m_i} t_{ij}\right)} \propto \prod_{i=1}^{k} \lambda_i^{n_i} e^{-\lambda_i w_i} \propto \prod_{i=1}^{k} L_i(\lambda_i) \tag{1}$$

Where $L(\lambda_i) \propto \lambda_i^{n_i} e^{-\lambda_i w_i}$ is an individual likelihood function, and $T_i = \sum_{j=1}^{n_i} t_{ij} + \sum_{j=n_1+1}^{m_i} t_{ij}$ is then a sufficient statistic for the estimation of the parameter λ_i.

The individual posterior function is given by

$$g(\lambda_i|t_i) = \frac{\lambda_i^{n_i} \exp(-\lambda_i T_i) f(\lambda_i)}{\int_0^\infty {}_i^{n_i} \exp(-\lambda_i T_i) f(\lambda_i) d\lambda_i}, \tag{2}$$

where $f_i(\lambda_i)$ is the individual prior distribution.

In this chapter we consider the use of Gamma distribution as our prior with p.d.f. as follows:

$$f(\lambda_i; \alpha_i, \beta_i) = \frac{1}{\Gamma(\alpha_i)\beta_i^{\alpha_0}} \lambda_i^{\alpha_i - 1} e^{-\lambda_i/\beta_i} \tag{3}$$

The Gamma distribution is one of some other priors such as Uniform, Non-informative, and Inverted Gamma distributions commonly used as prior for the Exponential posterior distribution. The main reason for using Gamma distribution as a prior is its mathematical tractability.

Substitute Eq. (3) into (2) and then the individual posterior distribution become:

$$g(\lambda_i|t_i; n_i, \alpha_i, \beta_i) = \frac{1}{\Gamma(\alpha_i + n_i)(\beta_i/(\beta_i t_i + 1))^{\alpha_i + n_i}} \lambda_i^{\alpha_1 + n_i - 1} e^{-\lambda_i(t_i + 1/\beta_i)} \tag{4}$$

For $0 < \lambda_i < \infty$

4 The Failure Rate Estimations

The failure rate estimation is the individual mean of posterior mentioned in Eq. (5) is as follows:

$$E(\lambda_i|t_i, n_i, \alpha_i, \beta_i) = \frac{\beta_i(\alpha_i + n_i)}{\beta_i t_i + 1} \tag{5}$$

With variance is as follows:

$$Var(\lambda_i|t_i; n_i, \alpha_i, \beta_i) = \frac{\beta_i^2(\beta_i + n_i)}{(\beta_i t_i + 1)^2} \tag{6}$$

5 Mean Time to Failure Estimation

Since the mean time to failure of individual posterior distribution, $MTTF_i$, is $1/E(\lambda_i|t_i, n_i, \alpha_i, \beta_i)$ or $\beta_i t_i + 1/\beta_i(\alpha_i + n_i)$, and then $MTTF$ for the system can be derived as follows:

$$MTTF_{System} = \frac{1}{\sum_{i=1}^{k} \frac{\beta_i(\alpha_i + n_i)}{\beta_i t_i + 1}} \tag{7}$$

6 Reliability Estimation

The reliability estimation can be obtained by first transforming posterior distribution in Eq. (5) to become:

$$g(R_i|t_i; n_i, \alpha_i, \beta_i) = \frac{1}{\Gamma(\alpha_i + n_i) t_i^{\alpha_i + n_i}} (t + 1/\beta_i)^{\alpha_i + n_i} (-\ln r)^{\alpha_i + n_i - 1} r^{(t + 1/\beta_i)/t_i - 1} \quad 0 < r_i < 1 \tag{8}$$

The individual posterior mean can be calculated as given by:

$$E(R|t_i; n_i, \alpha_i, \beta_i) = \frac{1}{[1 + t_i/(t + 1/\beta_i)]^{\alpha_i + n_i}} \tag{9}$$

With variance is as follows:

$$Var(R|t_i; n_i, \alpha_i, \beta_i) = \frac{1}{[1 + 2t_i/(t + 1/\beta_i)]^{\alpha_i + n_i}} - \frac{1}{[1 + t_i/(t + 1/\beta_i)]^{2\alpha_i + 2n_i}} \tag{10}$$

The reliability of the system can be derived from Eq. (10) as follows:

$$R_{System} = \prod_{i=1}^{k} E(R|t_i;\, n_i,\, \alpha_i,\, \beta_i)$$

$$= \prod_{i=1}^{k} \frac{1}{[1 + t_i/(t + 1/\beta_i)]^{\alpha_i + n_i}} \tag{11}$$

7 Conclusions

In this paper we investigate the exponential distribution as a posterior function and a Gamma distribution as its prior as our competing risks model. Other Exponential models with some other priors such as Uniform, Inverted Gamma, and Implied distributions are also open for the future investigation. It is also open for the cases of Two-parameter Exponential. The Exponential distribution is widely used in reliability analysis for a constant hazard rate and is one of some other common failure models as well as Weibull, exponential, Normal, Log-normal, and Inverse Gaussian models. This distribution is also known as the lower rank of the Weibull distributions

The estimations of this exponential model for its individual posterior function and for the system are developed. First by developing the likelihood function of the system and it shows that it is as a multiplicative form of individual distribution. Take a note here that we assume that each risk acts independently. The posterior distribution for each risks then developed together with its parameter of estimates such as failure rate and reliability estimations. It is followed by the estimation of the individual and system's MTTF and the reliability of the competing risks model. Based on these estimations, others such as the estimations for reliable or design life for a specific reliability, the probability of survival between a specific interval of time, the net probability if only one risk is present and the crude/partial crude probabilities if one or more risk is eliminated from the system while other risks remain are also open to be developed.

It should be noted that the choice of the time to failure distribution functions depend on the choice of its reliability measures that is meaningful and useful to the problem of interest. Different pattern of operating life of a device may required to have a specific probability of successfully performing its required function.

The author realizes that for all descriptions above of this model need a real or simulated data to test and prove the validity of the model. But the limitation of time to obtain or to conduct a simulation for such a data has put this matter for further investigation without eliminating the importance of this model. Another important measures of this analysis is to include the dependent risks in our model for further research.

References

1. Martz, H.F., Waller, R.A.: Bayesian Reliability Analysis. Krieger Publishing Company, Malabar (1991)
2. Fu, J., Tang, Y., Guan, Q.: Objective bayesian analysis for recurrent events in presence of competing risks. Qual. Technol. Quant. Manag. **11**(3), 265–279 (2014)
3. Do, G., Kim, Y.J.: Analysis of interval censored competing risk data with missing causes of failure using pseudo values approach. J. Stat. Comput. Simul. **87**(4), 631–639 (2016)
4. Liu, F., Shi, Y.: Inference for a simple step-stress model with progressively censored competing risks data from Weibull distribution. Commun. Stat.- Theory Methods **46**(14), 7238–7255 (2017)

5. Shen, Y., Xu, A.: On the dependent competing risks using Marshall-Olkin bivariate Weibull model: parameter estimation with different methods. Commun. Stat. – Theory Methods **47** (22), 5558–5572 (2018)
6. Liu, H.: Reliability and maintenance modeling for competing risk processes with Weibull inter-arrival shocks. Appl. Math. Model. **71**, 194–207 (2019)
7. Bhattacharjee, A.: Bayesian competing risks model: an application to breast cancer clinical trial with incomplete observations. J. Stat. Manag. Syst. **18**(4), 381–404 (2015)
8. Lu, T.: Bayesian nonparametric mixed-effects joint model for longitudinal-competing risks data analysis in presence of multiple data features. Stat. Methods Med. Res. **26**(5), 2407–2423 (2017)
9. Iskandar, I., Gondokaryono, Y.S.: Competing risk models in reliability systems, a Weibull distribution model with bayesian analysis approach. In: IOP Conference Series: Materials Science and Engineering, vol. 114, no. 1, pp. 012064 (2016)
10. Iskandar, I.: Competing risk models in reliability systems, a gamma distribution model with bayesian analyses approach. In: IOP Conference Series: Materials Science and Engineering, vol. 114, no. 1, pp. 012098 (2016)
11. Iskandar, I.: Competing risk models in reliability systems, an exponential distribution model with Bayesian analysis approach. In: IOP Conference Series: Materials Science and Engineering, vol. 319, no. 1, pp. 012069 (2018)
12. Iskandar, I., Razali, N.M.: Multi-mode failure models for attribute test data in reliability systems, a bayesian analysis approach using multi-nomial distribution model. In: Advanced Material Research, vol. 903, pp. 419–424 (2014)

Dump Truck Maintenance Contract Model Considering Operational Conditions (Load, Road Inclination and Environment Condition)

Fadhli Nishfi, Bermawi Priyatna Iskandar, and Rachmawati Wangsaputra(✉)

Industrial Engineering Program Study, Bandung Institute of Technology, Jl. Ganesha 10, Bandung, West Java 40132, Indonesia
{bermawi,rwangsap}@mail.ti.itb.ac.id

Abstract. A heavy-duty equipment, such as a dump truck operated in a mining industry experiences degradation caused by usage and operational condition. There are operational conditions influencing dump truck failure rate, such as load weight, road inclination and environmental conditions. In this research, dump truck failure is modeled using proportional hazard model (PHM) considering operational condition, age, and usage factors. Maintenance action is needed to control continuing failure to maintain the equipment always in good operation condition. A dump truck will need both preventive and corrective maintenance actions in order to be readily available for operation in mining industry. Considering that dump trucks are very complex and expensive, maintenance will be economical if it is conducted by an agent. In this research maintenance contract involved 2 parties, the owner and the agent. The agent offers 3 (three) options of maintenance contracts: (i) in house preventive maintenance conducting by the owner and corrective maintenance conducted by the agent (ii) both preventive and corrective maintenance conducted by the agent, (iii) preventive maintenance by the agent and in-house corrective maintenance by the owner, and determination of the optimal contract price for each option. Next the owner will choose the best option based on several options offered by the agent. The model used for decision is formulated by Nash game theory.

Keywords: Operational condition · Proportional hazard model (PHM) · Maintenance contract

1 Introduction

Equipment like a dump truck, in mining industry is used to transport mine from mining area to a processing area. This process happens continuously as the dump truck will degrade and at the end it will fail. This failure is influenced by usage level, usage time, and operation condition. These factors are important to be considered as heavy operation condition in Indonesia is heavy [1]. It is stated operational heavy and complex can influence the system life significantly causing the failure level increase [2]. There are two types of maintenance action: preventive maintenance consisting of a serial

M. N. Osman Zahid et al. (Eds.): iMEC-APCOMS 2019, LNME, pp. 342–348, 2020.
https://doi.org/10.1007/978-981-15-0950-6_52

action to control degradation rate and reduce the occurrence of failure, and secondly corrective maintenance, which is maintenance action conducted to put back a failed system [3].

Dump truck operated in a mining industry is a very heavy and expensive equipment; therefore, it will be economical to conduct the maintenance to the agent due to the resource at the owner side. For an owner, the availability of a dump truck is a prime key for achieve production target and getting optimal revenue; therefore, agent will offer several options for maintenance contract, for example partially maintenance or full maintenance. Maintenance contract where dump truck failure is modeled by accelerated failure time (AFT) considering the level of usage [4]. In mining industries, dump truck failure happens sooner because influenced by the operational condition. Failure and reliability of equipment in mining industries has been modeled to consider operational condition using proportional hazard model (PHM) to determine the availability of component of an equipment not for contract maintenance [5, 6]. This research suggests a maintenance contract model considering operational condition factors which are load weight, road inclination and environmental condition. Those factors never been used in previous model. The model is a proportional hazard failure model. The decision of the contract maintenance is formulated based on Nash game theory.

2 Methodology

2.1 Maintenance Contract

Heavy equipment like dump truck is a repairable equipment. Each equipment like dump truck has an interval period $[0, L)$, where L is a dump truck maintenance contract limit. Agent will offer three options for the contract maintenance for maximum benefits. The three contracts offered are as follow:

- Option-1: When an equipment fails in interval $[0, L)$, the owner will choose to conduct in house-PM and CM will be conducted by the agent, and the owner will be charged by the contract value C_{k_a} for the CM during the period $[0, L)$.
- Option-2: When the equipment fails in the interval $[0, L)$, all the maintenance activities, both for PM and CM, will be conducted by the agent with price of the contract C_G during the period $[0, L)$. Agent will pay the penalty if the availability is below the target.
- Option-3: When the equipment fails in the interval $[0, L)$, the owner will choose in-house CM and PM activity will be conducted by the agent with the contract price C_{p_a} for the PM action.

Furthermore, the following tables display the decision variables and parameters of the model being used in this research (Tables 1 and 2).

Table 1. Definitions of decision variables in the model.

Symbol	Decision variable	Symbol	Decision variable
C_{k_a}	Contract value of option-1	k	Number of PM
C_G	Contract value of option-2	δ	Level of PM
C_{p_a}	Contract value of option-3		

Table 2. Definitions of parameters in the model.

Symbol	Parameters	Symbol	Parameters
O_n	Option (n = 1, 2, 3)	C_{pe}	Penalty Cost
θ	PHM coefficient model	η	Target waktu downtime
Y	Operation Condition	$E[D(L)]$	Downtime expectation time [0, L)
L	Maintenance Contract Duration	$E[N(L)]$	Ekspektasi jumlah m. repair [0, L)
τ	Interval between PM	$L-E[D(L)]$	Availability of dump truck [0, L)
K	Dumptruck revenue	$g(d)$	Downtime distribution function
C_{k_n}	CM cost for option-n	$h_1(t)$	Failure function before PM
C_{f_n}	PM fixed cost for option-n	$h_1(v(t))$	Failure function after PM
C_{v_n}	PM variable cost for option-n	α	Weibull Dist. Parameter Scale
$J_{a_n}(k, \delta)$	CM total cost CM option-n	β	Weibull Dist. Parameter Shape
$J_{b_n}(k, \delta)$	PM total cost PM option-n	λ	Parameter of exponential parameter
$J_n(k, \delta)$	Total cost PM+CM option-n	$Xmax$	Penalty maximum time

2.2 Equipment and Repairement

The failure of the dump truck happens randomly and will increase due to the time of use and is affected by operation condition such as load, road inclination, and the environment condition in the industry [6], modeled by a blackbox approach. If T is a random variable stating the first time of failure with distribution function $F_0(t)$, then failure acceleration with first time distribution function is $h_0(t)$ and it increases along the increment of time t. The influence of operational condition is considered as covariant for PHM model making the failure rate model as follow:

$$h_1(t, Y) = h_0(t)\exp\left(\sum\nolimits_{i=1}^{m} \theta_i Y_i\right) \tag{1}$$

Owner will choose the best option based on contracts offered by agent for fulfill the equipment availability target where Y_i, i = 1, 2, …m is operational condition (co-variate) influencing the degradation level of dump truck and θ_i, i = 1, 2, … m is regression coefficient from m covariate Y. $h_0(t)$ is a baseline hazard function built from the first failure distribution. The failure acceleration is influenced by PM with imperfect maintenance. The effect of imperfect PM is formulated in Eq. 1, where it is assumed that PM will reduce the dump truck lifetime to the age younger than t to virtual age $v(t)$, that will make $v(t) < t$. The maintenance action is conducted at period $[0, L)$ with interval time PM τ, 2τ, …, $k\tau$,, where number of PM $1 \leq j \leq k$, thus the virtual age at time $j\tau$ is $v_j = j\delta\tau$. The age of dump truck will be younger and equal to $(1 - \delta)\tau$ for each PM maintenance, which make virtual age of dump truck after PM between j and $(j + 1)$ is:

$$v(t) = t - (1 - \delta)j\tau, j\tau \leq t \leq (j + 1)\tau \tag{2}$$

Where $0 \leq \delta \leq 1$, making the failure rate after PM formulated by $h_1(v(t))$ with total expectation of number of failures with minimal repair during the period $[0, L)$.

$$\begin{aligned} E[N] = \bar{N}(k, \delta) &= \begin{bmatrix} \int_0^\tau h_1 dt + \\ \sum_{j=0}^{k} \int_{j\tau}^{(j+1)\tau} h_1(v(t))dt \end{bmatrix} \\ &= \begin{bmatrix} \int_0^\tau h_1 dt + \\ \sum_{j=0}^{k} \int_{j\tau}^{(j+1)\tau} h_1(t - (1 - \delta)j\tau)dt \end{bmatrix} \end{aligned} \tag{3}$$

Furthermore, the total CM cost, total PM cost, and total maintenance cost can be formulated as the following Eqs. 4–6:

$$J_{a_n}(k, \delta) = C_{k_n}\bar{N}(k, \delta) \tag{4}$$

$$J_{b_n}(k, \delta) = \left((C_{f_n}k) + C_{v_n}k\tau(1 - \delta)\right) \tag{5}$$

$$J_n(k, \delta) = J_{a_n}(k, \delta) + J_{b_n}(k, \delta) \tag{6}$$

Dump truck availability will be determined by downtime, which is assumed to follow a distribution pattern and downtime is repairment time add by waiting time. Expectation downtime in interval [0, L) is influenced by number of failure and the duration of downtime (repairement time + waiting time), making expectation time for the downtime in interval [0, L) is stated as follows:

$$E[D(L)] = E[N(L)].E[d] \tag{7}$$

Where $E[N(L)]$ is expectation of number of failure in interval $[0, L]$ and $E[d]$ is duration of downtime with $1/\lambda$.

3 Numerical Example

3.1 PHM Failure Model

The numerical example based on hypothetic case with the scenario in PT. KPC. The scenario for this mining industry is $\alpha = 5.5$ and $\beta = 2.3$ where the change of the load (Y1) 351 kg = 1, where the road inclination is medium (Y2) 8%–10% = 0 and the environmental condition is low (Y3) = − 1 model hazard rate can be defined as follow:

$$h_1(t) = \left(\frac{\beta t}{\alpha^{\beta}}\right)^{\beta-1} e^{(\theta_1(Y_1)+\theta_2(Y_2)+\theta_3(Y_3))} \tag{8}$$

$$h_1(t) = \begin{bmatrix} \left(\frac{5.5t}{2.3^{5.5}}\right)^{5.5-1} \\ e^{(0,155(1)+0.272(0)+0.111(-1))} \end{bmatrix} \tag{9}$$

Where the dump truck failure data and operational condition data is generated by random number shown in Table 5. Therefore, the failure will increase considering all factors in operational conditions.

3.2 Optimal Maintenance Contract

The determination of optimal contract value for both parties using Nash game theory with objective getting a win-win solution decision between the owner and the agent. Therefore, owner can choose an optimal solution based on maximum profit expectation. Table 3 shows contract value, which maximizes profit for both options.

Table 3. Total benefit of owner and agent with Nash Game Theory.

Option	k*	δ*	Contract price (IDR million)	Owner's profit (IDR million)	Agent's profit (IDR million)
Option-1	1	0.5	5497.1	4091.0	4091.0
Option-2	1	0.7	6360.5	4180.5	4180.5
Option-3	1	0.3	5224.9	4084.9	4084.9

Using the Nash theory, it is found that the benefit expectation for each party are IDR 4.091 million for option-1, IDR 4.180 million for option-2, and IDR 4,084 million for option-3. From owner point of view, owner will prefer option-2 because it has more benefit compare to option-1 and option-3. Benefit expectation for option-2 is IDR 4.180 million with the contract value of IDR 6,360 million.

3.3 Parameter Change

The change in parameter conducted is the change in parameter φ showing the action of partial maintenance in option-1 and option-3 partly will be different with the maintenance action conducted by agent as in option-2. The capability partial maintenance in option-1 and option-3 is formulated with $\varphi > 1$. Full maintenance in option-2 is formulated with $\varphi = 1$ (Table 3). The effect of parameter φ for option-1 and option-3 show that the increment of φ will results δ^* getting smaller and this explains if conducting the partial maintenance as in option-1 and 3 the benefit will be lower than that in option-2. In possible solution by comparing option-1 and 3, the owner tends to choose option-1. If the reliability parameter is changed, the better solution is option 3 because will get more benefit compare to option-1. The next parameter changed is maintenance cost, conducted at in-house by owner for option-1 and 3 (Tables 4 and 5).

Table 4. Change on PM cost for option-1 for k* = 1 and $\varphi = 1.1$

δ*	Cost CM	Cost PM		Cg (Rp million)	Owner and agent's benefit
		Fix	Var		
0	300	150	100	5312.7	4470.3
0	300	210	140	5162.7	4320.3
0.1	300	255	170	5151.8	4208.2
0.1	300	270	180	5117.2	4173.7
0.3	300	285	190	5307.4	4142.5
0.5	300	315	210	5497.0	4090.9
0.7	300	330	220	5735.1	4071.9
0.7	300	345	230	5992.2	4057.8
1	300	390	260	6108.7	4034.2
1	300	450	300	6078.7	4004.2

Table 5. The change of CM cost in option-3 for k* = 1 and $\varphi = 1.1$

δ*	Cost CM	Cost PM		Cpa (Rp million)	Owner and agent's benefit
		Fix	Var		
1	150	300	200	4897.8	4597.8
1	210	300	200	4690.4	4390.4
1	255	300	200	4534.8	4234.8
0.7	270	300	200	4848.0	4188.0
0.5	285	300	200	5048.6	4148.6
0.3	315	300	200	5224.9	4084.9
0.2	330	300	200	5318.1	4058.1
0.1	345	300	200	5414.0	4034.0
0	390	300	200	5468.9	3968.9
0	450	300	200	5384.7	3884.7

Table 4 shows that the higher the PM cost at in-house in option-1 then the expectation of the benefit of each party will decrease. The lower the OM cost both benefit for both parties will increase. Table 5 shows that the higher the CM cost conducting at in-house as in option-3 the smaller the benefit expectation for both parties and vice versa. It can be concluded that if the in-house maintenance action increase then the owner decision will be choosing option-2. In comparison, the decision will change if maintenance action cost at in-house is smaller than in outsourcing maintenance action (agent).

4 Conclusion

This research studies how factors of operational condition influence the degradation level of dump truck, in order to determine maintenance contract between owner and agent using Nash Game Theory formulation. The dump truck failure is modeled using Proportional Hazard Model (PHM), that models the increasing rate of dump truck failure as a function of operational condition. Maintenance is proposed to control the rate of dump truck failure where agents offer maintenance option and owner chooses them based on maximum profit for both parties. Future potential research may include maintenance contract selection using other model, such as Stackelberg's game theory.

Acknowledgement. This work is funded by Ministry of Research, Technology and Higher Education of the Republic of Indonesia through the scheme of PDUPT 2019.

References

1. Dymasius, A., Wangsaputra, R., Iskandar, B.P.: Analysis of maintenance service contracts for dump trucks used in mining industry with simulation approach. In: IOP Conference Series: Materials Science and Engineering, vol. 114, no. 1 (2016)
2. Murthy, D.N.P., Jack, N.: Extended warranties, maintenance service and lease contracts modeling and analysis for decision-making. Springer, London (2014)
3. Iskandar, B.P., Cakravastia, A., Pasaribu, U.S., Husniah, H.: Performance-based maintenance contract for a fleet of dump trucks used in mining industry. In: Proceedings of 2014 2nd International Conference on Technology, Informatics, Management, Engineering and Environment, TIME-E 2014, pp. 117–122 (2014)
4. Barabadi, A.: Reliability and spare parts provision considering operational environment: a case study. Int. J. Perform. Eng. **8**(5), 497–506 (2012)
5. Gharahasanlou, A.N., Ataei, M., Khalokakaie, R., Ghodrati, B., Jafarie, R.: Tire demand planning based on reliability and operating environment. Int. J. Min. Geo-Eng. **50**(2), 239–248 (2016)
6. Husniah, H., Iskandar, B.P.: Maintenance contract with imperfect preventive maintenance. J. UNS **14**(1), 53–58 (2015)

Intelligent Condition Based Maintenance Using Adaptive Resonance Theory-2 Neural Network

R. Wangsaputra[1(✉)], H. Husniah[2], and Prasidhi Artono[1]

[1] Department of Industrial Engineering, Institut Teknologi Bandung, Jl. Ganesha 10, Bandung 40132, Indonesia
rwangsap@mail.ti.itb.ac.id, prasidhiartono@gmail.com

[2] Department of Industrial Engineering, Langlangbuana University, Jl. Karapitan no 116, Bandung 40261, Indonesia
henniehusniah@gmail.com

Abstract. Connector clamp is an essential component which effects transformer 150/20 kV performance in an electricity substation. Previous research developed a Statistical Process Control (SPC) chart combine with a Back Propagation Neural Network (BPNN) for determining the condition of connection clamp. Connector clamp condition assessment is designed using clamp and conductor's temperature as parameters at certain load values to shape limit-1 (3σ) and limit-6 (1.645σ). Its condition is necessary to be controlled effectively and efficiently. This paper improves the level of neural network. This research develops an Adaptive Resonance Theory-2 Neural Network with advantages that can renew the knowledge by adding new data for learning.

Keywords: Condition based maintenance · Adaptive Resonance Theory-2 Neural Network · Polynomial regression · Statistical process control and transformer connection clamp

1 Introduction

The idea of condition-based maintenance is to perform maintenance action based on actual condition and do the maintenance before the machine failure [1]. It has to be economically enough not too early to change the component but also not losing because of the failure. Unanticipated breakdown of equipment will cause more losses in economy and human sources. To reach this objective, condition monitoring has to be performed for the equipment, processes of manufacturing and operations to support the maintenance decision. The main principle used in condition-based maintenance is accuracy in assessing the condition whether it needs maintenance or not [2]. To be able to assess the condition accurately: (i) to accurately measure the parameter (ii) to accurately assess the condition based on parameter values. In doing so, it is very important to find a threshold where it is critical to prevent failure of equipment. Previous research an SPC and Back Propagation Neural Network (BPNN) are used to decide whether maintenance action is needed or not with a connector clamp for a

M. N. Osman Zahid et al. (Eds.): iMEC-APCOMS 2019, LNME, pp. 349–354, 2020.
https://doi.org/10.1007/978-981-15-0950-6_53

transformer used as a research object. This research increases the capability of the neural network to be able to learn new data in order to tune the knowledge by learning new condition.

2 Literature Study

2.1 Condition Based Maintenance

Condition based Maintenance (CBM) is a maintenance technique based on actual condition of a component [8]. Three functions of CBM are: (i) measuring the parameter condition, (ii) analyzing the condition of the parameter and (iii) determination of parameter condition limit.

2.2 Statistical Process Control (SPC)

Figure 1 show a control chart for connector clamp which is normally distributed with limit 1 (3σ) and limit-6 (1.645σ). The condition area is divided into 3 (three) regions: region-9 good condition, region-6 good mediate condition and region-1 bad condition. Using the control limit, the connector clamp is monitored for determining the next action needed based on the actual condition.

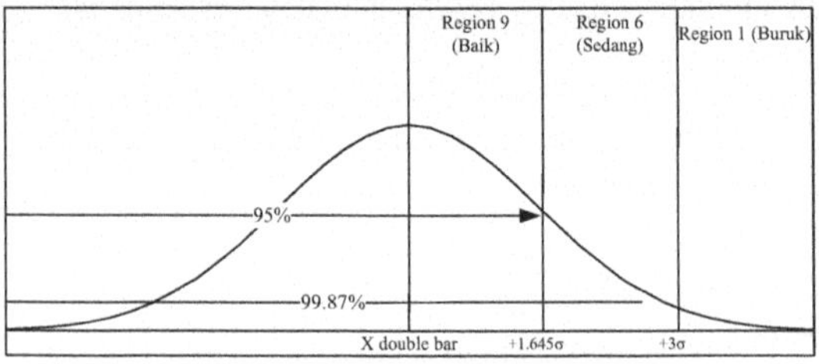

Fig. 1. Regions in a control chart

2.3 Load vs Temperature

Load is the electrical load caused by the flow of electron through a connector clamp [6]. Not all of those electrons can be flown by the connector clamp and the conductor which in turn converted into heat energy. Therefore, load and temperature have a very high correlation. Figure 2 shows the dependency between load in a component and temperature in that component [1].

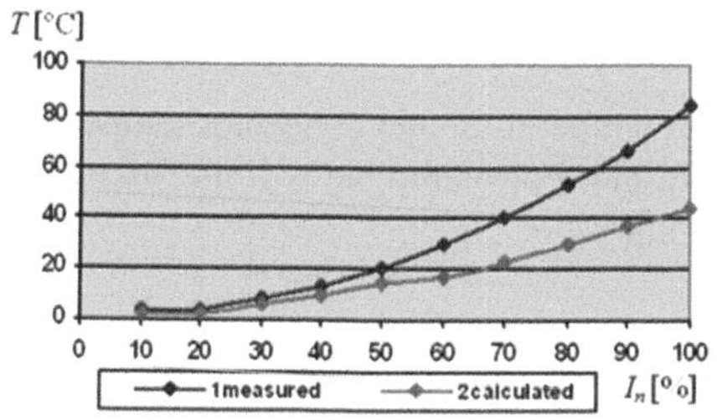

Fig. 2. Load versus temperature [7]

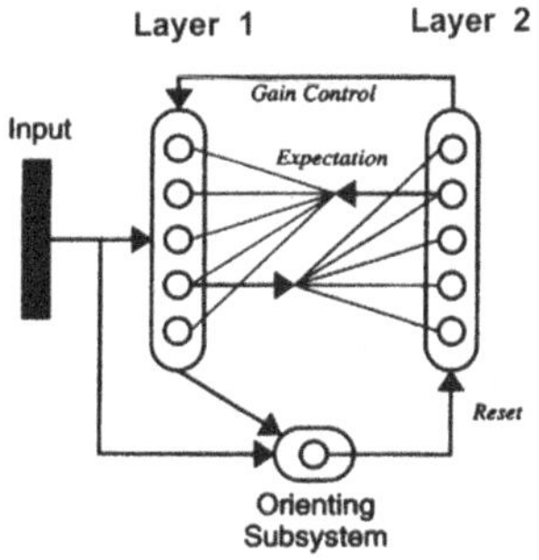

Fig. 3. ART-2 architecture

2.4 ART-2 NN for CBM

Adaptive Resonance Theory-2 Neural Network (ART-2 NN) is a neural network in which the learning process is based on resonance level between inputs and the knowledge matrix [4]. There are 2 (two) layers: Layer-1 and Layer-2 (see Fig. 3). The function of Layer-1 is to normalize the input; while the function of Layer 2 is to classify the input based on the current Long-Term Memory (LTM) and enhance the contrast by choosing a neuron which receives the maximum signal. The learning process is to adjust the LTM with the input pattern currently classified. ART2 NN is used to determine whether both clamp and conductor temperature show indicator whether the connector clamp is good or not. After teaching stage, the weight will be fixed and is able to determine whether those 2 (two) parameters result an emergency, checked right away, checked within 30 days, excellent condition or good condition.

3 Methods

Research method used consists of: (i) developing several pair conditions relating maintenance need for the clamp connection, (ii) train an ART-2 NN, (iii) testing the developed ART-2NN by comparing the SPC condition and ART-2 NN results.

1. For generating several conditions, it is collected for actual conditions. All condition in electricity company are collected
2. Design stage of intelligent condition-based maintenance. Three steps of the designing ICBM are: (i) observing the thermovision, calculating the limit and modelling the control chart. Observing using thermovision resulting load value, clamp temperature and conductor temperature. The calculation for the limit resulting the limit-6 and limit-1 for both the clamp and the conductor. Based on the output at design stage then it is conducted analysing and implementation in process.
3. Based on output, the analysing of the condition is conducted.

4 SPC and ART-2 Design

4.1 Statistical Process Control Chart

Table 1 shows a limit-1 and limit-6 for clamp temperature and connector temperature. Limit-1 and limit-6 is calculated based on Eq. (1). For clamp we will use upper limit and for the connector we will use the lower limit.

Table 1. Limit value for clamp and connector temperature

Load (A)	Clamp temperature			Connector temperature		
	Avg.	Limit 1	Limit 6	Avg.	Limit 1	Limit 6
1012.5	35.19	36.75	36.05	28.85	27.99	28.38
1087.5	36.80	38.03	37.48	29.00	27.68	28.28
1162.5	38.43	39.20	38.85	30.34	29.88	30.09
1237.5	39.80	41.11	40.52	30.98	30.09	30.49
1312.5	40.69	41.60	41.19	31.89	31.59	31.73
1387.5	42.54	43.49	43.06	33.79	33.00	33.35
1462.5	43.26	43.71	43.50	34.78	34.45	34.60

Due to the relationship between load and temperature, the SPC is generated for each specific load [5]. A second order polynomial regression is used to generate the SPC, where the regression polynomial regression (PR) will represent the relation between load and temperature for each limit value. The reason using second order PR is due to the determination coefficient, F significant and the means square error (MSE) value; the best comparing to when utilizing using linear regression and exponential.

Table 2. Limit value for clamp temperature

Klem	
Nilai Rata-Rata	$y_{9klem} = -0.000009139 I^2_{shooting} + 0.04068 I_{shooting} + 3.389$
Nilai Batas 1	$y_{1klem} = -0.00000966 I^2_{shooting} + 0.04013 I_{shooting} + 5.905$
Nilai Batas 6	$y_{6klem} = -0.000009425 I^2_{shooting} + 0.04038 I_{shooting} + 4.769$
Konduktor	
Nilai Rata-Rata	$y_{9kd} = 0.0000159 I^2_{shooting} - 0.0256 I_{shooting} + 38.34$
Nilai Batas 1	$y_{1kd} = 0.00001575 I^2_{shooting} - 0.02388 I_{shooting} + 35.74$
Nilai Batas 6	$y_{6kd} = 0.00001582 I^2_{shooting} - 0.02465 I_{shooting} + 36.91$

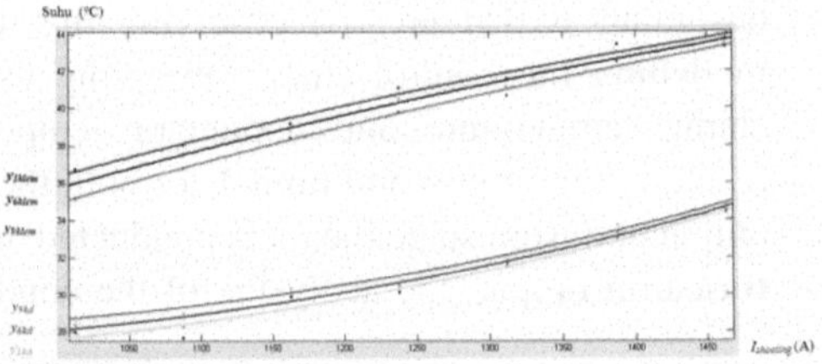

Fig. 4. Control limit based on polynomial regression

Table 2 shows a polynomial equation for the control for clamp temperature and conductor temperature.

Figure 4 shows control limit generated for inspection of clamp condition with polynomial on.

4.2 ART-2 NN CBM

The idea of using an Adaptive Resonance Theory-2 Neural Network (ART-2 NN) for CBM in this research is that in real time there will be a pair of condition where doesn't match with the condition-maintenance action matrix but the operator decided to use it as a reference either need action-1, action-2, action-3 or action-4. Two inputs used are clamp temperature and conductor temperature and also the information of the conductor temperature. ART-2 NN is one of neural network method in artificial intelligence which can classify a nonlinear relationship with specific training [9]. Figure 5 shows a network architecture for the ART-2.

The lack of previous ART-2 NN developed is on the assumption that both are regression, while in fact it is more complicated than that [7]. In this research we design the ART-2 NN architecture using 2 (two) inputs which is clamp temperature and conductor temperature and 1 (one) output which is the condition of the clamp connector either good or not.

5 Implementation

Table 3 shows the condition-maintenance action matrix.

Table 3. The initial condition vs maintenance action

<table>
<tr><td colspan="3" rowspan="3"></td><td colspan="6">CONDUCTOR CONDITION (9-6-1)</td></tr>
<tr><td colspan="2" rowspan="2">Below Limit-1</td><td colspan="4">Above Limit-1</td></tr>
<tr><td colspan="2">Between Limit 1-6</td><td colspan="2">Above Limit-6</td></tr>
<tr><td rowspan="6">CLAMP CONDITION (1-6-9)</td><td rowspan="5">Above Limit-6</td><td rowspan="2">Above Limit-1</td><td>Emergency</td><td></td><td>Emergency</td><td></td><td>Emergency</td><td></td></tr>
<tr><td></td><td>Need to be check right away</td><td></td><td></td><td></td><td></td></tr>
<tr><td rowspan="3">Between Limit 1-6</td><td></td><td rowspan="3">Need to be check right away</td><td colspan="2" rowspan="3">Checked within 30 days</td><td colspan="2" rowspan="3">Good Condition</td></tr>
<tr><td></td></tr>
<tr><td></td></tr>
<tr><td colspan="2">Below limit-6</td><td></td><td>Need to be check right away</td><td colspan="2">Good Condition</td><td colspan="2">Excellent Condition</td></tr>
</table>

The ART-2 NN architecture is as follow:

$a = 10$; $init\_W^{1:2} = 0.5$; $c = 0.850000$; $b = 10$; $init\_W^{2:1} = 0.0$;
$d = 0.800000$; $nonL2 = 150$; $\theta = -1.0$; $towu = 0.000001$; $nep = 4$; $\alpha = 0.9$;
$\rho = 0.999000$; $e = 0.0000001$

Data will be plot with axis coordinate show its relative position compare to the control chart limit and also the condition determined by the condition, if the operator want to add new data, it will added and the neural network will be learn to get new weight. The result of percentage error is 93,53%, meaning that from 100 experiment there are still 6,37% where the ART-2 NN is mistaken in assess the condition of the connector clamp.

6 Conclusion

This paper develops an ART-2 with 2 (two) inputs which are conductor temperature and clamp temperature with 4 (four) outputs. Connector clamp condition monitored based on clamp temperature and conductor temperature using thermovision. The ART-2 NN based CBM can function more realistic and can be implemented successfully. The data acquisition is conducted continuously and the ART-2 NN can be continuously learn and update the matrix value the weight for better work in the future. It is good that the system will increase the efficiency and effectivity of the Main Sub Station.

Acknowledgement. This work is funded by Kemenristekdikti - the Government of the Republic of Indonesia - through the scheme of PDUPT 2019.

References

1. Apandi, D.H.: Pembentukan model condition assessment transformator 150/20 KV sebagai tahap inisiasi implementasi condition-based maintenance (CBM) di PT PLN persero. Tugas Akhir Sarjana TI-ITB, Bandung (2010
2. Ben-daya, M., Deffuaa, S.O.: Maintenance, Modeling, and Optimization. Klwer Academic Publisher, Boston (2000)
3. Dileo, M., Manker, C., Cadick, J.: Condition Based Maintenance. Cadick Corporation, Texas (1999)
4. Hagan, M.T., Demuth, H.B., Beale, M.: Neural Network Design. PWS Publishing Company, Boston (1996)
5. Montgomery, D.C.: Statistical Quality Control: A Modern Introduction. Wiley, New York (2009)
6. Sebök, M.: Diagnostics of electric equipments by means of thermovision. University of Zllna, Slovakia (2009)
7. Cox, L., Wangsaputra, R.: Manufacturing assistant. New Mexico State University, Las Cruces (2002)
8. Wu, B., Tian, Z., Chen, M.: Condition based maintenance optimization using neural network based health condition prediction. Qual. Reliabil. Eng. Int. **29**(8), 1151–1163 (2010)
9. Yam, R.C.M., Tse, P.W., Li, L., Tu, P.: Intelligent predictive decision support system for condition-based maintenance. Int. J. Adv. Manuf. Technol. **17**(5), 383–391 (2001)

Nash Game Theory Leasing Contract Model of New and Recondition Complex Equipment

Mochamad Azka Harish, Andi Cakravastia(✉), and Bermawi P. Iskandar

Industrial Engineering and Management Post Graduate Program, Faculty of Industrial Technology, Bandung Institute of Technology, Bandung, Indonesia
{andi,bermawi}@mail.ti.itb.ac.id

Abstract. Complex condition of equipment maintenance lead many companies today considers outsourcing maintenance activities either to original equipment (OEM) or third-party maintenance company (agent). Objective of this research is to simultaneously consider two maintenance policies from perspectives of equipment lessor and lessee. The lessor offers two choices of equipment: (i) new equipment and (ii) reconditioned equipment. Lessee is going to compare these two options to maximize total profit. Nash game theory is applied to derive decision from both perspectives.

Keywords: Maintenance · Complex equipment · Outsource · Game theory

1 Introduction

1.1 A Subsection Sample

Complex equipment characterizes by high capital investment, rapid technology advancement, and high obsolete rate. Maintenance of complex equipment requires high skilled worker and high maintenance cost. In-house maintenance could be economically not feasible [1]. This complex condition, lead many companies today consider to outsource maintenance activities either to original equipment (OEM) or third-party maintenance company (agent).

The above situation inspires research dedicated in the area of equipment leasing. Researches in this area until today are considering two conditions of equipment: (i) New Equipment and (ii) Recondition Equipment. There were some previous efforts dedicated in the area of leasing contract for new equipment. Jaturonnatee et al. focused on maintenance action and impact on contract term on optimal maintenance strategy from perspective of equipment lessor [1]. Pakpahan and Iskandar develop leasing contract model to optimize contract value from perspective of lessee and standard performance level of the lessor [2]. Hamidi et al. consider two perspectives. Lessor has to define optimal maintenance policy and lessee decides optimal leasing period and usage rate to maximize total profit [3]. For the case of recondition equipment, Pongpech et al., consider strategy of upgrading used equipment to improve equipment reliability before it is leased to lessee [4].

M. N. Osman Zahid et al. (Eds.): iMEC-APCOMS 2019, LNME, pp. 355–361, 2020.
https://doi.org/10.1007/978-981-15-0950-6_54

Contribution of this research is to simultaneously consider these two policies from perspectives of equipment lessor and lessee. The lessor offers two choices of equipment: (i) new equipment and (ii) reconditioned equipment. Lessee is going to compare these two options to maximize total profit.

2 Mathematical Model

Notations of the mathematical models are described as follow:

P_{ph}	Revenue per output
$N(L)$	Number of downtime during leasing period
$E(D_b)$	Expected downtime option-1
$E(Dr)$	Expected downtime option-2
$E(TC_{tb})$	Expected additional cost option-1
$E(TC_{tr})$	Expected additional cost option-2

Decision variables:

δ	Rate of PM
k	Number of PM
H_{sb}	Lease price of new equipment
H_{sr}	Lease price of reconditioned equipment
Y	Number of equipment output

Parameters:

L	Period of equipment leasing.
H	Maximum output level
C_t	Additional cost
C_f	Cost of CM
$J_k^b(\delta, k)$	Total cost of CM option-1
$J_k^r(\delta, k)$	Total cost of CM option-2
C_0	Fixed cost PM
C_1	Variable cost PM
$J_p^b(\delta, k)$	Total cost PM option-1
$J_p^r(\delta, k)$	Total cost PM option-2

2.1 Failure Model

AFT (Accelerated Failure Time) model is applied to show relationship of usage rate to reliability of equipment, shown in Eq. (1)

$$\alpha_y = \left(\frac{y_0}{y}\right)^{\gamma} \alpha \tag{1}$$

Crow model is applied to estimate parameter α based on downtime data of new and reconditioned equipment. Two parameters of Weibull function applied in this research are shown in Eq. 2.

$$F(t) = 1 - e^{\left(\frac{t}{\alpha y}\right)^{\beta}} \tag{2}$$

2.2 Maintenance Model

Decreasing of age of equipment (t) to virtual age v(t) after maintenance on time $i\tau$ is $v_i = i\delta\tau$ for $i = 0, 1, \ldots, k$. Equipment age decrease $(1-\delta)\tau$ (Fig. 1) with $\tau = \frac{L}{k+1}$, so virtual age of equipment after maintenance is shown in Eq. 3.

$$v(t) = t - (1 - \delta)i\tau, \text{ for } i\tau \leq t < (i+1)\tau \tag{3}$$

Expected number of downtime during leasing period can be seen in Eq. (4)

$$E[N(\delta, k)] = \int_0^{\tau} h_0(t)dt + \sum\nolimits_{i=0}^{k} \int_{i\tau}^{(i+1)\tau} h_0(v(t))\, dt \tag{4}$$

Total CM cost from Eq. (4) can be formulated in Eq. (5)

$$J_k(\delta, k) = C_f E[N(\delta, k)] \tag{5}$$

2.3 Additional Cost Model

Lessee will be charged additional cost if output (Y) per period exceeds maximum output level (η). Expected additional cost during leasing period is given in Eq. 6.

$$E(TC_t) = C_t(max(0, Y - \eta)) \tag{6}$$

2.4 Lessor Expected Profit

Lessor expected profit for option-1 and option-2 are given is Eqs. 7 and 8, respectively.

$$E(\theta_1(\delta, k)) = (H_{sb}L + E(TC_{tb})) - J_k^b(\delta, k) - J_p^b(\delta, k) \tag{7}$$

$$E(\theta_2(\delta, k)) = (H_{sr}L + E(TC_{tr})) - J_k^r(\delta, k) - J_p^r(\delta, k) \tag{8}$$

2.5 Lessee Expected Profit

Lessee expected profit for option-1 and option-2 are shown in Eqs. 9 and 10, respectively.

$$E(\pi_1(Y)) = (P_{ph}Y(L - E(D_b))) - (H_{sb}L + E(TC_{tb})) \tag{9}$$

$$E(\pi_2(Y)) = (P_{ph}Y(L - E(D_r))) - (H_{sr}L + E(TC_{tr})) \tag{10}$$

2.6 Decision Problem

Decision problem from perspective of lessor and lessee of the equipment can be formulated as a Nash game theory model.

Based on Eqs. (7) and (9), decision model for option-1 is given in Eq. 11.

$$H_{sb} = \frac{1}{2L}\left(P_{ph}Y(L - E(D_b)) + J_k^b(\delta, k) + J_p^b(\delta, k) - 2E(TC_{tb})\right) \tag{11}$$

Following Eqs. (8) and (10), we can have decision model for option-2 as shown in Eq. 12.

$$H_{sr} = \frac{1}{2L}\left(P_{ph}Y(L - E(D_r)) + J_k^r(\delta, k) + J_p^r(\delta, k) - 2E(TC_{tr})\right) \tag{12}$$

3 Optimization

Solution of the model will be derived in the following steps:

Step 1: Define δ, $0 \leq \delta \leq 1$ (0, 0.1, 0.2, 0.3, 0.4, 0.5, 0.6, 0.7, 0.8, 0.9, 1) for $k = 1$.
Step 2: Calculate total CM cost from Eq. (5).
Step 3: Calculate PM cost from Eq. (6).
Step 4: Calculate minimum total maintenance cost. Therefore we can have δ^* for $k = 1$.
Step 5: Define $k = 2$ and repeat step 1, 2, 3, and 4. Calculate δ^* for $k = 2$.
Step 6: Define $k = 3$ and repeat step 1, 2, 3, and 4. Calculate δ^* for $k = 3$.
Step 7: Compare $k = 1$, $k = 2$, and $k = 3$ to minimize total maintenance costs k^*.
Step 8: Apply Nash game theory to define leasing contract value for option-1 and option-2.
Step 9: Calculate expected additional cost, Max(0, $Y-\eta$). If $Y > \eta$ then lessee is charged additional cost $Ct(Y-\eta)$.
Step 10: Calculate expected total profit for lessor and lessee. The next stage, from perspective of lessee, is to select contract option that maximizes total profit and optimal output level.

4 Numerical Example

Parameters of both options for numerical example are depicted in Tables 1, 2 and 3 are showing results of δ * and k* for option-1 and option-2. Under option-1 the lowest maintenance cost is 2,400,800 with $\delta^* = 0.7$ and $k^* = 1$. Under option-2 the lowest maintenance cost is 2,672,200 with $\delta^* = 0.6$ and $k^* = 1$. Total profit for lessor and lessee is depicted in Table 4.

Table 1. Parameter

	Option-1	Option-2
α	19,498	15,855
α_y	17,557	14,277
β	2,385	2,240
L	12	12
C_f	1,000,000	1,000,000
C_0	2,000,000	2,000,000
C_1	46,000	68.000
C_t	95	95
η	5000	5000
P_{ph}	150	130
Y	5500	5500

Table 2. δ* and k* of option 1

k*	δ*	Total maintenance cost
1	0.7	2,400,800
2	0.8	4,401,600
3	0.9	6,402,400

Table 3. δ* and k* of option 2

k*	δ*	Total maintenance cost
1	0.6	2,672,200
2	0.7	4,673,600
3	0.8	6,673,600

Table 4. Lessor and lessee profit

	Option 1	Option 2
Contract value	414,800	368,803
Lessor's profit	2,624,300	1,800,934
Lessee's profit	2,624,300	1,800,934

Total profits are 2,624,300 for option 1 and 1,800,934 for option 2, respectively. In this case option-1 is more preferable to the lessee than option-2. Subsequent step is to define Y^* for option-1 and option-2, shown in Figs. 1 and 2, respectively. If lessee is going to choose option-1, then number of output Y^* that maximize both profit is 5,700 with total profit 2,633,026. If lessee choose option-2 then Y^* is 5200 with total profit 1,812,408.

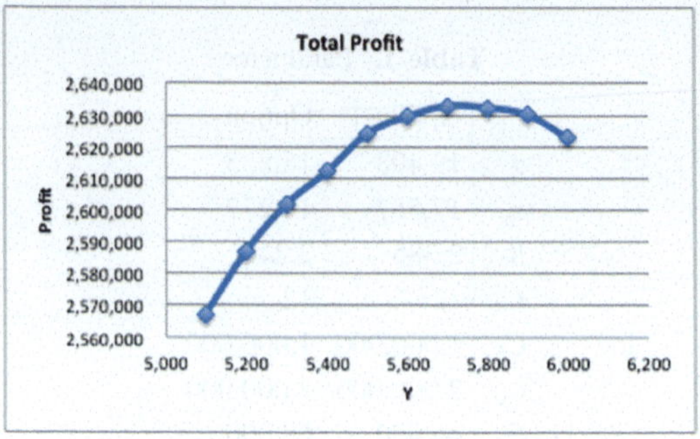

Fig. 1. Total profit under option 1 policy

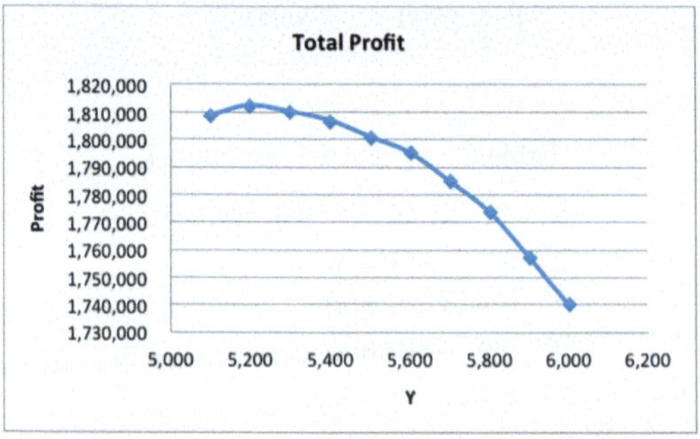

Fig. 2. Total profit under option 2 policy

5 Concluding Remarks

This paper proposes a leasing contract model that simultaneously considers new and reconditions equipment from perspective of lessor and lessee of the equipment. Applying Nash game theory model can derive solution of the decision-making problem from perspective lessor and lessee. Future research can be directed into considering penalty for breakdown frequency.

References

1. Jaturonnatee, J., Murthy, D.N.P., Boondiskulchok, R.: Optimal preventive maintenance of leased equipment with corrective minimal repairs. Eur. J. Oper. Res. **174**, 201–215 (2006)
2. Pakpahan, E.K.A., Iskandar, B.P.: Performance based lease contract involving discrete. J. Eng. Appl. Sci. **11** (2016)
3. Hamidi, M., Liao, H., Szidarovszky, F.: Non-cooperative and cooperative game-theoretic models for usage-based lease contracts. Eur. J. Oper. Res. **255**(1), 163–174 (2016)
4. Pongpech, J., Murthy, D.N.P., Boondiskulchok, R.: Maintenance strategies for used equipment under lease. J. Qual. Maint. Eng. **12**, 52–67 (2006)

Materials

Refining the Composition of Recycled Spent Lubricants Mixed with Alumina Nanofluids for Machining Purpose

Lim Syh Kai, Nurrina Rosli, and Ahmad Razlan Yusoff(✉)

Faculty of Manufacturing Engineering, Universiti Malaysia Pahang, 26600 Pekan, Pahang, Malaysia
fal0007.ump@gmail.com, razlan@um.edu.my

Abstract. Machining lubricant is used in the manufacturing industry for lubrication and cooling which are crucial in operations such as grinding and cutting process. However, machining lubricant lose efficiency in months due to thermal degradation and contamination. To recycle the used machining lubricants while enhancing the performance, the mixture of nanofluids (N) based Al_2O_3 with better thermal properties to synthesize with the recycled spent lubricants (RSL). This study suspended Al_2O_3 nanofluids in five base ratios of recycled spent lubricants (i.e. 40:60, 45:55, 50:50, 55:45 and 60:40) by ultrasonic homogenies. The two main parameters in cooling rate performance are thermal conductivity and dynamic viscosity. The thermal conductivity and dynamic viscosity are then measured at temperature range of 30 to 60 °C. The highest enhancement of thermal conductivity in 60:40 (N:RSL) was evaluated to be 18.5% higher than the 40:60 (N:RSL) base fluid at the temperature of 60 °C. However, the enhancement of dynamic viscosity was measured to be 2.4% for 60:40 (N:RSL) at 60 °C temperature. Therefore, this study recommends the use of recycled spent lubricants based Al_2O_3 as cutting fluid in the ratio of 60:40 for application in machining operations.

Keywords: Spent lubricants · Thermal conductivity · Dynamic viscosity

1 Introduction

Machining lubricants also named as metalworking fluids are considered as an accessory in a manufacturing process in order to increase the life of the cutting tool and productivity. The surface quality of the mechanical parts and cutting tools are crucial in evaluating the productivity of a manufacturing industry. In 2018, the amount of machining lubricants used in manufacturing industry was reported nearly 38 Mtonnes with an increase of 1.2% compared with the year of 2005 [1]. It is reported that European Union alone is consuming about 320 Mtonnes of machining lubricants within one year and two-thirds of it needs to be disposed due to the degradation of thermal efficiency of machining lubricants [2]. It is expected that cost of machining lubricants, including procurement, preparation, operation and the disposal are approximately 17% of the total manufacturing costs of a product. Therefore, recycling of spent lubricant can reduce manufacturing cost and solve waste disposal problems. In 2016, Liang et al.

M. N. Osman Zahid et al. (Eds.): iMEC-APCOMS 2019, LNME, pp. 365–370, 2020.
https://doi.org/10.1007/978-981-15-0950-6_55

[3] conducted an investigation on the tribological behavior of in-situ exfoliated graphene for water-based lubricants and found that graphene-enhanced lubricants improved the surface quality and coolants durability. Syam Sundar et al. [4] experimentally investigated the thermal conductivity of Al_2O_3 in 20:80 EG/water mixtures compared with 60:40 EG/water base fluids. They found that the thermal conductivity was significantly increased compared to the base fluid with higher ratio of ethylene glycol. The augmentation of thermal conductivity increased the heat transfer coefficient value as stated from Kole and Dey [5]. The average enhancement of heat transfer coefficient was up to 22% in their studied.

Few studies are available regarding to the nanofluids as lubricants in machining process. The nanofluid based Al_2O_3 is proposed as a surfactant to synthesize with recycling spent lubricant in order to enhance the performance of machining lubricant with better thermal properties. The present study focuses on thermal conductivity and dynamic viscosity of Al_2O_3 nano-mixtures in five difference ratios of 40:60, 45:55, 50:50, 55:45 and 60:40 as base fluids for measured temperature of 30 to 60 °C. The optimum volume ratio of nano-mixtures need to be determined by considering the optimum properties values of dynamic viscosity and thermal conductivity in bulk temperature condition as machining lubricants for application in machining purpose.

2 Experimental Procedures

2.1 Preparation of Nano-lubricants

Nanofluid was prepared by dispersing the nanoparticles in the mixture of water to ethylene glycol fluid. The aluminum oxide, Al_2O_3 nanoparticles in powder form have an average particle diameter of 13 nm size are suspended in nanofluid by using two-step method proposed by Yu et al. [6]. The nanofluid was mixed in five different ratios of recycled spent lubricants as base fluids, which were 40:60, 45:55, 50:50, 55:45 and 60:40, respectively. The ultrasonic homogenizer was employed to improve the dispersion stability of nanoparticles in the mixture. The mixture of Al_2O_3 machining lubricant is synthesized by using a magnetic stirrer and sonicated in ultrasonic bath for two hours, followed the studies by Azmi et al. [6, 7]. The stability of the nano-lubricant is observed and found to be stable for more than two months.

2.2 Thermal Conductivity Measurement

KD2 Pro thermal property analyzer (Decagon Devices, USA) measured the thermal conductivity of the Al_2O_3 nano-lubricant. The thermal properties of liquids and solids were determined by the device using the transient line heat source. A single needle sensor (KS-1) in the range of 0.002 to 2.00 W/m K was used, and the sensor was validated by measuring the thermal conductivity of the verification liquid such as glycerine that was provided by the supplier. The measured value of glycerine at 25 °C was 0.286 W/m K, which agreed with the calibrated data of 0.285 W/m K and within ±0.4% accuracy [6]. The validation process of the sensor was checked before each measurement of thermal conductivity. A water bath of WNB7L1 model by Memmert was used to maintain a constant temperature of the sample with an accuracy of 0.1 °C [8]. Thermal conductivity of four different ratios of Al_2O_3 nano-lubricants were measured within the temperature range of 30 to 60 °C. The consistency of data measurement can be ensured; a minimum of three values were taken for every base ratio at a specific temperature, and the average of the three values was analyzed.

2.3 Dynamic Viscosity Measurement

Dynamic viscosity of Al_2O_3 nano-lubricants in five different ratios of recycled spent lubricants as base fluids were measured by Brookfield (low viscosity digital viscometer) LVDV-III Ultra Programmable Rheometer. The device is equipped with a RheoCal program in personal computer for data measurement and collection. The viscometer measures the dynamic viscosity of solutions that range from 1 to 6000,000 mPa•s by utilizing an ultra-low adapter. The spindle connected to the viscometer was used to measure the viscosity of nano-lubricant [9]. The viscometer drives a spindle that is immersed in nano-lubricant. Spindle rotation created a viscous drag of the fluid opposite to the spindle, which is measured by the deflection of the calibrated spring. The adapter in this experiment has a provision for temperature circulation of bath fluid. The viscosity of different base ratios of nano-lubricants based Al_2O_3 started from 30 to 60 °C at an interval of 5 °C. Each measurement was conducted three times to generate a reliable data, and the average value of the three values was considered for analysis.

3 Results and Discussion

3.1 Thermal Conductivity of Nano-Lubricants Based Al_2O_3

The thermal conductivity of recycled spent lubricants mixed with different ratios of Al_2O_3 nanofluids are shown in Fig. 1(a). As observed from the figure, thermal conductivity is increased with the increasing of temperature and nanofluids percentages, followed the behavior of nanofluids thermal conductivity as stated by Syam Sundar et al. [4] and Lim et al. [6]. However, there is small fluctuation of thermal conductivity occurs at temperature 35 °C where the percentage of nanofluids 40% higher than 60% nano-lubricants. In overall, all the nano-lubricants with the percentage of 60% exhibited the maximum thermal conductivity. The observation seems to be related to the Brownian motion and convection heat transfer of nanoparticles where the increases of nanofluid ratio tends to increase the collision of particles with molecules of base liquid at high temperature [10].

3.2 Dynamic Viscosity of Nano-Lubricants Based Al_2O_3

The dynamic viscosity measurements of nano-lubricants based Al_2O_3 were conducted in temperature range from 30 to 60 °C as presented in Fig. 1(b). At the base fluids of

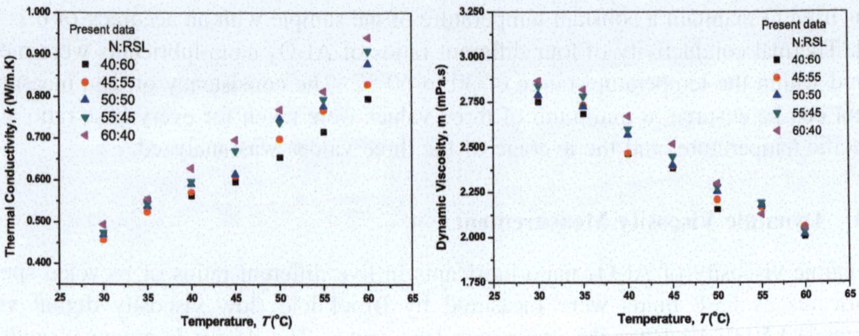

(a) (b)

Fig. 1. The results of nano-lubricants for different ratios of nanofluid (N) and recycled spent lubricant (RSL) (a) Thermal conductivity (b) Dynamic viscosity

60:40 (N:RSL), the dynamic viscosity enhancement percentage was 39.7% at 60 °C compared with the 40:60 (N:RSL) base fluids only 27.1%. The dynamic viscosity of the nano-lubricants increased as the volume concentration of nanofluid increased. In addition, each volume concentrations have similar increment pattern of viscosity. According to Syam Sundar et al. [11], who found similar pattern of viscosity, it is caused by shear resistance given by the particles onto the fluid. However, the dynamic viscosity of the nanofluid diminished exponentially as the temperature increased. As mentioned by Usri et al. [12] in their research paper, the temperature weakening the inter-particle and inter-molecular adhesion forces.

3.3 Optimum Thermal Properties of Nano-Lubricants for Machining

The optimum ratio of nano-lubricants at 60:40 (N:RSL) was measured for suitable dynamic viscosity and thermal conductivity in temperature range from 30 to 60 °C, as presented in Fig. 2. From the temperature of 45 °C, the thermal conductivity of nano-lubricants with base ratio 60:40 (N:RSL) started to above the average value, while the dynamic viscosity decreased to break even value. At 45 °C bulk temperature of machining coolants, the augmentation of thermal conductivity was approximately 46.7% compared with initial temperature of 30 °C. It was established that the optimum value was obtained using nano-lubricant base 60:40 (N:RSL) at the temperature of 45 °C. Omolayo et al. [13] reported that machining lubricant temperature was above 40 °C during machining process of aluminium alloys. Moreover, viscosity was vital to the internal friction force in a fluid and resistance to the molecule displacement in relative motion during exposed to shear stress [14].

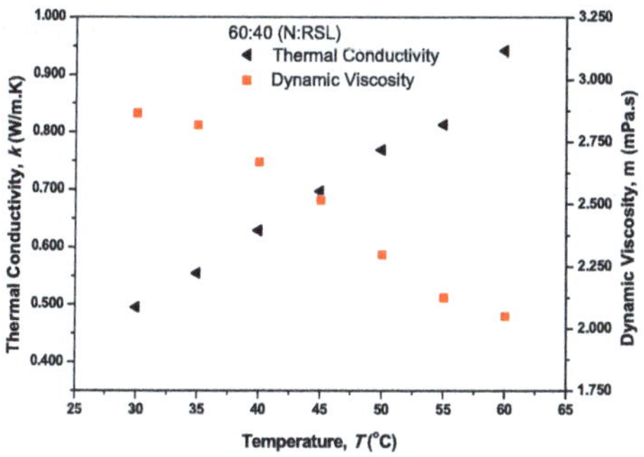

Fig. 2. Thermal properties variation of 60:40 (N:RSL) nano-lubricants with different temperatures

4 Conclusion

In this paper, the thermal conductivity and dynamic viscosity of Al_2O_3 base nano-recycled spent lubricants were investigated. The experimental analysis found that thermal conductivity increased with the increment of temperature and percentages of nanofluid as base fluids. However, the dynamic viscosity of nano-lubricants exponentially decreases with the increase of temperature and the decrease of nanofluid percentages as base fluid.

The thermal conductivity enhancement was determined to be greater than the enhancement of dynamic viscosity. The highest thermal conductivity enhancement increased by 18.5% compared to the 40:60 (N:RSL) base fluid. However, the enhancement of dynamic viscosity exceed 39.7% which is not giving benefit to the heat transfer. Therefore, the use of nano-lubricant of 60:40 (N:RSL) based Al_2O_3 is recommended for application as machining lubricants to reduce cutting temperatures. Further investigations and experiments on the performance of the machining processes such as milling and turning process by using nano-recycled spent lubricants as coolants are required to extend the present work.

Acknowledgments. The financial support by Ministry of Education Malaysia and Universiti Malaysia Pahang under Fundamental Research Grant Scheme of FRGS/1/2016/TK03/UMP/0 2/6 and Internal Research Grant of RDU 190348 are gratefully acknowledged.

References

1. Debnath, S., Reddy, M.M., Yi, Q.S.: Environmental friendly cutting fluids and cooling techniques in machining: a review. J. Cleaner Prod. **83**, 33–47 (2014)
2. Abdalla, H.S., Baines, W., McIntyre, G., Slade, C.: Development of novel sustainable neat-oil metalworking fluids for stainless steel and titanium alloy machining. Part 1. Formulation development. J. Adv. Manuf. Technol. **34**, 21–33 (2007)
3. Liang, S.S., Shen, Z.G., Yi, M., Liu, L., Zhang, X.J., Ma, S.L.: In-situ exfoliated graphene for high-performance water-based lubricants. Carbon **96**, 1181–1190 (2016)

4. Syam Sundar, L., Venkata Ramana, E., Singh, M.K., Sousa, A.C.M.: Thermal conductivity and viscosity of stabilized ethylene glycol and water mixture Al_2O_3 nanofluids for heat transfer applications: an experimental study. Int. Commun. Heat Mass Transfer **56**, 86–95 (2014)
5. Kole, M., Dey, T.K.: Thermophysical and pool boiling characteristic of ZnO-ethylene glycol nanofluids. Int. J. Therm. Sci. **62**, 61–70 (2012)
6. Lim, S.K., Azmi, W.H., Yusoff, A.R.: Investigation of thermal conductivity and viscosity of Al_2O_3/water–ethylene glycol mixture nanocoolant for cooling channel of hot-press forming die application. Int. Commun. Heat Mass Transfer **78**, 182–189 (2016)
7. Azmi, W.H., Sharma, K.V., Sarma, P.K., Mamat, R., Anuar, S., Dharma Rao, V.: Experimental determination of turbulent forced convection heat transfer and friction factor with SiO_2 nanofluid. Exp. Therm. Fluid Sci. **51**, 103–111 (2013)
8. Zakaria, I., Azmi, W.H., Mohamed, W.A.N.W., Mamat, R., Najafi, G.: Experimental investigation of thermal conductivity and electrical conductivity of Al_2O_3 nanofluid in water–ethylene glycol mixture for proton exchange membrane fuel cell application. Int. Commun. Heat Mass Transfer **61**, 61–68 (2015)
9. Azmi, W.H., Sharma, K.V., Sarma, P.K., Mamat, R., Najafi, G.: Heat transfer and friction factor of water based TiO_2 and SiO_2 nanofluids under turbulent flow in a tube. Int. Commun. Heat Mass Transfer **59**, 30–38 (2014)
10. Hussein, A.M., Sharma, K.V., Bakar, R.A., Kadirgama, K.: A review of forced convection heat transfer enhancement and hydrodynamic characteristics of a nanofluids. Renew. Sustain. Energy Rev. **29**, 734–743 (2014)
11. Syam Sundar, L., Venkata Ramana, E., Singh, M.K., Sousa, A.C.M.: Viscosity of low volume concentrations of magnetic Fe_3O_4 nanoparticles dispersed in ethylene glycol and water mixture. Chem. Phys. Lett. **554**, 236–242 (2012)
12. Usri, N.A., Azmi, W.H., Mamat, R., Abdul Hamid, K.: Viscosity of aluminium oxide (Al_2O_3) nanoparticle dispersed in ethylene glycol. Appl. Mech. Mater. **660**, 735–739 (2014)
13. Omolayo, M.I., Sunday, T.O., Ojo, P.B., Sunday, A.A., Imhade, P.O., Esther, T.A.: The effects of lubricants on temperature distribution of 6063 aluminium alloy during backward cup extrusion process. J. Mater. Res. Technol. **8**(1), 1175–1187 (2019)
14. Amiril, S.A.S., Rahim, E.A., Syahrullail, S.: A review on ionic liquids as sustainable lubricants in manufacturing and engineering: recent research, performance, and applications. J. Cleaner Prod. **168**, 1571–1589 (2017)

Fatigue and Harmonic Analysis of a Diesel Engine Crankshaft Using ANSYS

Aisha Muhammad[1,2(✉)], Mohammed A. H. Ali[1], and Ibrahim Haruna Shanono[3,4]

[1] Faculty of Mechanical and Manufacturing Engineering, Universiti Malaysia Pahang, Gambang, Malaysia
ayshermuhd@gmail.com

[2] Department of Mechatronics Engineering, Bayero University Kano, Kano, Nigeria

[3] Faculty of Electrical and Electronics Engineering, Universiti Malaysia Pahang, Gambang, Malaysia

[4] Department of Electrical Engineering, Bayero University Kano, Kano, Nigeria

Abstract. The performance and reliability of an internal combustion engine depend on the robustness of the engine crankshaft as it is one of its important moving component. An internal combustion engine is affected by the vibration activities of a crankshaft. This vibration performance is hard to determine due to the complex nature of the crankshaft. To avoid failure of an engine crankshaft, calculation of the crankshaft strength becomes an importance component. In this paper, fatigue, failure and harmonic response analysis of an internal combustion engine by Finite Element method (FEM) using ANSYS workbench were carried out. The analysis is carried out in ANSYS static structural, mechanical solver on a structural steel material. Result from the transient analysis indicates presence of stress at the crankpin journals and with the appearance of deformation at its crank nose. The failure analysis safety factor is high in the Gerber theory being a structural steel material.

Keywords: Crankshaft · FEM · ANSYS · Harmonic · Stress · Deformation · Fatigue

1 Introduction

A crankshaft is used to provide rotary motion when subjected to a number of continuous loads and stress. The idea behind the principle of a crankshaft is changing the input of devices (pumps, compressors, generators) from a sudden displacement to rotary form [1]. The crankshaft is hammered by the power impulse whenever the engine is running generating torsional vibration. In the past, the stress in a crankshaft is calculated using the methods of beam and frame model. The limitation to these methods is that they have a finite number of nodes. Recently, the advancement in computer technology has provide the means of calculating and analyzing the stress in a crankshaft using the method of Finite Element Method (FEM). The part of the structure that was modified as a result of the subjected stress due to harmonically changing loads

M. N. Osman Zahid et al. (Eds.): iMEC-APCOMS 2019, LNME, pp. 371–376, 2020.
https://doi.org/10.1007/978-981-15-0950-6_56

can be determined by carrying out the harmonic analysis of the structure. The three different approaches that can be used to carryout harmonic analysis includes: Full, Reduced, and Mode superposition (commonly used approach) [2–5]. Majority of materials have fatigue limit also known as endurance limit. Those having a stress level below this limit are said to have an infinite fatigue life. The three most common use theory for fatigue analysis to evaluate the mean stress on fatigue life include: Soderberg, Goodman and Gerber failure theory [2]. Analysis of a crankshaft have been conducted by numerous researchers using the ANSYS workbench. For calculating stress, or the effect of other components such as the flywheel on a crankshaft, harmonic analysis is used [6]. In [7], the dynamic twisting moment was determined using harmonic analysis. In another research, the transient analysis was carried out to determine the harmonic response for torsional deformation. Transient dynamic analysis for the steady state values of deformation, stress and stiffness on plates using the harmonic method of mode superposition was conducted and presented in [8]. In this paper, transient, fatigue failure and harmonic response analysis is carried out to determine the stress, deformation and modal analysis of a crankshaft. The analysis is carried out in ANSYS static structural, mechanical solver on a structural steel material.

2 Methodology

The design of the structure is done using Solid-Works software, which is then exported into ANSYS workbench for further analysis. Figure 1 below shows the sketched structure with the dimensions.

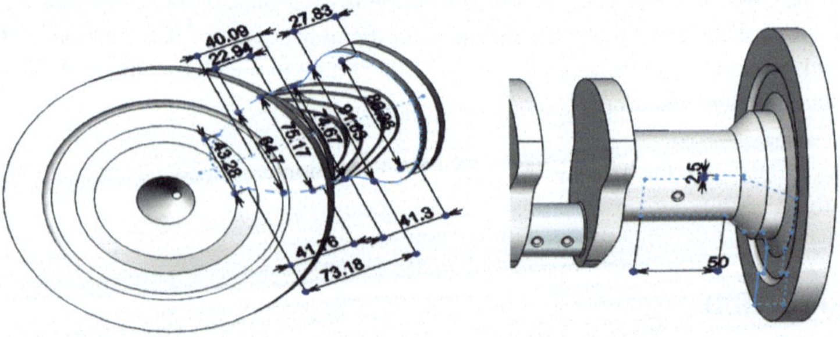

(a) (b)

Fig. 1. (a) Counter weight and flywheels (b) Main journals

Failures of engineering structures makes analysis to become very significant to ascertain safety of the structure because failure of the whole or part of the structure leads to high risk of life and financial loss [9, 10]. For an efficient and qualitative analysis of material, the material properties need to be correctly and carefully entered [11]. Depending on the aim of the analysis, some mechanical properties such as density, strength and coefficient of thermal expansion definition is optional [12]. Knowing and declaring the correct value of the material property is very useful for design analysis purpose [13, 14]. For this analysis, the material used is structural steel. The fatigue analysis of the crankshaft structure will be carried out for a moment of 100 Nm under a rotational velocity of 150 rmp.

In the presence of dynamic loading at stress level lower than the material yielding strength, fatigue normally occurs. Figure 2 shows the constraint for the failure analysis.

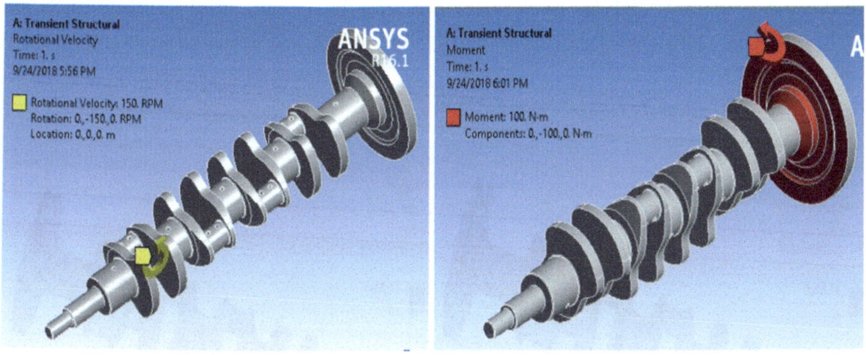

(a) (b)

Fig. 2. (a) Rotational velocity (b) Moment

The harmonic response analysis basically followed by the modal analysis. The frequency operation ranges and response to harmonically loads of the crankshaft is determined from the harmonic analysis. The modal analysis is the primary step to begin the harmonic analysis with the boundary conditions set as shown in Fig. 3.

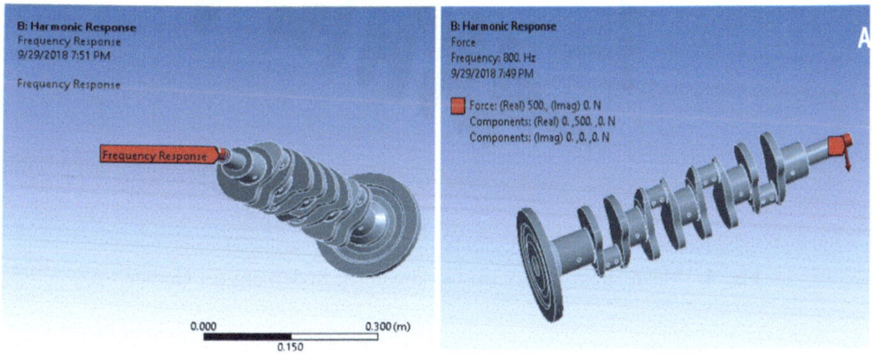

(a) (b)

Fig. 3. (a) Setting the frequency response (b) Force application

3 Result and Discussion

Figure 4 shows the safety factor result for the three empirical mean stress theories (Soderberg, Goodman, and Gerber) for an equivalent load ratio of 0 the by application and removal of the load (zero- base loading type). It shows Gerber theory having the highest safety factor followed by Goodman and then lastly the Soderberg theory. The material used for the Analysis been structural steel have high strength, toughness and ductile property. This clearly explained why the safety factor is higher for the Gerber mean stress theory.

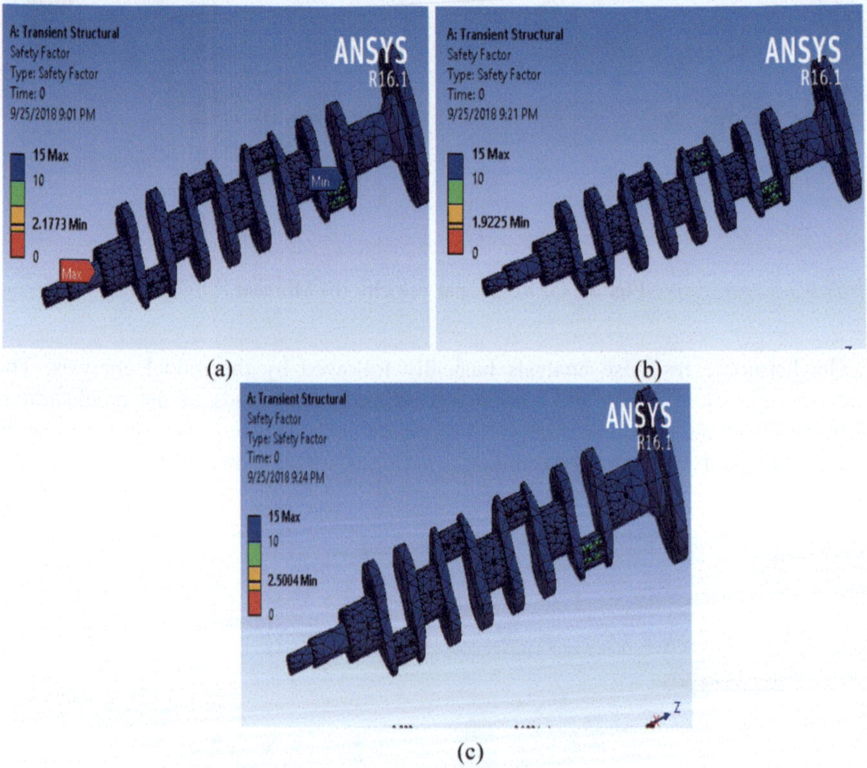

Fig. 4. Mean stress theory (a) Goodman (b) Soderberg (c) Gerber

The crankshaft lower order natural frequency has much effect on the system response as they generate larger vibrations. The total deformation for 4 first natural frequencies of the modal analysis are shown in Fig. 5. The frequency response indicates which frequency the crankshaft should operate without failing. For a linear frequency spacing type. The minimum and maximum range of frequencies for the analysis are 800 and 6800 Hz respectively, in an interval of 300. Figure 6 shows the fatigue life prediction for a single fully reversed loading.

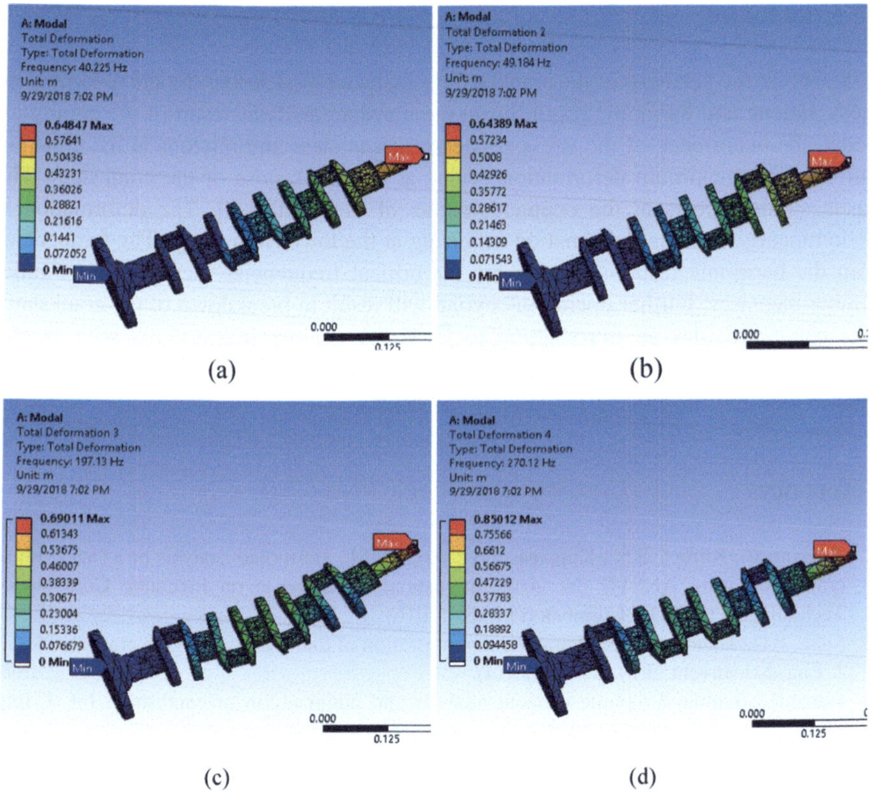

Fig. 5. (a) Total deformation (b) Total deformation 2 (c) Total deformation 3 (d) Total deformation 4

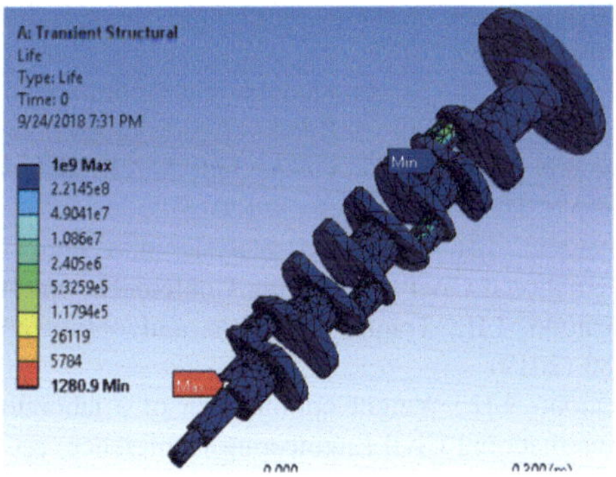

Fig. 6. Fatigue life

4 Conclusion

In this work, the behavior of the crankshaft is analyzed by determining the deformation, stress, fatigue and harmonic result. From the transient analysis result of the crankshaft model, the occurrence of the stress concentration is clear and obvious at the crankpin journals. The maximum deformation occurs at the crank nose of the crankshaft. The elastic strain appears at the counter weights of the crankshaft. The deformation is predominantly bending deformation occurring at the lower frequency. The frequencies from the harmonic response which are the critical frequencies increase as the order number increases. Further operation beyond will result in breakdown of the crankshaft. This paper provides an introduction to FEM for engineering systems with similar models and give a way for future research for dynamic analysis in multi-cylinder engines and optimization of the structural design.

References

1. Talikoti, B., Kurbet, S.N., Kuppast, V.V., Arvind, M.: Harmonic analysis of a two-cylinder crankshaft using ANSYS. In: 2016 International Conference on Inventive Computation Technologies (ICICT), Coimbatore, India (2016)
2. Shah, P.D., Bhabhor, K.K.: Parametric optimization of four cylinder engine crankshafts. Int. J. Eng. Sci. Invent. **3**(6), 38–43 (2014)
3. Garg, R., Baghla, S.: Finite element analysis and optimization of crankshaft. Int. J. Eng. Manag. Res. **2**(6), 26–31 (2012)
4. Yingkui, G., Zhibo, Z.: Strength analysis of diesel engine crankshaft based on PRO/E and ANSYS. In: Third International Conference on Measuring Technology and Mechatronics Automation, Shangshai, China (2011)
5. Chikalthankar, S.B., Nandedkar, V.M., Kaundal, S.K.: Finite element analysis approach for stress analysis of crankshaft under dynamic loading. Int. J. Sci. Eng. Res. **4**(2), 1–6 (2013)
6. Khasnis, V., Ukhande, M., Tilekar, G., Mane, R., Shegavi, G.: Crankshaft design optimization to improve dynamic balancing and fatigue strength. Int. J. Automot. Eng. Soc. Automot. Eng. Jpn. **6**(2), 59–66 (2015)
7. Montazersadgh, F.H., Fatemi, A.: Stress Analysis and Optimization of Crankshafts Subject to Dynamic Loading. The University of Toledo, Toledo (2007)
8. Talikoti, B., Kurbet, S.N., Kuppast, V.V., Arvind, M.: Modal analysis of a 2-cylinder crankshaft using ANSYS. IJERA **5**(12), 26–30 (2015)
9. Muhammad, A., Shanono, I.H.: Structural analysis of a knuckle joint using different materials. In: 1st International Civil Engineering Conference, pp. 100–106, Nigeria (2018)
10. Muhammad, A., Shanono, I.H.: Transient analysis and optimization of a knuckle joint. Kinetic **4**(2), 179–186 (2019)
11. Muhammad, A., Shanono, I.H.: Weight optimization of a laboratory stool using ANSYS workbench. In: 1st International Civil Engineering Conference, pp. 14–21, Nigeria (2018)
12. Barbero, E.J.: FEA of Composite Materials Using ANSYS. CRC Press, Boca Raton (2013)
13. Muhammad, A., Shanono, I.H.: Strength analysis and structural optimization of an I-shaped bracket. Niger. J. Technol. Res. **13**(2), 14–19 (2018)
14. Muhammad, A., Shanono, I.H.: Static analysis and optimization of a connecting rod. Int. J. Eng. Technol. Sci. (IJETS) **6**(1), 24–40 (2019)

Tensile Properties Comparison of Cassava Peel/Lycal, E-Glass 135/Lycal and Hybrid Cassava Peel+E-Glass 135/Lycal Composite with Hand Lay up Manufacturing Method

Lathifa Rusita Isna[1(✉)], Nur Mufidatul Ula[1], Syamsul Rizal[2], and Afid Nugroho[1]

[1] Aeronautic Technology Centre, Indonesian National Institute of Aeronautics and Space – LAPAN, Jl. Raya LAPAN Rumpin, Sukamulya, Rumpin, Bogor, Jawa Barat 16350, Indonesia
lathifarusita@gmail.com
[2] College of Aerospace Technology, Jl. Parangtritis KM.4,5, Druwo, Bangunharjo, Sewon, Bantul, Yogyakarta 55143, Indonesia

Abstract. The use of composite materials in UAV development has increased significantly. The chosen of composite material for the UAV structure is due to the advantage of composite both from the strength and the light side of material that good for improving structure performance because it can be reducing the weight of the structure. Nowadays, research in material not only focus on reducing weight but also focus on reducing undegradable waste to environment. Tensile test has been carried out for Cassava Peel/Lycal composite and E-Glass/Lycal Composite. The comparison results showed that the cassava/lycal composite has modulus elasticity average value (3445,3 ± 0.36 MPa) smaller than Hybrid Cassava Peel+E-Glass 135/Lycal Composite (4704.28 ±0,67 MPa) and E-Glass 135/Lycal Composite (5916,5 ± 1,24 MPa). This trend is happened also for tensile strength value with average 44,63 ± 5,11 MPa for Cassava Peel/Lycal, 143,01 ± 15,82 MPa for E Glass 135/Lycal composite and 109.68 ± 7,05 MPa for Hybrid Cassava Peel+E-Glass 135/Lycal Composite. This result showed that the strength of E Glass 135/Lycal Composite still much bigger than Cassava Peel/Lycal Composite with but not really higher than Hybrid Cassava Peel+E-Glass 135/Lycal Composite. It means the cassava composite can't be the alternative material for UAV skin but it can be used for another part inside the UAV that not directly touching from outside impact.

Keywords: Composite · UAV · Cassava Peel · E-glass · Hand lay-up

1 Introduction

Composite material is a material that widely used in Unmanned Aerial Vehicle (UAV). Composite materials has used for the main structures of aircraft such as wings, fuselage etc. [1]. The selection of this type of material is to reduce the weight of the aircraft structure.

M. N. Osman Zahid et al. (Eds.): iMEC-APCOMS 2019, LNME, pp. 377–383, 2020.
https://doi.org/10.1007/978-981-15-0950-6_57

E-glass composite with thermoset resin is the material that mostly used in unmanned aircraft. Nowadays, environmental friendly materials are needed and it is also very important for reducing the environmental harmness, enhancing the public awareness of environmental concerns and make new environment regulation for sustainability consumption [2]. Materials from natural fibers are needed to replace synthetic materials that are difficult to decompose. In a review of Natural Fiber Reinforced Polymer Composite (NFPCs) and Its Application, Mohammed and coworkers said that NFPCs was getting attention from many researcher due to the environmental advantages (eco-friendly and sustainability) [3]. Some specific design parameters that meet an airplane models specification are not available for public deployment. Specific design parameters such as mechanical properties (skin-core adhesion, flexural strength, impact resistance), Acoustical Properties (transmission loss), Resistance to cabin environment (vibration, humidity, fluid susceptibility, etc.) and weight are important information to design a structural material of airplane. There are some advantages of natural fiber if it chosen for aircraft structure such as replace the hazardous materials, Replace the non-renewable resources, and weight reduction that can be reduce fuel consumption/CO2 emissions [4].

Cassava peel fiber is one of some choices in this study that easy to obtain and easy to produce the composite using thermoset resin. Research that has been conducted by Ahmed Edirej et al., about the addition of cassava bagasse (CB) and cassava peel (CP) in bio-composite cassava films, the addition of both fibers increases tensile strength and modulus while reducing the breakthrough of bio-composite films [5]. Previous research has been conducted on the preparation and characterization of cassava starch/peel for composite film. Cassava starch was extracted and the peel was added as a fiber film to the composite. The addition of 6% cassava peel to the composite increased the elastic modulus and tensile stress up to 449.74 and 9.62 MPa [6].

The purpose of this study is to find out the comparison between the tensile strength of E-glass fiber composites, cassava peel fiber composites and hybrids both with lycal thermoset resin. The results of this comparison will be used as a database of material for making UAV at National Institute of Aeronautics and Space (LAPAN). The material database is used as reference for UAV's part manufacture where the strength of each part is adjusted to the type of UAV and its regulation [7, 8].

1.1 Composite Specimens

The specimens made by 3 sample plats of composite, the 1st plat was using cassava peel fiber with Lycal, the 2nd plat was using woven- E glass type WR135 with Lycal, and the 3rd plat was using Hybrid composite both Cassava Peel and E Glass WR 135 fiber with Lycal as Table 1 below:

Figure 1 shows the specimens before test. The Lycal resin was Lycal GLR 1011 (N) that mix with its hardener, this is a thermoset resin type. The ratio of the resin and fiber mixture was 60:40. The manufacturing method was a standard hand lay-up technique. The sample plats of composite were cured in room temperature for 24 h, and cut with geometry of 250 mm × 25 mm as mentioned in ASTM D3039 standard test.

Table 1. Sample information

Specimen code	Specimen content	Fiber orientation
A	Cassava Peel/Lycal	–
B	E-Glass 135/Lycal	[0/90]
C	Cassava Peel+E-Glass 135/Lycal	[0/90] E-Glass

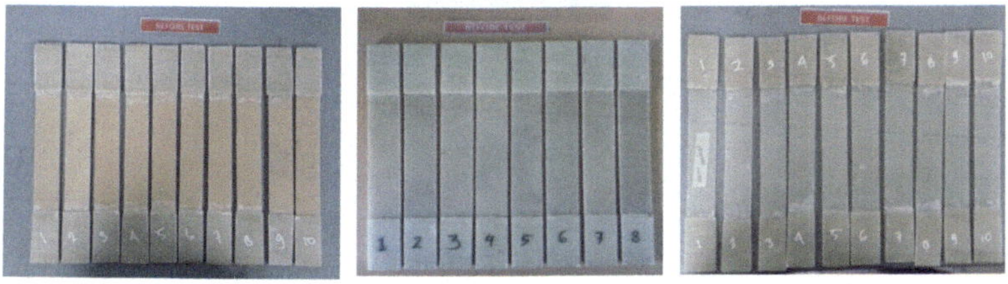

(A) (B) (C)

Fig. 1. Specimens before test

1.2 Tensile Testing Method

The testing method was all refer to ASTM D3039 standard tensile test of composite material. The test has done using Universal Testing Machine (UTM) Tensilon RTF - 2410 with crosshead speed of 2.0 mm/min. Figure 2 shows the research methodology, and the tensile test arrangement can be show as Fig. 3 below.

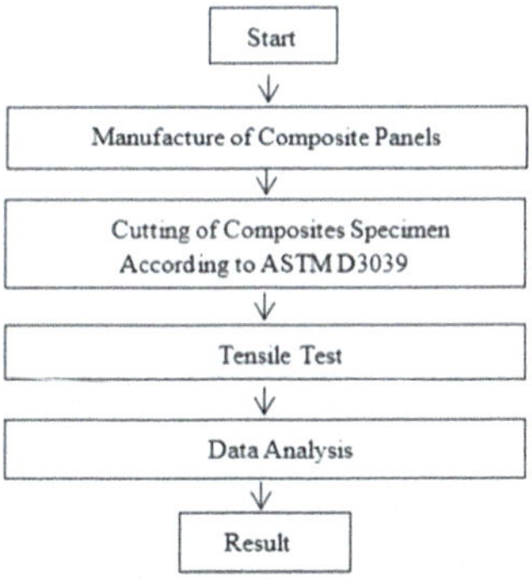

Fig. 2. Research methodology

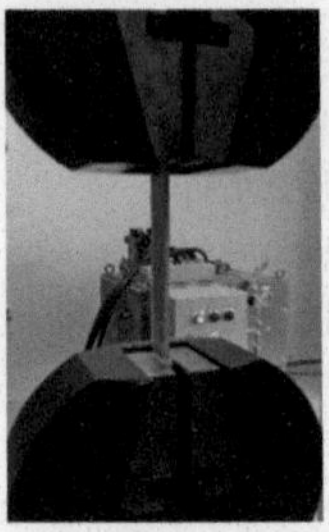

Fig. 3. Tensile test of composite material using universal testing machine

2 Data Analysis

The tensile test has been held using ASTM D3039 procedure standard [9]. This test is used to obtain the composite tensile stiffness and strength. The maximum stress (σ) of the specimen can be determined by the following equation:

$$\sigma i = Pi/A \quad (1)$$

The maximum strain (ε) on the mid-span of the specimen can be determined by the following equation:

$$\varepsilon i = \delta i/Lg \quad (2)$$

The modulus of elasticity E is comparison between stress and strain as following equation:

$$E^{chord} = \Delta\sigma/\Delta\gamma \quad (3)$$

3 Result and Discussion

The failure specimen after test are shown in Fig. 4. From the figure we can see that most of failures both for Cassava Peel/Lycal and E-glass WR135/Lycal sample occur at the top and bottom near the grip of the specimen, typical modes based on ASTM D3039 are LAT (Lateral, at grip/tab, Top) and LAB (Lateral, At grip/tab, Bottom). Different with other, the failure of Hybrid Cassava Peel+Eglass WR135/Lycal sample occur not only in the top/bottom near the grip side but also there are some delamination failure in the middle of the specimen gage between E-Glass and Cassava Peel fiber, it means there are two typical modes that represent this sample failure based on ASTM D3039 that is LAT (Lateral, at grip/tab, Top) and DGM (Edge Delamination, Gage, Middle). The Density of each specimen are 1.12 g/cm^3 for Cassava Peel/Lycal, 1.55 g/cm^3 for E-glass WR135/Lycal Composite and 1.32 g/cm^3 for Hybrid Cassava Peel+Eglass WR135/Lycal Composite. From this result, we can said that Cassava/Lycal composite has a lowest density than others.

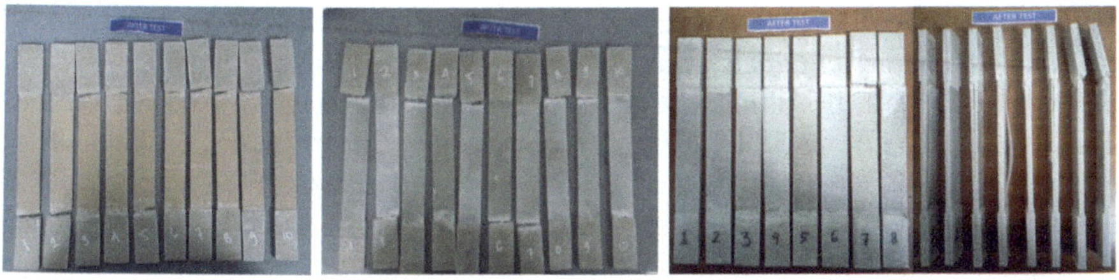

(A) (B) (C)

Fig. 4. The failure composite specimen after test

The results are given in Fig. 5. It shows that all of the sample failed in elastic mode.

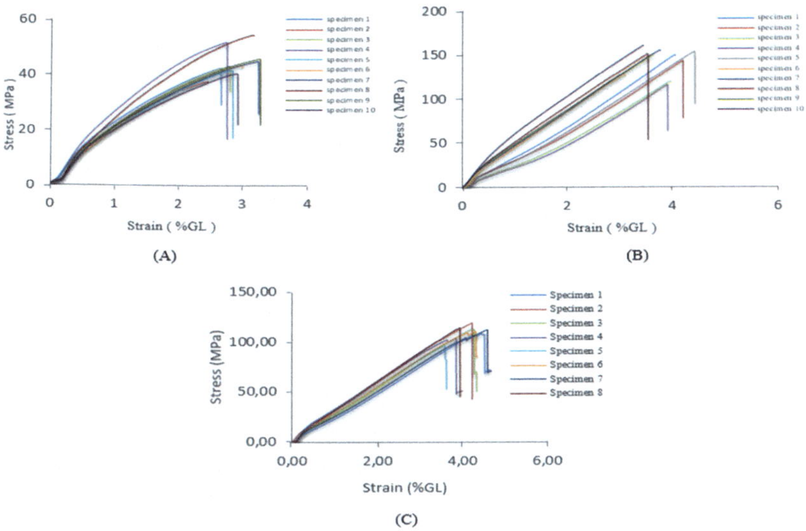

Fig. 5. Stress-strain graph of tensile test

Elastic modulus determines the film stiffness, so its indicates the high stiffness of a material [10]. From the result, we can see that the highest elastic modulus 3445.3 ± 0.36 MPa was recorded for Cassava Peel/Lycal, 5916.5 ± 1,24 MPa for Eglass WR135/Lycal and 4704.28 ± 0,67 MPa for hybrid cassava peel+E glass 135/Lycal composite.

The smallest strength value that occur in cassava peel composite is due to the poor adhesion of the hydrophilic group in its own fibers to the hydrophobic group in polymer matrix. The porous structure of most natural fibers causes water absorption, which leads to a change in the mechanical properties [11]. The hydrophilic character in the surface of Cassava peel is due to this fiber is principally made up of lignin and cellulose, it is refer to research by Syazwani *et al.* that demonstrated it from the FTIR spectra of each samples that consist of carboxyl (C = O) and hydroxyl (H = O) groups [12]. The hybrid composite has the middle strength value compare with other composite is due to the hydrophilic group of this composite has fewer than Cassava peel composite along with half of its compound consist of E-Glass fiber.

Refer to Table 2, the ultimate tensile strength (UTS) average of Cassava Peel+E-Glass 135/Lycal 44.63 ± 5,11 MPa, this result is still smaller than the tensile value of e-glass epoxy with 346.15 MPa where this type of composite has been applied by Nur et al. as LSU-02 NG LD wing skin. The result showed that its composite has failed when the load is 2.5 g [13]. Refer to FAR 23 for the normal category of UAV, the load of maneuver limits that must be met is between 1,52 to 3,8 g [7]. Further research is needed regarding the application of this material in UAV components.

Table 2. Tensile test result comparison

Data	A	B	C
	Average	Average	Average
Stress (Mpa)	44.63 ± 5,11	143.01 ± 15,82	109.68 ± 7,05
Elastic modulus (Mpa)	3445.3 ± 0.36	5916.5 ± 1,24	4704.28 ± 0,67
Break point strain (%GL)	2.91 ± 1.13	3.81 ± 1,39	4.14 ± 2.06
Density (g/cm^3)	1.12	1.55	1.32

The average stress value comparison shows in the following graph (Fig. 6):

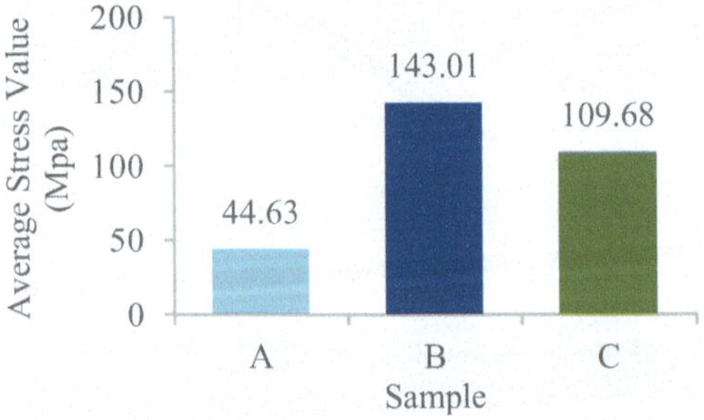

Fig. 6. Tensile strength average comparison

4 Conclusion

The comparison results showed that the Cassava Peel/lycal composite has modulus elasticity average value (3445,3 ± 0.36 MPa) smaller than Hybrid Cassava Peel+E-Glass 135/Lycal Composite (4704.28 ± 0,67 MPa) and E-Glass 135/Lycal Composite (5916,5 ± 1,24 MPa).

This trend is happened also for tensile strength value with average 44,63 ± 5,11 MPa for Cassava Peel/Lycal, 143,01 ± 15,82 MPa for E Glass 135/Lycal composite and 109.68 ± 7,05 MPa for Hybrid Cassava Peel+E-Glass 135/Lycal Composite. This result showed that the strength of E Glass 135/Lycal Composite still bigger than Cassava

Peel/Lycal Composite but not really higher than Hybrid Cassava Peel+E-Glass 135/Lycal Composite. It means the cassava composite can't be the alternative material for UAV skin but it can be used for another part inside the UAV that not directly touching from outside impact.

Acknowledgement. This research was supported and funded by Aeronautic Technology Centre – Indonesian National Institute of Aeronautics and Space (LAPAN – Indonesia). The authors would like to thank to Mr. Gunawan Setyo Prabowo and Mr. Agus Aribowo for the support to our research and the process of making this paper.

References

1. Hadi, B.K., Rofa, B.K.: Experimental tensile strength analysis of woven-glass/epoxy composite plates with central circular hole. J. Phys: Conf. Ser. **1005**(1), 012001 (2018)
2. May-Pat, A., Valadez-González, A., Herrera-Franco, P.J.: Effect of fiber surface treatments on the essential work of fracture of HDPE-continuous henequen fiber-reinforced composites. Polym. Test. **32**(6), 1114–1122 (2013)
3. Mohammed, L., Ansari, M.N.M., Pua, G., et al.: A review on natural fiber reinforced polymer composite and its applications. Int. J. Polym. Sci. (2015)
4. Martín, P.P.: New generation of materials for the aerospace industry : green aircraft interiors, United States of America (2012)
5. Edhirej, A., Sapuan, S.M., Jawaid, M., et al.: Effect of cassava peel and cassava bagasse natural fillers on mechanical properties of thermoplastic cassava starch: comparative study. In: C3 - AIP Conference Proceedings in Advanced Materials for Sustainability and Growth (2017)
6. Edhirej, A., Sapuan, S.M., Jawaid, M., Ismarrubie Zahari, N.: Preparation and characterization of cassava starch/peel composite film. Polym. Compos. **9**(5), 1704–1715 (2018)
7. Administration FA: federal aviation regulation, part 23-airworthiness standards: normal category airplanes (2017)
8. Ministry of Transportation R of I Civil Aviation Safety Regulation (CASR) part. 23 (2001)
9. ASTM D3039: Standard test method for tensile properties of polymer matrix composite materials. D3039-2017, pp. 1–13 (2017)
10. Edhirej, A., Sapuan, S.M., Jawaid, M., Zahari, N.I.: Tensile, barrier, dynamic mechanical, and biodegradation properties of cassava/sugar palm fiber reinforced cassava starch hybrid composites. BioResources **12**, 7145–7160 (2017)
11. Kaczmar, J.W., Pach, J., Kozlowski, R.: Use of natural fibres as fillers for polymer composites. Int. Polym. Sci. Technol. **34**, 45–50 (2018)
12. Mohd-Asharuddin, S., Othman, N., Mohd Zin, N.S., Tajarudin, H.A.: A chemical and morphological study of cassava peel: a potential waste as coagulant aid. In: MATEC Web Conference, vol. 103, p. 06012 (2017)
13. Ula, N.M., Marta, A., Wijaya, Y.G., Muksin: Wing static test of LSU-02 NGLD aircraft using Whiffletree method, p. 030025 (2019)

Dip-Coating Methods for Carbon Membrane Fabrication: Effects of Coating-Carbonization-Cycles on Hydrogen Separation Prepared from P84/NCC

Norazlianie Sazali[1,2(✉)], Mohd Syafiq Sharip[1], Haziqatulhanis Ibrahim[1], Ahmad Shahir Jamaludin[1], and Wan Norharyati Wan Salleh[3]

[1] Faculty of Mechanical Engineering, Universiti Malaysia Pahang, 26600 Pekan, Pahang, Malaysia
azlianie@ump.edu.my
[2] Centre of Excellence for Advanced Research in Fluid Flow (CARIFF), Universiti Malaysia Pahang, Lebuhraya Tun Razak, 26300 Gambang, Kuantan, Pahang, Malaysia
[3] Advanced Membrane Technology Research Centre (AMTEC), School of Chemical and Energy, Faculty of Engineering, Universiti Teknologi Malaysia, 81310 Skudai, Johor Darul Takzim, Malaysia

Abstract. In this paper depicts the manufacture and assessment of tubular carbon membrane equipped from BTDA-TDI/MDI (P-84) polyimide mixes with Nanocrystalline cellulose (NCC). Given earlier investigations, we planned the theory that tubular carbon membrane performance could impose constraints by controlling the carbonization conditions which directed with a heating rate of 3 °C/min, a final temperature of 800 °C and stabilization temperature of 300 °C. The principal purpose of this examination is to acquaint successful dip-coating strategies with produce superior tubular carbon membrane. The coating-carbonization cycles (1, 2, 3, and 4 times) have been considered. This methodology empowers quick and straightforward assessment of dip-coating techniques to yields high separation performance. Gas separation performance of the carbon membranes was adequately carried out by a single gas permeation experiment of H_2, and N_2 to explore the transport component in the carbon membrane separation process. In this case, the most elevated selectivity of 434.68 ± 1.39 for H_2/N_2; side by side with H_2 permeance of 1399.66 ± 5.22 GPU shall accomplish by employing two coating-carbonization-cycles.

Keywords: Coating-carbonization-cycle · P84 co-polyimide · Nanocrystalline cellulose (NCC) · Tubular carbon membrane · Hydrogen

1 Introduction

Membrane innovation has sociably attracted many reviews in consideration with the gas separation enterprises, for example, hydrogen recovery, air separation, olefin/paraffin separation, CO_2 capture, nature gas dehydration, and many other [1].

M. N. Osman Zahid et al. (Eds.): iMEC-APCOMS 2019, LNME, pp. 384–389, 2020.
https://doi.org/10.1007/978-981-15-0950-6_58

Carbon membrane innovation is in the process of being created promptly for these reasons. The requests on inorganic membrane have expanded because of the reduced temperature and chemical resistance of natural membrane. The hand-selected molecular sieving system of carbon membrane is especially helpful to accomplish great separation which is smooth between gases with practically relative molecular size [2]. Some other way, this approach turns out to be more energy conservation thus progressively efficient by contrasting with other gas separations technique. In this manner, a few specialists have participated in evolving new materials with a few manufacture techniques for carbon membrane [3]. One method is to manufacture the carbon membrane as a tubular aided membrane to limit imperfection and manage the cost of high gas separation performance. The fundamental focal point of this examination ponders to give better comprehension of the best dip-coating process parameters to control the morphology of the tubular aided carbon membrane. Through the author learning, dip-coating techniques are a standout amongst the most significant strategies for generation zone to supplant the conventional spin coating and spray coating methods. It has something to do with the fact of no impediment in supported estimate size, and the usage of polymers is low [4]. Hydrogen (H_2) is a decent fuel possibility for clean energy generation since it just generates water as side-effect upon burning. Filtration of H_2 from the output stream is mandatory for earlier utilisation. As for H_2/N_2 separation, if the molecular size of H_2 is small, H_2 is specialised in permeating through the membrane. In any case, the H_2 gas that penetrates through the membrane ought to be compacted again before it tends to utilise and this is a keen energy procedure [5]. In this paper, our target is to manufacture the PI/NCC-based carbon tubular membrane for both H_2 filtration and flue gas treatment. The tubular membrane design is suitable for industrialised gas separation, and mixing of NCC with PI which can enhance the gas separation performance [6]. As we all know, this is the initial endeavour of mixing NCC with PI to create carbon tubular layers for H_2/N_2 separation.

Dip-coating of porous tubular alumina ceramic with a solvent of PI-copolyimide mixes with nanocrystalline cellulose (NCC) was utilised to manufacture carbon membrane. As one of the writer philosophies, the manufacturing by a composite of NCC is the initial examination; henceforth bring the reproducible way of NCC itself. Past investigation proposed that cellulose is one of the natural linear polymers of b −(1/4) - D-glucopyranose that is owing novel highlights, for example, nontoxicity, high biocompatibility, and biodegradability [6]. Additionally, it is likewise, has excellent mechanical strength with durable adsorption manner. Truth be told, the one of a kind local property owe by the cellulose has pull in specialist in including cellulose substances as a supplement substance inside the mixing materials [7]. The hydrophobized cellulose inferred so as to keep away from muster of cellulose started from the surface hydroxyl cluster. Past investigation expressed that, NCC is one of the hydrophobized cellulose subordinates for lean or thin cellulose that frequently utilize. Amid membrane manufacture, it is essential to keep away membrane from any fractures and pinholes flaw so as to keep up great membrane selectivity. Therefore, the decision of material to aid this process is a crucial job. In this examination, tubular support produced from alumina ceramic is utilised because of its high physical quality, high gas diffusivity and tolerance to the high-temperature carbonization treatment amid membrane fabrication [8]. Although the vast majority of the analysts in this field would prefer to pick support materials with a deformity-free microporous. Several even on nanoporous surface layer to guarantee useful invention by steer clear of deformities generation, the uprising expense joined by this best quality will help in the long run keep down the capability of carbon membrane applications. In the present study, an affordable tubular macro porous Al_2O_3 is chosen as the membrane support material to acknowledge cost sparing in the fabrication process. Nevertheless, the issue of the vulnerability of carbon membrane established from the low-cost assist should initially suppress so as to guarantee reproducibility and versatility of the membrane. Dip-coating is considered the least difficult and most possibly applied manufacture method in the market. Numerous works are in progress to enhance this system [9]. Hence, this paper will review about dip-coating state because of less research on dip-coating process conditions.

2 Experimental Section

2.1 Materials

Polyimide BTDA-TDI/MDI (P-84), polyvinylpyrrolidone (PVP), and microcrystalline cellulose (MCC) which acquired from Sigma Aldrich whereas N-methyl-2-pyrrolidone (NMP) obtained from Merck (Germany). All synthetic compounds were utilized with no more than that of purification. Nanocrystalline cellulose (NCC) is presently not financially accessible and was equipped already in our investigation (7). Porous tubular ceramic assist (TiO_2) with 8 cm long, 13 mm in external diameter, 10 mm internal diameter with the standard pore size of 0.2 µm (porosity of 40-half) that bought from Shanghai Gong Tao Ceramics Co., Ltd.

2.2 Preparation of Carbon Membrane

Polymer mixture comprise of 15wt% of P-84 out of a NMP is set up under consistent mixing at 80 °C. 7wt% of NCC was bit by bit added to the mixture and keeps mixing until homogeneous mixture was acquired. The mixture was sonicated to eliminate trapped bubbles from the mixture and put aside for 12 h. The tubular aid was immersed into the polymeric mixture for 45 min and experienced 'aging' at 80 °C inside the fume hood. The membranes were then submerged in methanol for 2 h before being set inside an oven at 100 °C for 24 h to permit moderate expulsion of the solution. Carbon membranes were set up via carbonization procedure of the supported polymeric membranes. The supported polymeric membranes were positioned at the focal point of the Carbolite horizontal tubular furnace to experience carbonization operation. The carbonization procedure executed at various carbonization temperature of 800 °C under Argon condition (200 ml/min). The heating rate of 3 °C/min was connected all through the procedure. In the wake of finishing each heating cycle, the provided membrane was cooled generally to room temperature. Carbon membrane without supported was produce by employing an identical procedure for characterization purpose.

2.3 Pure Gas Permeation Measurements

The carbon tubular membrane was experimental in a gas penetration framework as portrayed in our past examination [10]. The carbon tubular membrane laid inside a tubular stainless-steel module of 14 cm long. The membrane was adjusted with elastic O-rings to enable the membrane to the position in the module without spillages. The porousness of four pure gases with various molecular sizes; H_2 (2.80), and N_2 (3.64) as a result carbon membrane created from various coating cycles at a feed pressure of 8 bar. On account of the molecular diameter who is nearly identical, the gas separation by gas compound containing H_2, CH_4, CO and CO_2 is an intricate issue. Thus, gain insight into developed of porous carbon structure is expected to manage the pore sizes in the membrane. The permeance, P/l (GPU) and selectivity, α of the membranes were determined and the selectivity characterized as the penetration proportion of fast gas permeation to moderate gas permeation.

3 Results and Discussions

3.1 Gas Permeation Measurements

Dip-coating procedures might have profoundly recommendable for reasonable applications because of their relative simplicity. The impact of coating parameters additionally affected the moving or transfer properties of the carbon membranes [10]. It expressed that fitting manipulation of the coating cycles created carbon membrane with high solidness while coating time improved the membrane selectivity and decreased the membrane's porousness. Amid dip covering process, the interfacial stress among polymer and aiding which diminishes the Tg of the polymer subsequently will restrain the development of complicates structure [7]. Later in the carbonization process, distinction in carbon microstructure are because of the contrast at homogeneous polymer surface. In the dip-coating technique, the gases permeability came to very nearly a consistent incentive after two coatings and can be considered that two times coating cycle are the best while though in film casting which required three coatings. Liu stated four successive dip-coating is essential to create their membranes [11] (Table 1).

Table 1. Gas separation performance of carbon membranes produced from different coating-carbonization cycles.

Carbon membrane			
Sample (CM-PI/NCC)	Permeance (GPU)		Selectivity
	H_2	N_2	H_2/N_2
1 cycle	1306.27 ± 3.22	3.05 ± 1.21	428.29 ± 1.87
2 cycles	1399.66 ± 5.22	3.22 ± 3.21	434.68 ± 1.39
3 cycles	1105.21 ± 2.39	2.71 ± 1.03	407.835 ± 1.76
4 cycles	988.91 ± 2.94	2.55 ± 2.71	387.81 ± 2.44

Increasing the number of coatings cycle to four brought about roughly no adjustment in permeance for dip-coating and film casting systems. Nonetheless, for the pouring technique, H_2 permeance was diminished relentlessly after four successive coatings [12]. The selectivity of the equipped membrane by dip-coating was higher than that for the film casting method a role as notified by past examination. The coating carbonization cycles have prompted an expansion of the membrane selectivity. Enlarging the number of coatings to four brought about roughly comparative permeance for dip-coating and film casting systems. Selectivity of the equipped membrane by dip-coating was higher due to better permeation of PI/NCC mixture into membrane surface pinholes. For the strategies, as the number of repeated coatings expanded, the exposed pores or deformities on the membrane surface were additionally clogged or blocked and thus, enhance production was attained. Besides, a correlation of film casting and dip-coating strategies demonstrated improve inclusion of deformities by the coating layer in the dip-coating procedure than that in film casting [13]. In respect to the optimum manufacture states, the aided carbon membrane shall examine so as to decide their gas permeances. Sad to say, all membrane may experience the small insignificant despite the fact that the membrane was seen to be without fracture or gap. The residue from the surrounding air could have debased with the aided membrane which brings about little pinholes and infinitesimal defects on the membrane.

4 Conclusions

From this examination, the impact of dip-coating conditions on PI/NCC carbon membrane was explored. As a result, there are couple of outcomes can be made:

1. The enlarging of carbonization-covering-cycles from 1 to 2 would upgrade pore configuration because of disintegration preceding carbonization method. Existence of NCC as added substance gives better pore auxiliary properties due to their nanocrystalline structure.
2. In this context, the carbon membrane carbonized at 800 °C demonstrates the most encouraging outcome to H_2/N_2 selectivity of 1399.66 ± 5.22 with H_2 permeance of 213.56 ± 2.17. Positive findings acquired in this examination show the capability of crystalline cellulose, particularly NCC, for thermally labile added substance warrants for further study.

Acknowledgement. The authors would also gratefully acknowledge the financial support from the Ministry of Education and Universiti Malaysia Pahang under the Fundamental Research Grant Scheme (Project Number: RDU1703253 and RDU191105).

References

1. Sołowski, G., Shalaby, M.S., Abdallah, H., Shaban, A.M., Cenian, A.: Production of hydrogen from biomass and its separation using membrane technology. Renew. Sustain. Energy Rev. **82**, 3152–3167 (2018)
2. Sazali, N., Salleh, W.N.W., Ismail, A.F., Kadirgama, K., Othman, F.E.C., Ismail, N.H.: Impact of stabilization environment and heating rates on P84 co-polyimide/nanocrystalline cellulose carbon membrane for hydrogen enrichment. Int. J. Hydrogen Energy **44**(37), 20924–20932 (2018)
3. He, X.: Techno-economic feasibility analysis on carbon membranes for hydrogen purification. Sep. Purif. Technol. **186**((Supplement C)), 117–124 (2017)
4. Li, P., Wang, Z., Qiao, Z., Liu, Y., Cao, X., Li, W., et al.: Recent developments in membranes for efficient hydrogen purification. J. Membr. Sci. **495**, 130–168 (2015)
5. Ismail, N.H., Salleh, W.N.W., Sazali, N., Ismail, A.F., Yusof, N., Aziz, F.: Disk supported carbon membrane via spray coating method: effect of carbonization temperature and atmosphere. Sep. Purif. Technol. **195**, 295–304 (2018)
6. Salleh, W.N.W., Ismail, A.F.: Carbon membranes for gas separation processes: recent progress and future perspective. J. Membr. Sci. Res. **1**(1), 2–15 (2015)
7. Sazali, N., Salleh, W.N.W., Ismail, A.F., Kadirgama, K., Othman, F.E.C.: P84 co-polyimide based-tubular carbon membrane: effect of heating rates on helium separations. Solid State Phenom. **280**, 308–311 (2018)
8. Sazali, N., Salleh, W.N.W., Ismail, A.F., Nordin, N.A.H.M., Ismail, N.H., Mohamed, M.A., et al.: Incorporation of thermally labile additives in carbon membrane development for superior gas permeation performance. J. Natural Gas Sci. Eng. **49**, 376–384 (2018)
9. Sazali, N., Salleh, W.N.W., Ismail, A.F., Ismail, N.H., Yusof, N., Aziz, F., et al.: Influence of intermediate layers in tubular carbon membrane for gas separation performance. Int. J. Hydrogen Energy **44**(37), 20914–20923 (2018)
10. Sazali, N., Salleh, W.N.W., Nordin, N.A.H.M., Ismail, A.F.: Matrimid-based carbon tubular membrane: effect of carbonization environment. J. Ind. Eng. Chem. **32**, 167–171 (2015)
11. Liu, R., Qiao, X., Chung, T.S.: The development of high performance P84 co-polymide hollow fibers for pervaporation dehydration of isopropanol. Chem. Eng. Sci. **60**, 6674–6686 (2005)
12. Ismail, N.H., Salleh, W.N.W., Sazali, N., Ismail, A.F.: Effect of intermediate layer on gas separation performance of disk supported carbon membrane. Sep. Sci. Technol. **52**(13), 2137–2149 (2017)
13. Ismail, N.H., Salleh, W.N.W., Sazali, N., Ismail, A.F.: The effect of polymer composition on CO_2/CH_4 separation of supported carbon membrane. Chem. Eng. Trans. **45**, 1465–1470 (2015)

Current Advances in Membranes for Competent Hydrogen Purification: A Short Review

Mohd Syafiq Sharip[1], Norazlianie Sazali[1,2](✉), Mohd Nizar Mhd Razali[1], Farhana Aziz[3], and Mohd Hafiz Dzarfan Othman[3]

[1] Faculty of Mechanical Engineering, Universiti Malaysia Pahang, 26600 Pekan, Pahang, Malaysia
azlianie@ump.edu.my
[2] Centre of Excellence for Advanced Research in Fluid Flow (CARIFF), Universiti Malaysia Pahang, Lebuhraya Tun Razak, 26300 Gambang, Kuantan, Pahang, Malaysia
[3] Advanced Membrane Technology Research Centre (AMTEC) School of Chemical and Energy, Faculty of Engineering, Universiti Teknologi Malaysia, 81310 Skudai, Johor Darul Takzim, Malaysia

Abstract. Hydrogen (H_2) has been recognized as the clean potent energy carrier to pacify the worldwide energy and environmental problem. The tremendously escalate of high-quality hydrogen urge the development of membrane technology for hydrogen separation that believe have high potential and advantages compare to the conventional hydrogen separation technique. In correlation, the growing of research field like chemical and material sciences, membrane sciences and technology, and advance membrane technology over the decade provides important information on the membrane technology for hydrogen separation. Moreover, the present review was set up as the preliminary study and guidance for the future research and development of membrane materials and membranes for hydrogen purification, and hence promote the development of sustainable and clean hydrogen energy.

Keywords: Hydrogen purifications · Gas separation · Materials characteristic · Sustainable energy

1 Introduction

In the present years, H_2 is a top-notch transporter for clean energy which draws in refreshed and regularly expanding concern around the globe generally because of the progression of power cells, ecological loads along with worldwide environmental alteration issues [1–3]. Enhancement and separation innovations are imperative in the thermochemical procedures of H_2 creation from petroleum products [4]. The membrane reactors show dependable affirmations in transferring the harmony at whatever point the transfer response of water–gas associated with converting monoxide into H_2 [5]. Membranes are additionally vital in the accompanying refinement of H_2. Generally,

M. N. Osman Zahid et al. (Eds.): iMEC-APCOMS 2019, LNME, pp. 390–395, 2020.
https://doi.org/10.1007/978-981-15-0950-6_59

there are two inorganic membrane classifications in the cleansing and generation of H_2: metal alloy and thick segment metal, including porous ceramic membrane. Sol-gel or hydrothermal method are utilized in manufacturing porous ceramic membrane for elevated consistency and toughness in high temperature, hydrothermal conditions and severe pollutants, for example, microporous membrane which exhibit affirmations in gas transfer response at high temperatures. Figure 1 shows Palladium-based membrane for H_2 purification [6].

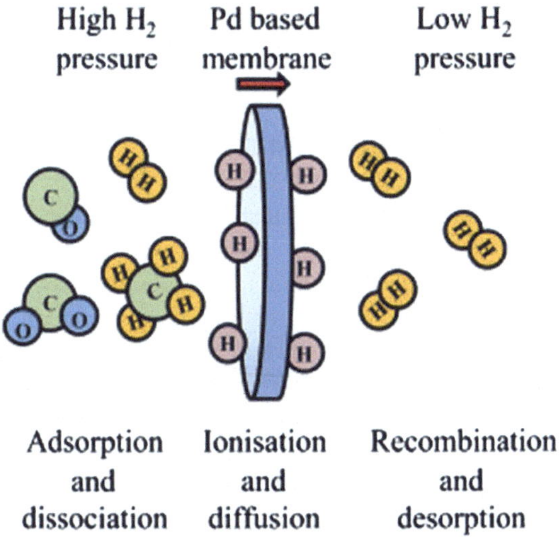

Fig. 1. Palladium-based membrane for H_2 purification [6].

Eminent importance of separation innovation in diverse parts of synthetic and energy enterprises is broadly recognized. In a way, a primary development will be accomplished by the advancing of technologies related to the separation of aerosolised species [7]. This may be because of a high fossil-fuel request in addition to the interest for several gases for pharmaceutical and industrial applications [8]. For instance, it determined that in the fossil-fuel derivative segment alone 41 of the established gas sources inside the US are sub-quality which should be updated through the expulsion of exorbitant carbonic acid gas, H_2S, N_2 with different polluting influences so as to satisfy the wellhead procedure or pipeline transmission necessities [9]. A comprehensive investigation ponders are carried out on current strategies paying little heed to different setup gas separation technologies present, for example, pressure swing sorption and fluid ingestion so as to detected procedures with abilities to offer expense-effective task and less complicated management [10]. Membranes innovation has been a possible option for numerous uses of gas separation mostly because of its multiple benefits, for example, little footprint, simple scale-up and high energy strength [11].

As indicated by the referred accounts, following in the year of 2002, the gas separation membrane innovation has turned into a $150million/year business with the prospect to develop in the coming years [5]. Therefore, a comprehensive discovery of

R&D membrane innovation development is required so as to conquer the hindrances. Beside high caliber, economically accessible membrane composed of a chemical compound, carbon membranes offer an utterly remarkable and engaging membranes class with recognized determinations along with excellent gas separation performance. Carbon membranes contribute striking attribute with excellent thermal and compound strength beside the capacity to exceed the permeability – selectivity exchange off. Carbon membranes have an inkling by taking the lead in acknowledgement and appearance of morphology in separation performance. This is bolstered by the porous structure which permits high penetrability as a result of high profitability while conservative size and segregation of molecules structure result in high selectivity which given by the molecular sieving system.

2 Membrane-Based H_2 Purification

As a relatively new and rapidly developing technology, membrane technology exhibits inherent advantages of energy-efficiency, cost-effective and environmental compatibility compared to conventional separation techniques. Moreover, membrane technology can be facilely coupled with other separation techniques to enhance the efficiency and economics of separation process. Nowadays, membrane technology has been widely used in water treatment, meanwhile it has also commercialized for air separation, natural gas sweetening and hydrogen recovery from ammonia purge gas. With the rapid development of hydrogen economy and membrane science, membrane-based gas separation technology shows great potential for the hydrogen purification market as well. Great demands for high-quality hydrogen products provide the driving force for research and development of advanced membrane materials and membranes for hydrogen purification [7].

2.1 H_2-Selective Membranes

The H_2-selective membranes were suggested as polymeric membrane by several researchers worldwide based on experimental process on the membrane technology for hydrogen purification before 2010. Across the modern era, the technology of membrane was believed has been developed significantly and the research on membrane increase their potential on application for the biohydrogen purification [12]. Unfortunately, there are still no extensive review in the hydrogen purification on the membrane. The uses of hydrogen basically being established worldwide as promising energy that can initially fulfill by having precise technique of separation for producing the high quality of hydrogen. The use of membrane technology in hydrogen purification was believe to be success based on the successful of the practice of advance membrane technique for separation using inorganic material, organic material and polymers that have lowest cost and great performance compare to the conventional separation techniques. The previous study shows that the low-cost membrane material such as PEO-based copolymers and polyvinylamine produce vast achievement in filtrated material and the membrane module such as Proteus™ and Polaris™ produce significant performance in remove CO_2 from the shifted gas. In relation, the hydrogen separation membrane currently being separating into five categories such as proton conducting membranes, porous membranes, polymeric membranes and dense metal membranes. In order to achieve positive result of membrane for hydrogen separation, there are several parameters must be taken care with caution such as the flux of membrane can be use, the perm selectivity and the temperature range. Moreover, the membrane also must have high chemical stability characteristic, high mechanical durability characteristic and low cost respectively. This characteristic was important to follow because membrane is the crucial component to act as barriers for the separation process take place. Figure 2 shows H_2 separation by syngas system [12].

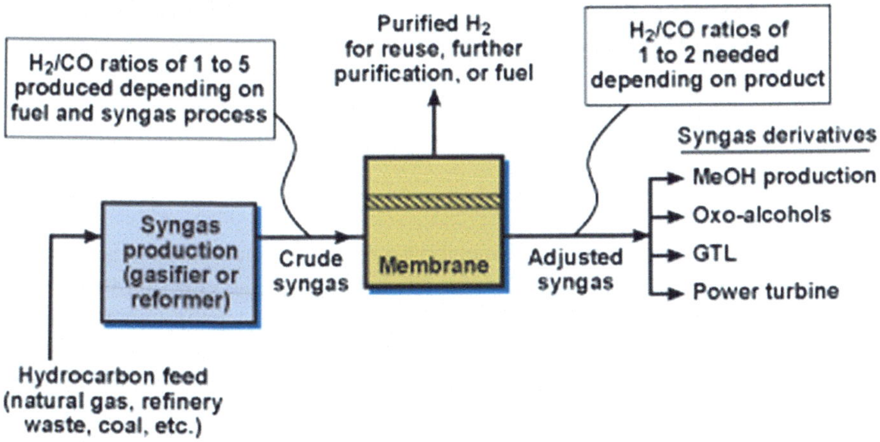

Fig. 2. H_2 separation by syngas system [12].

In current energy usage, the fossil fuel is still the dominant resource for fulfill daily basis energy for human in earth. However, the fossil fuel has been estimated to be execute for 5–20 years more. Even though there are plenty research and development of renewable energy technologies were setting up but still the use of renewable energy reclines the domination of fossil fuel as major energy resource in the world. Thus, the use of membrane for hydrogen separation is major step to utilize the fossil fuel for being more sustainable energy. The hydrogen was well known as with the characteristic of high gravity energy density (1.43 × 108 J/kg) and low greenhouse gas emission [9] that can benefit as the energy carrier and medium for the sustainable energy system. Nowadays, the interest for the hydrogen has been increase tremendously in chemical industries with 53 million metric tons of hydrogen worldwide was produced annually, and the hydrogen market valued at $88 billion in 2010 [13]. Figure 3 illustrates hydrogen as fuel system.

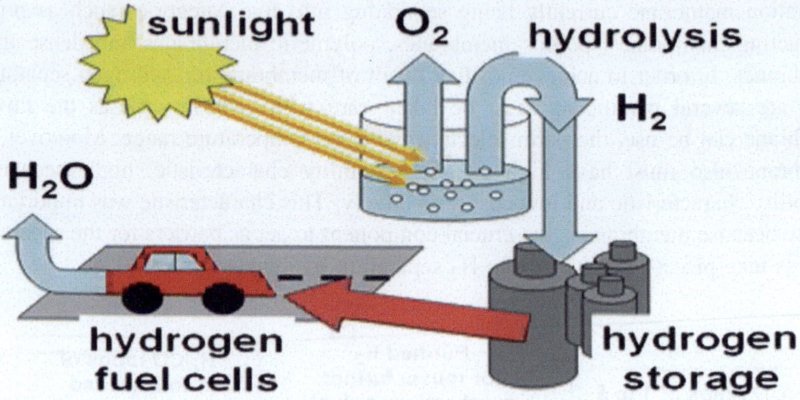

Fig. 3. H_2 as fuel system [13].

3 Conclusions

In completion, the further research on membranes for hydrogen separation is crucial to be carried forward. The feasibility study on the application, the effect on the temperature, pressure of membrane technologies and the material use really important factor to be set on as preliminary bases for the membrane technology development in hydrogen separation field [14–16]. The development of membrane technology these day and the literatures published contribute great exemplary in order to develop the meet requirement of practical application for membrane technology in hydrogen separation in future.

Acknowledgement. The authors would also gratefully acknowledge the financial support from the Ministry of Higher Education and Universiti Malaysia Pahang under the Fundamental Research Grant Scheme (Project Number: RDU1703253, RDU191105, and RDU190354).

References

1. He, X.: Techno-economic feasibility analysis on carbon membranes for hydrogen purification. Sep. Purif. Technol. **186**((Supplement C)), 117–124 (2017)
2. Sazali, N., Salleh, W.N.W., Ismail, A.F., Kadirgama, K., Othman, F.E.C., Ismail, N.H.: Impact of stabilization environment and heating rates on P84 co-polyimide/nanocrystaline cellulose carbon membrane for hydrogen enrichment. Int. J. Hydrogen Energy **44**(37), 20924–20932 (2018)
3. Sazali, N., Salleh, W.N.W., Ismail, A.F., Wong, K.C., Iwamoto, Y.: Exploiting pyrolysis protocols on BTDA-TDI/MDI (P84) polyimide/nanocrystalline cellulose carbon membrane for gas separations. J. Appl. Polym. Sci. **136**(1), 46901 (2018)
4. Sazali, N., Salleh, W.N.W., Ismail, A.F., Ismail, N.H., Aziz, F., Yusof, N., Hasbullah, H.: Effect of stabilization temperature during pyrolysis process of P84 co-polyimide-based tubular car-bon membrane for H_2/N_2 and He/N_2 separations. In: IOP Conference Series: Materials Science and Engineering, vol. 342, no. 1, p. 012027 (2018)

5. Sazali, N., Salleh, W.N.W., Izwanne, M.N., Harun, Z., Kadirgama, K.: Precursor selection for carbon membrane fabrication: a review. J. Appl. Membr. Sci. Technol. **22**(2) (2018)
6. Zhang, K., Way, J.D.: Palladium-copper membranes for hydrogen separation. Sep. Purif. Technol. **186**, 39–44 (2017)
7. Ismail, N.H., Salleh, W.N.W., Sazali, N., Ismail, A.F.: The effect of polymer composition on CO2/CH4 separation of supported carbon membrane. Chem. Eng. Trans. **45**, 1465–1470 (2015)
8. Haider, S., Lindbråthen, A., Lie, J.A., Andersen, I.C.T., Hägg, M.-B.: CO2 separation with carbon membranes in high pressure and elevated temperature applications. Sep. Purif. Technol. **190**, 177–189 (2018)
9. Lee, J., Kim, J., Hyeon, T.: Recent progress in the synthesis of porous carbon materials. Adv. Mater. **18**(16), 2073–2094 (2006)
10. Haider, S., Lindbråthen, A., Lie, J.A., Carstensen, P.V., Johannessen, T., Hägg, M.-B.: Vehicle fuel from biogas with carbon membranes; a comparison between simulation predictions and actual field demonstration. Green Energy Environ. **3**(3), 266–276 (2018)
11. Pereira, L.M.C., Vega, L.F.: A systematic approach for the thermodynamic modelling of CO2-amine absorption process using molecular-based models. Appl. Energy **232**, 273–291 (2018)
12. Burra, K.R.G., Bassioni, G., Gupta, A.K.: Catalytic transformation of H_2S for H_2 production. Int. J. Hydrogen Energy **43**(51), 22852–22860 (2018)
13. Hames, Y., Kaya, K., Baltacioglu, E., Turksoy, A.: Analysis of the control strategies for fuel saving in the hydrogen fuel cell vehicles. Int. J. Hydrogen Energy **43**(23), 10810–10821 (2018)
14. Sazali, N., Salleh, W.N.W., Ismail, A.F., Nordin, N.A.H.M., Ismail, N.H., Mohamed, M.A., et al.: Incorporation of thermally labile additives in carbon membrane development for superior gas permeation performance. J. Nat. Gas Sci. Eng. **49**, 376–384 (2018)
15. Ismail, N.H., Salleh, W.N.W., Sazali, N., Ismail, A.F.: Effect of intermediate layer on gas separation performance of disk supported carbon membrane. Sep. Sci. Technol. **13**, 2137–2149 (2017)
16. Ismail, N.H., Salleh, W.N.W., Sazali, N., Ismail, A.F., Yusof, N., Aziz, F.: Disk supported carbon membrane via spray coating method: effect of carbonization temperature and atmosphere. Sep. Purif. Technol. **195**, 295–304 (2018)

Microstructure and Mechanism of Silicanizing Process on Mild Steel Substrate Using Tronoh Silica Sand at 1000 °C for 4 H

Yusnenti Faziran Mohd Yunos[1(✉)] and Mohd Yusri Ibrahim[2]

[1] University of Technology PETRONAS, Tronoh, Perak, Malaysia
yusnenti@gmail.com
[2] Polytechnic Kuching Sarawak, Kuching Sarawak, Malaysia

Abstract. This paper presented a new route of producing coating onto a mild steel substrate called "Silicanizing" by using natural silica sand originated from Tronoh Silica Sand (TSS) at elevated temperatures. Influence of the silicanizing process parameters especially temperature and time on the silicanized coating layer's microstructure and mechanism were studied. Formation of Fe-SiO_2 via interdiffusion of Fe and SiO_2 were determined by the rate of grain boundary diffusion of SiO_2 through the growing Fe-SiO_2 phase shown at Silicanizing process temperature of 1000 °C for 4 h. Relevant silicanizing mechanism and chemical reaction involved in transforming TSS particles into thin film silicanized-coating on mild steel substrate is expected to be established.

Keywords: Silica sand · Coating · Mild steel

1 Introduction

The importance of coatings and the synthesis of new materials for industry have led to a tremendous increase in innovative coating processing technologies. Many of the surface treatment techniques, especially silicanizing can be applied economically and it is defined as imparting silica-like properties to a substrate material by addition or reaction of silica sand containing compounds, which results in formation of a thin surface film. Several metal oxide coatings including SiO_2 were reported for corrosion protection of metals [1–3]. This research has laid the foundation for the coating production of SiO_2 fine particles from Tronoh mineral resources (TSS) using silicanizing processes. Among methods of surface treatment, siliconizing is a common method that has been widely used in industry. This is because of its have advantages such as flexibility, simple, economical and suitable for the material especially TSS as powder coating as well as mild steel substrate. A novel method has been reported for coating surfaces with dense layer of silica in solutions [4]. A homogeneous coating could be formed with multiple layers of silica particles. Initially, small particles coated the surface and larger particles bound to them in the final step. The silica coatings acted as diffusion barrier layers and generated outstanding corrosion resistance despite the

M. N. Osman Zahid et al. (Eds.): iMEC-APCOMS 2019, LNME, pp. 396–401, 2020.
https://doi.org/10.1007/978-981-15-0950-6_60

thinness of the coatings [5]. These characteristics enable silica a good ceramic coated. The siliconizing process has been described as a heterogeneous solid/liquid reaction followed by a mechanism of diffusion of the reactant [6, 7]. Some researches attempted to use the mechanism to standardize the siliconizing process. The thermodynamics, chemical reaction and kinetics of the pack siliconizing process are strongly affected by the operating conditions and pack components. Generally, pack siliconizing can be considered as a self-generated process carried out in a thermodynamically closed system under isothermal conditions [8]. In this paper, the influence of the microstructure of "silicanizing" process during transformation was studied. Relevant silicanizing mechanism and chemical reaction involved in transforming TSS particles into thin coating on mild steel substrate is expected to be established.

1.1 Experimental Procedures

Material and Silicanizing Process
A rectangular mild steel specimen with dimension 18 mm × 18 mm × 5 mm were embedded in TSS. This was performed using fire clay jacket as crucible, which subjected to surface hardening treatment. For this, samples surface was ground with series of SiC paper of 600 grits. Subsequently, the sample were cleaned in acetone baths and dried in air. TSS powder with size below 150 μm was used in the silicanizing experiments. In the next step, the sample was subjected to heat treatment in furnace at elevated parameters.

2 Result and Discussion

2.1 Microstructure of Silicanizing Process

Figure 1 demonstrate SEM image of the coating layer by silicanizing process at 1000 ° C for 4 h. In addition, the surface morphology of coating structured by silicanizing process is presented in Fig. 1. The findings reveal that diverse structured layer experienced temperature point and time during the silicanizing process. The SiO_2 concentrations varied steadily across of the thin layer to the surface and diffused and deposited into substrate, which led to the formation of coating layers rich in SiO_2. The substrate was covered with SiO_2 with brown in colour and the thickness of the layer was almost uniform dimension. Figure 1(b) illustrates the bond Fe-SiO_2 interface which the interface is clear several micro-cracks are observed in the substrate beneath the bond Fe-SiO_2 interface. In particular, the pack mixture containing 80–90 wt% SiO_2 resulted in deposition of Fe-SiO_2 with an average thickness 500 μm. The as-formed coatings were very rough but more homogeneous, crystallize and had porosity.

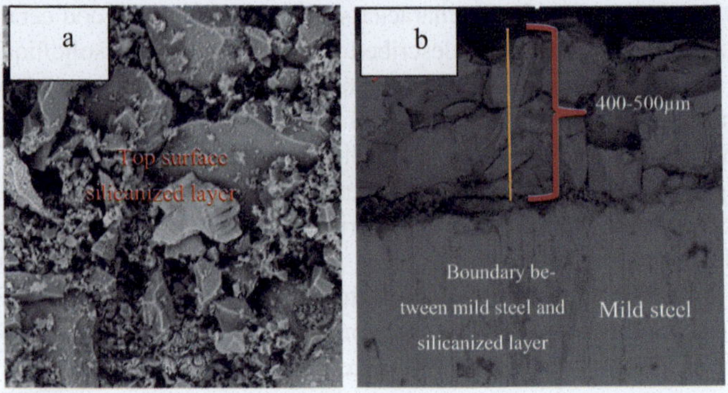

Fig. 1. SEM image of siliconized layer at 1000 °C for 4 h (a) top surface view (b) cross-sectional view.

Formation of the coating compounds is the result of reaction of the SiO_2 with the chemically deposited mild steel substrate. Table 1 demonstrates their chemical composition concentration. For an accurate identification of the as-formed structure, SEM-EDS analysis of several regions of the final coating was performed. The results demonstrated the formation of three different Fe-SiO_2 compounds. The area SEM-EDS scan-line denoted in Table 2 was found to contain 91.0 wt% Fe, 1.8 wt% Si. Furthermore, insignificant amounts of O_2 (less than 7.2 wt%) was found, which can be attributed to the amounts of oxygen in the mixture of SiO_2 pack powder.

Table 1. Chemical composition of silicanized layer at 1000 °C for 4 h

Element	Composition concentration %wt
Fe	47.4
Si	43.8
Al	3.16
K	1.83
P	1.22
Other	2.59

Table 2. wt% concentration silicanizing layer at 1000 °C for 4 h.

Element	Composition concentration %wt
Fe	91
Si	1.8
O	7.2

The aforementioned results are in agreement with the XRD pattern (Fig. 2) acquired from the same sample. By combining the results of SEM-EDS analysis (Table 2) and XRD characterization it was concluded that the substrates correspond with FeSi, Fe_2SiO_4, $FeSiO_3$ and SiO_2 respectively. Furthermore, deposition at 1000 °C for 4 h resulted in full transformation of SiO_2 into substrate. This process was continued because the as-formed compounds are very brittle and cross-sectional sample preparation for microscopic observation was impossible practically.

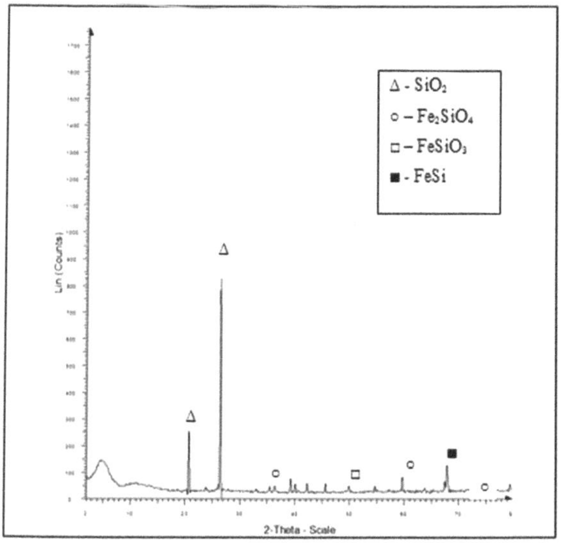

Fig. 2. XRD analysis of silicanized layer at 1000 °C for 4 h

In fact, as reported for mechanical properties of Fe and SiO_2, all four equilibrium phases in the Fe-SiO_2 system can be formed, depending on the Fe/SiO_2 atomic ratio [9]. The description of the FeSi binary and phases were combined into the description of a semi-stoichiometric phase Fe-SiO_2 to reflect the continuous solubility of this phase in the ternary Fe-Si-O system [10]. Furthermore, the diffusion of SiO_2 to this interface is accelerated by the presence of the ternary elements, Fe, Si and O. When coating powder (TSS) containing SiO_2 is used, the coating exhibited a microstructure of vertically oriented dendrites solidification growing on the interfacial layer. This structure resembles the columnar structure, which is useful for thermal barrier coatings. Moreover, the micro-cracks were mainly found between the dendrites and not in the main branches. Consequently, the microstructure is less prone to mechanical failure due to interconnected networks of cracks. This is because the thermal conduction coefficient of the steel was ten times higher than of the SiO_2 and the thermal gradient was more or less vertical in the coating. The effect of SiO_2 on the microstructure of the Fe-SiO_2 coating is demonstrated in the Fe-Si-O phase diagram [11]. This phase diagram represents a generalization of the equation with the result of current study [11].

2.2 Chemical Reaction of Silicanizing Process

The silicanizing process phenomenon is more pronounced for the Fe-containing alloys. The Fe diffusivity was very weak in SiO_2, where thermal SiO_2 oxide is a barrier for the diffusion of Fe into SiO_2. The result supported the possibility that SiO_2 can be formed through diffusion reaction between Fe and SiO_2. In the case of Fe-SiO_2, high temperature (1000 °C for 4 h) is considered as stable oxide phase. This could be attributed to two scenarios: (1) the high temperature Fe-SiO_2 phase decomposed during cooling into bcc and (2) Fe-SiO_2 did not form as a result of nucleation difficulties or increased surface energy. At the processing temperature and time, the pack power silicanizing and substrate react to form solid-state deposition.

The equilibrium is established in the pack and the substrate surface is driven by chemical potential gradients in the solid-state phase and surface reactions at the substrates to deposit the silicanized layer. This mechanism has strong dependence on pack powder silicanizing content. In this research, a large number SiO_2 were deposited and diffused layers of various composition deposited on Fe and silicanized at 1000 °C for 4 h. based on these results, the structural changes occurred as Fig. 3. The process is described as; (1) The fine-grain structure of as-deposited layers quickly re-crystalline and formed large grains of FeSi, Fe_2SiO_4 and $FeSiO_3$ through solid-state diffusion reaction; (2) Fe react with SiO_2 and nucleates grains of the FeSi, Fe_2SiO_4 and $FeSiO_3$ (silicates) along the silicanized layer interface. The orientation of the nucleated phase is random, considering the amorphous nature of SiO_2, the lack of structural relationship between body centered cubic (bcc) and silicates phases. The phase grains coalesce and a continuous layer formed separating the SiO_2 and thin coating (silicanized layer) and (3) The Fe-SiO_2 reaction released unbounded Si and O that diffused into unreacted silicanized thin layer.

Since solubility of oxygen in Fe is very low, precipitation of the phase transforming occurs throughout the thin layer, coarsening forms round particles of the oxide. Since migration of oxygen into substrate is much faster than migration of SiO_2, the rate-controlling step for growth Fe_2SiO_4, FeSi and $FeSiO_3$ could be the outward diffusion of SiO_2. A schematic diagram illustrating the formation of Fe_2SiO_4, FeSi and $FeSiO_3$ is shown in Fig. 3. The dashed line presents the interface between austenite and the oxide x^{os} and x^{so} are equilibrium silicon concentration in the oxide and mild steel respectively. x is the average SiO_2 concentration near the steel surface.

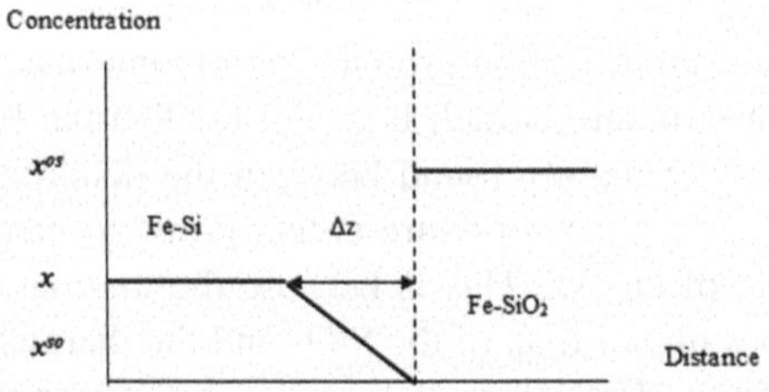

Fig. 3. The schematic diagram for the formation of Fe-SiO_2

3 Conclusion

The formation of TSS on mild steel is caused by silicanizing process (heat treatment) and produced thin coating layer (silicanized layer) due to high temperature and time. It was found that at high temperature, 1000 °C for 4 h, FeSi, $FeSiO_3$ and Fe_2SiO_4 including SiO_2 can be formed spontaneously on mild steel in furnace using fine particles TSS. The formation of silicanized layer of Fe-SiO_2 was predicted with respect to the thermodynamics supported with experiment data. To form Fe-SiO_2, SiO_2 needs to diffuse more comparing with the formation of FeSi, Fe_2SiO_4 and $FeSiO_3$. The prediction of thickness of the coating layer depending on process time and temperature is important for technological and industrial applications.

Acknowledgements. This work was supported by the Minister of Education's Fundamental Research Grant Scheme FRGS/2/2013/TK04/UTP/02/3. The authors would also like to acknowledge of assistance provided by Universiti Teknologi PETRONAS.

References

1. Broide, Z.S., Udovitsshii, V.I.: Prospects for the commercial use of coatings based on diffusion siliconizing. Metal Sci. Heat Treat. **23**(7), 489–491 (1981)
2. Othman, M., Yusnenti, F.M.Y., MohdYusri, I.: Siliconizing process of mild steel substrate by using Tronoh silica sand (TSS): an experimental investigation. Procedia CIRP **26**, 554–559 (2015)
3. Wang, Y., Wang, D., Yan, J.: Preparation and characterization of MoSi/MoB composite coating on Mo substrate. J. Alloys Compd. **589**, 384–388 (2014)
4. Yang, H., Li, Y., Tang, G., He, N., Zhang, Y.: Influence of duty cycle on composition and microstrucutre of siliconized layer using pulse electrodeposition. Adv. Mater. Res. **139–141**, 666–669 (2010)
5. Yusnenti, F.M.Y., Othman, M.: The mechanism and microstructure of silicanizing process on mild steel substrate in the medium of silica sand. ARPN J. Eng. Appl. Sci. **11**, 20 (2016). ISSN 1819-6608
6. Xiao, L.R., Cai, Z.G., Yi, D.Q., Li, Y., Lui, H., Huang, D.: Morphology, structure and formation mechanism of silicide coating by pack cementation process. Trans. Nonferrous Met. Soc. China **16**, 239–244 (2006)
7. Ugur, S., Azkan, O., Senol, Y., Saduman, S.: Kinetics of iron silicide deposited on AISI D2 steel by pack method. Metal **15**, 17.5 (2013)
8. Ren, Z., Hu, X., Xue, X.: Solid state reaction studies in Fe_3O_4-TiO_2 system by diffusion couple method. J. Alloys Compd. **580**, 182–186 (2013)
9. Yusnenti, F.M.Y., Othman, M., Mustapha, M., MohdYusri, I.: Silicanizing process on mild steel substrate by using Tronoh silica sand: microstructure, composition and coating growth. In: IOP Conference Series: Materials Science and Engineering, vol. 114, p. 012117 (2016)
10. Tahir, A., Othman, M.: Characterization and properties of aluminium-silica sand nanoparticles composites. Solid State Sci. Technol. **19**, 138–149 (2011). ISSN 0128-7389

Surface Roughness and Tool Wear in Edge Trimming of Carbon Fiber Reinforced Polymer (CFRP): Variation in Tool Geometrical Design

Syahrul Azwan Sundi[1,2](✉), R. Izamshah[2], M. S. Kasim[2], M. F. Jaafar[2], and M. H. Hassan[3]

[1] Faculty of Mechanical & Manufacturing Engineering Technology, Universiti Teknikal Malaysia Melaka (UTeM), Hang Tuah Jaya, 76100 Durian Tunggal, Melaka, Malaysia
syahrul.azwan@utem.edu.my
[2] Faculty of Manufacturing Engineering, Universiti Teknikal Malaysia Melaka (UTeM), Hang Tuah Jaya, 76100 Durian Tunggal Melaka, Malaysia
[3] School of Mechanical Engineering, Universiti Sains Malaysia, 14300 Nibong Tebal, Pulau Pinang, Malaysia

Abstract. Composite materials such as carbon fiber-reinforced plastics (CFRPs) for product structures have been steadily increasing due to the superior material properties such as high strength, low weight and corrosion resistance especially in the aerospace industry. This research initiated to investigate the effect of a various router or burrs tool geometrical feature towards surface roughness as well as tool wear during edge trimming process on a specific CFRP material. CFRP panel which measured in 3.25 mm thickness with 28 plies in total has been chosen to be the main study material. There were three various types of tool geometries made of un-coated carbide with the same diameter, 6.35 mm which vary in a number of flute and helix angle are successfully studied. Surface roughness measurement was taken using Mitutoyo Surftest SJ-410. Optical microscope, Nikon MM-800 is utilized to further observe the impact on the tool wear of the cutting edges. The result reveals that tool type 3 (T3) resulted in the minimized surface roughness with respect to the overall averaged Ra value whilst tool Type 1 (T1) achieved the highest surface roughness value. Meanwhile, the tool type 2 (T2) falls in the middle between the rests of two others type of tool. Observation of tool wear illustrated a clear fractured tool observed on the T1 compared to the T3 which the tooltip of the pyramidal cutting edge was still obviously intact.

Keywords: CFRP · Edge trimming · Surface roughness · Tool wear

M. N. Osman Zahid et al. (Eds.): iMEC-APCOMS 2019, LNME, pp. 402–408, 2020.
https://doi.org/10.1007/978-981-15-0950-6_61

1 Introduction

Carbon Fiber Reinforced Polymers (CFRP) are extensively used in today's aerospace industry due to their lightweight, good fatigues performance and high static strength. The behavior of composite such as its inhomogeneity and interaction with the cutting tool whilst machining is a complex phenomenon to be understood. Machining may possibly cause the quality of the machined composite part such as delamination, cracking, fiber pull-out, and burned matrices. The abrasive nature of the reinforcement fibers and the need to shear them neatly bring to the additional requirements and constraints on the selection of the best tool materials as well as geometry [1].

Prakash conducted an experiment to study the effect of various tool geometry. They discovered that trapezoidal geometry generated lower cutting force and moderate surface roughness with no delamination [2]. Haddad reported that burr tool seems to minimize defects on trimmed surface. Tool geometry and the cutting parameters which impacted the formation of the chip thickness were the two main factors discovered affecting the mass of harmful particles [3]. Gara found that the transverse roughness does not depend on cutting conditions but only on tool geometry. Contrary to the longitudinal roughness which was not only depending on the tool geometry but also the cutting conditions [4]. Sundi revealed that the tool geometrical feature especially the variation number of teeth or flute for router type tool may affecting the surface quality during edge trimming CFRP material [5]. Meanwhile, Sheikh-Ahmad et al. discovered that the wear of burr tools is best correlated with cutting temperature, delamination depth, and normal face [6].

This paper attempts to provide a more detailed investigations regarding the effects of various geometrical design mainly for router or burrs tool towards surface roughness and the impact on the tool wear in edge trimming process on a specific CFRP material.

2 Experimental Procedure

2.1 Material

The CFRP panel measured 3.25 mm in thickness and the type of fabric was unidirectional (UD). It has 28 number of plies in total which consist of 2 thin layer of glass/epoxy woven fabrics at the top and bottom of the CFRP laminate. The 26 unidirectional plies were made of carbon/epoxy prepreg. The stacking sequence was $[45/135/90_2/0/90/0/90/0/135/45_2/135]s$. The nominal fiber volume fraction is 60%. Table 1 illustrates the overall specification of CFRP material used in this work.

Table 1. CFRP details

Composite composition	No. of ply	Areal density	Fabric type	CPT/Ply
Carbon	26	203 g/m^3	Unidirectional	0.125
Glass	2	107 g/m^3	Woven	0.08
Total thickness (mm)				3.25

2.2 Cutting Tool Design

There were three types of router or burrs tool made of carbide (uncoated) and has diameter of 6.35 mm with different geometrical features have been chosen and investigated in this work (refer Fig. 1 and Table 2). Type 1 (T1) is defined as fine, Type 2 (T2) medium and Type 3 (T3) smooth. The machine used for to perform the edge trimming process was a Hass CNC Gantry Router – 3 Axis GR-510.

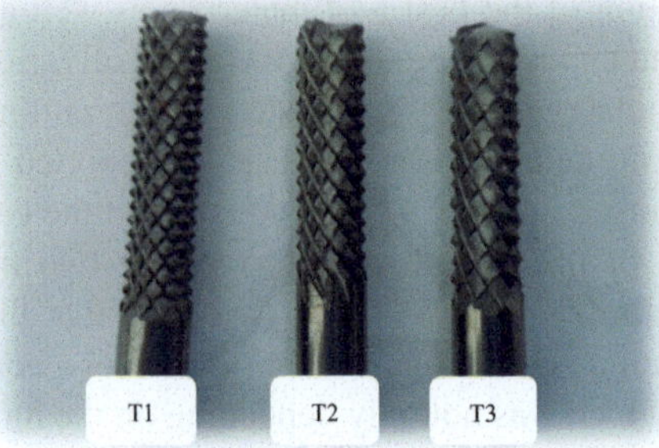

Fig. 1. Three different geometrical feature of router or burrs tool

Table 2. Router or burrs tool properties

	Diameter (mm)	Number of teeth	Number of helix		Angle of helix (°)		Length (mm)
			Right	Left	Right	Left	
T1	6.35	12	12	12	28	28	80.0
T2	6.35	10	11	10	26	26	75.0
T3	6.35	9	7	9	19	32	75.0

2.3 Machining Parameters

The range of cutting speed (V_c) (Table 3) applied for each type of chosen router tool was between 50 m/min to 150 m/min. Meanwhile, the cutting feed (f_z) was varied from 0.05 feed/rev to 0.15 feed/rev. Down-milling has been selected as the mode of machining configuration. In this work, the edge trimming process performed with 100% of tool diameter or step width (a_e) and the depth of cut (a_p) was taken in full thickness of the selected composite panel.

Table 3. Machining parameters

		Tool type	V_c (m/min)	Spindle speed, N (RPM)	Feed/Rev, F_z	Feed rate, V_f (mm/min)
T1	R1	Fine	50	2526	0.05	126
	R2	Fine	100	5053	0.1	505
	R3	Fine	150	7579	0.15	1137
T2	R1	Medium	50	2526	0.1	253
	R2	Medium	100	5053	0.15	758
	R3	Medium	150	7579	0.05	379
T3	R1	Smooth	50	2526	0.15	379
	R2	Smooth	100	5053	0.05	253
	R3	Smooth	150	7579	0.1	758

2.4 Surface Finish and Tool Wear Measurement

Surface finish of the workpiece was measured using surface roughness tester Surftest SJ-410 (Fig. 2). In this study, Ra (Arithmetical mean deviation) is used to measure the surface finish. Longitudinal surface roughness is evaluated in this work. There were 5 points of measurements taken on every machined surface and final average Ra is obtained to represent the result of surface finish on every specimen. On the other hand, Nikon MM-800 (Fig. 3) microscope is utilized to observe the tool wear for each type of tool geometrical design. The magnification range is 1x magnification to 100x magnification. Therefore, it helps on identifying the details during the tool wear observation.

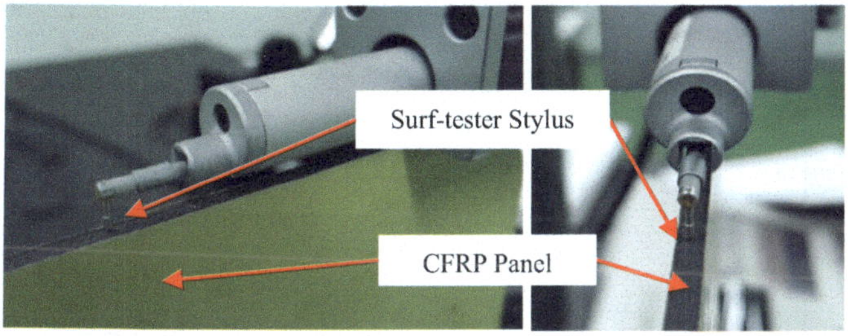

Fig. 2. Longitudinal surface roughness measurement by SJ-410 roughness tester

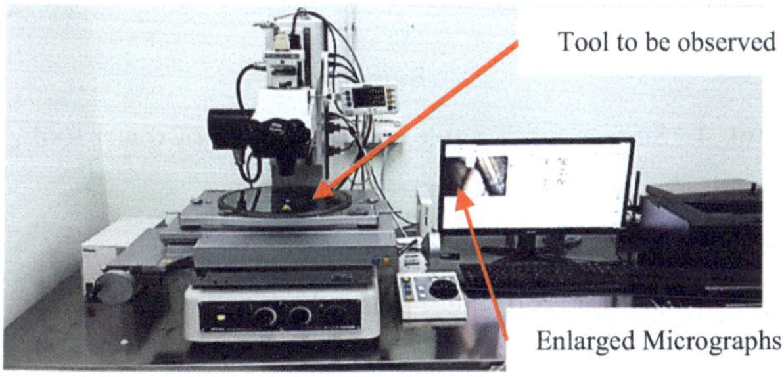

Fig. 3. Tool wear observation by Nikon MM-80 optical microscope

3 Result and Discussion

3.1 Surface Roughness Analysis

From Fig. 4, it is obviously seen that the Type 3 (T3) tool indicates the lowest averaged surface roughness value, Ra which ranged between 2.22 μm to 5.29 μm. Meanwhile, Type 1 (T1) tool shows the highest Ra values ranged between 2.86 μm to 19.36 μm. Type 2 (T2) tool falls in the middle between the rests of two others type of tool which stated the range of Ra value was between 3.91 μm to 5.28 μm. In general, T3 resulted the lowest surface roughness exhibited by the Run no. 1 (R1) which the machining parameter applied was cutting speed, V_c at 50 m/min and feed per tooth, f_z 0.15 mm/rev. In contrast, T1 has presented the highest surface roughness, Ra value indicated by the Run no. 3 (R3) which the machining parameter applied was cutting speed, V_c at 150 m/min and feed per tooth, f_z 0.15 feed/rev. An enormous difference of Ra value obtained by the R3 of T1 and T3 which is approximately (19.36 μm −3.78 μm = 15.58 μm or almost 400% or 4 times different) which provides an evidence that the difference in tool geometrical design impacted the trimmed surface quality. Both mentioned tools (T1 & T3) were having different geometrical design in details. T1 has more number of flutes compared to the T3 (refer Table 2). In addition, T1 has identical helix angle for both right and left flutes whilst T3 has different helix angle for both flutes.

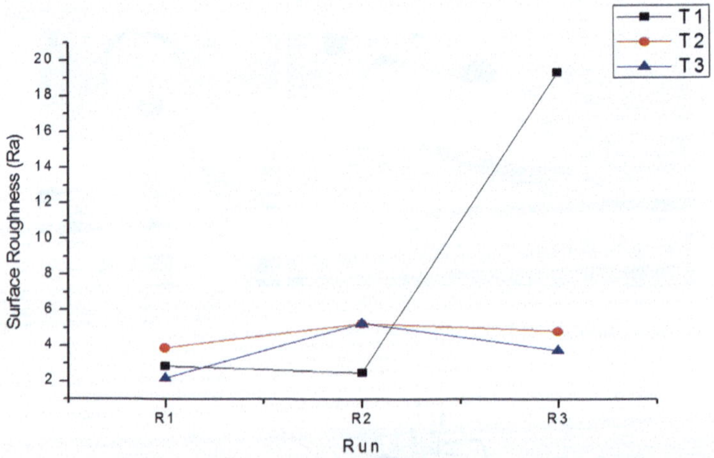

Fig. 4. Surface finish result for all the types of router tool

3.2 Tool Wear Observation

Figure 5 presents the overall and the enlarged micrographs for all selected tool geometries of router tool with the specified cutting condition applied. The main focus is given on the T1 and T3 at Run no. 3 (R3) which resulted the highest and the lowest surface roughness value. The cutting edge on the T1 exhibits severe worn off or also known as fractured tool. Meanwhile, the cutting edge on the T3 was still obviously intact. Hence, it could be summarized that the tool wear condition influenced the surface quality of the trimmed surface. Fractured tool condition on T1 produced worst surface finish whilst good cutting edge on T3 was the main reason led to better surface finish produced. This was believed due to the ability of the cutting edge to neatly cut the fibers which indirectly caused the formation of the chip thickness. Moreover, variation of feed rate applied may also impacted the tool wear generally. This finding is consistent with findings of past studies by Janardhan et al. which stated that the number of teeth fractured was influenced by the feed rate and cutting speed [7]. Therefore, an important conclusion could be drawn from the finding of this work that variation in the tool geometry design of router or burr tool may affecting the result of surface quality in edge trimming of a specific CFRP material.

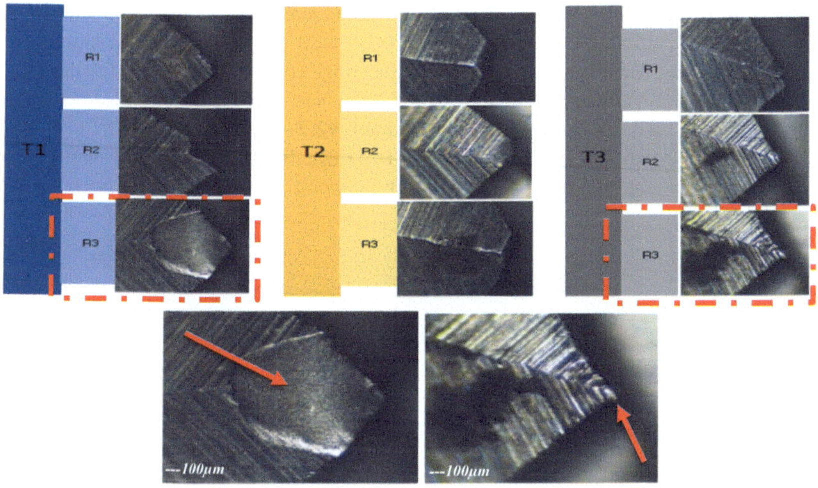

Fig. 5. Fractured tool observed of T1 (R3) (bottom-left); Tool tip or pyramidal edge tooth still intact T3 (R3) (bottom-right).

4 Conclusion

This paper presented results of surface roughness analysis and observation of tool wear on edge trimming of CFRP with three different tool geometries. Tool T3 resulted in the minimized surface roughness with respect to the overall averaged Ra value whilst tool T1 achieved the highest surface roughness value. In addition, clear fractured condition observed on the T1 compared to the T3 which the pyramidal cutting edge was still obviously intact.

Acknowledgement. Authors are grateful to Universiti Teknikal Malaysia Melaka for serving a platform to perform this study and Ministry of Education (MOE) for their financial support. (FRGS/2018/FKP-AMC/F00378).

References

1. Sheikh-Ahmad, J.Y.: Machining of Polymer Composites. Springer, New York (2009)
2. Prakash, R., Krishnaraj, V., Zitoune, R., Sheikh-Ahmad, J.: High-speed edge trimming of CFRP and online monitoring of performance of router tools using acoustic emission. Materials (Basel) **9**(10), 798 (2016)
3. Haddad, M., Zitoune, R., Eyma, F., Castanie, B.: Study of the surface defects and dust generated during trimming of CFRP: Influence of tool geometry, machining parameters and cutting speed range. Compos. Part A Appl. Sci. Manuf. **66**, 142–154 (2014)
4. Gara, S., Tsoumarev, O.: Effect of tool geometry on surface roughness in slotting of CFRP. Int. J. Adv. Manuf. Technol. **86**(1–4), 451–461 (2016)
5. Sundi, S.A., Izamshah, R., Kasim, M.S., Mohd Amin, A.T., Kumaran, T.: Influence of router tool geometry on surface finish in edge trimming of multi-directional CFRP material. In: IOP Conference Series: Materials Science and Engineering, vol. 469, p. 012026 (2019)
6. Sheikh-Ahmad, J.Y., Dhuttargaon, M., Cheraghi, H.: New tool life criterion for delamination free milling of CFRP. Int. J. Adv. Manuf. Technol. **92**(5–8), 2131–2143 (2017)
7. Janardhan, P., Sheikh-Ahmad, J., Cheraghi, H.: Edge trimming of CFRP with diamond interlocking tools edge trimming of CFRP with diamond interlocking tools. SAE Int. 2006-01-31 (2006)

Surface Roughness and Cutting Forces During Edge Trimming of Multi-directional Carbon Fiber Reinforced Polymer (CFRP)

Syahrul Azwan Sundi[1,2](✉), R. Izamshah[2], M. S. Kasim[2], M. F. Jaafar[2], and M. H. Hassan[3]

[1] Faculty of Mechanical and Manufacturing Engineering Technology, Universiti Teknikal Malaysia Melaka (UTeM), Hang Tuah Jaya, 76100 Durian Tunggal, Melaka, Malaysia
syahrul.azwan@utem.edu.my
[2] Faculty of Manufacturing Engineering, Universiti Teknikal Malaysia Melaka (UTeM), Hang Tuah Jaya, 76100 Durian Tunggal, Melaka, Malaysia
[3] School of Mechanical Engineering, Universiti Sains Malaysia, 14300 Nibong Tebal, Pulau Pinang, Malaysia

Abstract. Carbon fiber reinforced polymers (CFRPs) have found wide-ranging applications in numerous industrial fields especially in aerospace as well as automotive industries due to their excellent mechanical properties such as high strength, low weight and corrosion. The aim of the present work is to investigate the effect of machining parameters on surface quality and cutting forces in edge trimming process for a specific CFRP material. There were two variation of machining parameters focused in this work namely spindle speed (N) and feed per tooth (f_z). The range of spindle speed applied was of 2506 rpm (low), 5012 rpm (moderate), and 7518 rpm (high) speed whilst for feed per tooth; 0.05, 0.1, and 0.15 mm/rev. The CFRP panel measured 3.25 mm in thickness and the type of fabric was unidirectional (UD) with 28 number of plies in total has been chosen to be the main study material. Router or burr tool geometry made of uncoated tungsten carbide with a diameter of 6.35 mm was used to perform the edge trimming process. Surface roughness measurement was taken using Mitutoyo Surftest SJ-410. Kistler Type 9257B dynamometer is attached during the edge trimming process to record the cutting forces. The result reveals that the smallest value of the surface roughness (1.62 μm) is obtained by Run 1 (R1) and the highest surface roughness value (12.62 μm) exhibited by the R9. The resultant force exhibits no clear trend on the cutting forces for all the machining parameters applied. However, it is strongly believed that the thermal effect especially the low temperature of glass transition, *Tg* of the polymer matrix was the main reason affecting the result of the resultant force. Details results elaborated and discussed further in this manuscript.

Keywords: CFRP · Edge trimming · Surface roughness · Cutting forces

M. N. Osman Zahid et al. (Eds.): iMEC-APCOMS 2019, LNME, pp. 409–415, 2020.
https://doi.org/10.1007/978-981-15-0950-6_62

1 Introduction

In manufacturing of aero-structural composite material, milling and drilling are critical for finishing trimmed edges of panels or making accurate holes. However, due to their anisotropic, heterogeneous and brittle behavior, these materials are hard to be machined by comparing to the metals or isotropic materials.

Surface roughness is one of the measurable criteria which has a huge impact on dimensional accuracy, performance of mechanical properties of the machined composite as well as on the overall production costs. In achieving the desired surface roughness of the machined surface, it is compulsory to understand the mechanisms of material removal and the kinetics of machining processes. Machined or trimmed surface was found to be smoother and the cracks reach one or two of fiber diameters into the composites when the CFRP is machined parallel to the fiber [1]. Ucar [2] discovered that the chips removed from CFRP are in discontinuous scrap or powder forms because CFRP is brittle and the removed chips are fractured instead of being elastically and plastically deformed. The larger the feed rate the larger the cutting force due to the chip thickness [2]. Duboust [3] proved the surface roughness increased with machining distance generally following a steady trend. They also concluded that diamond coated tool with multiple cutting teeth or also known as burrs tool was able to produce a good quality surface although at high feed rate in comparison with polycrystalline diamond (PCD) tool [3]. In more recent work by Sundi et al. summarized that spindle speed was found to be the main influential factor towards surface finish in edge trimming a specific CFRP material [4].

It appears from the aforementioned investigations that numerous researches have been conducted to study the effects of tool type and cutting conditions towards surface roughness, delamination and cutting temperature. To the best of authors' knowledge, only few works thoroughly described the relationship between machining parameters towards the trimmed surface finish and the cutting forces in edge trimming of CFRP material. Therefore, this was the motivation behind the present study.

2 Experimental Procedure

2.1 Material

The CFRP panel measured 3.25 mm in thickness and the type of fabric was unidirectional (UD). It has 28 number of plies in total which consist of 2 thin layer of glass/epoxy woven fabrics at the top and bottom of the CFRP laminate. The 26 unidirectional plies were made of carbon/epoxy prepreg. The stacking sequence was $[45/135/90_2/0/90/0/90/0/135/45_2/135]s$. The nominal fiber volume fraction is 60%. Table 1 illustrates the overall specification of CFRP material used in this work.

Table 1. CFRP details

Composite composition	No of ply	Areal density	Fabric type	CPT/ply
Carbon	26	203 g/m^3	Unidirectional	0.125
Glass	2	107 g/m^3	Woven	0.08
Total thickness (mm)				3.25

2.2 Cutting Tool Design

The type cutting tool used in this study was router or burr tool made of tungsten carbide (uncoated) and diameter 6.35 mm (Fig. 1). Both right and left helix angles were 30°. Number of tooth was 10 flutes for both right and left helix.

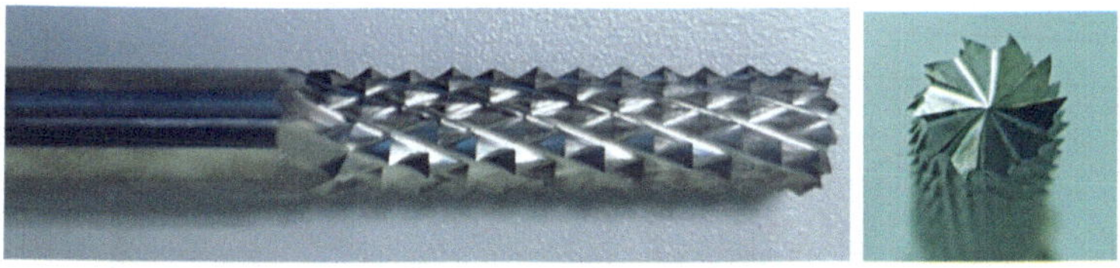

Fig. 1. Geometrical feature of router or burrs tool

The machine used for to perform the edge trimming process was a Hass CNC Gantry Router – 3 Axis GR-510.

2.3 Machining Parameters

There were two variation of machining parameters focused in this work namely spindle speed (N) and feed per tooth (f_z). The range of spindle speed applied was of 2506 rpm (low), 5012 rpm (moderate), and 7518 rpm (high) speed whilst for feed per tooth; 0.05, 0.1, and 0.15 mm/rev. Taguchi method (Orthogonal Array L9) one of the statistical techniques was adopted to plan the overall experiment. Table 2 represents the machining parameters applied in this work. Down milling has been selected as the mode of machining configuration. Total travel distance of each run was 260 mm.

2.4 Surface Finish and Cutting Forces

Surface roughness plays a major role in specifying surface finish of machined CFRP material in production. Machining processes can rapidly change the surface layer, which results in a change of mechanical properties of the composite. Cutting speed, feed rate, cutting force as well as tool geometry were the examples of factors affecting the surface quality. Surface roughness tester; Surftest SJ-410 (Fig. 2) is used to measure the surface finish of the workpiece. In this study, Ra (Arithmetical mean deviation) is referred to measure the surface roughness. There were 5 points of measurement taken on every machined surface and final average Ra is obtained to represent the result of surface finish on every specimen. Meanwhile, Kistler Type 9257B dynamometer is attached during the edge trimming process to record the cutting forces.

Table 2. Machining parameters

Run (R)	Cutting speed, V_c	RPM	Feed per tooth, F_z (mm/rev)	V_f mm/min
1	50	**2506**	0.05	**125**
2	100	**5012**	0.15	**752**
3	50	**2506**	0.1	**251**
4	100	**5012**	0.05	**251**
5	100	**5012**	0.1	**501**
6	150	**7518**	0.1	**752**
7	150	**7518**	0.05	**376**
8	50	**2506**	0.15	**376**
9	150	**7518**	0.15	**1128**

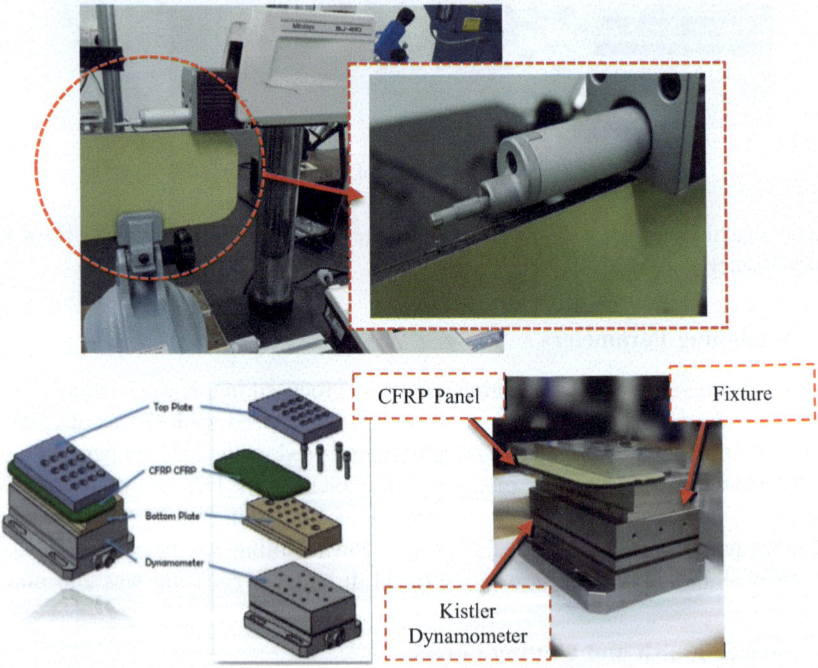

Fig. 2. Longitudinal surface roughness measurement by SJ-410 roughness tester (top). Fixture and Kistler Type 9257B dynamometer arrangement during edge trimming process (bottom)

3 Result and Discussion

3.1 Surface Roughness Analysis

Figure 3 presents the result of averaged longitudinal surface roughness, Ra values for Run 1 (R1) until Run 9 (R9). It is obviously seen that the smallest value (1.62 μm) of the surface roughness is obtained by Run 1 (R1) which the spindle speed applied was

2506 rpm and feed rate at 125 mm/min. Meanwhile, the highest surface roughness value (12.62 μm) exhibited by the R9 which has the highest setting of spindle speed, 7518 rpm and the highest feed rate, 1128 mm/min. Looking at the two sets of identical feed rate (R3&R4 = 251 mm/min, R7&R8 = 376 mm/min) it could be summarized that an increase in spindle speed improves the value of surface roughness. This finding is inconsistent with the work by M. Haddad et al. which revealed that the quality of the machined surface is mainly affected by the cutting speed followed by cutting distance [5].

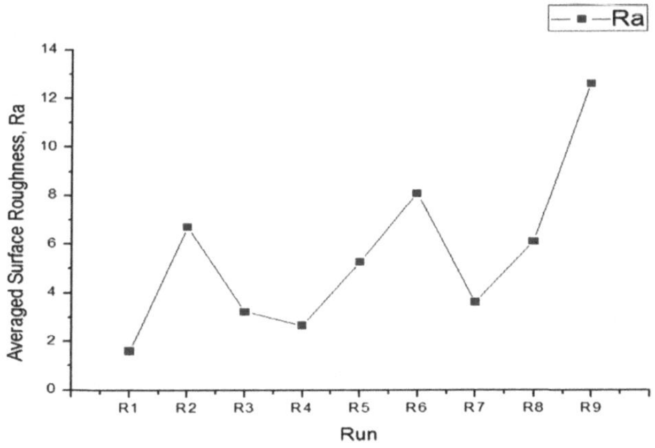

Fig. 3. Surface roughness result for all runs (R) (averaged Ra values)

3.2 Cutting Forces Analysis

The cutting force generated during the edge trimming process due to various cutting parameters was measured using the Kistler type 9257B dynamometer. The Eq. (1) is used for calculating the resultant cutting forces.

$$\mathrm{Fr} = \sqrt{(Fx)2 + (Fy)2 + (Fz)2} \tag{1}$$

Fr = Resultant Force
Fx = Force acting along x-axis
Fy = Force acting along y-axis
Fz = Force acting along z-axis

Figure 4 illustrates the result of resultant forces for Run 1 (R1) until Run 9 (R9). The smallest resultant force is clearly indicated by R9 (16.18 N) which has the highest setting of spindle speed, 7579 rpm and the highest feed rate, 1137 mm/min. Meanwhile, the highest resultant force is obtained by R3 (18.69 N) which the spindle speed applied was 2506 rpm and feed rate at 251 mm/min. Overall, the resultant force exhibits no clear trend on the cutting forces for all the machining parameters applied. These results were contradicted by the study carried out by Zám et al. whom considered, feed rate has the most significant effect on the cutting force, followed by the cutting width [6]. However, it is strongly believed that the smallest resultant force presented by R9 which had the highest spindle speed and feed rate was due to the thermal effect especially the low temperature of glass transition, *Tg* of the polymeric matrix. Heat evacuated through the workpiece remains 16% less regardless of cutter type, spindle speed, or feed. This is mainly attributed to the CFRP's poor thermal conductivity [7] The main concern was when this temperature exceeded, the mechanical properties of the overall composites will absolutely inconsistent [8]. Therefore, further investigation on the cutting temperature during machining or trimming process is highly recommended in the future work in order to gain better understanding on the relationship between cutting temperature and cutting forces.

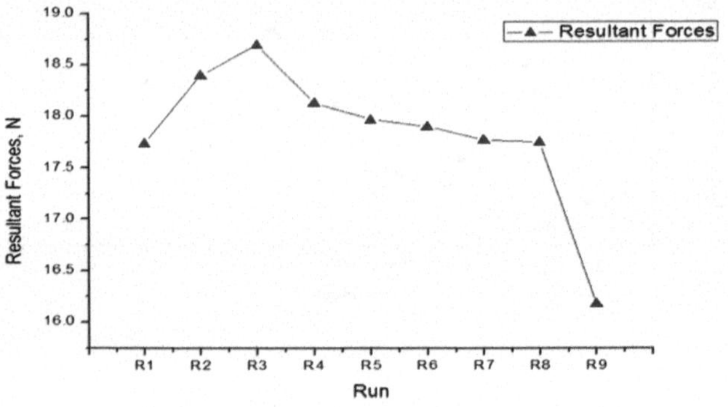

Fig. 4. Resultant forces result, N

4 Conclusion

This paper presented results of surface roughness analysis and cutting forces on the edge trimming of CFRP composite with burrs tool geometry. The smallest value of the surface roughness (1.62 μm) is obtained by Run 1 (R1). Besides, the highest surface roughness value (12.62 μm) exhibited by the R9. The resultant force exhibits no clear trend on the cutting forces for all the machining parameters applied. However, the thermal effect especially the low temperature of glass transition, *Tg* of the polymer matrix was believed the main reason affecting the result of the resultant force.

Acknowledgement. Authors are grateful to Universiti Teknikal Malaysia Melaka for serving a platform to perform this study and Ministry of Education (MOE) for their financial support. (FRGS/2018/FKP-AMC/F00378).

References

1. Koplev, A., Lystrup, A., Vorm, T.: The cutting process, chips, and cutting forces in machining CFRP. Composites **14**(4), 371–376 (1983)
2. Ucar, M., Wang, Y.: End-milling machinability of CFR laminated composite. J. Adv. Mater. **37**(4), 46–52 (2003)
3. Duboust, N., Melis, D., Pinna, C., Ghadbeigi, H., Collis, A., Ayvar-Soberanis, S., Kerrigan, K.: Machining of carbon fibre: optical surface damage characterisation and tool wear study. Procedia CIRP **45**, 71–74 (2016)
4. Sundi, S.A., Izamshah, R., Kasim, M.S., Abdullah, M.K.A.: Effect of machining parameters on surface quality during edge trimming of multi-directional CFRP material: Taguchi method. In: IOP Conference Series: Materials Science and Engineering, vol. 469, p. 012095 (2019)
5. Haddad, M., Zitoune, R., Eyma, F., Castanié, B.: Machinability and surface quality during high speed trimming of multi directional CFRP. Int. J. Mach. Mach. Mater. **13**(2/3), 289–310 (2013)
6. Balázs, S.M., Geier, N.: Machinability analysis of carbon fibre-reinforced plastics (CFRP) using compression tools. Manuf. Technol. **1**(3), 5–10 (2018)
7. Sheikh-Ahmad, J.Y., Almaskari, F., Hafeez, F.: Thermal aspects in machining CFRPs: effect of cutter type and cutting parameters. Int. J. Adv. Manuf. Technol. **100**(9–12), 2569–2582 (2018)
8. Schorník, V., Daňa, M., Zetková, I.: The influence of the cutting conditions on the machined surface quality when the CFRP is machined. Procedia Eng. **100**, 1270–1276 (2015)

A Brief Review on Utilization of Hybrid Nanofluid in Heat Exchangers: Theoretical and Experimental

Haziqatulhanis Ibrahim[1], Norazlianie Sazali[1,2](✉), Ahmad Shahir Jamaludin[1], Wan Norharyati Wan Salleh[3], and M. H. D. Othman[3]

[1] Faculty of Mechanical Engineering, Universiti Malaysia Pahang, 26600 Pekan, Pahang, Malaysia
azlianie@ump.edu.my
[2] Centre of Excellence for Advanced Research in Fluid Flow (CARIFF), Universiti Malaysia, Lebuhraya Tun Razak, 26300 Gambang, Kuantan, Pahang, Malaysia
[3] Advanced Membrane Technology Research Centre (AMTEC), School of Chemical and Energy, Faculty of Engineering, Universiti Teknologi Malaysia, 81310 Skudai, Johor Darul Takzim, Malaysia

Abstract. Heat exchanger is important for cooling and heating in industrial sectors. Current working fluid widely used for heat transfer in heat exchangers is water, which requires large space in a plant. Researchers has found an engineered, nanosized colloidal suspensions named nanofluids that have high potential in replacing water due to its superior thermal properties. Whilst nanofluid shows a promising heat transfer enhancement in heat exchanger, the dispersion of two different types of nanoparticles in a base fluid are expected to have a better efficiency of heat transfer. This paper compiled the studies done by various re- searchers in implementing hybrid nanofluids as working fluid in heat exchangers and its limitations.

Keywords: Nanotechnology · Hybrid nanofluid · Heat exchanger · Heat transfer

1 Introduction

In order to achieve maximum thermal efficiency of heat transfer, numerous studies were carried out concerning the enhancement of heat transfer [1–3]. With addition to current globalization activity, environment is experiencing temperature rises and thus, energy savings has become an important theme in researchers' scope of work. Most of industrial sector employed the concept of heat transfer in their facilities. For decades, heat exchangers have been used widely in industries such as petrochemical industry, food processing industry and manufacturing industry. Therefore, it is essential to optimize energy usage of the heat exchangers.

Heat exchanger is a device that allows exchange of heat between two fluids until thermal equilibrium is achieved without mixing the fluids. In a heat exchanger,

M. N. Osman Zahid et al. (Eds.): iMEC-APCOMS 2019, LNME, pp. 416–422, 2020.
https://doi.org/10.1007/978-981-15-0950-6_63

working fluids are used to transfer heat from or to applied fluids. Applied fluids in this context are fluids that needs to be cooled or to be heated depends on its functional usage. Conventional fluid used to absorb or transfer heat from applied fluids are water, which then results in drawback of needing large size of heat exchanger to accommodate higher heat transfer. After several attempts, a new type of working fluid called nanofluid was discovered in 1995 [4–6]. The suspension of small, nano-sized particles into working fluid has flourished attentions since then due to its superior thermophysical properties. To date, there are many literatures reporting on various types of nanofluids for enhancement of heat transfer in heat exchanger [5, 7, 8]. However, to author's knowledge, there are scarce to none that focuses on application of hybrid nanofluid in heat exchangers. This paper addressed the preparation method of hybrid nanofluids and studies done by past researchers in the recent 5 years.

2 Preparation Method of Hybrid Nanofluids

Mono nanofluid or only known as nanofluid is a single type of nanoparticles that were incorporated in a base fluid while hybrid nanofluid is mixture of two or more types of nanoparticles in different proportions, fused into base fluids at required volume concentration [9]. Two benefits of hybrid nanofluid compared to single type nanofluid are that they exhibit better thermal properties and have superior rheology [6]. There are two approaches established in preparing the hybrid nanofluids; one step method and two step method.

In single step approaches, synthesize of nanoparticles and dispersion of the particles in base fluids are conducted simultaneously. Corresponds to its name, nanofluids from this method is accomplished in only a single step. For two step method, nanoparticles will be processed separately and dispersed in a base fluid [10]. Figures 1 and 2 illustrates the flow of one step method and two step method, respectively. Depending on the suitability between nanoparticles and base fluids, most of researchers commonly used two step method in their studies as it produced large scale of nanoparticles at low cost [7, 10]. However, the usage of surfactant or dispersant are partially required in two step method to depress agglomeration due to forces between independent nanoparticles [10].

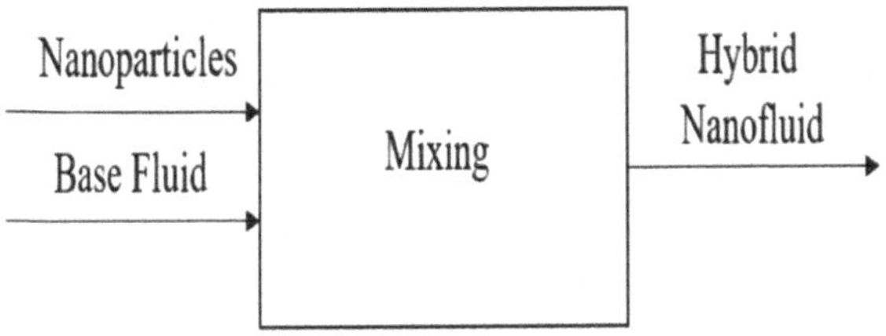

Fig. 1. Illustration for one step approach

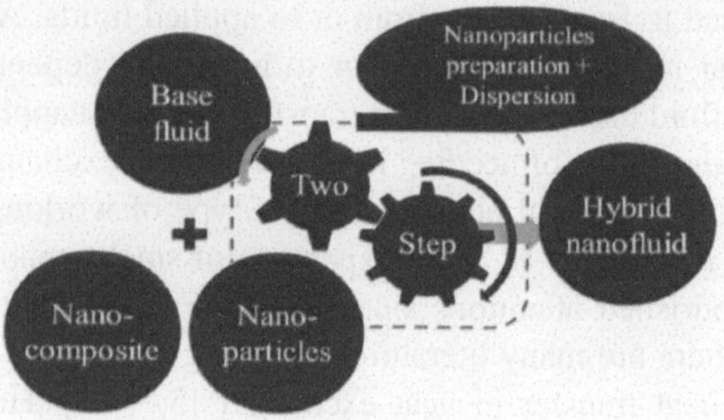

Fig. 2. Two step method illustration [6]

Surfactants helps to reduce agglomeration and increase the stability of the suspension. It can be classified into four categories; (1) Non-ionic surfactant, (2) Anionic surfactant, (3) Cationic surfactant and (4) Amphoteric surfactant, and is chosen following its compatibility [6].

3 Hybrid Nanofluids in Heat Exchangers

Following the demand in energy-efficient and economical heat transfer devices comes the idea of utilizing hybrid nanofluids in heat exchangers. Performance of nanofluids are affected by various parameters, in which thermos-physical properties (i.e. Thermal conductivity, viscosity, specific heat and density) of the hybrid nanofluids itself plays an important role. Other crucial parameters include temperature, particle concentration and particle size [6, 7]. Researchers conducted various studies in both numerical and experimental way in order to evaluate the performance of hybrid nanofluids when operated in heat exchangers. Few commercially available heat exchangers include plate heat exchanger (PHE) and shell and tube heat exchanger (STHE).

Allahyar *et al.* [11] in their experimental work investigated thermal performance of both hybrid alumina-silver nanocomposite dispersed in distilled water and mono alumina nanofluid employed in a coiled-type heat exchanger. The experiment was conducted at constant wall temperature and under laminar flow. They observed that heat transfer rate when operated using hybrid nanofluids of 0.4% volume concentration were 31.58% higher compared to when using distilled water as the working fluid. Moreover, they noted that increase in particles concentration led to an increment in heat transfer rate. On the other hand, Huang *et al.* [12] studies the behavior of hybrid water-based alumina with multi-walled carbon nanotubes in a corrugated plate heat exchanger (demonstrated in Fig. 3). From their findings, heat transfer coefficient for hybrid nanofluids are slightly higher than single alumina/water nanofluid and water itself with addition that it has smaller pressure drop, denoting its energy-efficiency. In plate heat exchanger, strong turbulence due to liquid flow inside the narrow-corrugated channels helps to enhance the heat transfer [13–15].

Recently, similar study on water-based alumina-multiwalled carbon nanotubes (MWCNT) nanofluids operated in plate heat exchanger was conducted by Bhattad

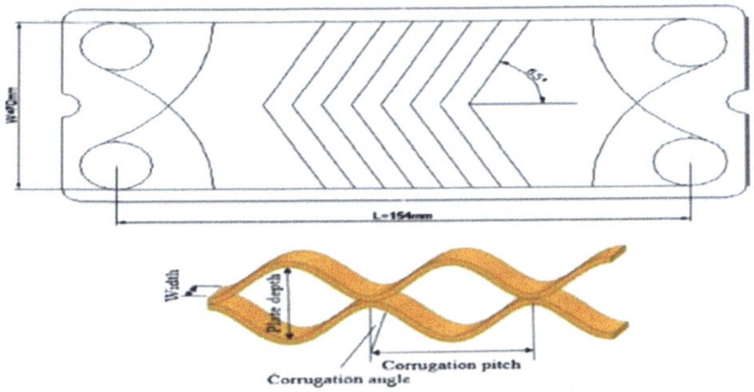

Fig. 3. Schematic plate heat exchanger used by Huang *et al.* [12].

et al. [16] using simulation CFD software. The results from their study was in conform with aforementioned study done by Huang *et al.* [12], where heat transfer coefficient increased when hybrid nanofluid was used. Three parameters; inlet temperature of nanofluid, flow rate and concentration of nanofluid that affects rate of heat transfer were evaluated in their literature. Aluminium nitride nanoparticles dispersed in ethylene glycol utilized in a double pipe heat exchanger was experimentally investigated by Hussein [17] under laminar flow. He suggested that heat transfer efficiency may augmented up to 160% when lower volume fractions of nanofluids is used. Furthermore, their results show that increment in flow rate decreases the friction factor, contradict with when volume concentration was increased. Therefore, low volume concentration is more favorable as it will produce low friction factor.

Arsan *et al.* [18] performed experimental work in shell and tube model heat exchanger using magnesium and aluminium nanoparticles dispersed in distilled water with different volume fraction ranging from 5 vol.% to 25 vol.%. They explained that by increasing nanoparticles volume incorporated in the base fluids, the performance of heat transfer in shell and tube heat exchanger increased. Next, effect on heat transfer and pressure drop in plate heat exchanger was investigated experimentally and numerically by Kumar and Tiwari [19]. Similar to previous paper, they also varied their volume concentration of nanofluids (Titanium dioxide and MWCNT in distilled water) from 0 vol.% to 1.50 vol.%. Numerical analysis was done via CFD software, employing the discrete phase model of plate heat exchanger. They found out that their experimental results were closely match with results from simulation except for cold water side pressure drop. However, they concluded that simulation study could help to predict the application of real scale heat exchanger, specifically the plate heat exchanger. Table 1 below addressed the findings from various researcher.

Table 1. Summary of past literatures

Study mode	Heat exchanger	Nanoparticles	Base fluid	Findings	Ref
Experimental	Coiled type heat ex-changer	Aluminium and silver	Distilled water	Hybrid nanofluid at 0.4 vol.% showed the best performance of heat transfer	[11]
	Plate heat ex-changer	Aluminium oxide and MWCNT	Distilled water	Hybrid nanofluid possessed highest heat transfer coefficient with low pressure drop	[12]
	Shell and Tube heat exchanger	Magnesium and Aluminium	Distilled water	Heat transfer rate enhanced when hybrid nanofluids volume fraction increased	[18]
Simulation	Plate heat ex-changer	Aluminium oxide and MWCNT	Distilled water	Discrete phase model gives better prediction than homogenous model and hybrid nanofluids showed excel- lent heat transfer performance	[16]
	Plate heat exchanger	Titanium dioxide and MWCNT	Distilled water	-Usage of hybrid nanofluid showed positive augmentation on heat transfer -Drawback in terms of fluid viscosity	[19]

4 Conclusion

In conclusion, in depth study need to be done in order to implement the usage of hybrid nanofluids in heat exchanger. This paper presented a mini review of recent studies on heat transfer enhancement in a heat exchanger. Based on various literatures, it was known that hybrid nanofluid does exhibits a superior performance for heat transfer but due to its complexity, it still has not been utilized in a real scale heat exchanger [20]. Type of heat exchanger, type of nanoparticles, volume concentration and size of nanoparticles need to be taken into consideration as it influences the performance of heat transfer in a heat exchanger.

Acknowledgement. The authors would gratefully acknowledge the financial support from the Ministry of Higher Education and Universiti Malaysia Pahang under the Fundamental Research Grant Scheme (Project Number: RDU1703253 and RDU191105). The authors would also like to show our gratitude to Associate Professor Dr Agus Saptoro from Curtin University Malaysia for his insight in this topic.

References

1. Sukarno, D.H.: Challenges for nanofluid applications in heat transfer technology. J. Phys: Conf. Ser. **795**(1), 12020 (2017)
2. El Bécaye Maïga, S., Tam Nguyen, C., Galanis, N., et al.: Heat transfer enhancement in turbulent tube flow using Al2O3 nanoparticle suspension. Int. J. Numer. Methods Heat Fluid Flow **16**(3), 275–292 (2006)
3. Kumar, P., Pandey, K.M.: Effect on heat transfer characteristics of nanofluids flowing under laminar and turbulent flow regime – a review. In: IOP Conference Series: Materials Science and Engineering vol. 225, no. 1, p. 12168 (2017)
4. Taherian, H., Alvarado, J.L., Languri, E.M.: Enhanced thermophysical properties of multiwalled carbon nanotubes based nanofluids. Part 1: critical review. Renew. Sustain. Energy Rev. **82**, 4326–4336 (2018)
5. Suganthi, K.S., Rajan, K.S.: Metal oxide nanofluids: review of formulation, thermo-physical properties, mechanisms, and heat transfer performance. Renew. Sustain. Energy Rev. **76**, 226–255 (2017)
6. Babar, H., Ali, H.M.: Towards hybrid nanofluids: preparation, thermophysical properties, applications, and challenges. J. Mol. Liq. **281**, 598–633 (2019)
7. Sajid, M.U., Ali, H.M.: Thermal conductivity of hybrid nanofluids: a critical review. Int. J. Heat Mass Transf. **126**, 211–234 (2018)
8. Bahiraei, M., Rahmani, R., Yaghoobi, A., Khodabandeh, E., Mashayekhi, R., Amani, M.: Recent research contributions concerning use of nanofluids in heat exchangers: a critical review. Appl. Therm. Eng. **133**, 137–159 (2018)
9. Phanindra, Y., Kumar, S.D., Pugazhendhi, S.: Experimental investigation Al2O3 & Cu/Oil hybrid nano fluid using concentric tube heat exchanger. In: Materials Today: Proceedings 5 (5, Part 2), pp. 12142–12150 (2018)
10. Ranga Babu, J.A., Kumar, K.K., Srinivasa Rao, S.: State-of-art review on hybrid nanofluids. Renew. Sustain. Energy Rev. **77**, 551–565 (2017)
11. Allahyar, H.R., Hormozi, F., ZareNezhad, B.: Experimental investigation on the thermal performance of a coiled heat exchanger using a new hybrid nanofluid. Exp. Therm. Fluid Sci. **76**, 324–329 (2016)
12. Huang, D., Wu, Z., Sunden, B.: Effects of hybrid nanofluid mixture in plate heat exchangers. Exp. Therm. Fluid Sci. **72**, 190–196 (2016)
13. Khan, T.S., Khan, M.S., Chyu, M.-C., Ayub, Z.H.: Experimental investigation of single-phase convective heat transfer coefficient in a corrugated plate heat exchanger for multiple plate configurations. Appl. Therm. Eng. **30**(8), 1058–1065 (2010)
14. Wajs, J., Mikielewicz, D.: Influence of metallic porous microlayer on pressure drop and heat transfer of stainless steel plate heat exchanger. Appl. Therm. Eng. **93**, 1337–1346 (2016)
15. Nilpueng, K., Wongwises, S.: Experimental study of single-phase heat transfer and pressure drop inside a plate heat exchanger with a rough surface. Exp. Therm. Fluid Sci. **68**, 268–275 (2015)
16. Bhattad, A., Sarkar, J., Ghosh, P.: Discrete phase numerical model and experimental study of hybrid nanofluid heat transfer and pressure drop in plate heat exchanger. Int. Commun. Heat Mass Transf. **91**, 262–273 (2018)

17. Hussein, A.M.: Thermal performance and thermal properties of hybrid nanofluid laminar flow in a double pipe heat exchanger. Exp. Therm. Fluid Sci. **88**, 37–45 (2017)
18. Arsan, R., Aravindh, M., Prabhu, N., Ramalingam, D.: Experimental investigation on thermal performance of shell and tube heat exchanger using hybrid nanofluid. Int. J. Adv. Res. Sci. Eng. **7**(2), 512–519 (2018)
19. Kumar, D., Tiwari, A.K.: CFD simulation of plate heat exchanger using hybrid nanofluid. Int. J. Mech. Eng. Technol. **9**(9), 1411–1418 (2018)
20. Sazali, N., Salleh, W.N.W., Izwanne, M.N., Harun, Z., Kadirgama, K.: Precursor selection for carbon membrane fabrication: a review. J. Appl. Membr. Sci. Tech. **22**(2), 131–144 (2018)

A Review on Effectiveness of Numerous Technologies by Utilizing Hydrogen

Mohd Syafiq Sharip[1], Norazlianie Sazali[1,2](✉), Haziqatulhanis Ibrahim[1], Ahmad Shahir Jamaludin[1], and Farhana Aziz[3]

[1] Faculty of Mechanical Engineering, Universiti Malaysia Pahang, 26600 Pekan, Pahang, Malaysia
azlianie@ump.edu.my

[2] Centre of Excellence for Advanced Research in Fluid Flow (CARIFF), Universiti Malaysia Pahang, Lebuhraya Tun Razak, 26300 Gambang, Kuantan, Pahang, Malaysia

[3] Advanced Membrane Technology Research Centre (AMTEC), School of Chemical and Energy, Faculty of Engineering, Universiti Teknologi Malaysia, 81310 Skudai, Johor Darul Takzim, Malaysia

Abstract. Numerous technologies for H_2-rich gaseous mixture production has been know through studies. However, most approaches do not satisfy the purity level for certain process such as fuel cell technology that demands a high-purity hydrogen supply (>99.99 vol.%). Therefore, purification technology to purify hydrogen is taken into consideration in order to meet purity requirement by certain applications, and it is a key factor to efficient hydrogen supply. In addition, CO_2 capture in pre-combustion of integrated gasification combined cycle (IGCC) power plants include the separation of H_2-CO_2 to generate electricity at a lower hydrogen purity. Advantage of hydrogen gaseous are its low volumetric energy density compared to other materials and the storage is essential for wider applications using hydrogen as well as hydrogen purification.

Keywords: Hydrogen purifications · Fuel cell technologies · Hydrogen storage · Selective membrane

1 Introduction

Abundance and longer lifetime of natural gas has initiated the idea of using it as primary feedstock in most petrochemical processing industries. Compared to other energy sources available, natural gas is the cleanest sources, generating low pollution than fuel. According to literature, it is expected that worldwide natural gas consumption will reach 182 trillion cubic feet in 2030, increasing from 95 trillion cubic feet in 2003 [1]. However, raw natural gas contains impurities such as methane (CH_4), and other hydro- carbons such as ethane (C_2H_6), propane (C_3H_8), butane (C_4H_{10}), CO2 coexists and other impurities such as hydrogen sulphide (H_2S), sulphur dioxide (SO_2) and nitrogen (N_2) are also present [2]. Therefore, it needs to undergo purification prior to the usage to prevent and minimized the fouling effect, corrosion, as well as pipeline

M. N. Osman Zahid et al. (Eds.): iMEC-APCOMS 2019, LNME, pp. 423–428, 2020.
https://doi.org/10.1007/978-981-15-0950-6_64

rupture [3]. In addition, calorific value and transportability of natural gas would increase when carbon dioxide is removed from the gas mixture [4]. This separation can be achieved by employing thin barriers called membranes to remove the unwanted components and the technique was proven to be economically and technically excellent by past researchers [5]. Currently, performance H_2-selective membranes have been widely studied for separation of gas mixture H_2-CO_2. Molecules of H_2 are known to possess smallest kinetic diameter compared to other types of gases molecules, thus, it has a large diffusion co- efficient which is suitable for permeation through membranes. Despite the reported achievements of hydrogen, gas separation via membrane technology still require lot of intensive investigations before being able to practice in a large-scale industrial application [6]. This is due to low quality hydrogen production from the reported H_2–CO_2 separation that unable to satisfy the requirement. To improve competitively of membrane-based separation in terms of technology and economy, few challenges that need to be addressed by both scientific and engineering community has been outlined below [7]:

I. Materials' ability to develop a perfect, crack-free thin layer on porous supports or a self-standing thin film with enough mechanical strength, is the most essential factor in order to reach successful practical application, but are rarely investigated [8]. Even though the chosen materials possess a high permeability, it is unable to be practiced in industrial sector if it cannot be easily synthesized into a thin-selective layer. Therefore, fabrication of thin membrane with large area need to be handled alongside the advancements in advanced membrane materials. It should be noted that pre-requisite for a membrane to be successfully applied commercially are the stability of its separation performance under different practical conditions (i.e. pressure, temperature and impurities).

II. Development in materials science since the past decades has also shows improvement in performance of conventional membrane materials via structural optimization. Introduction to several advanced materials for novel membrane fabrication include metal organic frameworks (MOFs), graphene-based materials, thermal rear-ranged (TR) polymers, polymers of intrinsic microporosity (PIMs), ionic liquids (ILs) and functionalized polymers, which are specialized for hydrogen purification and produced from thermochemical and biotechnological method. Membranes for H_2–CO_2 separation can be classified as H_2-selective membranes or CO_2-selective membrane depending on its favorable gas species permeations. Alternatively, the gas mixture can be separated according to type of membrane materials, specifically inorganics membranes; polymeric and mixed matrix membranes. The only well- known H_2-selective membrane is molecular sieving mechanism for microporous membranes which resulted in high-efficiency gas transport. To achieve molecular sieving effects, complete control of pore structures need to be achieved first [9]. It includes pore size, shape of pores and pore size distribution. MOFs shows all the aforementioned requirements and thus, became the potential candidate for molecular sieving H_2-selective membranes. It is predicted that investigations for new fabrication approaches can produced a highly molecular-sieving MOF membrane.

2 Fuel Cell Development

Fuel cells advancement as well as environmental strains for example climate change concern lead to increasing global attention of hydrogen as clean energy carrier of high quality in the recent years [10, 11]. Purification and separation technologies are crucial in thermochemical operations of transforming fossil fuels into hydrogen. Membrane reactors demonstrate exceptional potential in equilibrium shift involving water–gas shift reaction in forming hydrogen from carbon monoxide. Membranes are also significant in hydrogen purification. There are two classification of inorganic membranes for hydrogen purification and production which are porous ceramic as well as metal alloys and dense phase metal membranes. Hydrothermal technique and sol-gel are usually used in preparing the porous ceramic membranes with qualities such as high durability and stability in elevated temperature, severe contamination as well as hydrothermal surroundings [12]. Particularly, microporous membranes exhibit potentials in water gas shift reaction at elevated temperatures. Registry of membrane technology importance and relevance to recent creation of zero-emission power technologies is necessary in order to study the vital concerns involved with these membranes according to economic and technical benefits as well as shortcomings. Hydrogen economy concept can be defined as hydrogen utilization as the primary energy carrier. This concept is already well-known among the policy makers and futurists for several decades. Promising prospective brought by hydrogen is widely recognized for nearly two centuries. Hydrogen was used as fuel in the first combustion engine build by Isaac de Rivaz in 1805 [13] Fig. 1 illustrates hydrogen as fuel system [10].

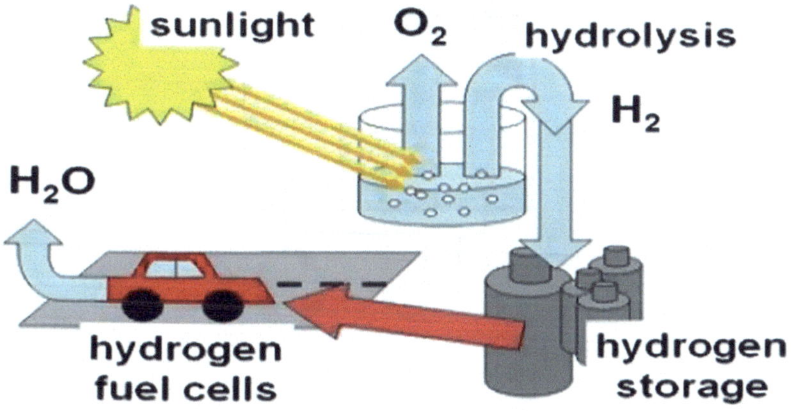

Fig. 1. H2 as fuel system [10].

Nevertheless, steam and petroleum preceded hydrogen in powering engines until now. Implications of shifting to hydrogen economy are seriously considered by various countries. Increasing attention gained by hydrogen is attributed to its potential in solving two main concerns faced by the global economies which are achieving independence in energy while reducing the effect of economic activity on the environment [11]. Realization of hydrogen economy needs to be preceded by the development of four crucial technologies which are:

1. Cost effective hydrogen manufacture in carbon restricted system of global energy.
2. This area comes with a few trials such as hydrogen production from fossil fuels with inclusion of carbon sequestration as well as growing employment of renewable sources.
3. Hydrogen storage and purification technologies with abilities to separate as well as refine hydrogen streams according to subsequent usage and storage systems essentials. US DOE target of 6.5 wt% needs to be achieved in developing practical and efficient hydrogen storage devices.
4. An effective, broadly accessible with sound management hydrogen distribution and delivery structures.
5. Effective fuel cells as well as other energy transformation technologies utilizing hydrogen (Fig. 2).

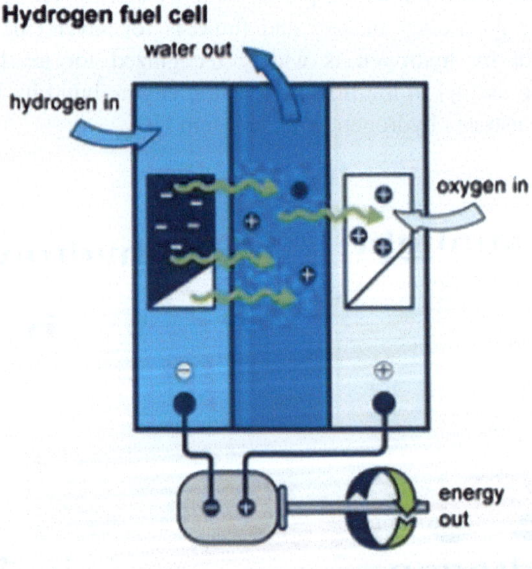

Fig. 2. H2 as fuel cell [13].

3 Generation of H_2 and Purification Needs

Separation and purification technology are very crucial for H_2 derived from fossil fuels. In the process, water-gas shift reaction occurs and carbon monoxide is converted to H_2 inside a membrane reactor. In this context, membrane reactor shows a great potential in shifting the equilibrium. It should be noted that membranes also play an important role in purification mechanism of H_2. Economical way to produce H_2 is by stream reforming, where steam reacted with hydrocarbons and nickel are added to accelerate the process [14, 15]. Example of competitive separating technology for H_2 from the streams is amine absorption, pressure swing adsorption (PSA), and membrane separation [16]. Among the list, gas separation using membrane system are more cost-effective than PSA in comparison to the relative capital investment and unit recovery cost [17]. Selective removal of H_2 from the reaction system shifted the reaction equilibrium to the products side, resulting in higher conversion of CH_4 to H_2 and carbon dioxide, and it can be obtained at lower temperatures. A simulation study which employed Pd-based disk membranc with 100 μm of thickness proved the enhancement of steam reforming performance in a real membrane catalytic system [18, 19]. From their study, the production of H_2 was increased but at high temperature range of 700 or 800 °C. However, available commercial membrane for Pd-based membrane are too thick and does not work effectively at temperature suggested. Few properties of a good membrane reactors include high separation selectivity, high gas permeance, and the durability and stability of the membrane itself. There are two inorganic membrane classes for production and purification of H_2 which are dense phase metal, metal alloys and ceramics, and porous ceramic membranes [20].

4 Conclusion

Currently hydrogen-based technology is used in various gas separation processes such as H_2 recovery, H_2 purification, as well as H_2 storage. Although H_2 have been extensively applied for gas separation process and fuel cell development, these technologies are less competitive as compared to the conventional processes concerning to their performance and the durability. Few challenges that need to be addressed has been outlined in order to achieve a high-quality gas separation performance. Purification technology is essential, and it is a key factor to efficient hydrogen supply.

Acknowledgement. The authors would also gratefully acknowledge the financial support from the Ministry of Education and Universiti Malaysia Pahang under the Fundamental Research Grant Scheme (Project Number: RDU1703253, RDU191105, and RDU190354).

References

1. Yeo, Z.Y., Chew, T.L., Zhu, P.W., Mohamed, A.R., Chai, S.P.: Conventional processes and membrane technology for carbon dioxide removal from natural gas: a review. J. Nat. Gas Chem. **21**, 282–298 (2012)
2. Scholes, C.A., Stevens, G.W., Kentish, S.E.: Membrane gas separation applications in natural gas processing. Fuel **96**, 15–28 (2012)
3. Chua, M.L., Xiao, Y.C., Chung, T.S.: Modifying the molecular structure and gas separation performance of thermally labile polyimide-based membranes for enhanced natural gas purification. Chem. Eng. Sci. **104**, 1056–1064 (2013)

4. Ahmad, A.L., Adewole, J.K., Leo, C.P., Ismail, S., Sultan, A.S., Olatunji, S.O.: Prediction of plasticization pressure of polymeric membranes for CO2 removal from natural gas. J. Membr. Sci. **480**, 39–46 (2015)
5. Adewole, J.K., Ahmad, A.L., Ismail, S., Leo, C.P.: Current challenges in membrane separation of CO2 from natural gas: a review. Int. J. Greenh. Gas Control **17**, 46–65 (2013)
6. Lu, G.Q., Diniz da Costa, J.C., Duke, M., Giessler, S., Socolow, R., Williams, R.H., et al.: Inorganic membranes for hydrogen production and purification: a critical review and perspective. J. Colloid Interface Sci. **314**(2), 589–603 (2007)
7. Sazali, N., Norharyati, W., Md. Nordin, N.A.H., Mohamed, M., Ismail, A., Yusof, N., et al.: Influence of Carbonisation temperature on gas permeation properties of matrimid-based carbon membrane (2017)
8. Sazali, N., Salleh, W.N.W., Ismail, A.F.: Carbon tubular membranes from nano-crystalline cellulose blended with P84 co-polyimide for H2 and He separation. Int. J. Hydrog. Energy **42**(15), 9952–9957 (2017)
9. Sazali, N., Salleh, W.N.W., Izwanne, M.N., Harun, Z., Kadirgama, K.: Precursor selection for carbon membrane fabrication: a review (2018)
10. Hames, Y., Kaya, K., Baltacioglu, E., Turksoy, A.: Analysis of the control strategies for fuel saving in the hydrogen fuel cell vehicles. Int. J. Hydrog. Energy **43**(23), 10810–10821 (2018)
11. Ma, C., Yu, J., Wang, B., Song, Z., Xiang, J., Hu, S., et al.: Chemical recycling of brominated flame retarded plastics from e-waste for clean fuels production: a re-view. Renew. Sustain. Energy Rev. **61**, 433–450 (2016)
12. Wei, Q., Wang, F., Nie, Z.-R., Song, C.-L., Wang, Y.-L., Li, Q.-Y.: Highly hydrothermally stable microporous silica membranes for hydrogen separation. J. Phys. Chem. B **112**(31), 9354–9359 (2008)
13. Schemme, S., Breuer, J.L., Samsun, R.C., Peters, R., Stolten, D.: Promising catalytic synthesis pathways towards higher alcohols as suitable transport fuels based on H2 and CO2. J. CO2 Util. **27**, 223–237 (2018)
14. Acar, C., Dincer, I.: The potential role of hydrogen as a sustainable transportation fuel to combat global warming. Int. J. Hydrog. Energy (2018, in press)
15. Ismail, N.H., Salleh, W.N.W., Sazali, N., Ismail, A.F.: Effect of intermediate layer on gas separation performance of disk supported carbon membrane. Sep. Sci. Technol. **52**(13), 2137–2149 (2017)
16. Ismail, N.H., Salleh, W.N.W., Sazali, N., Ismail, A.F., Yusof, N., Aziz, F.: Disk supported carbon membrane via spray coating method: effect of carbonization temperature and atmosphere. Sep. Purif. Technol. **195**, 295–304 (2018)
17. Tseng, H.-H., Wang, C.-T., Zhuang, G.-L., Uchytil, P., Reznickova, J., Setnickova, K.: Enhanced H2/CH4 and H2/CO2 separation by carbon molecular sieve membrane coated on titania modified alumina support: effects of TiO2 intermediate layer preparation variables on interfacial adhesion. J. Membr. Sci. **510**(Suppl. C), 391–404 (2016)
18. Sridhar, S., Smitha, B., Shaik, A.: Pervaporation-based separation of methanol/MTBE mixtures—a review. Sep. Purif. Rev. **34**(1), 1–33 (2005)
19. Lu, H., Zhu, L., Wang, W., Yang, W., Tong, J.: Pd and Pd–Ni alloy composite membranes fabricated by electroless plating method on capillary α-Al2O3 substrates. Int. J. Hydrog. Energy **40**(8), 3548–3556 (2015)
20. Wu, W., Yang, Q., Su, B.: Centimeter-scale continuous silica isoporous membranes for molecular sieving. J. Membr. Sci. **558**, 86–93 (2018)

Phosphorus/Nitrogen Grafted Lignin as a Biobased Flame Retardant for Unsaturated Polyester Resin

Salman Farishi, Annisa Rifathin(✉), and Benni F. Ramadhoni

Center for Polymer Technology, Agency for Assessment and Application of Technology, South Tangerang, Indonesia
annisa.rifathin@bppt.go.id

Abstract. Nowadays, polymers are widely used for various applications but have some disadvantages such as poor fire retardancy and thermal stability. However, some types of flame retardants for polymer have negative impacts on health and environment. Lignin, the second most highly abundant material from biomass, is one of the potential flame retardant for polymer. In this study, biobased flame retardant for unsaturated polyester (UP) was successfully fabricated from lignin using safer material. Lignin was functionalized using polyethylene imine and diphosphorous pentoxide to improve its fire retardancy and thermal stability. The chemical structure of functionalized lignin was characterized using FTIR and the results show that phosphate and nitrogen groups were successfully grafted on lignin. Based on UL 94 test results, we concluded that functionalized lignin using polyethylene imine and diphosphorous pentoxide can reduce burning rate of UP by 31.98% while the neat one reduced only as high as 22.87%. The TGA results also show that the functionalized lignin increases thermal stability of UP higher than the unfunctionalized one.

Keywords: Lignin · Flame retardant · Unsaturated polyester

1 Introduction

The combustibility of polymer is a challenge. This weakness is improved by adding flame retardant [1, 2]. Widely used flame retardant for polymers include brominated flame retardant (BFR). Unfortunately, some types of BFR such as polybrominated diphenyl ether (PBDE) which is a persistent organic pollutants (POPs), are known to be toxic [3, 4]. Other types of BFR are also reported to be potentially toxic in living things and can pollute the environment [5]. Therefore, innovation is needed to find alternative flame-retardant materials that are safe and environmentally benign as substitute for BFR.

Lignin is a second most abundant biopolymers after cellulose. It has been reported that more than 50 million tons of lignin are produced globally in a year as a by-product of pulp and paper industry. But less than 2% are used as chemical products and the leftovers are considered as waste [6, 7]. Therefore, lignin is a promising material, thanks to its abundancy as well as chemical modifiable functional groups to produce certain properties [8, 9].

Some studies suggest that the addition of nitrogen and phosphate groups to lignin can improve flame retardancy of lignin by forming char and inert gases. Nitrogen-

M. N. Osman Zahid et al. (Eds.): iMEC-APCOMS 2019, LNME, pp. 429–434, 2020.
https://doi.org/10.1007/978-981-15-0950-6_65

composed flame retardant will form inert gases (ammonia, nitrogen) that slow down the combustion. Furthermore, phosphate assists the formation of char that influences flame retardancy through slowing down the heat transfer, preventing the entry of oxygen, and polymer degradation [10, 11]. Some groups reported successful functionalization of lignin using polyethylene imine (PEI) and diethyl phosphite [11]. Unfortunately, the later chemical is listed as a hazardous material [12].

At present, there is limited information about efficient and safer route to functionalized lignin to be used as flame retardant. Therefore, in this study, lignin is functionalized using safer materials i.e. polyethylene imine (PEI) and diphosphorous pentoxide (P_2O_5) using modified method of Liu et al. [11]. The resulted material was used as flame retardant in unsaturated polyester.

2 Experimental Methods

2.1 Materials

Alkalinated lignin (Tokyo Chemical Industry, Japan Ltd) was used as precursor. Polyethylene imine (PEI) (Tokyo Chemical Industry, Japan Ltd) and diphosphorus pentoxide (P_2O_5) (Merck, Germany Ltd) were used as nitrogen and phosphorous sources respectively. Formaldehyde and tetrahydrofuran (THF) (Merck, Germany Ltd), were used as reaction media. Sodium hydroxide (NaOH) (Merck, Germany Ltd) was used as pH adjuster. Hydrochloric acid (HCl) (Merck, Germany Ltd) were used as pH adjuster and precipitator. All reagents used were analytical grade.

Unsaturated polyester resin, Yucalac 157 BQTN (Justus Kimiaraya, ID Ltd), was used as base polymer while methyl ethyl ketone peroxide, Mepoxe (Justus Kimiaraya, ID Ltd), was used as catalyst.

2.2 Methods

Modified method of Liu et al. was used to synthesis PN-lignin [11]. The synthesis was divided into 2 steps, i.e. functionalization reaction with nitrogen using PEI and functionalization with phosphate using P_2O_5 consecutively.

N-Lignin: At first, 42 g of lignin and 10 g each of formaldehyde and PEI were mixed for 5 h in 330 ml distilled water at 50 °C and mixing speed 80 rpm. 20% NaOH solution was added to adjust pH to 10. Following 5 h mixing, HCl solution was added into the mixture to adjust pH to 3 or 4 and to precipitate the supernatant. Supernatant was washed and neutralized using distilled water. The resulted powder, namely N-Lignin, was dried overnight in an oven at 80 °C. 37.2 g of N-lignin were collected.

PN-Lignin: Secondly, 36 g of N-lignin powder were dissolved in 168 ml of THF and stirred. 12.4 g of P_2O_5 were added to the solution [13]. Reaction was kept for 7 h at 70 °C and mixing speed 80 rpm. As the reaction ended, the solvent was evaporated using vacuum oven and the resulted powder was washed with distilled water several times until pH 7 was reached. The powder was dried overnight in an oven at 80 °C. Resulted powder was labelled as PN-Lignin.

Characterization of Functionalized Lignin: A FTIR spectrometer Tensor 27 with Hyperion 2000 was used to investigate the chemical composition of PN-lignin. TGA was performed with TGA/SDTA 851 (Mettler Toledo). 20 mg of sample was heated from 50 °C to 900 °C at heating rate of 10°C/min under nitrogen atmosphere. Peak temperature of mass loss (T_{onset}) and mass of resulted char were determined.

Composite Preparation: The composite specimen was prepared by mixing 12.5% of flame retardant with unsaturated polyester resin using mechanical stirrer. 1% of catalyst was added to accelerate reaction.

Characterization of Composite Samples: Flammability test was conducted according to UL-94 standard method. All samples with dimension 125 mm × 14 mm × 5 mm were tested in vertical setting and exposed to flame for 10 s. The burning rate was determined using Eq. 1.

$$Burning\ rate\ \left(mm^3/second\right) = \frac{volume}{time} \tag{1}$$

Thermal stability of samples was analysed using TGA/SDTA 851 (Mettler Toledo). All samples were heated from 50 °C to 900 °C at heating rate of 10 °C/min under nitrogen atmosphere. Onset temperature of decomposition and mass of resulted char residue were determined.

3 Results and Discussion

3.1 Characterization of Functionalized Lignin

Figure 1 shows the FTIR spectra of neat lignin, N-lignin, and PN-lignin. For neat lignin, a broad absorption band at around 3400 cm^{-1} is attributed to the stretching vibration of aromatic and aliphatic OH groups.

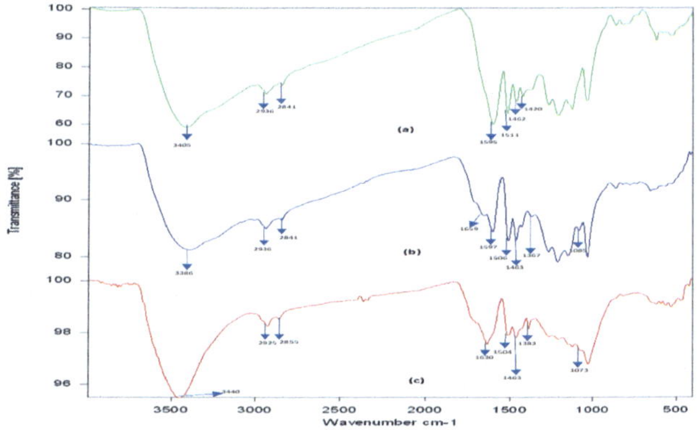

Fig. 1. FTIR spectra of (a) Neat Lignin, (b) N-Lignin, (c) PN-Lignin

The peaks around 2900 cm^{-1} are assigned to C-H stretching vibration. For neat lignin, a broad absorption band at around 3400 cm^{-1} is attributed to the stretching vibration of aromatic and aliphatic OH groups, and the peaks around 2900 cm^{-1} are assigned to C-H stretching vibration. The absorption band at 1595–1420 cm^{-1} belongs to the stretching vibration of aromatic rings of lignin. After functionalization by grafting nitrogen (N-Lignin), several new absorption peaks of nitrogen group formed, which are 1659 cm^{-1} which belongs to -NH; 1367 cm^{-1} of aromatic amine; and 1085 cm^{-1} of C-N. This suggested that PEI chain has been grafted on lignin. As a PN-lignin, the additional peaks of 1630 cm^{-1} (P = O) and 1073 cm^{-1} (P-O-C and PN) are formed which indicate the presence of a phosphate group [11, 14–16]. In addition, there was reduction in the intensity of -OH group (3400 cm^{-1}) of functionalized lignin compared to neat lignin. This indicated the conversion of the free hydroxyl groups in lignin to phosphate nitrogen group during reaction [16]. Therefore, the FTIR results demonstrated that phosphate and nitrogen were successfully grafted in lignin.

3.2 TGA Analysis of Functionalized Lignin

The TGA analysis was performed to obtain residue and thermal stability of functionalized lignin. As shown in Fig. 2a, PN-lignin had a lower initial decomposition temperature, which is 170 °C, than neat one at 277 °C. As reported in previous study, the lower initial decomposition temperature of functionalized lignin, the earlier char produced that can effectively protect the materials from heat and flame. Previous studies also reported that the addition of a phosphate group to lignin will increase the amount of char formed [11, 16]. However, this study found that the number of chars formed in PN-lignin almost similar as the neat one. This is probably due to the addition of PN groups to lignin is not yet optimum.

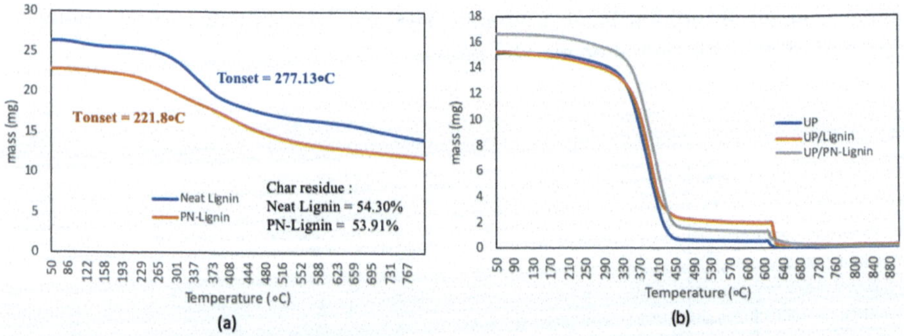

Fig. 2. TGA Thermogram (a) Effect of functionalization lignin on thermal properties, (b) Effect of neat lignin and functionalized lignin on decomposition temperature of unsaturated polyester resin

3.3 Flame Retardancy and Thermal Stability Properties

Composite specimens of unsaturated polyester/lignin (UP/lignin) and unsaturated polyester/PN-lignin (UP/PN-lignin) were successfully manufactured. The flammability of samples was tested using UL-94, with the burning rate of each samples reported in Table 1. Burning rate of UP was 64 mm^3/second, while UP/neat lignin and UP/PN-lignin were 49 mm^3/second, and 44 mm^3/second, respectively. In general, the addition of neat and functionalized lignin decreased the burning rate of unsaturated polyester resin. The addition of neat lignin reduced burning rate of UP by as low as 22.87%, while grafted nitrogen/phosphate lignin (PN-lignin) succeeded to reduce burning rate of UP by as much as 31.98%. Although neat and PN-lignin had similar char residue, the fire retardancy of the later was higher than of the former. This is probably caused by the effect of more rapid char produced by PN-lignin compared with neat one, as described above. Therefore, optimization grafting PN on lignin will improve its flame retardancy properties.

Table 1. Flammability and thermal stability properties of UP/Lignin samples

Sample	Burning rate (mm^3/second)	Tonset (°C)
UP	64.18	326
UP/Neat Lignin	49.50	330
UP/PN Lignin	43.66	335

Figure 2b and Table 1 showed the thermal decomposition behaviour of UP, UP/lignin, and UP/PN-lignin composites in nitrogen environment. The TGA results show that UP started to decompose at 326 °C while UP/lignin was at 330 °C. A slightly increased about 4 °C compared with UP. The highest increment was obtained by UP/PN lignin (335 °C), which 9 °C higher than UP. This indicates the addition of lignin and PN-lignin can improved the thermal stability of UP.

4 Conclusion

Functionalized lignin (PN-lignin) was successfully fabricated using Polyethylene imine and Diphosphorus pentoxide to raise flame retardancy and thermal stability of Unsaturated polyester resin. Compared to neat lignin, the addition of the same amount of PN-lignin to UP further enhances its flame retardancy properties. The TGA results show that PN-Lignin can increased thermal stability of UP much higher than neat lignin. However, the results of study show that the amount of char residue from neat and PN-lignin is not significantly different.

References

1. Sorathia, U., Rollhauser, C.M., Hughes, W.A.: Improved fire safety of composites for naval applications. Fire Mater. **16**(3), 119–125 (1992)
2. Sorathia, U., et al.: Screening tests for fire safety of composites for marine applications. Fire Mater. **25**(6), 215–222 (2001)
3. Stapleton, H.M., Allen, J.G., Kelly, S.M., et al.: Alternate and new brominated flame retardants detected in U.S. house dust. Environ. Sci. Technol. **42**, 6910–6916 (2008)
4. Mackintosh, S.A., Wallace, J.S., Gross, M.S., Navarro, D.D., et al.: Review on the occurrence and profiles of polybrominated diphenyl ethers in the Philippines. Environ. Int. **85**, 314–326 (2015)
5. Alaee, M., Arias, P., Sjodin, A., Bergman, A.: An overview of commercially used brominated flame retardants, their applications, their use patterns in different countries/regions and possible modes of release. Environ. Int. **29**(6), 683–689 (2010)
6. Gosselink, R.J.A., de Jong, E., Guran, B., Abächerli, A.: Co-ordination network for lignin—standardisation, production and applications adapted to market requirements. Ind. Crops Prod. **20**(2), 121–129 (2004)
7. Laurichesse, S., Avérous, L.: Chemical modification of lignins: towards biobased polymers. Prog. Polym. Sci. **39**(7), 1266–1290 (2014)
8. Alzagameem, A., Khaldi-Hansen, B.E., Buchner, D., Larkins, M., et al.: Lignocellulosic biomass as source for lignin-based environmentally benign antioxidants. Molecules **23**, 2664 (2018)
9. Hobbs, C.E.: Recent advances in biobased flame retardant additives for synthetic material. Polymers **11**, 224 (2019)
10. Yu, Y., Fu, S., Song, P., Luo, X., Jin, Y., Lu, Y., Wu, Q.: Functionalized lignin by grafting phosphorus-nitrogen improves the thermal stability and flame retardancy of polypropylene. Polym. Degrad. Stab. **97**, 541–546 (2012)
11. Liu, L., Qian, M., Song, P., Huang, G., Yu, Y., Fu, S.: Fabrication of Green lignin based flame retardants for enhancing the thermal and fire retardancy properties of polypropilene/wood composites. ACS Sustain. Chem. Eng. **4**(4), 2422–2431 (2016)
12. Mathews, R.J.: A comparison of the australia group list of chemical weapon precursors and the CWC schedules of chemicals. Chem. Weapons Conv. Bull. **21**, 1–3 (1993)
13. Mendis, G.P., Weiss, S.G., Korey, M., et al.: Phosphorylated lignin as a halogen-free flame retardant additive for epoxy composites. Green Mater. **4**(4), 150–159 (2016)
14. Wang, S., Zhang, Y., Chen, H., Dai, X., Zou, Z., Si, N.C.: Ultrasonic-assisted synthesis of aminated lignin by a Mannich reaction and its decolorizing properties for anionic azo-dyes. RSC Adv. **4**, 28156–28164 (2014)
15. Yue, X., Chen, F., Zou, X.: Improved interfacial bonding of PVC/Wood flour composites by lignin amine modification. BioResources **6**(2), 2022–2034 (2011)
16. Yu, Y., Fu, S., Song, P., Luo, X., Jin, Y., Lu, F., Wu, Q., Ye, J.: Functionalyzed lignin by grafting phosphorus-nitrogen improves thermal stability and flame retardancy of polypropilene. Polym. Degrad. Stab. **97**, 541–546 (2012)

Effect of Glass Fibers and Aramid Fiber on Mechanical Properties of Composite Based Unmanned Aerial Vehicle (UAV) Skin

Benni F. Ramadhoni[1(✉)], Ara Gradiniar Rizkyta[1], Atik Bintoro[2], and Afid Nugroho[2]

[1] Balai Teknologi Polimer, Tangerang Selatan 15314, Indonesia
benni.ramadhoni@bppt.go.id
[2] Pusat Teknologi Penerbangan, Bogor 16350, Indonesia

Abstract. The use of carbon-based fiber that can disrupt radar signal transmission on the flights especially on UAV remains problem. To address the problem, composites based UAV skin using E-glass, S-glass and aramid fibres as well as epoxy resin has been successfully manufactured by vacuum infusion process. Full factorial design of experiment (DoE) and analysis of variance (ANOVA) was performed using Minitab 16 to analyse the effects of layer variations on specific tensile, compression and bonding shear. Fiber configuration was divided into two layers, namely layer 1–9 and 10–18. In general, mechanical properties i.e. tensile strength, bonding shears and compression strength increased compared with epoxy resin properties. Tensile strength and modulus increased by 484% and 204% respectively, while bonding shear and compressive strength increased by 24226% and 94% respectively. The overall result indicated that the best properties of composites-based UAV skin were obtained by the utilization of aramid fiber.

Keywords: Composite · Mechanical properties · UAV

1 Introduction

Property of composite depends on the behaviour of its constituents as well as of the interfaces between the fiber and resin [1]. Development of composite containing more than one type of reinforcement is motivated by the desire to combine advantageous features of reinforcement to improve performance as well as to reduce weight and cost [2].

In aerospace applications, especially Unmanned Aerial Vehicle (UAV), thermoset is more widely used than thermoplastic, due to its easier to fill in the fiber [3]. Among thermoset resins, epoxy is the most commonly used resin, due to its high chemical resistance, excellent dimensional stability, good fiber adhesion and performance under wet conditions.

To increase flight time of UAVs, lightweight materials are needed, namely S-glass, E-glass, aramid, carbon fiber and graphite. More specifically, carbon fiber is the most

M. N. Osman Zahid et al. (Eds.): iMEC-APCOMS 2019, LNME, pp. 435–440, 2020.
https://doi.org/10.1007/978-981-15-0950-6_66

suitable material for UAV due to its high specific strength-to-weight [4]. However, the use of carbon based fiber can disrupt radar signal transmission on the flights [5].

Processing methods that can be used for manufacturing composite for UAV include vacuum infusion, vacuum bag, hand lay up, resin transfer moulding (RTM), prepreg lay-up and filament winding with oven or autoclave curing [6, 7]. However, hand lay-up and pre-preg lay-up show low reproducibility while RTM and filament winding require sophisticated machine. In contrary, vaccum infusion and vaccum bag require only simple machine, facile set up but manage to produce good reproducibility, and improving fibres-to-resin ratio with minimized of voids so that improve mechanical properties compared with hand lay-up process [8].

In this research, we conducted a study to compare aramid, E glass and S glass fibers as well as hybrid arrangement thereof on the mechanical properties of UAV skin, manufactured using epoxy resin and vacuum infusion process. Main and interaction effect analyses on specific mechanical properties were carried out using ANOVA in Minitab 16.

2 Materials and Methods

2.1 Materials

Epoxy Renlam LY 5138-2and Ren Hy 5138(Huntsman, Indonesia) were used as the matrix and hardener respectively. E-glass EW 130 and S-glass SW220B-90A (JustusKimiaraya, Indonesia) as well as aramid (Coats Rejo, Indonesia) were used as the reinforcement.

2.2 Manufacturing Method

All composites were manufactured using vacuum infusion except for control sample. Control sample was made in advanced from unsaturated polyesther (UP) and a layer of E-glass using hand lay up method.

Table 1 list all layer combinations. In a typical experiment, eighteen layers of fibers were arranged and enclosed by bagging film that connected to inlet and outlet flow tube. Epoxy resin in a container was pumped through a tube to wet the entire laminate. The specimen was placed at room temperature (25 ± 2 °C) up to 3 h for curing.

2.3 Measurement Testing

The cured specimen was conditioned at 23 ± 2 °C and 50 ± 5% humidity over 40 h prior to measurements. Density was obtained using analytical balance following the method in ASTM D792. A typical size specimen was weighed in air (m_{air}) and in a solution ($m_{dissolved}$) and both masses were recorded and the density was calculated. Tensile strength and modulus test were carried out according to ASTM D3039. The tests were carried out at rate of 2 mm/minute. Bonding shear test was carried out in

Table 1. List of fiber combination

Formula	Layer 1–9	Layer 10–18
1	Aramid	Aramid
2	S-glass	S-glass
3	E-glass	E-glass
4	S-glass	Aramid
5	Aramid	S-glass
6	S-glass	E-glass
7	E-glass	S-glass
8	Aramid	E-glass
9	E-glass	Aramid

accordance with ASTM D5868 using epoxy resin as adhesive. Shear rate was 13 mm/minute. Compression test was carried out in accordance with ASTM 6641. Compression rate was 1.3 mm/minute. All properties were divided by density to obtain specific properties.

3 Results and Discussion

Main Effect Analysis of Mechanical Properties

Main effect analysis analyze the effect of isolated parameter on property of interest. Figure 1 (a) shows that specific tensile strength of layer 1-9 and 10-18 was mainly affected by aramid fiber, while E-glass fiber had the least effect. These results were consistent with specific tensile strength value of formula 1 (260.66 MPa cm^3/g) and formula 2 (184.16 MPa cm^3/g) in Table 2. E-glass fiber has the lowest strength compared to S-glass and aramid fiber.

On the other hand, Fig. 1 (b) reveals that aramid and E-glass fiber positively effected specific tensile modulus of layer 1-9 and 10-18 respectively, while S-glass fiber gave the least influence. These followed the results of formula 1, 2 and 3 i.e. 6.81 GPa cm^3/g, 6.67GPa cm^3/g and 4.93 GPa cm^3/g respectively. Figure 1 (c) demonstrates that aramid fiber had the slightest effect on specific compressive strength, while E-glass withstood compressive stress superiorly. Figure 1 (d) shows that specific bonding shear was affected mostly by aramid fiber in layer 1-9 and S-glass fiber in layer 10-18, while E-glass and aramid fiber had minor effect on specific bonding shear in layer 1-9 and 10-18 respectively. Compared to S-glass and E-glass fibers, Aramid fiber not only has the highest specific strength, but also the highest tensile modulus. As well as aramid fiber has high stiffness caused by its symmetrical internal structure [9].

Interaction Effect Analysis of Mechanical Properties

Interaction effect analysis analyze the effect of combined parameter. Figure 2 (a) shows that regardless of aramid fiber has the most effect on specific tensile strength, utilization of whole aramid fiber at 18 layers (formula 1), did not resulted in the highest specific tensile strength. The highest specific tensile strength (321.9 MPa cm^3/g) was obtained

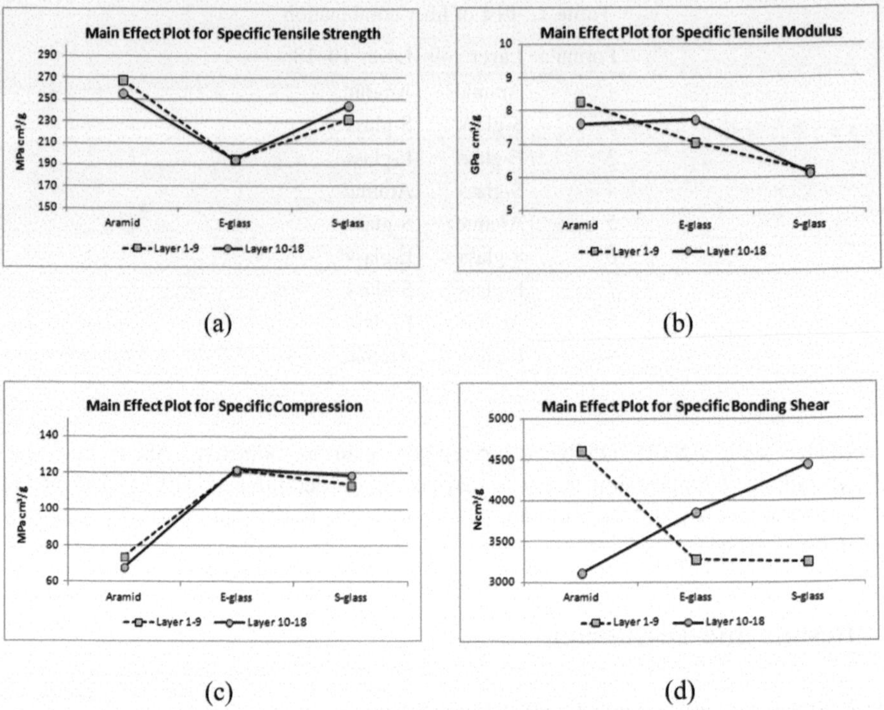

Fig. 1. Main effect analysis of specific (a) tensile strength (b) tensile modulus(c) compression (d) bonding shear

Table 2. Results of density measurement and mechanical test of composites

Formula	Density	Specific tensile strength (MPa cm3/g)	Specific tensile modulus (GPacm3/g)	Specific compressive strength (MPa cm3/g)	Specific bonding shear (N cm3/g)
1	1.103	260.66	6.81	61.13	3932.91
2	1.679	184.16	6.67	169.65	3817.15
3	1.696	233.01	4.93	129.92	4117.33
4	1.348	282.00	7.83	82.43	3206.23
5	1.404	321.90	7.35	92.08	5390.31
6	1.699	178.96	5.79	128.13	3238.38
7	1.723	176.82	6.07	134.54	3823.56
8	1.295	219.84	10.67	69.60	4519.69
9	1.276	222.88	8.35	61.40	2194.00
Control	1.747	150.51	3.03	55.84	921.00

by combining aramid and S-glass (formula 5). This is also true on applying merely whole E-glass for all 18 layers (formula 3). Whole E-glass fiber reinforced composite did not become the lowest specific tensile strength. The lowest specific tensile strength

(176.8 MPa cm^3/g) was obtained by combining E-glass and S-glass fiber (formula 7). Similar with specific tensile strength, the highest specific tensile modulus was not resulted from applying whole aramid fiber in every layer. The highest specific tensile modulus (10.7 GPa cm^3/g) was obtained by combining aramid fiber in layer 1–9 with E-glass fiber in layer 10–18 (formula 8). On the other hand, applying S-glass fiber in layer 1–9 with E-glass in layer 10–18 (formula 6) gave the lowest specific tensile modulus (2.6 GPa cm^3/g). Figure 2(c) depict that whole E-glass fiber arrangement (formula 2) showed the best specific compressive strength (169.65 MPa cm^3/g), while whole aramid arrangement (formula 1) gave the smallest one (61.13 MPa cm^3/g). Figure 2(d) shows that combination between aramid and S-glass fiber (formula 1) gave the best bonding shear (5390.31 N cm^3/g), while combination between E-glass and aramid fiber (formula 9) gave the lowest one (2194.00 N cm^3/g).The interaction between two different fibers was responsible for all those results.

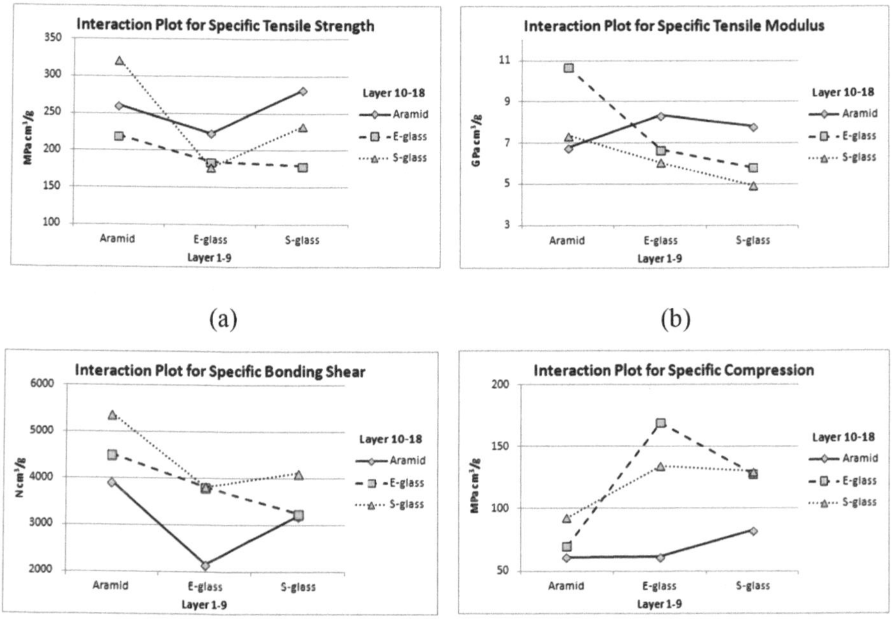

(a) (b)

(c) (d)

Fig. 2. Interaction effect analysisof specific (a) tensile strength (b) tensile modulus(c) bonding shear (d) compression

4 Conclusion

Composites based UAV skin has been successfully manufactured by vacuum infusion process using epoxy resin as matrix as well as E-glass, S-glass and aramid fibres as the reinforcements. The result show that specific tensile strength and specific modulus was mainly affected by aramid fiber. However aramid fiber had the slightest effect on specific compressive strength, while E-glass withstood compressive stress superiorly. The same results also occur at interaction effect analysis of compression that E-glass fiber arrangement showed the highest value.

References

1. Karger-Kocsis, J., et al.: Interphase engineering with nanofillers in fiber-reinforced polymer composites. Interface/interphase in Polymer Nanocomposites, pp. 71–102 (2017)
2. Muhammed, O.S., Hussein, A.K., Abdel-Rahim, R.H.: Effect of filler type on some physical and mechanical properties of carbon fibers/polyester composites. Eng. Tech J. **31 Part (A)** (15) (2013)
3. He, X., Li, Y., Nie, M.: Root-like glass fiber with branched fiber prepared *via* molecular self-assembly. R. Soc. Chem. **6**, 45492–45494 (2016)
4. Goh, G.D., Agarwala, S., Goh, G.L.: Challenges and potential, aerospace science and technology. Additive Manufacturing in Unmanned Aerial Vehicle (UAVs), pp. 45492–45496 (2016)
5. Fan, Y., Yang, H., Liu, X.: Preparation and study on radar absorbing materials of nickel-coated carbon fiber and flake graphite. J. Alloy. Compd. **461**, 490–494 (2008)
6. Maley, A.J.: An investigation into low-cost manufacturing of carbon epoxy composite and a novel "mouldless" technique using the Vacuum Assisted Resin Transfer Moulding (VARTM) method. Department of Mechanical and Aerospace Engineering Carleton University Ottawa, Ontario, Canada (2008)
7. Wang, P., Drapier, S., Molimard, J.: Numerical and experimental analysis of resin infusion manufacturing processes of composite materials. J. Compos. Mater. **46**(13), 1617–1631 (2011)
8. Mohd Yuhazri, Y., Phongsakon, P.T., Sihombing, H.: A comparison process between vacuum infusion and hand lay-up method toward Kenaf/Polyster composites. Int. J. Basic Appl. Sci. IJBAS-IJENS **10**(03) (2010)
9. Jogi, S.A., et al.: Evaluation of impact strength of epoxy based hybrid composites reinforced with E-Glass/Kevlar 49. Mehran Univ. Res. J. Eng. Technol. **36**(4) (2017)

Surface Roughness of Laser Modified Die Surface Change Under Thermal Cyclic Loading

Annie Lau Sheng[1], Izwan Ismail[1,2,3(✉)], Fazliana Fauzun[1], and Syarifah Nur Aqida[1,2]

[1] Faculty of Mechanical & Manufacturing Engineering, University Malaysia Pahang, 26600 Pekan, Malaysia
izwanismail@ump.edu.my
[2] Automotive Engineering Centre, University Malaysia Pahang, 26600 Pekan, Malaysia
[3] Centre of Excellence for Advanced Research in Fluid Flow (CARIFF), University Malaysia Pahang, 26300 Gambang, Malaysia

Abstract. Thermal fatigue crack results in decrease of die service life has brought to a significant loss in the die casting industry due to the high cost of die. Laser surface modification is proposed in this study to enhance the surface properties of die and increase the resistance of thermal crack. In this study, the changes of surface roughness of laser modified die surface were investigated during thermal cyclic loading. H13 tool steel samples were sectioned while the surface of the samples were laser modified at different parameters. Laser modified samples were subjected to thermal cyclic loading by continuously heating in molten aluminium and cooling in water bath at respective temperature range of 850–900 °C and 27 °C. The results of surface roughness and morphology were obtained using a display optical microscope after 3000, 4000 and 5000 cycles of thermal cyclic loading. The surface roughness of laser modified tool steel with hardness properties of 745 HV experienced the least changes throughout the thermal cyclic loading process. The minimal changes of surface roughness on the laser modified die reduces the formation of the oxide layer and thus reduces the thermal fatigue of die casting dies.

Keywords: Die casting · Surface roughness · Thermal crack

1 Introduction

High pressure die casting process is capable of producing sharply defined and net shape dimensional, metal part with smooth or textured surface. Aluminium die-casting acts as a vital role especially in the automotive industry due to its light weight and good formability characteristic. By adjusting and determining the parameter of the aluminium die casting process, the quality of die-cast part produced can be improved. However, the quality of tool steels in the die casting process decreases and the lifetime is limited due to thermal fatigue.

M. N. Osman Zahid et al. (Eds.): iMEC-APCOMS 2019, LNME, pp. 441–447, 2020.
https://doi.org/10.1007/978-981-15-0950-6_67

During operation, the tool steel dies are exposed to high-pressure peaks of melt and temperature gradients. Temperature gradients between the surface and the die core are causing thermal stresses, which leads to the accumulation of local plastic strains on the die surface and finally causes the surface cracks [1]. Other failures due to thermal include heat checking, wear, plastic deformation and corrosion.

The enhancement of tool steel surface can be carried by using laser surface modification which increases the surface hardness of the material and improve the resistance against thermal fatigue. Surface modification using laser can produce small grain microstructure via rapid solidification. Through this method, the tribological and mechanical properties of the material can be improved [2]. The surface roughness (Ra) is a critical factor that causes cracking or wear occur. In many applications, a suitable surface roughness value is required to avoid premature failure from surface-initiated cracking [3].

In this study, the changes of surface roughness in laser modified tool steels which is subjected to thermal cyclic loading were investigated. The as-received and laser modified AISI H13 with different surface roughness were subjected to thermal cyclic loading test.

2 Methodology

The materials used in this study is AISI H13 tool steel (5.64 wt% C, 1.37 wt% O, 0.85 wt% Si, 1.03 wt% V, 8.84 wt% Cr, 1.51 wt% Mo and Fe balance). AISI H13 tool steel is commonly used for aluminium die casting tool due to its high hardness and resistance of thermal fatigue [4, 5]. The chromium, molybdenum and vanadium in the AISI H13 tool steel are the main alloying ingredients for the hot working tool steel [6]. AISI H13 tool steels were machined into a rectangular shape with a dimension of 56 mm 33 mm. The surface of test specimens treated with different laser surface modification to increase the thermal fatigue resistance. The thermal cyclic loading test which includes cyclic immersion test was started when the temperature of molten aluminium alloy reaches 850 °C. When the temperature of molten aluminium is more than its melting point (630 °C), it can prevent the sticking of molten aluminium to the test specimens. The test specimens were immersed in molten aluminium alloy for 11.2 s followed by a rotation and immersion in water at 27 °C for 11.2 s. After that, the test specimens were rotated back to its original position. The total time taken per cycle was 31 s which include 4.3 s for each rotation time.

Before the thermal cyclic loading test, the average surface roughness of test specimens was measured by using a Mahr MarSurf PS1 surface roughness tester. The evaluation length and cut-off length were set at 4.0 mm and 0.8 mm respectively which complete the ISO 4288 standard. The surface roughness of test specimens was measured after 3000, 4000 and 5000 cycles. The changes of surface roughness for all samples before and after the thermal cyclic loading test were determined.

3 Results and Discussion

The surface roughness (Ra) of laser modified samples were affected by laser parameters. The samples used in this study were as-received AISI H13, laser modified AISI H13 with low surface roughness (9.0 μm) and laser modified AISI H13 with high surface roughness (42.3 μm). The laser modified sample with surface roughness of 9.0 μm produced hardness properties of 745 HV while the modified sample with surface roughness of 42.3 μm experienced hardness properties reduction down to 591 HV. The surface roughness and hardness of the as-received AISI H13 were 1.3 μm and 202 HV respectively. After the laser surface modification, the surface hardness of the samples was increased approximately 3 times compared to the as-received sample. Table 1 summarises the surface roughness results at different cycles of thermal fatigue test for all samples.

Table 1. Comparison of surface roughness before and after thermal fatigue test.

Samples cycles	Surface roughness, Ra (μm)			
	Initial	3000	4000	5000
As-received AISI H13	1.3	4.9	5.4	6.8
Laser modified AISI H13 with low surface roughness	9.0	5.9	6.3	7.4
Laser modified AISI H13 with high surface roughness	42.3	5.9	7.3	8.2

Overlapping of laser spot was the significant parameter that affects surface roughness of laser modified surface. The surface roughness decreased when overlapping of laser spots decreased. The previous work by Qin & Chen [7] obtained a similar trend which justifies this statement. On the other hand, the hardness increased when overlapping decreased. Low overlapping lead to grain refinement of microstructure occurred. This was due to the low interaction time between laser beam and the surface of sample. Decreasing surface temperature caused rapid solidification and results in finer grains formation. This was also the reason why laser surface modification process enhances the surface properties which increases the hardness properties and resistance of thermal fatigue [8, 9].

The changing of surface roughness (Ra) after 3000, 4000 and 5000 cycles of thermal fatigue (TF) test for as-received AISI H13 and laser modified AISI H13 were shown in Fig. 1. From the experiment, the surface roughness of all samples increased when the number of cycles was increased.

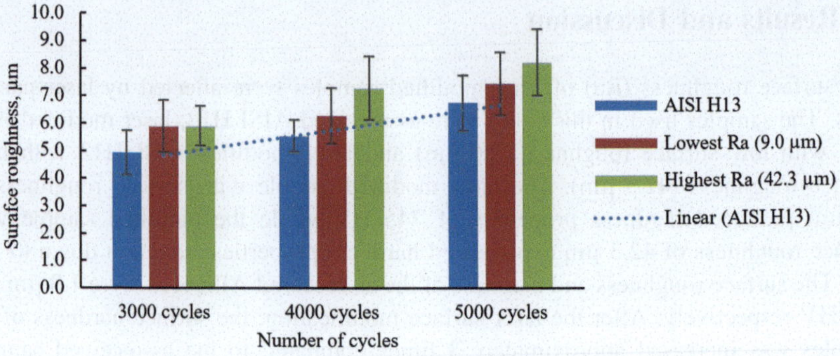

Fig. 1. Surface roughness changes after thermal fatigue test.

Thermal fatigue crack and oxide layer formation after the thermal fatigue test caused the surface roughness of samples increased. As reported by previous work, the increment of immersion time leads to crack increase [10]. Immersion time stands for the number of cycles of thermal fatigue test. When the number of cycles increased, the thermal fatigue crack also increased. The presence of oxygen and high-temperature environment caused the formation of oxide layer on the surface of the sample. Such condition hastens the surface crack on the samples. In addition, the high temperature in aluminium die casting environment also caused the AISI H13 tool steel to undergo thermal softening trend and accelerated crack growth.

Figure 2 shows the rate of change of surface roughness for all samples after subjected upto 3000, 4000 and 5000 cycles of thermal fatigue test. The surface roughness of as-received AISI H13 undergoes a huge changing which was more than 200% after 3000 cycles of thermal fatigue test. The rate of change for as-received AISI H13 was 273.31% which increased from 1.3 μm (before TF test) to 4.9 μm (after 3000 cycles). The surface roughness keeps increasing gradually by 12.42% and 21.96% after 4000 and 5000 cycles respectively. On the contrary, the surface roughness of all laser modified samples decreased after 3000 cycles and increased after 4000 and 5000 cycles. The rate of change for laser modified sample with lowest Ra was reduced to −34.37% which mean surface roughness decreased from 9.0 μm to 5.9 μm. Laser modified sample with highest Ra also had −86.06% of the rate of change after 3000 cycles. The surface roughness decreased from 42.3 μm to 5.9 μm. After 4000 and 5000 cycles, the rate of change for laser modified sample with highest Ra was higher than the laser modified sample with lowest Ra. When compare as-received AISI H13 and laser modified samples with lowest and highest Ra, the rate of surface roughness change was: laser modified AISI H13 with lowest Ra < laser modified AISI H13 with highest Ra < as-received AISI H13.

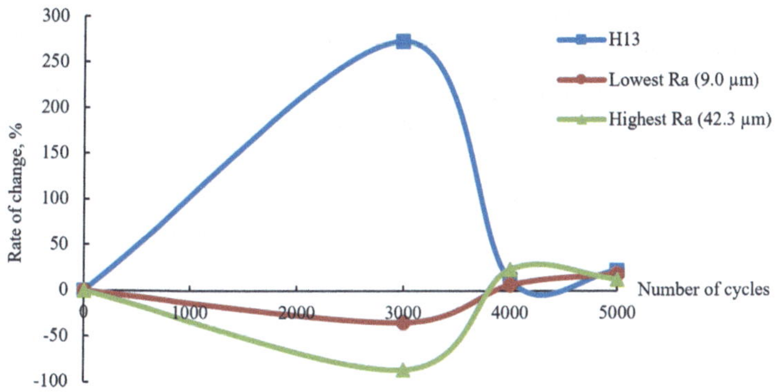

Fig. 2. Rate of surface roughness change after 3000, 4000 and 5000 cycles of thermal fatigue test.

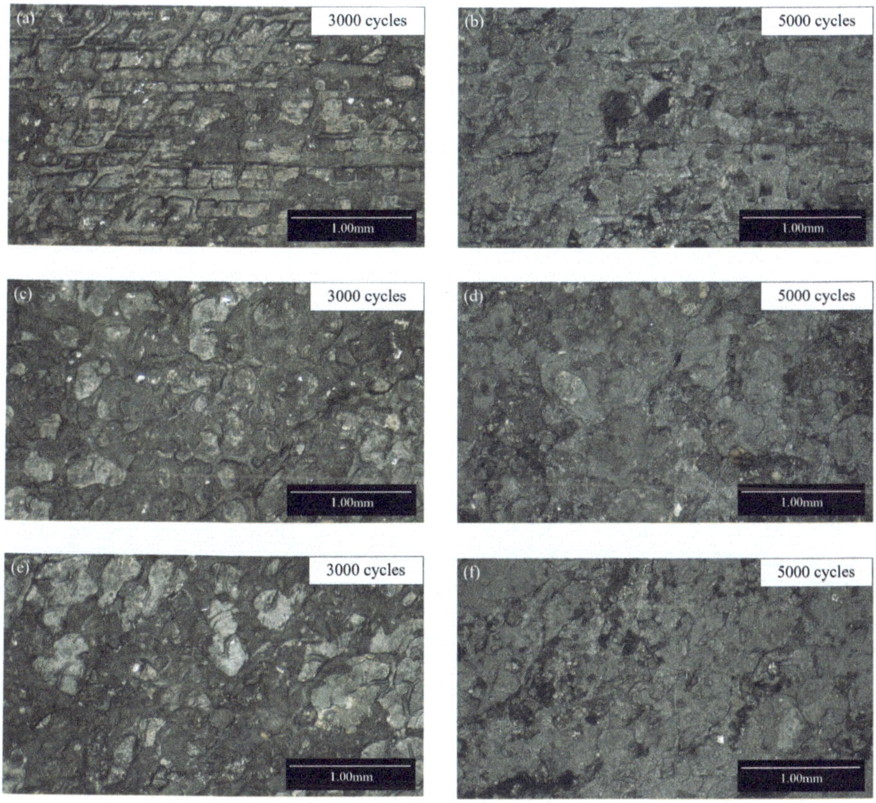

Fig. 3. Surface morphology after 3000 and 5000 cycles of thermal fatigue test: (a)–(b) AISI H13; (c)–(d) laser modified sample with lowest Ra; (e)–(f) laser modified sample with highest Ra.

The as-received AISI H13 has the highest rate of surface roughness change among all the samples. This is because as-received AISI H13 has the lowest hardness and weakest resistance of thermal fatigue among the samples. Thermal fatigue crack and wear occurred on the surface of as-received AISI H13 were more than laser modified samples as shown in Fig. 3. Improvement in the hardness properties of laser modified samples due to the refinement of grain in microstructure enhances the resistance of thermal fatigue and wear. Hence, the change of surface roughness in laser modified samples was insignificant compared to the as-received AISI H13.

4 Conclusion

In this study, the AISI H13 tool steel was treated by laser surface modification process and changes of surface roughness of laser modified die surface were investigated during thermal cyclic loading. The following conclusions are drawn:

(a) In comparison, laser modified AISI H13 tool steel undergo refinement of grain which increases the hardness and resistance of thermal fatigue crack. This can reduce the rate of surface roughness change.
(b) By minimising the change of surface roughness, suitable parameters of laser surface modification process like low overlapping need to be applied to increase the life time of die casting dies.
(c) Laser modified AISI H13 tool steel with low surface roughness and high hardness has excellent surface properties toward the high-temperature environment.

Acknowledgements. The authors would like to acknowledge the Ministry of Education Malaysia (MOE) for the funding of this research under the Fundamental Research Grant Scheme (FRGS/1/2016/TK03/UMP/02/4) RDU160141.

References

1. Muhič, M., Tušek, J., Kosel, F., Klobčar, D., Pleterski, M.: Thermal fatigue cracking of die-casting dies. Metalurgija **49**(1), 9–12 (2010)
2. Aqida, S.N., Maurel, M., Brabazon, D., Naher, S., Rosso, M.: Thermal stability of laser treated die material for semi-solid metal forming. Int. J. Mater. Form. **2**(1), 761–764 (2009)
3. Mumtaz, K., Hopkinson, N.: Top surface and side roughness of Inconel 625 parts processed using selective laser melting. Rapid Prototyping J. **15**(2), 96–103 (2009)
4. Lu, Y., Ripplinger, K., Huang, X., Mao, Y., Detwiler, D., Luo, A.A.: A new fatigue life model for thermally-induced cracking in H13 steel dies for die casting. J. Mater. Process. Technol. **271**, 444–454 (2019)
5. Bailey, N.S., Katinas, C., Shin, Y.C.: Laser direct deposition of AISI H13 tool steel powder with numerical modeling of solid phase transformation, hardness, and residual stresses. J. Mater. Process. Technol. **247**, 223–233 (2017)
6. Telasang, G., Dutta Majumdar, J., Padmanabham, G., Manna, I.: Structure-property correlation in laser surface treated AISI H13 tool steel for improved mechanical properties. Mater. Sci. Eng., A **599**, 255–267 (2014)

7. Qin, Q., Chen, G.X.: Effects of parameters on surface roughness of metal parts by selective laser melting. Adv. Mater. Res. **834–836**, 872–875 (2013)
8. Cabeza, M., Castro, G., Merino, P., Pena, G., Román, M.: Laser surface melting: a suitable technique to repair damaged surfaces made in 14 Ni (200 grade) maraging steel. Surf. Coat. Technol. **212**, 159–168 (2012)
9. Jia, Z.X., Liu, Y.W., Li, J.Q., Liu, L.J., Li, H.L.: Crack growth behavior at thermal fatigue of H13 tool steel processed by laser surface melting. Int. J. Fatigue **78**, 61–71 (2015)
10. Abdulhadi, H., Ahmad, S., Ismail, I., Ishak, M., Mohammed, G.: Experimental investigation of thermal fatigue die casting dies by using response surface modelling. Metals (Basel) **7**(6), 191 (2017)

Tensile Properties of Hybrid Woven Glass Fibre/PALF Reinforced Polymer Composite

Mawarnie Ismail[1,2(✉)], M. R. M. Rejab[1], J. P. Siregar[1], Zalinawati Muhamad[1,3], and Ma Quanjin[1]

[1] Faculty of Mechanical & Manufacturing Engineering, Universiti Malaysia Pahang, 26600 Pekan, Pahang, Malaysia
Mawarnie85@gmail.com

[2] Politeknik Mukah, 96400 Mukah, Sarawak, Malaysia

[3] Politeknik Sultan Mizan Zainal Abidin, 23000 Dungun, Terengganu, Malaysia

Abstract. The world today is experiencing a significant global warming problem. One of the causes of the problem is due to the use of excessive and uncontrollable material. Replacing the material with more environmentally friendly is a priority today, especially from wasted natural materials such as pineapple leaf fibre (PALF). Therefore, the aim of the study is on an investigation of tensile properties from composites made by reinforcing woven glass fibre and short fibre PALF into epoxy resin. The fibre contains a fixed glass fibre with 20 wt% and PALF contains with a range of 5 wt%, 10 wt% and 15 wt%. The architecture of hybrid composite comprises of bottom and top are woven glass fibre meanwhile middle layer is short fibre of PALF. Epoxy resin was poured into it, then hand lay-up technique was used to spread the epoxy evenly. Cold compression method was used to cure the composite. The results show that 10 wt% of PALF and 20 wt% woven glass fibre have shown the best tensile strength and stiffness. This concluded that the best filler of PALF is at 10 wt%.

Keywords: Pineapple leaf fibre (PALF) · Woven glass fibre · Hybrid composite · Tensile properties

1 Introduction

Nowadays, request on an alternative material is highly demanded especially in automotive, aircraft, marine and offshore industries. Increasing demand for environmentally friendly materials, rising prices of petroleum-based plastics, increasing depletion rates and pressing environmental regulations have created an increasing interest in composites [1]. The conception of natural resources has gained importance in recent years. Research and engineering interest have been shifting from monolithic materials to natural-glass fibre reinforced polymeric materials. Natural fibres such as wood fibre composites are being used in a large number of applications in decks, docks, window frames and moulded panel composites, meanwhile car bumper manufactured by kenaf/glass epoxy material, banana reinforced composites apply at under-floor protection for passenger car [2].

M. N. Osman Zahid et al. (Eds.): iMEC-APCOMS 2019, LNME, pp. 448–454, 2020.
https://doi.org/10.1007/978-981-15-0950-6_68

Recently, natural fibre reinforced polymer matrix composites has been gaining importance all over the world because of their environmental friendliness and light-weight [3]. Based on their origin, natural fibres can be generally classified to bast fibres such as jute, flax, hemp, kenaf, and mesta; leaf fibres such as pineapple, sisal, henequen, and screw pine; and seed or fruit fibres such as coir, rice, cotton, and oil palm [4]. The important advantages of using these natural fibres are availability, easy biodegradability, renewable and cost-effective [5]. Natural fibres may play a major role in developing biodegradable composites to resolve the current ecological and environmental problems. Natural fibres are lighter and cheaper, but they have low mechanical properties than glass fibres. The use of hybrid fibres may solve this issue. Most of the studies on natural fibres are concerned with single reinforcement. The addition of natural fibre to the glass fibre can make the composite hybrid relatively cheaper and easier to use.

Hybrid composites refer to a combination of two or more different materials combined in a common matrix. Hybridization of two or more different materials is found to be an effective approach to design materials with different requirements and applications. Result from hybrid composite can acquired new properties, which are determined by chemical and individual components, structure and interface of the different components [6]. Problems of high-cost synthetic fibres and low mechanical properties of natural fibres can be solved by blending them in a common matrix. By having a hybrid composite, what lacks in one material can be complemented by the strength of the other material [7]. E-Glass has 2.58 g/cm^3 of density, 3445 MPa for tensile strength and 72.5 GPa for tensile modulus [8]. Meanwhile, the properties of PALF are 1.2–1.53 g/cm^3 for density, 290.61 MPa for tensile strength, 5.83 GPa for Tensile Modulus [9] and cellulose content 70–80% [10]

In this research, the hybrid composites used are glass fibre as a synthetic fibre meanwhile pineapple leaf fibre (PALF) as a natural fibre. Fibre loading is one of the important factors affecting the strength of composite material as it determines the load transfer within the composite sample, thanks to the fibre-matrix bonding. Hence, this study is focused on investigating the effect of different fibre loading on tensile properties of glass fibre/pineapple leaf fibre (PALF) hybrid composite.

2 Materials and Methods

2.1 Material Preparation

The PALF is supplied from Acheh, Indonesia in long fibres form. PALF is easily growing in Malaysia and Indonesia as agricultural products. The synthetic fibre used in this research is E-Glass mat woven fibre supplied by Salju Bistari Sdn. Bhd. Epoxy Amite 100 purchased from Mecha Solve Engineering Sdn Bhd, Malaysia, was used as a matrix. A long fibre of PALF was cut to short fibre around 15 cm length, then used crusher machine to grind the fibre become shorter which is around 1–20 mm length. The short fibre then sieved using a sieve machine to characterize the fibre size distribution. Most of PALF had around 3–14 mm length. Drying process of PALF took placed in Vacuum Oven for 24 h at 80 °C to remove moisture in fibre, then stored in the vacuum bag to avoid from the air inlet.

2.2 Material Fabrication

The manufacturing process of the sample begins with a weight distribution of PALF, Glass Fibre and Epoxy resin. Table 1 showed each weight distribution of PALF, Glass Fibre and Epoxy. The variable weight distribution of PALF was 5 wt%, 10 wt% and 15 wt%, while glass fibre was consistent with 20% throughout the combination. For example, sample GP1 represents Glass Fibre (G), PALF (P) and 1 is the percentage of natural fibre.

Table 1. Summary of weight distribution of PALF, glass fibre and epoxy resin

Sample	Glass fibre wt%	PALF wt%	Epoxy wt%	Glass fibre (g)	PALF (g)	Epoxy resin (g)	Total (g)
G2	20	–	80	76	–	182.4	182.4
GP1	20	5	75	76	9.12	171.0	256.1
GP2	20	10	70	76	18.5	159.6	254.1
GP3	20	15	65	76	27.8	148.2	252.0

Firstly, the woven mat of glass fibre was cut to 20 wt% which is equivalent to 4 layers and according to mould which in size of 260 mm × 192 mm. Then, short fibres of PALF were distributed equally on top of one layer of the woven mat, then covered with another one layer of glass fibre. Mould was prepared and wiped with a special wax remover before pouring the epoxy resin. The ratio between epoxy and hardener was 3:1 by weight, and was stirred with a magnetic stirrer at 200 rpm for 3 min to avoid cross-epoxy crossbar. After that, the mixing resin was left in a container for 5 min to ensure no trapped air bubbles. Next, the mixture was poured into mould evenly before a layer of E-Glass fibre mat is placed then poured epoxy again. An E-Glass mat contains on short PALF was placed repeated with epoxy. Next, a glass fibre placed and lastly poured epoxy before curing. Figure 1 shows the sequence of layering between the glass fibre and PALF. The warp and weft directions of the fabric were neglected based on the assumption that the fabrics have similar properties in each direction. This due to glass fibre woven mats are balance plies (0°/90°) [11]. After fabrication, the mould was cured under 5 psi load under cold press machine for at least 8 h. Next, the plates were cut according to ASTM D638 by using a CNC milling machine.

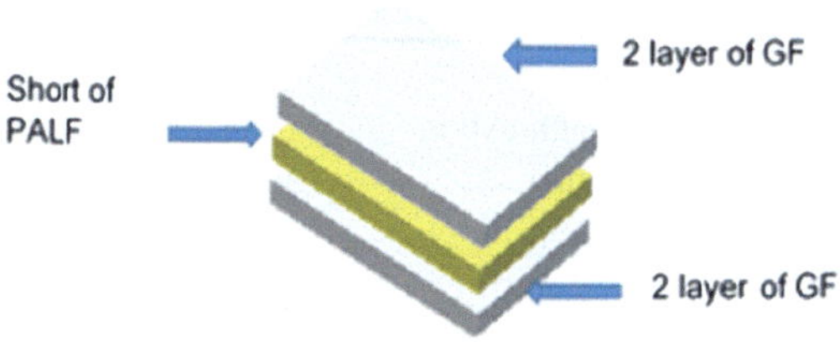

Fig. 1. Sequence layering of glass fibre and PALF

2.3 Tensile Test

The tensile test was conducted according to the 638 type IV Dumbell shape American Society for Material Testing (ASTM) with Universal Testing Machine (Instron Co. 3369). The test was performed at a loading rate of 2 mm/min at room temperature as shown in Fig. 2(a), as load and displacement data were recorded to analyse Young Modulus and elongation at break. Five specimens for each sample were examined based on ASTM 638 recommendation, as shown in Fig. 2(b). All the specimens were tightened hardly to prevent slippage during testing.

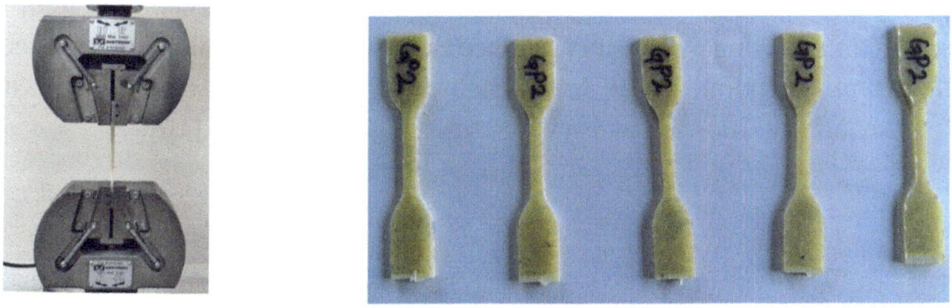

(a) (b)

Fig. 2. Photos of (a) specimen test for tensile (b) five specimens each variant.

3 Results and Discussion

The specimens were maintained under the same experimental conditions to prevent data variation. When each specimen reached its maximum elongation and break, the data was recorded e.g., load, tensile stress and tensile strain. Table 2 shows the average of maximum load, tensile strength, tensile modulus and elongation at break.

Table 2. Summary of result from tensile tests

Sample	Max. load (kN)	Tensile strength (MPa)	Tensile modulus (GPa)	Elongate a t break (%)
G20	2.209	113.963	1.463	0.42
GP1	3.066	165.140	1.585	0.60
GP2	3.447	175.680	1.723	0.75
GP3	3.392	136.708	1.285	0.73

Tensile strength of PALF/Glass fibre reinforced hybrid composite presented in Fig. 3. Based on this bar chart, the highest tensile strength is GP2 at weight distribution of PALF 10% with 175.68 MPa meanwhile the lowest is solely Glass fibre with 20% 113.963 MPa. It was observed that tensile strength increases with fibre loading up to 10% PALF/20% glass fibre but decreases with further loading of PALF. This result slightly higher than Glass/Jute/Pineapple hybrid composite reported by [12] with value 78 MPa at 0.42 weight distribution for highest tensile strength and lowest 40 MPa at 0.18. But, looking at 15 wt% of PALF the tensile strength decrease about 22.2% from GP2. This situation is due to poor interfacial bonding between reinforcement and matrix happened, and also due to access amount of PALF and less amount of matrix cause the mixing process not happened equally. Tensile strength of natural fibre depend on cellulose content, so when the cellulose content high the tensile strength becomes increased, contradicting with increase content of hemicellulose decreased tensile strength. So, at this point, PALF had a high cellulose content with 70–80% made this hybrid, especially at 10 wt% of PALF have a higher strength [10].

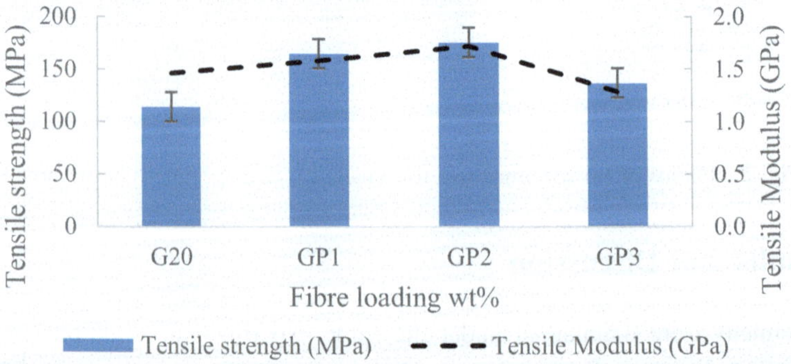

Fig. 3. Effect of PALF loading on tensile properties of the hybrid composite.

The effect of PALF loading on tensile modulus of hybrid composite is presented in Fig. 3. The addition of PALF in this hybrid composite increased the tensile modulus until optimum at 10 wt% PALf/20 wt% GF with value 1.72 GPa. Further loading of PALF at 15wt% PALF/20 wt% GF decreased the modulus with value 1.28 GPa. This result quite high compared to glass/jute/pineapple hybrid composite with 0.9 GPa the highest Tensile Modulus at 0.42 volume fraction and 0.5 GPa for the lowest at 0.18 volume fraction [12].

4 Conclusion

The effect of PALF weight distribution on hybrid glass fibre/PALF epoxy composite to obtain the tensile strength, tensile modulus and elongation at break were investigated. It can be concluded that, addition of PALF made composite became more strength and stiffness. In contrast, the strength becomes less if too much inclusion of PALF. Approximately 54% increased of strength when added with PALF and 26% increased of stiffness from solely glass fibre. The best fibre loading to be added in hybrid composite was at 10 wt% of PALF either at tensile strength, tensile modulus or elongation at break. This proves that natural fibre plays an important role as a filler to enhance mechanical properties of woven glass fibre. Synthetic fibres have better properties but natural fibre reinforced composites showed better performance. Natural fibre reinforced composites also have a better impact on the environment as compared to the synthetic fibre reinforced composites. Natural fibre will help to improvethe climate without hampering the green globalization.

Acknowledgement. The authors are grateful to the Ministry of Higher Education and Universiti Malaysia Pahang for funding this research with PGRS180319.

References

1. Sun, G., Tong, S., Chen, D., Gong, Z., Li, Q.: Mechanical properties of hybrid composites reinforced by carbon and basalt fibers. Int. J. Mech. Sci. **148**, 636–651 (2018)
2. Sanjay, M.R., Arpitha, G.R., Naik, L.L., Gopalakrishna, K., Yogesha, B.: Applications of natural fibers and its composites: an overview. Nat. Res. **7**(3), 7 (2016)
3. Vignesh, P., Venkatachalam, G., Shankar, A.G., Singh, A., Pagaria, R., Prasad, A.: Studies on tensile strength of sugarcane fiber reinforced hybrid polymer matrix composite. Mater. Today: Proc. **5**(5), 13347–13357 (2018)
4. Arjmandi, R., Hassan, A., Zakaria, Z.: Lignocellulosic Fibre and Biomass-Based Composite Materials. Woodhead Publishing, Malaysia (2017)
5. Magarajan, U., kumar, S.D., Arvind, D., Kannan, N., Hemanandan, P.: A comparative study on the static mechanical properties of glass fibre vs glass-jute fibre polymer composite. Mater. Today: Proc. **5**(2), 6711–6716 (2018)
6. Bachtiar, D., Siregar, J.P., bin Sulaiman, A.S., bin Mat Rejab, M.R.: Tensile properties of hybrid sugar palm/kenaf fibre reinforced polypropylene composites. Appl. Mech. Mater. **695**, 155–158 (2015)
7. Zin, M.H., Abdan, K., Norizan, M.N.: Structural Health Monitoring of Biocomposites Fibre-Reinforced Composites and Hybrid Composites. Woodhead Publishing, Malaysia (2019)
8. Kulkarni, A.A., Vanakudre, V.M.: Preparation and analysis of hybrid epoxy composites. IJESC **7**(6) (2017)

9. Asim, M., Jawaid, M., Abdan, K., Ishak, M.: Effect of pineapple leaf fibre and kenaf fibre treatment on mechanical performance of phenolic hybrid composites. Fibers Polym. **18**(5), 940–947 (2017)
10. Djafari Petroudy, S.R.: Advanced High Strength Natural Fibre Composites in Construction. Woodhead Publishing, Malaysia (2017)
11. Siregar, J.P., Jaafar, J., Cionita, T., Jie, C.C., Bachtiar, D., Rejab, M.R.M., Asmara, Y.P.: The effect of maleic anhydride polyethylene on mechanical properties of pineapple leaf fibre reinforced polylactic acid composites. Int. J. Precis. Eng. Manuf.-Green Technol. **6**(1), 101–112 (2019)
12. Reddy, M.I., Kumar, M.A., Raju, C.R.B.: Tensile and flexural properties of jute, pineapple leaf and glass fiber reinforced polymer matrix hybrid composites. Mater. Today: Proc. **5**(1), 458–462 (2018)

Modification of Layered Structure in Manganese Oxide Nanorods for Electrode of Supercapacitor

Radhiyah Abd Aziz[1(✉)] and Rajan Jose[2]

[1] Faculty of Manufacturing Engineering, Universiti Malaysia Pahang, 26600 Pekan, Pahang DarulMakmur, Malaysia
radhiyah@ump.edu.my
[2] Faculty of Industrial Science and Technology, Universiti Malaysia Pahang, 26600 Pekan, Pahang DarulMakmur, Malaysia

Abstract. One-dimensional of layered structure manganese oxide (MnO_2) has been synthesized via hydrothermal route without further heat treatment. Varying the hydrothermal route parameter has an effect on the layered structure of MnO_2. α-MnO_2 obtained in this study has been physicochemical characterized using X-Ray diffraction (XRD), Fourier Transmission Infra-Red (FTIR), Field Emission Electron Microscopy (FESEM), Transmission Electron Microscopy (TEM) and Selected Area Electron Diffraction (SAED). Its electrochemical property has been evaluated by performing cyclic voltammetry using potentiostat. From the electrochemical analysis, it was shown MnO_2 with larger layered structure having high specific (*Cs*) compared to that of MnO_2 with smaller layered structure. This high C_S is originated from a combination of electrochemical double layer and pseudo-capacitance storage mechanisms. Besides, the appearance of water molecules within the layered spacing assisted the cations diffusion process.

Keywords: A-MnO_2 · Hydrothermal · Supercapacitor · Layered structure · Nanorods

1 Introduction

One-dimensional (nanorods, nanowires, and nanofibers) nanostructure of manganese dioxide (MnO_2) have made extreme research interests over the years due to their superior properties (e.g., catalytic, optical, magnetic, electrical and electrochemical) [1, 2]. Their main importance for energy storage devices have been intensively investigated as promising electrode materials in primary/secondary batteries and electrochemical capacitors. It is due to their superior electrochemical properties, low-cost, environmentally friendly and facile method of preparation [3, 4]. Several routes have been used to synthesize MnO_2, such as chemical precipitation [1], cathodic electrodeposition [2, 5], solvothermal [6], sonochemistry [7], and hydrothermal methods [8–10]. Powerful synthesis approach of hydrothermal method is used to synthesize different forms of MnO_2 and it offers various benefits (mild conditions of pH and temperature, and variety of precursors) [10]. Crystalline structure and morphology of

M. N. Osman Zahid et al. (Eds.): iMEC-APCOMS 2019, LNME, pp. 455–460, 2020.
https://doi.org/10.1007/978-981-15-0950-6_69

the MnO_2 is strongly affected their electrochemical properties. MnO_2 exists in various polymorphic forms (α, β, γ and δ), which are different in the spatial arrangement of basic octahedral [MnO_6] units. α-MnO_2 have is received special attention as cathode materials for lithium batteries [5], since the large MnO_2tunnels existing in the crystalline lattice of α-MnO_2 are believed to facilitate the accommodation and transportation of inserting lithium ions.

Synthesis of various MnO_2 nanostructures via a facile and mild hydrothermal route without using any template and surfactant is reported here. Both δ-MnO_2 and α-MnO_2 nanostructures were synthesized based on the hydrothermal reaction of $MnSO_4$ and $KMnO_4$ in aqueous medium and at mild temperatures. Effects of hydrothermal synthesis conditions on the evolution of structural morphology and phase transformation of MnO_2 nanostructures were investigated. The modification of layered structure by changing the hydrothermal parameters have been reported and showed the benefits of layered structure for energy storing mechanism.

2 Methodology

In a typical synthesis, 4.225 g of $MnSO_4$ was mixed with 2.37 g $KMnO_4$ by dissolving it in 40 ml distilled water. The mixture was stirred under magnetic stirring, with 200 ml total volume of the solution. Then the solution was charged into a Teflon(PTFE)-lined autoclave, tightly sealed and oven-heated at different temperature (100 °C, 120 °C, and 140 °C) for 24 h, and then cooled to room temperature naturally. The precipitate was centrifuged and repeatedly washed with deionized water, then dried at 100°C for one hour to obtain the final product. The prepared samples are labelled as MnO_2-a, MnO_2-b and MnO_2-c for 100 °C, 120 °C, and 140 °C, respectively.

3 Characterization

Crystal structure of the final sample was studied by X-ray diffraction (XRD) technique using the Rigaku Miniflex II X-ray diffractometer employing Cu Kα radiation (λ = 1.5406 Å). Morphology and microstructure of the materials were studied by scanning and transmission electron microscopic techniques. The shape and size of the samples were obtained using a field emission scanning electron microscope (JSM-7800F (JEOL, Japan).

For electrochemical evaluation, a working electrode was prepared by mixing the active material with polyvinylidene fluoride (PVDF) (Sigma Aldrich, USA) and carbon black (Super P conductive, Alfa Aesar, UK) in a mass ratio of 85:5:10 and was stirred in N-methyl-2-pyrrolidinone to achieve a homogeneous mixture. The resulting slurry was then pasted on a nickel foam substrate (current collector) and the electrode was dried at 70 °C for 12 h. The dried electrode was then pressed using a pelletizer at 5 tonne. A three electrode system was used to evaluate the electrochemical performance by cyclic voltammetry (CV) using a potentiostat (Autolab PGSTAT 30, Eco Chemie B.V., The Netherlands) employing NOVA 1.9 software in 1 M KOH aqueous electrolyte at room temperature. A platinum rod and a saturated Ag/AgCl electrode were used as the counter and the reference electrodes, respectively. The kinetics at the electrode were studied using electrochemical impedance spectroscopy (EIS) using the AutolabPotentiostat.

4 Discussion

Figure 1(a) shows the XRD recorded for the hydrothermally synthesized MnO_2 nanostructures at different experiment parameter as mentioned in methodology section. The XRD analysis was done in the range of 10°–80° at 2θ° with scan rate of 0.02°/min. It can be seen from the XRD pattern of MnO_2-a (Fig. 1(a)) is indexable to a tetragonal α-MnO_2 phase. The coexisting of potassium ion in the MnO_2 is arising from one of the precursors used to prepare the MnO_2. However, in this study, the characteristic peak represent potassium ion was not observed for all samples. The amount of potassium ion may be very negligible to be detected using XRD. While, XRD pattern of MnO_2-b (Fig. 1(a)) is attributed to the body centered tetragonal, which are in agreement with the standard values (JCPDS 44-1041). MnO_2-b showing an interlayer spacing of 6.6 nm attributed by (110) peak at around 2θ = 12.4°. For layered MnO_2 was formed as the characteristic peaks of the layered structure clearly appear around 2θ = 12.4°. It can be seen that MnO_2-a has larger interlayer spacing of 6.9 nm compared to that of MnO_2-b (6.7 nm). There are several humps in the XRD pattern of the MnO_2-a which are attributed to the presence of water in the layered structure.

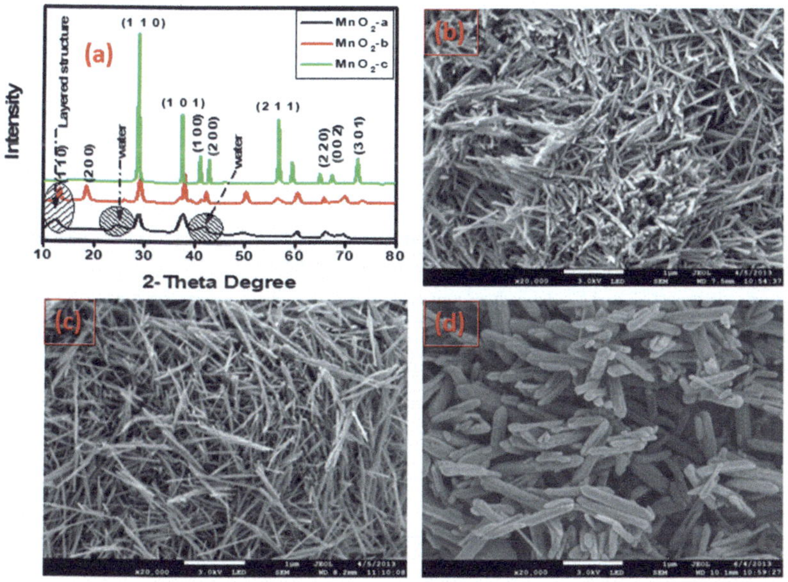

Fig. 1. (a) XRD pattern of the sample, and FESEM images of sample (b) MnO_2-a, (c) MnO_2-b and (d) MnO_2-c.

The size and morphology of each sample were examined by FESEM. All of samples composed of nanorods structure but with different diameter and length. FESEM picture in Fig. 1(b) and (c) shows that the sample MnO_2-a and MnO_2-b is composed of dense bare shape nanorods within the diameter range of 20–30 nm and with various of length. Whereas, sample MnO_2-c (Fig. 1(d)) is having larger diameter compared to other samples ($\sim$200 nm) as a result of high temperature during hydrothermal treatment. This high temperature hydrothermal treatment induced more fusion process between MnO_2 particle and forming larger diameter nanorods.

Figure 2(a) shows the CV profiles at scan rate of 100 mV/s for MnO_2–a, MnO_2-b and MnO_2-c in 1 M of KOH. All samples showed CV profile with the quasi-rectangular shape, which indicating the pseudocapacitance nature of MnO_2. It was almost symmetric between the cathodic and the anodic processes for sample MnO_2-a, which pertained that the samples had been an excellent electrode material for super-capacitor in the alkaline electrolyte. Almost non-symmetrical shape of CV was observed for sample MnO_2-b and MnO_2-c reflecting to the less reversibility of electrode materials. Besides, it can be seen that the area under the curves of CV of MnO_2-a is larger compared to that of MnO_2-b and MnO_2-c, demonstrating the highest of *Cs*.

At this higher scan rates, the ions were intercalated probably only on the surface of the electrode, whereas at lower scan rate, the ions could diffuse into the inner active sites as well, because lower scan rates provide longer lengths of time permitting better intercalation of the ions with the active sites. However, from the CV shape of sample MnO_2-a at this higher scan rate, it can be seen that there are redox peaks appeared which corresponded to the Mn transtition of Mn^{4+} and Mn^{3+}. This could be attributed to the presence of water molecules in the interlayer spacing of the sample (as shown in Fig. 1(A)), whereby leads to the succeeding electrochemical ion diffusion, intercalation/de-intercalation, and transport process, resulting in the redox process of Mn between Mn^{4+} and Mn^{3+}. Thus, higher *Cs* of 678 F/g was obtained by sample MnO_2-a and followed by 196F/g and 91 F/g for sample MnO_2-b and MnO_2-c, respectively as shown in Fig. 2(b). Redox peak is also proven by the non-triangular charge-discharge curve as shown in Fig. 2(c). However, it was observed that all samples are having similar charge-discharge curve since this analysis has been done at low current density. Thus, all samples have longer time for intercalation/de-intercalation process, contributing to the redox reaction. Two mechanisms have been generally proposed for the charge storage process of MnO_2; (1) intercalation of protons from electrolytes ($C^+ = H^+$, Li^+, Na^+, and K^+) in the bulk of electrode material upon reduction, and then, followed by de-intercalation upon oxidation [11]. And (2) surface adsorption of electrolyte cations (C^+) on MnO_2 [11].

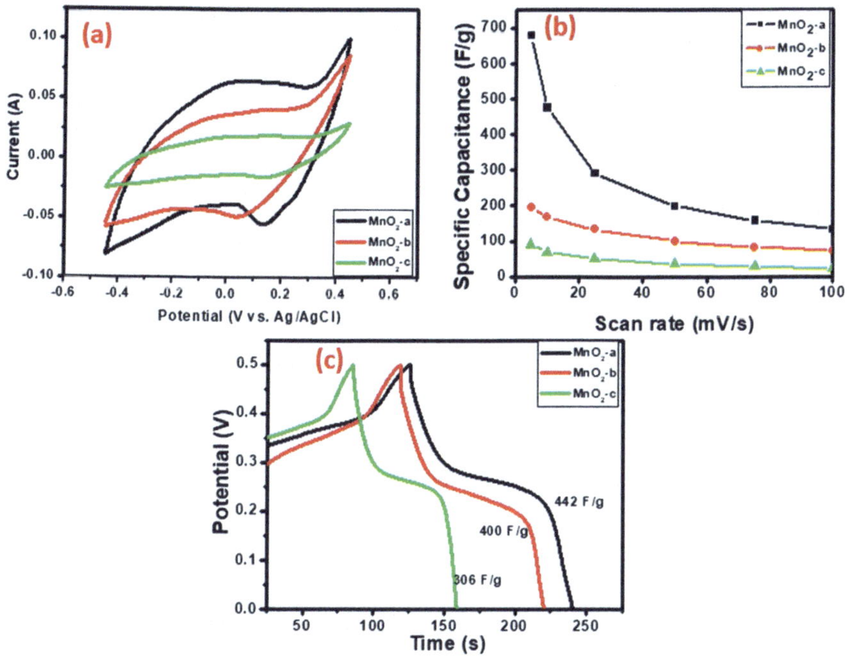

Fig. 2. (a) Cyclic voltammetry of sample at scan rate of 100 mV/s, (b) *Cs* versus scan rate, and (c) Charge-discharge curve at 0.1 A/g current density.

5 Conclusion

In summary, simple and free-template hydrothermal route has been used to prepare MnO_2 with different interlayer spacing and crystal structure but with similar one-dimensional nanorods with various sizes. microstructure. Sample with bigger interlayer spacing (MnO_2-a) exhibited high *Cs* due to the better ion diffusion and intercalation/de-intercalation for redox reaction to be occurred even at high scan rate of cyclic voltammetry. Besides, the presence of water molecules in the interlayer spacing facilitates better ion diffusion process.

Acknowledgement. This project is supported by the Research and Innovation Department of the Universiti Malaysia Pahang through grant number RDU1703156.

References

1. Li, Y., Xie, H., Wang, J., Chen, L.: Preparation and electrochemical performances of α -MnO_2 nanorod for supercapacitor. Mater. Lett. **65**, 403–405 (2011)
2. Yousefi, T., Nozad, A., Hossein, M., Aghazadeh, M.: Facile synthesis of α-MnO one-dimensional (1D) nanostructure and energy storage ability studies. J. Solid State Chem. **190**, 202–207 (2012)

3. Zhang, Y., Mo, Y.: Preparation of MnO_2 electrodes coated by Sb-doped SnO_2 and their effect on electrochemical performance for supercapacitor. Electrochim. Acta **142**, 76–83 (2014)
4. Hao, J., Zhong, Y., Liao, Y., Shu, D., Kang, Z.: Face-to-face self-assembly graphene/MnO_2 nanocomposites for supercapacitor applications using electrochemically exfoliated graphene. Electrochim. Acta **167**, 412–420 (2015)
5. Nagarajan, N., Humadi, H., Zhitomirsky, I.: Cathodic electrodeposition of MnO x films for electrochemical supercapacitors. Electrochim. Acta **51**, 3039–3045 (2006)
6. Yue, G.H., Yan, P.X., Yan, D., Qu, D.M., Fan, X.Y., Wang, M.X., Shang, H.T.: Solvothermal route synthesis of single-crystalline α-MnO_2 nanowires. J. Cryst. Growth **294**, 385–388 (2006)
7. Zolfaghari, A., Ataherian, F., Ghaemi, M., Gholami, A.: Capacitive behavior of nanostructured MnO_2 prepared by sonochemistry method. Electrochim. Acta **52**, 2806–2814 (2007)
8. Wang, H., Qian, D., Lu, Z., Li, Y., Cheng, R., Zhang, W.: Facile synthesis of a -MnO2 nanorods and their electrochemical performances. J. Cryst. Growth 1–8 (2007)
9. Cao, G., Su, L., Zhang, X., Li, H.: Hydrothermal synthesis and catalytic properties of α and β-MnO_2 nanorods. Mater. Res. Bull. **45**, 425–428 (2010)
10. Ji, Z., Dong, B., Guo, H., Chai, Y., Li, Y., Liu, Y.: A facile hydrothermal synthesis and growth mechanism of novel hollow β-MnO_2 polyhedral nanorods. Mater. Chem. Phys. **136**, 831–836 (2012)
11. Toupin, M., Brousse, T., Be, D.: Charge storage mechanism of MnO_2 electrode used in aqueous electrochemical capacitor. Chem. Mater. **16**, 3184–3190 (2004)

Investigation on the Effect of Build Orientation and Heat Treatment on Tensile Strength and Fracture Mechanism of FDM 3D Printed PLA

Nanang Fatchurrohman[1(✉)], Nurul Najihah Najlaa Noor Hamdan[1], Mebrahitom Asmelash Gebremariam[1], and Kushendarsyah Saptaji[2]

[1] Faculty of Mechanical and Manufacturing Engineering, Universiti Malaysia Pahang, 26600 Pekan, Pahang, Malaysia
fatchurrohman@ump.edu.my

[2] Faculty of Engineering and Technology, Mechanical Engineering Department, Sampoerna University, Jakarta, Indonesia

Abstract. Three-dimensional (3D) printing is one of the many popular types additive manufacturing. Current FDM product has low tensile strength due to the printing orientation that affect to the low bonding layer by layer inside the material. Furthermore, experimental work of FDM using different printing orientation are still limited. The aim of this investigation is to characterize the effect of build orientation and heat treatment on the mechanical performance of PLA samples manufactured using fused deposition modelling (FDM) - 3D printer. Specimens were fabricated according to ASTM-D638 type IV. The next investigation was to analyse the effect of build orientation and heat treatment on the printed specimens. Tensile tests were carried out to determine the mechanical response of the printed specimens. The highest result for ultimate strength and yield strength achieved by heat-treated on-edge orientation, 47.84 MPa and 43.94 MPa respectively while the highest elastic modulus is untreated upright orientation, 8.96 GPa. The results showed that different orientations effect the behaviour of tensile strength and yield strength of the 3D printed PLA. Heat treatment process effected the layer bonding of the specimen as it strengthens the bonding between the layer. In addition, the results have highlighted different fracture behaviour for the upright orientation, on-edge and flat orientations.

Keywords: FDM · 3D printing · PLA · Build orientation · Heat treatment

1 Introduction

Nowadays, polylactic acid (PLA) is one of the most commonly used material for 3D printing or known as fused deposition modelling (FDM). The variety of the products that can be produced with 3D printers is growing steadily such as prototypes, moulds, tools, organ in human body parts, prosthetics, toys and furniture [1]. 3D printing can be used to perform product development which involves conceptual design selection,

M. N. Osman Zahid et al. (Eds.): iMEC-APCOMS 2019, LNME, pp. 461–465, 2020.
https://doi.org/10.1007/978-981-15-0950-6_70

which is an activity engaging with numerous types of data including technical-customer specifications and current design developments [2–4]. The 3D printer works by placing the thin lines of molten material side by side and on top of each other to create a printed product. The printers using this technique are fed with a filament of thermoplastic printing material. This filament is melted through an extrusion die connected to a three-dimensional positioning system. The nozzle is controlled to perform a selective material deposition that produces the geometry of the desired component [5]. Precise and simpler design methods are required to accurately perform the static assessment of 3D printing materials. Therefore, 3D printing is an "additive" process that allows more complex-shaped components to be manufactured more effectively than traditional "subtractive" technologies, with a remarkable degree of accuracy in both shape and dimensions [6].

FDM strategy is exceptionally convincing because of its relationship to desktop 3D printers. 3D printing frames a 3D geometry by collecting singular layers of expelled thermoplastic fibre, for example PLA, which have melting temperatures sufficiently low for utilize in melt expulsion in open air non-committed offices [6]. PLA is a biodegradable thermoplastic aliphatic polyester derivative from renewable assets. PLA was one of the highest utilization volumes among bioplastics around the world [8]. However, PLA brittleness and low thermal resistance limit its applications [9]. Current FDM product has low tensile strength due to the printing orientation that affect to the low bonding layer by layer inside the material. Therefore, in this investigation, the objective is to fabricate 3D printed PLA test specimen using FDM using different orientations and heat treatment. The next objective is to perform tensile test and fracture analysis on the specimen. The final objective is to analyse the effect of building orientation and heat treatment on the tensile strength and yield strength.

2 Methodology

Material for this investigation PLA filament 1.75 mm Makerbot was utilised. Test specimen dimension was produced according to ASTM D638 Type IV. This is the standard for polymer/PLA tensile testing. The 3D model of the specimen was designed in a CAD software and transferred to MakerBot software to prepare for the G-codes 3D printing. Figure 1 shows the specimens build orientations during 3D printing.

Each build orientation consisted of five specimens. The total were 30 specimens. 15 out of 30 specimens (5 specimens for each build orientation) undergo heat treatment process using a vacuum oven. The heat treatment temperature was set to 65 °C as this is the glass transition temperature for PLA [10]. It took 30 min for the oven to reach 65 °C from room temperature. Once it reached the desired temperature, test specimen was put in the oven and the temperature remained constant at 65 °C for 15 min. After 15 min of heat treated, test specimen was rested in the oven for 60 min to allow the specimen to slowly cool before proceeding to tensile testing.

The tensile test was performed to express the mechanical and deformation properties of a specimen under a parallel stress at a given rate. The specimens were loaded until they are fractured. The best possible load of the specimens was calculated at the end of the tensile test. The equipment then automatically records all data prior to export

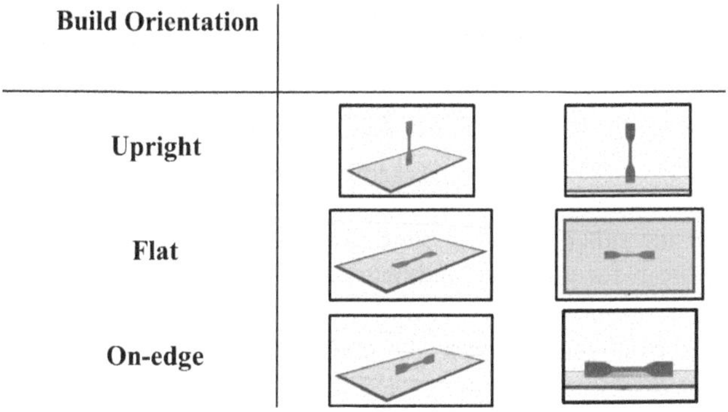

Fig. 1. Isometric view of specimens build orientation.

to provide a graphical representation of the test progress. Fracture analysis was performed by examining cracked specimens. This was to investigate the fracture modes of the untreated and heat-treated test specimens.

3 Result and Discussion

The build orientation significantly influenced the mechanical properties, the ductility and the failure behavior. On-edge and flat orientations showed the highest values for maximum tensile strength, while the upright orientation resulted in the lowest. The results in Table 1 showed a brittle behavior for upright orientation.

Table 1. Average yield strength and ultimate strength.

Build orientation	Heat treatment (65 °C)	Yield strength (MPa)	Ultimate strength (MPa)	Elastic modulus (GPa)
Up-right	Untreated	18.2091	20.3515	8.9698
	Heat-treated	18.6833 (+2.5%)	24.5583 (+17%)	5.3642 (−40%)
Flat	Untreated	30.3996	41.6246	3.0794
	Heat-treated	41.2957 (+26.4%)	45.7108 (+8.9%)	3.1401 (+1.9%)
On-edge	Untreated	41.5920	46.0643	3.2186
	Heat-treated	43.9184 (+5.3%)	47.8352 (+3.7%)	3.3328 (+3.4%)

On-edge and flat orientations, however, showed a ductile behavior with significant plastic deformation. In particular, on-edge samples showed the value of the maximum tensile strain at break as more layers were drawn longitudinally. This result was in agreement with previous finding [8].

Heat-treated specimen was still a ductile material due to a small increasing percent of yield strength therefore both untreated and treated specimens are easy to be plastically deformed. The reason of higher ultimate strength and highest decreasing of elastic modulus for heat-treated specimen as much as 40% for the heat treated specimen is due to the inter-layer bond of the specimen. Next, flat orientation shows the highest increasing percentage of yield strength which is 26.4%. It shows that heat-treated specimens were easy to deform a plastic deformation compared to untreated specimen. Untreated specimen is more ductile as it has a trans-layer failure. From Table 1, ultimate strength and elastic modulus of flat orientation increased 8.9% and 1.9% respectively. On the other hand, on-edge orientation has the smallest differences of mechanical responses for untreated and heat-treated specimen. On-edge orientation has the most optimum results compared to upright and flat orientation. This is because heat treatment only gives a small effect to the inter-layer and trans-layer of the specimen.

The heat treated specimens, on-edge and flat orientations showed the highest values for the maximum tensile strength, while the upright orientation gave the lowest. The maximum yield strength and tear strength for upright orientation are 18.68 MPa and 24.56 MPa, which is lower than for edge orientation and flat.

Figure 2 shows the cross section of untreated and heat-treated specimens after tensile tested. Untreated specimens have void in their printing layer as showed in circle A, B and C for each build orientation. With heat treatment, number of void is reduced as showed in circle A', B' and C'. For upright orientation, the samples were pulled parallel to the layer deposition direction and the load was perpendicular to their fibers, resulting in failure of the inter-layer bond. In the case of edge and flat orientation, the specimens were pulled perpendicular to the layer deposition direction, and therefore, the fibers were pulled parallel to the loading direction, resulting in failure of the trans-layer. The heat treatment strengthened the bond between the layer, thus the inter-layer and trans-layer were stronger. The current findings were in agreement with previous research [11].

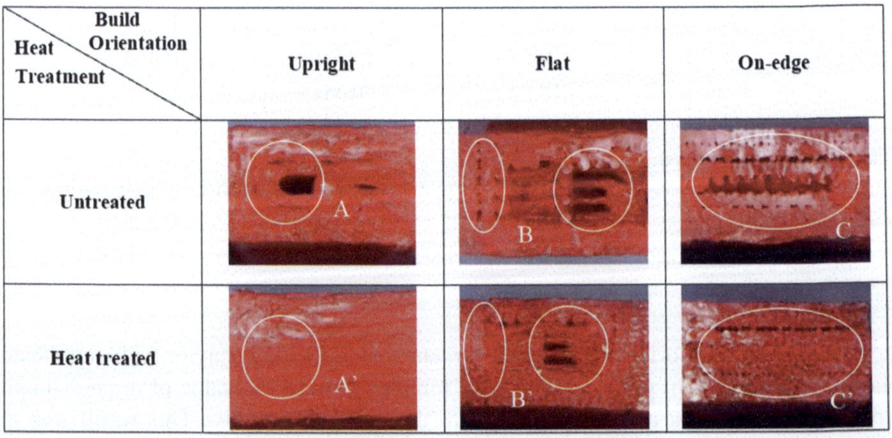

Fig. 2. Fracture analysis of the specimens.

4 Conclusion

This research was accomplished with all the objectives had been achieved. From this investigation, it can be concluded that the upright orientation, on-edge and flat orientations have affected the behaviour of tensile strength and yield strength of the 3D printed PLA. Heat treatment process affected the layer bonding of the specimen as it strengthened the bonding between the layer. In addition, the results have highlighted different fracture behaviour for different printing orientations. Future investigation would be suggested to test the specimens using fatigue test, which is to observe the lifetime of the specimens against fatigue loading.

Acknowledgement.. The authors gratefully appreciated Universiti Malaysia Pahang for the financial support through Internal Research Grant RDU180324.

References

1. Rayna, T., Striukova, L.: From rapid prototyping to home fabrication: how 3D printing is changing business model innovation. Technol. Forecast. Soc. Chang. **102**, 214–224 (2016)
2. Fatchurrohman, N., Sulaiman, S., Sapuan, S.M., Ariffin, M.K.A., Baharudin, B.T.: A new concurrent engineering–multi criteria decision making technique for conceptual design selection. In: Applied Mechanics and Materials, vol. 225, pp. 293–298 (2012)
3. Lu, B., Li, D., Tian, X.: Development trends in additive manufacturing and 3D printing. Engineering **1**(1), 85–89 (2015)
4. Marini, C.D., Fatchurrohman, N., Azhari, A., Suraya, S.: Product development using QFD, MCDM and the combination of these two methods. In: IOP Conference Series: Materials Science and Engineering, vol. 114, no. 1, pp. 012089. IOP Publishing (2016)
5. Ferreira, R.T.L., Amatte, I.C., Dutra, T.A., Bürger, D.: Experimental characterization and micrography of 3D printed PLA and PLA reinforced with short carbon fibers. Compos. B Eng. **124**, 88–100 (2017)
6. Ahmed, A.A., Susmel, L.: Additively manufactured PLA under static loading: strength/cracking behaviour vs. deposition angle. Procedia Struct. Integrity **3**, 498–507 (2017)
7. Fayazbakhsh, K., Movahedi, M., Kalman, J.: The impact of defects on tensile properties of 3D printed parts manufactured by fused filament fabrication. Mater. Today Commun. **18** (December 2018), 140–148 (2019)
8. Chacón, J.M., Caminero, M.A., García-Plaza, E., Núñez, P.J.: Additive manufacturing of PLA structures using fused deposition modelling: effect of process parameters on mechanical properties and their optimal selection. Mater. Des. **124**, 143–157 (2017)
9. Shojaeiarani, J., Bajwa, D.S., Stark, N.M.: Green esterification: a new approach to improve thermal and mechanical properties of poly (lactic acid) composites reinforced by cellulose nanocrystals. J. Appl. Polym. Sci. **135**(27), 46468 (2018)
10. Fargo3D Printing Homepage. https://www.fargo3dprinting.com/annealing-makes-3d-prints-better/. Accessed 19 July 2019
11. Wang, G., Zhang, D., Li, B., Wan, G., Zhao, G., Zhang, A.: Strong and thermal-resistance glass fiber-reinforced polylactic acid (PLA) composites enabled by heat treatment. Int. J. Biol. Macromol. **129**, 448–459 (2019)

Influence of Glass Fiber Content on the Flexural Properties of Polyamide 6-Polypropylene Blend Composites

Nurizzathanis Mohamad Kusaseh[1],
Dewan Muhammad Nuruzzaman[1(✉)],
Mohammad Asaduzzaman Chowdhury[2],
A. K. M. Asif Iqbal[1], Noor Mazni Ismail[1], Nanang Fatchurrohman[1],
and Chan Shin Yi[1]

[1] Faculty of Mechanical and Manufacturing Engineering, University Malaysia Pahang, 26600 Pekan, Pahang Darul Makmur, Malaysia
dmnuruzzaman@ump.edu.my

[2] Department of Mechanical Engineering, Dhaka University of Engineering and Technology, Gazipur, Gazipur 1700, Bangladesh

Abstract. Recently, polymer blend composites are receiving significant attention to fulfill the growing demand from automotive industries because of their improved properties as compared to that of pure polymer blends. In this research work, considering 0%, 3%, 6% and 9% glass fiber content, four different compositions of polyamide 6 (PA6)-polypropylene (PP)-glass fiber composites such as, 80%PA6+20%PP, 77%PA6+20%PP+3%GF, 74%PA6+20%PP+6%GF and 71%PA6+20%PP+9%GF were prepared using an injection molding machine. Different process parameters were taken into consideration in order to produce dog bone shaped specimens free of defects. All flexural tests were performed according to ASTM standard. Results showed that flexural yield strength steadily improves as fiber content increases. Test data showed that flexural stiffness or modulus is lowest for pure blend and it gradually improves with addition in glass fiber. Flexural strength is lowest for pure blend and it gradually develops with increase in fiber content. On the other hand, it was observed that flexural strain is hardly impacted by fiber content.

Keywords: Flexural properties · Polyamide · Polypropylene · Glass fiber

1 Introduction

Thermoplastic polymer materials are reinforced by different types materials in order to strengthen mechanical properties and these polymer composites can be prepared by different manufacturing techniques. During past decade, research efforts were applied on dissimilar composite materials to examine mechanical or tribological properties [1–10]. Different types of reinforcing and matrix materials, different processing methods as well as different operating parameters are responsible for properties of these composites. Fiber reinforced polyamide-6 composite was prepared by application of injection process and flexural properties were examined in detail [11]. Experimental data revealed

M. N. Osman Zahid et al. (Eds.): iMEC-APCOMS 2019, LNME, pp. 466–471, 2020.
https://doi.org/10.1007/978-981-15-0950-6_71

that modulus and flexural strength of reinforced polyamide-6 were much increased than that for polyamide-6 without reinforcement. Experimental data revealed that strength properties and elastic modulus of composites were markedly affected by fiber content and fiber orientation [12]. During injection process, parameters such as fill time, temperature etc. were theoretically examined for pure polyamide material [13]. It was noticed that strain rate was important parameter for properties of composites [14]. Experiments were performed to examine properties of reinforced nylon [15]. It was apparent that modulus and strength of composite were markedly influenced by fiber content. Polypropylene-epoxy based composites were prepared using E-glass fiber reinforcement and it was proved that type of resin has marked influence on impact strength [16]. Research efforts were applied to examine flexure properties for nylon 6-polypropylene based blend composites [17, 18]. It was found that flexure strength, flexure modulus and impact strength of composites were influenced by fiber content.

In this study, four types of polyamide 6-polypropylene based composites were prepared considering 0%, 3%, 6% and 9% glass fiber reinforcement. How the flexural properties such as flexural yield strength, flexure modulus, flexure strength, flexural strain etc. were influenced by fiber content that were examined in detail.

2 Experimental

Four types composite specimens were prepared by the application of injection molding process. Polyamide 6-polypropylene based composite specimens were prepared considering 0%, 3%, 6% and 9% glass fiber reinforcement. Four types of specimens of various compositions such as, 80%PA6+20%PP, 77%PA6+20%PP+3%GF, 74%PA6+20%PP +6%GF and 71%PA6+20%PP+9%GF were processed in conformity with ASTM standard. Considering suitable process parameters, different types of dog-bone shaped composite specimens were prepared. Moreover, temperatures at different sections/zones of barrel such as, rear section and nozzle section were varied depending on the composition of composite. During injection process, temperature control was very critical in order to produce defect-free specimen. After successful preparation of composite specimens, flexural tests were performed to examine flexural properties of these composites. Using a universal test machine, three point bending flexural experiments were performed using crosshead speed of 3 mm/min according to ASTM standard and Bluehill 2 software was used to generate all data. The detailed flexural characteristics such as, flexural yield strength, flexural modulus, flexural strength, flexural strength at break and flexure strain at break were examined for different types of composites.

3 Results and Discussion

Flexural experiments of glass fiber reinforced PA6-PP blend composites were carried out and at slope threshold 0.2%, flexural yield strength results of composites are shown in Fig. 1. At this stress level, material started to deform plastically means that non-reversible deformation occurred at this stress level. Experimental data revealed that pure blend 80% PA6+20%PP started to deform permanently at reasonably low stress level of 38.13 MPa. It is apparent that due to addition of 3%GF, composite showed an improvement in yield strength of 40.88 MPa which is 7.2% greater than that of pure blend. Results showed that 74%PA6+20%PP+6%GF composite exhibited moderately improved yield strength of 46.06 MPa which is nearly 21% higher than that of pure blend. Due to addition of 9% glass fiber, composite showed further improvement in yield strength to 50.03 MPa. It is noticeable that due to more fiber content, flexural yield strength improved steadily and deformation of composite started plastically at reasonably greater stress level.

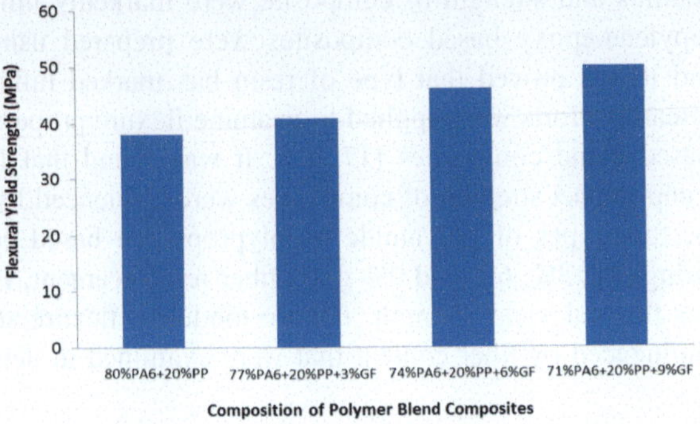

Fig. 1. Flexural yield strength of PA6-PP-GF composites.

The experimental results of flexural modulus are shown in Fig. 2. These results are measure of stiffness of composites to resist bending deformation. The obtained result reveals that 80%PA6+20%PP blend has low modulus of 1.1 GPa. Due to the addition of 3%GF, composite shows little improved modulus of 1.26 GPa, which is 14.5% more than modulus for pure blend. 74%PA6+20%PP+6%GF composite shows 32.7% increased modulus of 1.46 GPa as compared to modulus of pure blend. Finally, due to addition of 9%GF, the composite shows 44.5% improved modulus of 1.59 GPa. From these results, it is evident that resistance to deformation or stiffness improves gradually as fiber content rises.

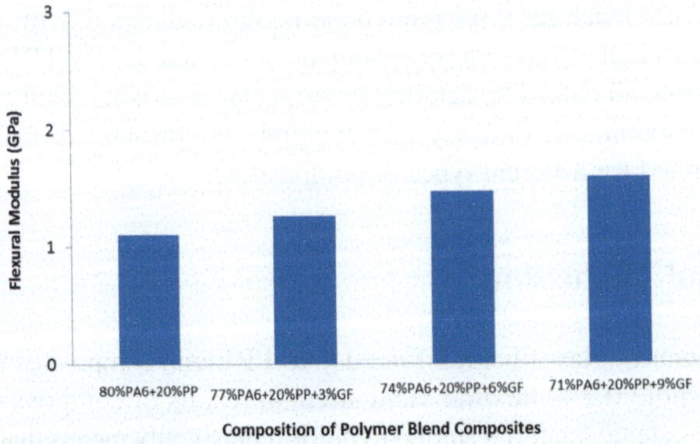

Fig. 2. Flexural modulus of PA6-PP-GF composites.

Flexural strength of PA6-PP-GF composites is presented in Fig. 3 and in experiments, the composites experienced maximum stress at extreme fiber. Experimental data show that 80%PA6+20%PP has flexural strength of 38.85 MPa. With 3% glass fiber addition, composite shows an improved strength of 41.43 MPa which is about 6.6% higher than the strength of pure blend. The obtained result reveals that because of increased fiber content 6%, flexural strength shows 21% increased strength of 46.97 MPa as compared to that for pure blend. Experimental result also reveals that 9% GF composite has reasonably high strength of 51.26 MPa and this is about 32% higher than strength of 80%PA6+20%PP. From these results, it is obvious that flexural strength improves gradually as fiber content increases up to 9%. Figure 4 represents flexural strength at break of the composites. Obtained data show that pure blend has the breaking strength of 38.55 MPa. Due to addition of 3%, 6% and 9% glass fiber, the composite shows the strength of 41.01, 46.42 and 50.11 MPa respectively. These results almost correspond to maximum flexural strengths of composites (Fig. 3), means that, the extreme fiber (outermost layer) of the composite reached to breaking point instantly after reaching the maximum strength under flexure loading.

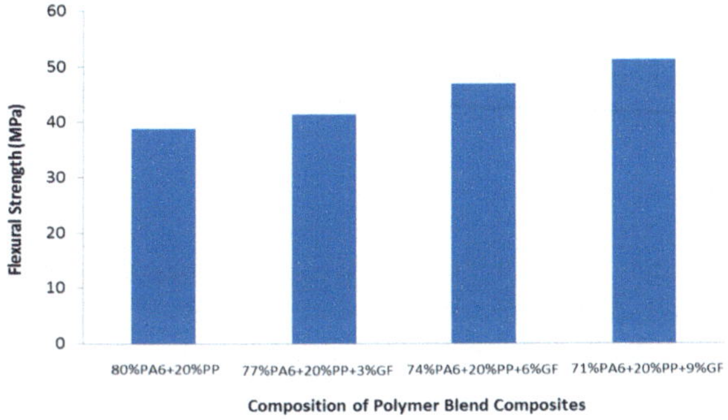

Fig. 3. Flexural strength of PA6-PP-GF composites.

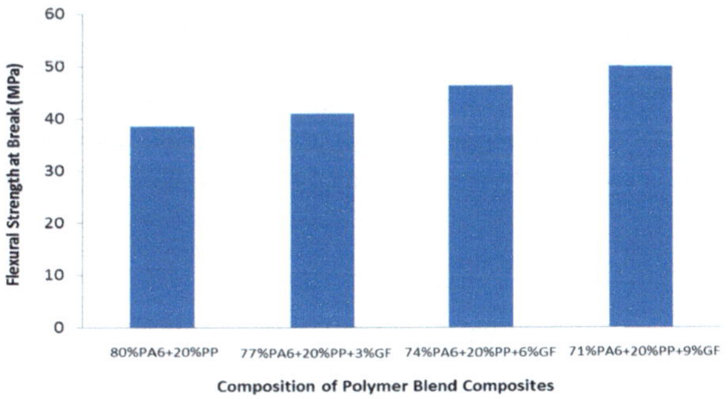

Fig. 4. Flexural strength at break of PA6-PP-GF composites.

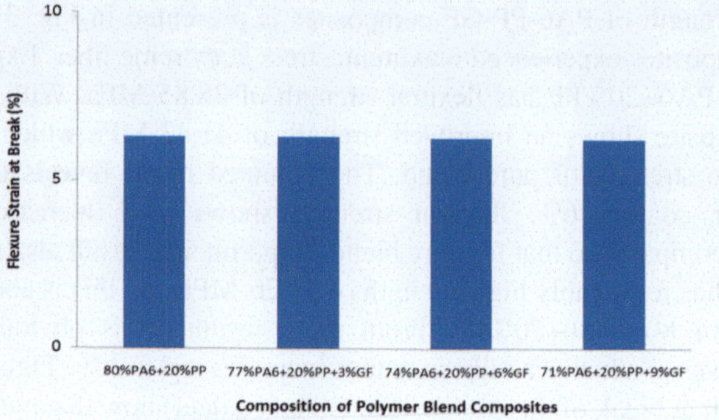

Fig. 5. Flexure strain at break of PA6-PP-GF composites.

Figure 5 represents flexure strain at break (slope threshold 0.2%) of composites. During experiment, this unit deformation (%) was at outermost layer of composite. Experimental data revealed that pure blend has flexure strain of 6.32% at breaking point and composites have 6.31%, 6.27% and 6.23% strain for fiber content of 3%, 6% and 9% respectively. It is distinctly clear that these strain results at breaking point are hardly influenced by fiber content up to 9%.

4 Conclusions

Using injection molding, polyamide 6-polypropylene based composites were successfully prepared considering 0%, 3%, 6% and 9% glass fiber contents. Three point flexural properties of these composites were investigated in detail. It was found that pure blend exhibits low flexural yield strength which improves steadily with increased content of fiber. Pure blend shows low flexural modulus and it rises gradually with increased fiber content. Experimental data also show that pure blend has minimum flexural strength and it rises gradually due to increased fiber content. It is found that flexural strengths almost correspond to breaking strengths. Furthermore, it is evident that flexural strain is hardly influenced by fiber content up to 9%.

Acknowledgements. This work was carried out under the research grant RDU1703131 and the authors are grateful to University Malaysia Pahang (UMP) for financial support. The authors also acknowledge the assistance by technical staffs of faculty of mechanical and manufacturing engineering.

References

1. Landesmann, A., Seruti, C.A., Batista, E.D.M.: Mechainal properties of glass fiber reinforced polymers members for structural applications. Mater. Res. **18**, 1372–1383 (2015)
2. Braga, R.A., Magalhaes, P.A.A.: Analysis of the mechanical and thermal properties of jute and glass fiber as reinforcement epoxy hybrid composites. Mater. Sci. Eng., C **56**, 269–273 (2015)

3. Chaichanawong, J., Thongchuea, C., Areerat, S.: Effect of moisture on the mechanical properties of glass fiber reinforced polyamide composites. Adv. Powder Technol. **27**, 898–902 (2016)
4. Feldmann, M.: The effects of the injection moulding temperature on the mechanical properties and morphology of polypropylene man-made cellulose fibre composites. Compos. Part A: Appl. Sci. Manuf. **87**, 146–152 (2016)
5. Unterweger, C., Bruggemann, O., Furst, C.: Effects of different fibers on the properties of short-fiber-reinforced polypropylene composites. Compos. Sci. Technol. **103**, 49–55 (2014)
6. Hartl, A.M., Jerabek, M., Lang, R.W.: Effect of fiber orientation, stress state and notch radius on the impact properties of short glass fiber reinforced polypropylene. Polym. Test. **43**, 1–9 (2015)
7. Nuruzzaman, D.M., Rahaman, M.L., Chowdhury, M.A.: Friction coefficient and wear rate of polymer and composite materials at different sliding speeds. Int. J. Surf. Sci. Eng. **6**, 231–245 (2012)
8. Mortazavian, S., Fatemi, A.: Effects of fiber orientation and anisotropy on tensile strength and elastic modulus of short fiber reinforced polymer composites. Compos. Part B: Eng. **72**, 116–129 (2015)
9. Ou, Y., Zhu, D.: Tensile behavior of glass fiber reinforced composite at different strain rates and temperatures. Constr. Build. Mater. **96**, 648–656 (2015)
10. Ou, Y., Zhu, D., Zhang, H., Huang, L., Yao, Y., Li, G., Mobasher, B.: Mechanical characterization of the tensile properties of glass fiber and its reinforced polymer (GFRP) composite under varying strain rates and temperatures. Polymers **8**, 196 (2016)
11. Nagakura, M., Tanimoto, Y., Nishiyama, N.: Effect of fiber content on flexural properties of glass fiber-reinforced polyamide-6 prepared by injection molding. Dent. Mater. J. **36**, 415–421 (2017)
12. Bajracharya, R.M., Manalo, A.C., Karunasena, W., Lau, K.: Experimental and theoretical studies on the properties of injection moulded glass fibre reinforced mixed plastics composites. Compos. Part A: Appl. Sci. Manuf. **84**, 393–405 (2016)
13. Nuruzzaman, D.M., Kusaseh, N., Basri, S., Oumer, A.N., Hamedon, Z.: Modeling and flow analysis of pure nylon polymer for injection molding process. In: IOP Conference Series: Materials Science and Engineering, vol. 114, pp. 1–7 (2016)
14. Elanchezhian, C., Ramnath, B.V., Hemalatha, J.: Mechanical behaviour of glass and carbon fibre reinforced composites at varying strain rates and temperatures. Procedia Mater. Sci. **6**, 1405–1418 (2014)
15. Nuruzzaman, D.M., Iqbal, A.K.M.A., Oumer, A.N., Ismail, N.M., Basri, S.: Experimental investigation on the mechanical properties of glass fiber reinforced nylon. IOP Conference Series: Materials Science and Engineering vol. 114, pp. 1–7 (2016)
16. Arikan, V., Sayman, O.: Comparative study on repeated impact response of E-glass fiber reinforced polypropylene & epoxy matrix composites. Compos. Part B: Eng. **83**, 1–6 (2015)
17. Kusaseh, N.M., Nuruzzaman, D.M., Ismail, N.M., Hamedon, Z., Azhari, A., Iqbal, A.K.M. A.: Flexure and impact properties of glass fiber reinforced nylon 6-polypropylene composites. In: IOP Conference Series: Materials Science and Engineering, vol. 319, pp. 1–6 (2018)
18. Nuruzzaman, D.M., Kusaseh, N.M., Chowdhury, M.A., Rahman, N.A.N.A., Oumer, A.N., Fatchurrohman, N., Iqbal, A.K.M.A., Ismail, N.M.: Experimental investigation on flexure and impact properties of injection molded polypropylene-nylon 6-glass fiber polymer composites. In: IOP Conference Series: Materials Science and Engineering, vol. 342, pp. 1–7 (2018)

Effect of Delamination in Drilling of Natural Fibre-Reinforced Composite

Suraya Hamirudin Husin, Nurul Mohd Helmi, Nanang Fatchurrohman, Mebrahitom A. Gebremariam, and Azmir Azhari(✉)

Faculty of Manufacturing and Mechatronics Engineering Technology, Universiti Malaysia Pahang, 26600 Pekan, Malaysia
azmir@ump.edu.my

Abstract. The use of natural fibre-reinforced composites is continuously getting more attention worldwide mainly due to their environmental friendliness. The present work investigates the effect of drilling parameters namely spindle speed and feedrate on the delamination damage of Kenaf fibre-reinforced composite laminates. Two different laminate thicknesses were fabricated (i.e. 4 and 6 mm). It was found that the spindle speed did not significantly affect the delamination damage. Whereas, increasing the feedrate resulted in a higher delamination factor possibly because of a higher thrust force. Meanwhile, the different in laminate thicknesses did not notably influence the delamination damage during the drilling process. The results of the present work are useful to be used for future reference in the study on the drilling of natural fibre-reinforced composite.

Keywords: Natural fibre reinforced composite · Drilling process · Push-out delamination

1 Introduction

Reinforced plastics composite material is a combination of various sources of reinforcement materials and plastics which acts as a binder. As a result, it can withstand a stronger load as the strength and other properties are improved. Reinforcements are used to provide the strength and rigidity at a lower cost. In recent years, there is a tendency to use composite materials in many industries due to its various advantages. However, majority of reinforcement materials or fibres are made from non-biodegradable materials. Lately, there is a huge interest of using natural fibre-reinforced composites (NFRC) due to the environmental and sustainability aspects of engineering materials. NFRC also offer numerous benefits which includes low cost, biodegradability, eco-friendly nature and relatively good mechanical properties [1].

Amongst various machining processes, drilling is a commonly employed technique in industry because of the requirement to assemble components in mechanical pieces and structures. The drilling of composite laminates is significantly affected by the tendency of these materials to delaminate under the action of drilling forces (thrust force) which may result in fibre pull-out and thermal damages especially the polymeric

M. N. Osman Zahid et al. (Eds.): iMEC-APCOMS 2019, LNME, pp. 472–477, 2020.
https://doi.org/10.1007/978-981-15-0950-6_72

matrix [2]. Furthermore, Capello [3] concludes that the delamination is the most critical damage since it can severely impair the performance of the machined components.

Various studies were done to examine the effect of drilling on composite materials. Hocheng and Tsao [4] conducted a study in order to determine the best drill tool, method and operating conditions. They found that the application of special drill bits, pilot hole and back-up plate may help to prevent or reduce the delamination. Maleki et al. [5] found that the HSS twist drill was found to generate a lower thrust force compared to the CoroDrill 854 and CoroDrill 856 thus producing less delamination. In terms of the effect of machining parameters, the feed rate and cutting speed are seen to make the largest contribution to the delamination factor [6]. During machining of four different fibre-reinforced composites namely glass, hemp, jute and banana, Babu et al. [7] concluded that the use of high cutting speed and low feed rate favour the minimum delamination for all composite materials. Therefore, it is interesting to make a further study on the effect of drilling parameters namely spindle speed and feedrate on the delamination during drilling process of Kenaf fibre-reinforced composite laminate.

2 Methodology

Two different thicknesses of composite laminates were fabricated using hand lay-up technique. Woven glass fibre and chopped strand mat Kenaf fibre were used as reinforcement materials. Their properties are shown in Table 1. The matrix use as the binder of the reinforcement was epoxy resin. Epoxy is one of the thermosetting polymers that has been used widely because of its quality and low cost. The properties of the epoxy resin are shown in Table 2. The fibre fabrics were cut into square of 200 mm × 200 mm. The layers were properly stacked with different number of plies as to produce laminates with two thicknesses. Natural Kenaf fibres were layered between the upper and lower plies of glass fibres as to provide a more rigid shape to the laminate. The orientation of glass fibre fabric was kept constant all the time during the lay-up process and they are considered as bidirectional laminates [0°/90°]. Meanwhile, plies of chopped Kenaf fibres as shown in Fig. 1(a) were placed in between the layers. One and two plies of Kenaf fibre were used to produce 4 and 6 mm thicknesses of composite laminates respectively. The open mould concept was used during the hand lay-up process where the load applied was 10 kg. The final fibre volume fraction of the laminate was calculated to be around 0.8.

Table 1. Properties of reinforcement fibres [1].

Fibre	Density (g/cm^3)	Tensile strength (MPa)	Specific tensile strength (MPa)	Elastic modulus (GPa)	Specific elastic modulus (GPa)
Kenaf	1.45	930	641	53	3.5
Glass fibre	2.5	2000–3000	800–1400	70	29

Table 2. Properties of epoxy resin [6].

Density (g/cc)	Elastic modulus (GPa)	Tensile strength (MPa)	Compressive strength (MPa)	Elongation (%)	Cure shrinkage (%)
1.1–1.4	3–6	35–100	100–200	1–6	1–2

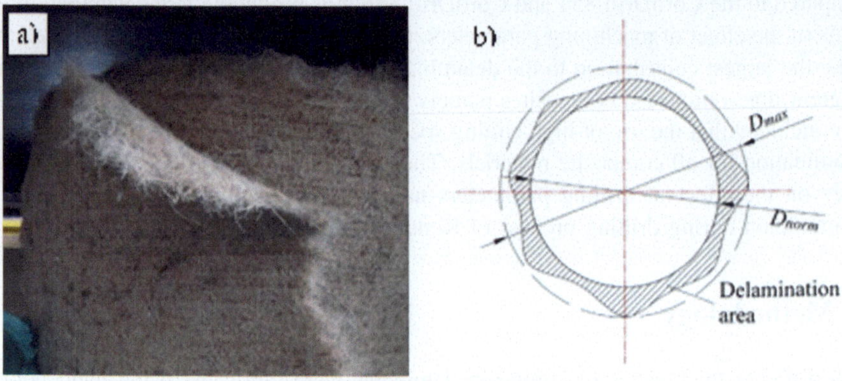

Fig. 1. (a) Kenaf strand mat fibre, (b) Illustration of the maximum diameter, D_{max} and original diameter, D_{norm} [10].

Vertical CNC milling machine was used for the drilling experiment. Diameter of the drilling tools was fixed for all sets of experiment at 10 mm using HSS drill bit. All experiments were conducted without coolant. The machining parameters and their respective levels are shown in Table 3. These parameters were chosen as they are commonly used parameters in the drilling process of composite materials [8, 9]. The design of experiments (DoE) was planned using a full-factorial approach which required nine (9) total number of experiments as presented Table 4.

Table 3. Machining parameters and their respective levels.

Parameter	Spindle speed (RPM)	Feedrate (mm/min)
Level 1 (Low)	800	25
Level 2 (Medium)	1000	46
Level 3 (High)	1200	58

The delamination of drilled holes at the exit of the tool was evaluated using an optical video measuring system for the calculation of delamination factor. The average delamination factor, F_d, for each drilled is calculated as a ratio of the maximum diameter of the damage area, D_{max}, to the drill diameter, D_{norm}, as illustrated in Fig. 1 (b) [10].

Table 4. Experimental layout and results.

Trial no.	Spindle speed (RPM)	Feedrate (mm/min)	F_d for 4 mm thickness	F_d for 6 mm thickness
1	800	25	1.2068	1.2077
2	800	46	1.2402	1.2094
3	800	58	1.2969	1.2892
4	1000	25	1.263	1.2341
5	1000	46	1.2531	1.2851
6	1000	58	1.2027	1.2889
7	1200	25	1.1634	1.2166
8	1200	46	1.2404	1.2102
9	1200	58	1.2628	1.2149

3 Results and Discussions

The results of delamination factors for both laminates thicknesses at different spindle speed and feedrate are displayed in Table 4. The effect of feedrate and spindle speed on the delamination factor is shown in Fig. 2. It is to note that the delamination factor was measured at the exit of the hole. This is due to the fact that more significant delamination occurs at the exit of the hole or push-out delamination as compared to the peel-up delamination (hole entry) [5]. In general as observed from Fig. 2, increasing the feedrate produces a higher delamination for both thicknesses of laminates. This can be attributed to the increase of thrust force with an increase of feedrate during the drilling process. A higher friction between the cutting edges of the drill bit with the fibre particularly at a higher feedrate may accelerate the increase of the thrust force. Furthermore, the increase in thrust force is due to the fact that the shear area is elevated while increasing the feedrate. As a result, a higher thrust force tends to cause a bigger damage to the hole thus increasing the amount of push-out delamination. In contrast, Maleki et al. [5] observed unexpectedly a reverse pattern where they found a decrease in the delamination size with a rise in feedrate. They attributed it to the higher heat generation for a lower feedrate which increases matrix softening thus increasing the delamination size. However, a similar finding with the present study was observed when using drill bit CoroDrill 854. Furthermore, it has been widely reported that the drill bit type largely contributes to the effect of delamination size [5, 6, 11]. In case of the spindle speed, there is no obvious trend on its effect in causing the delamination damage as shown in Fig. 2. This similar effect has been reported previously where the effect of spindle speed is small or negligible compared to the feed rate on the delamination damage [1, 6]. Similarly, there is no clear trend on the delamination damage for different laminate thicknesses (i.e. 4 and 6 mm) as shown in Fig. 2.

The optical images of push-out delamination damage are shown in Figs. 3 and 4 for both laminate thicknesses respectively. It is obvious that for both laminate thicknesses, the severity of uncut fibres and delamination damage is more significant while drilling at a higher feed rate. Due to the difference in the nature of fibres and in the mechanisms

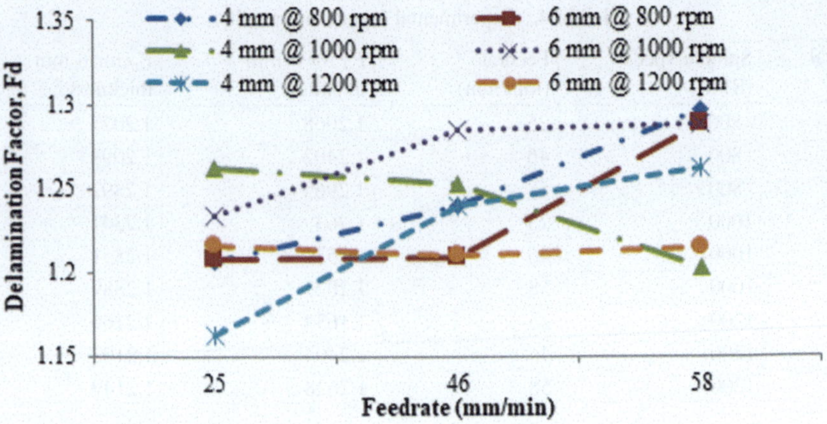

Fig. 2. Effect of delamination factor for different thicknesses of laminates at different cutting speeds and feedrates.

of material removal in natural fibre-reinforced composites, it is relatively difficult to establish the optimum drilling parameters. Furthermore, as previously mentioned, the delamination damage for both thicknesses is not significantly differed from one to another. Possibly, the different in thickness for both laminates is very small (i.e. 2 mm) hence no considerable effect on the delamination damage.

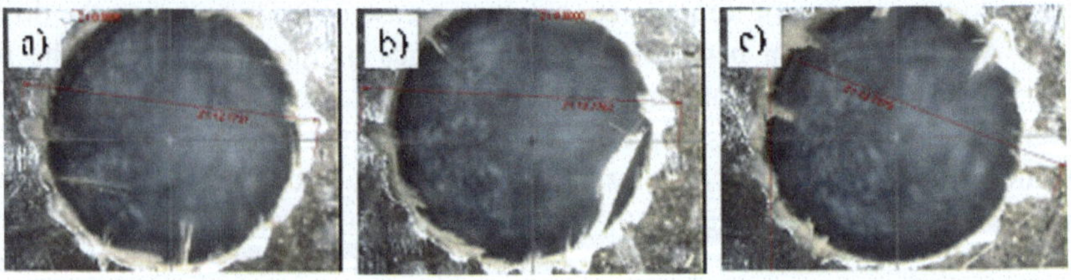

Fig. 3. Push-out delamination for the 4 mm thickness laminate at spindle speed of 1200 rpm and feedrate of, (a) 25 mm/min, (b) 46 mm/min, and (c) 58 mm/min.

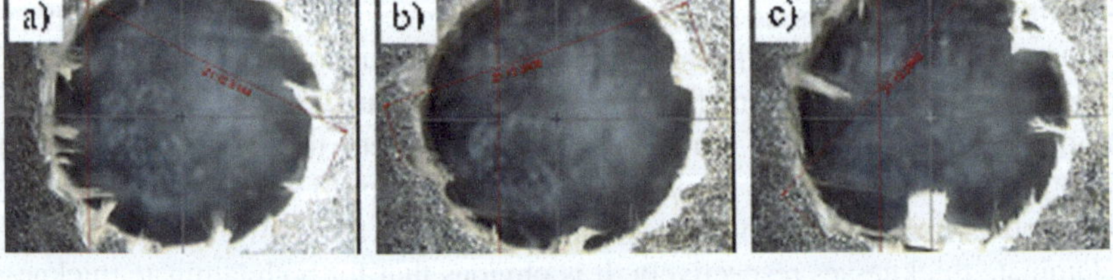

Fig. 4. Push-out delamination for the 6 mm thickness laminate at spindle speed of 1200 rpm and feedrate of, (a) 25 mm/min, (b) 46 mm/min, and (c) 58 mm/min.

4 Conclusion

Based on the findings in the present work, it can be concluded that the drilling parameters can affect the delamination damage of natural fibre-reinforced composite. The spindle speed did not significantly influence the delamination damage. Meanwhile, the feedrate had greatly contributed to the delamination damage especially at a higher feedrate. In term of laminate thickness, it can be concluded that it has no significant effect on the delamination damage.

Acknowledgement. Authors would like to gratefully acknowledge the financial support from the Universiti Malaysia Pahang through RDU1703313.

References

1. Lotfi, A., Li, H., Dao, D.V., Prusty, G.: Natural fiber–reinforced composites: a review on material, manufacturing, and machinability. J. Thermoplast. Compos. Mater. (2019, online)
2. Ogawa, K., Aoyama, E., Inoue, H., Hirogaki, T., Nobe, H., Kitahara, Y., Katayama, T., Gunjima, M.: Investigation on cutting mechanism in small diameter drilling for GFRP (thrust force and surface roughness at drilled hole wall). Compos. Struct. **38**, 343–350 (1997)
3. Capello, E.: Workpiece damping and its effect on delamination damage in drilling thin composite laminates. J. Mater. Process. Technol. **148**, 186–195 (2004)
4. Hocheng, H., Tsao, C.: The path towards delamination-free drilling of composite materials. J. Mater. Process. Technol. **167**, 251–264 (2005)
5. Rezghi Maleki, H., Hamedi, M., Kubouchi, M., Arao, Y.: Experimental investigation on drilling of natural flax fiber-reinforced composites. Mater. Manuf. Processes **34**, 283–292 (2019)
6. Vigneshwaran, S., Uthayakumar, M., Arumugaprabu, V.: Review on machinability of fiber reinforced polymers: a drilling approach. Silicon **10**, 2295–2305 (2018)
7. Babu, G.D., Babu, K.S., Gowd, B.U.M.: Effect of machining parameters on milled natural fiber-reinforced plastic composites. J. Adv. Mech. Eng. **1**, 1–12 (2013)
8. Geng, D., Liu, Y., Shao, Z., Lu, Z., Cai, J., Li, X., Jiang, X., Zhang, D.: Delamination formation, evaluation and suppression during drilling of composite laminates: a review. Compos. Struct. **216**, 168–186 (2019)
9. Upputuri, H.B., Nimmagadda, V.S.: Optimization of drilling process parameters used in machining of glass fiber reinforced epoxy composite. Mater. Today: Proc. (2019, in press)
10. Abdul Nasir, A.A., Azmi, A.I., Lih, T.C., Shuaib, N.A.: Experimental study towards determination of critical feed for minimization of delamination damage in drilling flax natural fibre composites. Procedia CIRP **77**, 191–194 (2018)
11. Azmir, M.A., Nair Sivasankaran, P., Hamedon, Z.: Experimental study on drilling process of CFRP composite laminate. Mater. Sci. Forum **638**, 927–932 (2010)

Investigation on Microstructure and Hardness of Aluminium-Aluminium Oxide Functionally Graded Material

Dewan Muhammad Nuruzzaman[1](✉), A. K. M. Asif Iqbal[1], Maziyana Marzuki[1], Mohammad Asaduzzaman Chowdhury[2], Noor Mazni Ismail[1], Muhammad Ihsan Abdul Latiff[1], Md. Mustafizur Rahman[1], and Mebrahitom Asmelash Gebremariam[1]

[1] Faculty of Mechanical and Manufacturing Engineering, Universiti Malaysia Pahang, 26600 Pekan, Pahang Darul Makmur, Malaysia
dmnuruzzaman@ump.edu.my

[2] Department of Mechanical Engineering, Dhaka University of Engineering and Technology, Gazipur, Gazipur 1700, Bangladesh

Abstract. This study investigated the microstructure and hardness of aluminium-aluminium oxide (Al-Al_2O_3) functionally graded material (FGM). Preparation of metal-ceramic functionally graded material was carried out following powder metallurgy (PM) route. Four-layered aluminium-aluminium oxide (Al-Al_2O_3) graded composite structure was processed using 0%, 5%, 10% and 15% (from first layer to fourth layer) aluminium oxide as ceramic concentration. A cylindrical steel die was used for the fabrication process of the FGM green compact. The green compact was prepared by applying cold pressing technique using a hydraulic press. The sintering process was implemented at 600 °C sintering temperature and 3 h sintering time using 2-step cycle. Microstructural characterization of the sample was conducted layer by layer using high resolution optical microscopy (OM). Hardness of the sample was also performed layer by layer using Vickers microhardness tester. The obtained results revealed that there is a uniform ceramic particle distribution within the metallic phase. From the microstructural observation it was clear that smooth transition occurred from one layer to next layer and each interface was distinct. It was also observed that there is a steady increase in layer hardness with the increase in ceramic concentration.

Keywords: Microstructure · Hardness · Aluminum · Aluminium oxide

1 Introduction

For diverse engineering applications, new generation engineering material known as functionally graded material (FGM) possesses multifunctional characteristics in one component. Within FGM, each composition gradually changes from one side to other side along thickness direction. For desired properties, FGM can be processed by combining the properties of metal and ceramic. For preparation of these graded materials, different processing methods, such as powder metallurgy, thermal spraying, vapor deposition,

M. N. Osman Zahid et al. (Eds.): iMEC-APCOMS 2019, LNME, pp. 478–483, 2020.
https://doi.org/10.1007/978-981-15-0950-6_73

centrifugal casting etc. were followed by researchers [1–3]. Powder metallurgy is a method to achieve the desired properties and sintering is an important process to produce homogeneous or inhomogeneous structure free of defects [4–6]. Different combinations of graded composite materials were investigated theoretically and experimentally [7–10]. The properties of these composite structures were markedly influenced by sintering parameters, material composition, layer number, applied load, sintering environment etc. Multi-layered Graded composite materials were prepared by the application of laser melting deposition method and it was reported that there is strong bonding between substrate and deposited material [11]. Graded composite panels prepared by aluminum/high-density polyethylene were analysed theoretically and experimentally [12]. It was concluded that experimental results were agreed well with theoretical predictions. Characteristics of aluminium based composite systems were analyzed and it was revealed that hardness of outer parts of composites is more than that of inner parts or middle parts [13]. Moreover, strength of outer region was greater than strength of inner region. Al/SiC graded composite material was manufactured by centrifugal casting method and the characteristics such as, microstructure, hardness etc. were analyzed [14]. Experimental results showed that properties of the composites were varied by applied load, sliding speed etc. Six-layered nickel-alumina functionally graded composite structure was prepared by pressureless sintering and microstructural observation confirmed that gradation of composite took place from first layer to sixth layer [15].

In this study, graded composite specimen of aluminium-aluminium oxide (Al-Al_2O_3) was prepared considering 0%, 5%, 10% and 15% ceramic concentration. The graded composite structure was prepared through powder metallurgy method and 2-step sintering cycle. The properties such as microstructure and hardness were examined layer by layer.

2 Experimental

Four-layered Al-Al_2O_3 graded composite specimen was prepared considering 0%, 5%, 10% and 15% ceramic (Al_2O_3) concentration and through powder metallurgy method. In the fabrication process, three major steps were involved such as, mixing/blending of powders, compacting and sintering of specimen. At first stage, based on molecular weight of aluminium and aluminium oxide powders, the constituent of two materials for each layer composition was measured. After that, by allowing sufficient time, the constituents of two powder materials for each layer were sieved, mixed and blended in order to achieve good homogeneous mixture for each layer. During powder stacking process, the composition of powder was made in the order from 85%Al+15%Al_2O_3, 90%Al+10%Al_2O_3, 95%Al+5%Al_2O_3 and 100%Al+0%Al_2O_3 at the end of a cylindrical steel die (diameter 30 mm) made of alloy steel. The cold compaction of specimen was carried out by applying 30 ton (294.3 kN) uniaxial load manually using a hydraulic press. After application of loading, sufficient time was allowed so that the specimen (green compact) became at the stage of stable condition. Then, the specimen was carefully ejected from the die so that no wear or tear of specimen surface occurred. At this stage, the green compact was very fragile and had low cohesive strength. The graded composite specimen was sintered by a sintering furnace using a two-step sintering profile. The sintering process was carried out at 600 °C sintering temperature and 3 h sintering time. The sintered specimen was prepared for microstructural study. Characterization of specimen was performed layer by layer using metallurgical microscope. Finally, hardness of the sample was measured using a Vickers micro-hardness tester. Hardness measurements were carried out layer by layer for a dwell time 15 s using 300 gf (2.94 N) loading. Along the longitudinal axis (thickness direction) of the specimen, average hardness of 10 readings (within each layer) was considered as hardness value for each layer.

3 Results and Discussion

The microstructure of four-layered Al-Al_2O_3 graded composite specimen was examined using a high resolution optical microscope. Characterization of sintered specimen was carried out for each metal-ceramic graded layer to determine whether ceramic particle distribution within the metallic phase is uniform or not. At first step, the sintered specimen was cut along the longitudinal direction and then it was mounted by cold mounting process. After grinding and polishing of the mounted specimen, microstructural observations were carried out and the resulting micrographs/images for all layers are presented in Fig. 1(a)–(d).

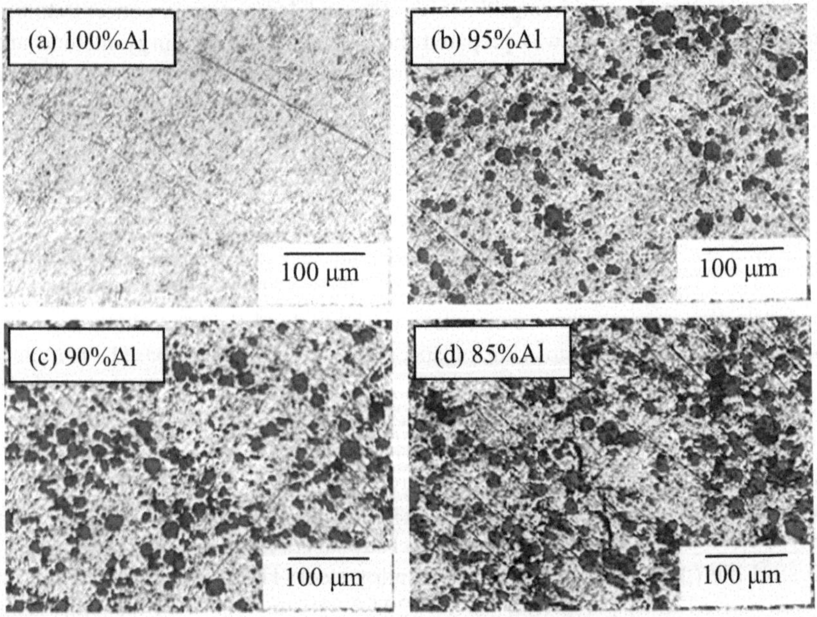

Fig. 1. Microstructures of Al-Al_2O_3 graded composite layers (a) 100% Al - 1st layer (b) 95%Al +5%Al_2O_3 - 2nd layer (c) 90%Al+10%Al_2O_3-3rd layer and (d) 85%Al+15%Al_2O_3-4th layer.

The microstructure of Fig. 1(a) consists only pure Al for first layer whereas, Fig. 1 (b), (c) and (d) consist of 95%Al+5%Al_2O_3, 90%Al+10%Al_2O_3 and 85%Al+15% Al_2O_3 for second, third and fourth layers respectively. It was noticed that the lighter phase is Al matrix and the darker phase is Al_2O_3. From these micrographs, it is distinctly evident that there is almost uniform dispersion of Al_2O_3 particles within Al matrix for second, third and fourth layers. It is also apparent that Al_2O_3 ceramic percentage is increased from second layer until fourth layer (5% to 15%).

Figure 2(a)–(c) show the interfaces for first-second layer, second-third layer and third-fourth layer by AA′, BB′ and CC′ respectively. In relation to these interfaces, no cracking was observed in between the layers or across any interface means that good powder stacking process, good preparation of each layer, good compaction process and good metal-ceramic bonding. It can also be observed that within graded structure, there is continuous and graded microstructure across each interface of adjacent two layers, as reported in previous investigation [8]. Moreover, it is distinctly evident that all interfaces are nearly straight and parallel to each other which establish that in order to prepare the graded composite specimen, all processes were well implemented.

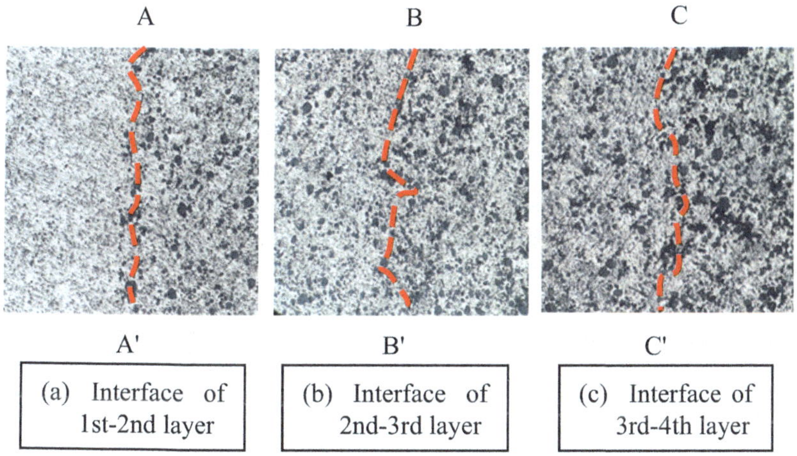

Fig. 2. Four layered Al-Al_2O_3 graded composite: Micrographs of interface: (a) 1st-2nd layer interface (AA′) (b) 2nd-3rd layer interface (BB′) (c) 3rd-4th layer interface (CC′).

Figure 3 shows the hardness variation for each layer and test results revealed that average hardness of first layer (100%Al) was 31.54 HV. From the indentation results, average hardnesses of second, third and fourth layers were measured as 32.08 HV, 32.52 HV and 32.94 HV. It is very clear that hardness value is steadily increased at each layer when the ceramics content is increased at the corresponding layer. These obtained results are well agreed with the results obtained by previous investigation [8]. The increment of hardness for each layer was reasonably consistent as aluminium oxide particles were well dispersed within aluminium matrix. Furthermore, it is distinctly evident that hardness increases layer by layer of the graded structure very consistently.

It is convinced that because of high hardness of ceramic particles and good interfacial bonding between metal and ceramic particles, continuous improvement in hardness occurred along the gradient direction of composite specimen [11, 16].

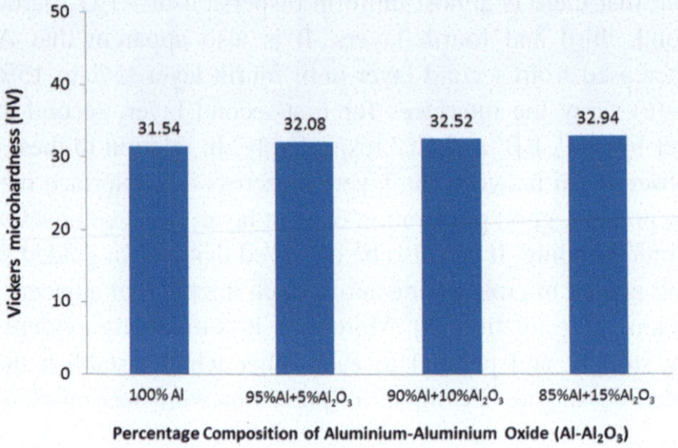

Fig. 3. Hardness variation in each layer of four-layered Al-Al_2O_3 graded composite structure.

4 Conclusions

Four-layered aluminium-aluminium oxide graded composite structure was prepared considering 0%, 5%, 10% and 15% ceramic concentration. Microstructural observations revealed that there is uniform dispersion of ceramic particles within the matrix for second, third and fourth layers of the graded composite structure. No cracking was observed in between the layers or across any interface. Results also show that there is continuous and gradient microstructure across each interface of adjacent two layers. It is understood that all interfaces are nearly straight and parallel to each other. It is believed that preparation processes of graded composite structure were well implemented. Moreover, experimental data show that there is continuous improvement in hardness along the gradient direction of the composite structure.

Acknowledgement. This research work was a preliminary investigation in relation to fundamental research grant scheme (FRGS) RDU190141 and the authors are grateful to Universiti Malaysia Pahang (UMP) for financial support. The authors also acknowledge the assistance by technical staffs of faculty of mechanical and manufacturing engineering.

References

1. Khoddami, A.M., Sabour, A., Hadavi, S.M.M.: Microstructure formation in thermally-sprayed duplex and functionally graded NiCrAlY/Yttria-stabilized zirconia coatings. Surf. Coat. Technol. **201**, 6019–6024 (2007)
2. Watanabe, Y., Hattori, Y., Sato, H.: Distribution of microstructure and cooling rate in Al-Al_2Cu functionally graded materials fabricated by a centrifugal method. J. Mater. Process. Technol. **221**, 197–204 (2015)

3. Wu, K., Scheler, S., Park, H.S., Porada, M.W.: Pressureless sintering of ZrO_2-$ZrSiO_4$/NiCr functionally graded materials with a shrinkage matching process. J. Eur. Ceram. Soc. **33**, 1111–1121 (2013)
4. Sun, L., Sneller, A., Kwon, P.: Fabrication of alumina/zirconia functionally graded material: From optimization of processing parameters to phenomenological constitutive model. Mater. Sci. Eng., A **488**, 31–38 (2008)
5. Nuruzzaman, D.M., Kamaruzaman, F.F.B.: Processing and mechanical properties of aluminium-silicon carbide metal matrix composites. In: IOP Conference Series: Materials Science and Engineering, vol. 114, pp. 1–7 (2016)
6. Shahrjerdi, A., Mustapha, F., Bayat, M., Sapuan, S.M., Majid, D.L.A.: Fabrication of functionally graded hydroxyapatite-titanium by applying optimal sintering procedure and powder metallurgy. Int. J. Phys. Sci. **6**, 2258–2267 (2011)
7. Tahvilian, L., Fang, Z.Z.: An investigation on thermal residual stresses in a cylindrical functionally graded WC-Co component. Mater. Sci. Eng., A **557**, 106–112 (2012)
8. Erdemir, F., Canakci, A., Varol, T.: Microstructural characterization and mechanical properties of functionally graded Al2024/SiC composites prepared by powder metallurgy techniques. Trans. Nonferrous Met. Soc. China **25**, 3569–3577 (2015)
9. Jamaludin, S.N.S., Basri, S., Hussain, A., Al-Othmany, D.S., Mustapha, F., Nuruzzaman, D. M.: Three-dimensional finite element modeling of thermomechanical problems in functionally graded hydroxyapatite/titanium plate. Math. Probl. Eng. **2014**, 1–20 (2014)
10. Jamaludin, S.N.S., Basri, S., Mustapha, F., Nuruzzaman, D.M., Latiff, M.I.A., Ismail, N.M.: Phase contamination characterization of stepwise-built functionally graded hydroxyapatite/titanium (HA/Ti) sintered under various atmospheres. Mater. Sci. Forum **889**, 90–95 (2017)
11. Zhang, C.H., Zhang, H., Wu, C.L., Zhang, S., Sun, Z.L., Dong, S.Y.: Multi-layer functional graded stainless steel fabricated by laser melting deposition. Vacuum **141**, 181–187 (2017)
12. Lin, Q., Chen, F., Yin, H.: Experimental and theoretical investigation of the thermo-mechanical deformation of a functionally graded panel. Eng. Struct. **138**, 17–26 (2017)
13. Radhika, N., Raghu, R.: Development of functionally graded aluminium composites using centrifugal casting and influence of reinforcements on mechanical and wear properties. Trans. Nonferrous Met. Soc. China **26**, 905–916 (2016)
14. Prabhu, T.R.: Processing and properties evaluation of functionally continuous graded 7075 Al alloy/SiC composites. Arch. Civil Mech. Eng. **17**, 20–31 (2017)
15. Latiff, M.I.A., Nuruzzaman, D.M., Basri, S., Ismail, N.M., Jamaludin, S.N.S., Kamaruzaman, F.F.: Preparation and characterization of 6-layered functionally graded nickel-alumina (Ni-Al2O3) composites. In: IOP Conference Series: Materials Science and Engineering, vol. 342, pp. 1–8 (2018)
16. Kamaruzaman, F.F, Nuruzzaman, D.M., Ismail, N.M., Hamedon, Z., Iqbal, A.K.M.A., Azhari, A.: Microstructure and properties of aluminium-aluminium oxide graded composite materials. IOP Conference Series: Materials Science and Engineering, vol. 319, pp. 1–6 (2018)

The Effect of MAPE Compatibilizer Agent on the Tensile Strength of Recycled PET/HDPE Plastic Composite

Nik Ruqiyah Nik Hassan[1], Noor Mazni Ismail[1(✉)], Dewan Muhammad Nuruzzaman[1], Noraini Mohd Razali[1,2], and Suriati Ghazali[2]

[1] Faculty of Manufacturing Engineering, Universiti Malaysia Pahang, 26600 Pekan, Malaysia
drmazni@ump.edu.my

[2] Faculty of Chemical and Nature Resources Engineering, University Malaysia Pahang, Lebuhraya Tun Razak, 26300 Gambang, Kuantan, Pahang, Malaysia

Abstract. This paper reports the effect of compatibilizer agent, polyethylene-grafted-maleic anhydride (MAPE) on the tensile strength of recycled polyethylene terephthalate (rPET)/high density polyethylene (HDPE) plastic composite. The 10% rPET/90% HDPE plastic composites with and without the addition of compatibilizer agent MAPE were prepared by using hot pressing machine. The tensile test was performed using universal testing machine. The result shows that the tensile strength of rPET/HDPE composite with the addition of MAPE increased up to 156.3 MPa, about 6 times higher than pure HDPE (24.26 MPa). The increased tensile strength seen in rPET/HDPE/MAPE composite can be suggested due to the compatibilizer agent MAPE increased the interfacial bonding between rPET and HDPE, which consequently increased the strength of the composite. The finding was supported by the SEM analysis done on the fractured surface and FTIR analysis presenting the intermolecular bonding of rPET/HDPE composite.

Keywords: Polyethylene Terephthalate (PET) · High Density Polyethylene (HDPE) · Polyethylene-grafted-maleic anhydride (MAPE)

1 Introduction

The increasing number of plastic waste has become one of the major environmental issues due to the plastics are non-biodegradable. One of the most common plastic waste is PET because it is frequently used as commercial plastic in our daily lives as well as in industries [1]. Many researchers nowadays focus to reduce the PET waste by recycling the used PET and turn it into useful product. However, it is not easy and there are many challenges to face when working with recycled PET or any other plastics. These include the incompatibility between the recycled plastic and the matrix, which could affect the mechanical properties of the composite [2, 3], and also the properties of the recycled plastic itself prone to downgrade due to the degradation and decomposition [4]. On another note, HDPE is one of the thermoplastics that has average

M. N. Osman Zahid et al. (Eds.): iMEC-APCOMS 2019, LNME, pp. 484–489, 2020.
https://doi.org/10.1007/978-981-15-0950-6_74

mechanical properties with the tensile strength in a range of 23.00 MPa to 29.5 MPa [5, 6]. Even there are limited researches investigate the use of recycled PET, particularly as a filler in HDPE matrix and the use of MAPE compatibilizer in rPET/HDPE composite, there are common researches reported that the mechanical properties of polymer could be improved by introducing additive or compatibilizer agent in polymer matrix [7, 8]. Therefore, in this research, the experiment focuses more on the effect of rPET as a filler and MAPE as a compatibilizer in rPET/HDPE composite. The effect of both rPET filler and MAPE compatibilizer in improving the tensile strength of HDPE were studied accordingly.

2 Experimental

In this experiment, three different types of rPET/HDPE composites were prepared by using hot-press machine. The composites are rPET/HDPE/0%MAPE, rPET/HDPE/5% MAPE, and rPET/HDPE/10%MAPE. The rPET acted as a filler in HDPE matrix, and MAPE acted as a compatibilizer. The rPET plastic bottles were first collected and ground into flakes size (Fig. 1). Then, 10% of rPET flakes were mixed with the 90% HDPE pellet. 5% of MAPE were added into rPET/HDPE/5%MAPE sample, and 10% of MAPE were added into rPET/HDPE/10%MAPE sample. No MAPE was added into rPET/HDPE/0%MAPE sample. The hot pressing temperature was set at 135 °C, which was the melting temperature of HDPE. After the hot pressing process, the tensile test was performed by using universal testing machine (UTS) based on ASTM D638 standard. The test was conducted with a crosshead speed of 2 mm/min. The average of five readings were taken for each type of rPET/HDPE composite. The characteristics of the composite were obtained from the sample characterization using tensile test, scanning electron microscope (SEM) and fourier-transform infrared spectroscopy (FTIR).

Fig. 1. Raw materials: HDPE pellet and rPET flakes

3 Result and Discussion

Figure 2 shows the tensile strength of pure HDPE and rPET/HDPE composites. It was observed that the rPET/HDPE composite without the addition of MAPE has lower tensile strength (7.41 MPa) than pure HDPE (24.26 MPa). The lower strength most probably due to poor interfacial bonding between rPET and HDPE, acted as a weak spot in the composite. Meanwhile, the tensile strength of samples with the addition of 5% MAPE and 10% MAPE increased significantly up to 116.9 MPa and 156.3 MPa, respectively. High tensile strength can be suggested due to strong interfacial bonding between rPET and HDPE improved by MAPE compatibilizer.

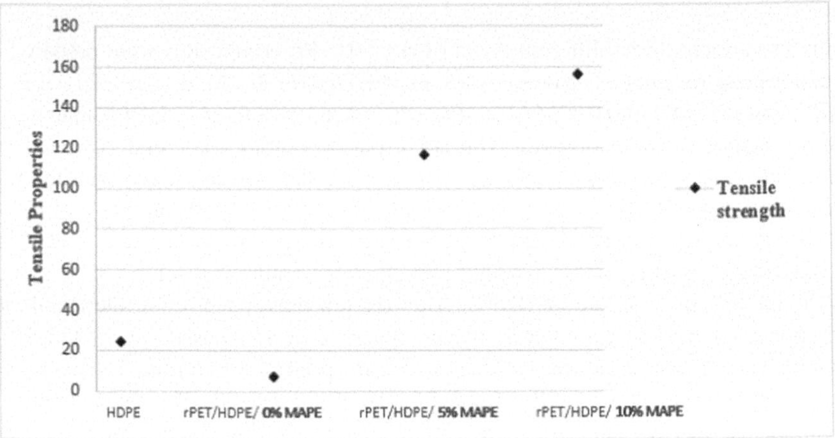

Fig. 2. Tensile strength of HDPE and rPET/HDPE composites

Figure 3 presents the SEM images of composite fractured surface after tensile test. Figure 3(a) displays that the rPET/HDPE sample without the addition of MAPE has poor interfacial adhesion with clear gaps seen in between rPET and HDPE phases. The poor interfacial adhesion has caused the tensile strength to decrease lower than the pure HDPE. The weak filler-matrix interface between rPET and HDPE also eased the pull out of rPET when stress applied during tensile test. In Fig. 3(b), the SEM images of rPET/HDPE sample with the addition of 5% MAPE shows that the interfacial adhesion between rPET and HDPE was significantly improved. The intact surface and significant white line tear exposed the fact that HDPE was also pulled out during the tensile test, indicating strong adhesion between the HDPE and RPET [9], as thus increased the strength of the composite. Similar observation seen in rPET/HDPE sample with the 10% MAPE as shown in Fig. 3(c). Particularly, the higher percentage of MAPE allowed more sites in rPET functional group to be coupled and bonded with HDPE, resulting in high tensile strength as observed in tensile test. The last figure which is Fig. 3(d) shows the chemical elements of rPET/HDPE composite screened using energy-dispersive X-ray (EDX), demonstrates different morphology of rPET and HDPE.

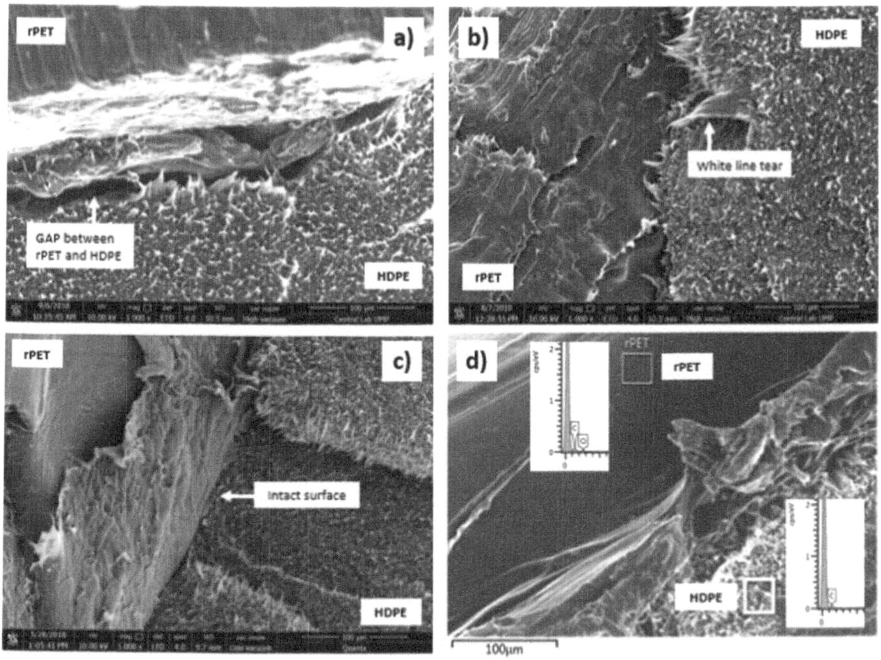

Fig. 3. SEM images of (a) RPET/HDPE/0%MAPE, (b) RPET/HDPE/5%MAPE, (c) RPET/HDPE/10%MAPE, and (d) RPET/HDPE/10%MAPE EDX analysis

The intermolecular reaction between rPET and HDPE, with and without the addition of MAPE, were investigated using FTIR. The results were shown in Fig. 4. From the Fig. 4(a), it can be observed that the main peaks of pure HDPE arise from the vibration characteristics of C-H ethylene group, the asymmetric and symmetric stretching at 2915 cm^{-1} and 2847 cm^{-1}, the bending vibrations at 1471 cm^{-1} and 1462 cm^{-1}, and rocking vibration at 716 cm^{-1}. Following the result, as can be seen in Fig. 4(b), the addition of rPET into HDPE was confirmed with the noticeable peaks arise at 1712 cm^{-1}. The peak at 1712 cm^{-1} attributed to the C = O stretching of ester group belongs to rPET [10]. Meanwhile, Fig. 4(c) and (d) show the FTIR spectrums for the composite rPET/HDPE with the addition of 5% MAPE and 10% MAPE, respectively. The intensity of C = O peak at 1712 cm^{-1} if compared to the C-H stretching peaks at 2915 cm^{-1} and 2847 cm^{-1} was found to decrease with the addition of MAPE, suggesting the coupling reaction occurred in rPET/HDPE composite. Notably, these changes support the stronger interfacial bonding and enhanced tensile strength seen in the rPET/HDPE/MAPE composite.

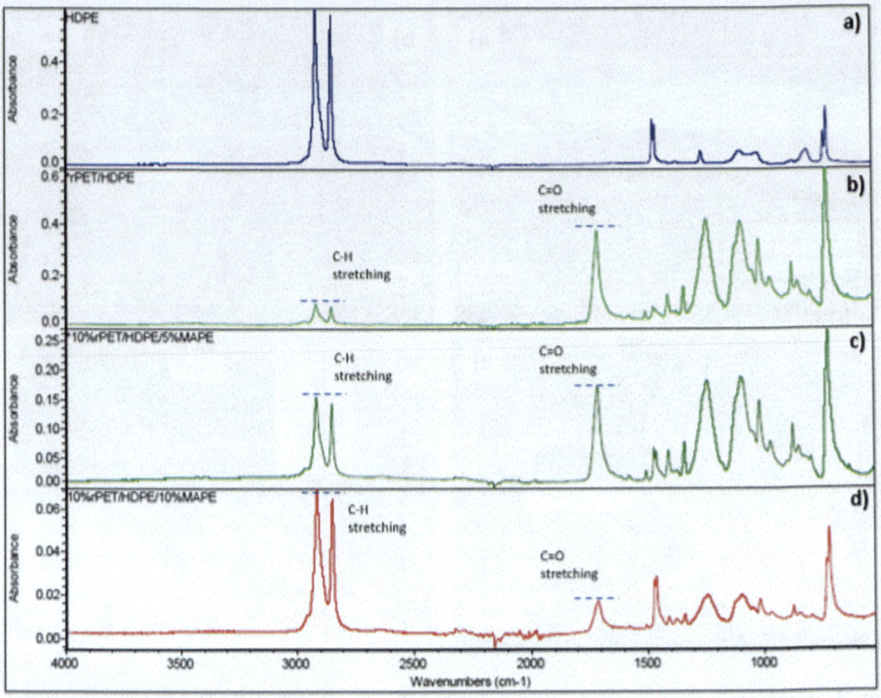

Fig. 4. FTIR spectrums of (a) HDPE, (b) RPET/HDPE/0%MAPE, (c) RPET/HDPE/5%MAPE, and (d) RPET/HDPE/10%MAPE.

4 Conclusion

Based on the experimental results discussed in this research, it can be concluded that the presence of MAPE compatibilizer in rPET/HDPE composite has significantly improved the interfacial bonding between HDPE and rPET phases. It increased the compatibility of rPET/HDPE composite and enhanced the tensile strength of the composite as shown in tensile result. The result was also supported by SEM and FTIR which confirmed strong interfacial bonding formed between the rPET filler and HDPE matrix.

Acknowledgement. The authors would like to express gratitude to Universiti Malaysia Pahang for the financial support of this research through RDU130386 and RDU170390 grants.

References

1. Williams, P.T., Slaney, E.: Analysis of products from the pyrolysis and liquefaction of single plastics and waste plastic mixtures. Resour. Conserv. Recycl. **51**, 754–769 (2007)
2. Eichhorn, S.J., Baillie, C.A., Zafeiropoulos, N., et al.: Current international research into cellulosic fibres and composites. J. Mater. Sci. **36**, 2107–2131 (2001)

3. Pracella, M., Chionna, D., Anguillesi, I., Kulinski, Z., Piorkowska, E.: Functionalization, compatibilization and properties of polypropylene composites with Hemp fibres. Compos. Sci. Technol. **2**, 554–574 (2006)
4. Fechine, G.J.M., Souto-Maior, R.M., Rabello, M.S.: Structural changes during photodegradation of poly (ethylene terephthalate). J. Mater. Sci. **37**, 4979–4984 (2002)
5. Hamid, F., Akhbar, S., Halim, K.H.K.: Mechanical and thermal properties of polyamide 6/HDPE-g-MAH/high density polyethylene. Procedia Eng. **68**, 418–424 (2013)
6. Adhikary, K.B., Pang, S., Staiger, M.P.: Dimensional stability and mechanical behaviour of wood-plastic composites based on recycled and virgin high-density polyethylene (HDPE). Compos. Part B Eng. **39**, 807–815 (2008)
7. Fu, S., Feng, X., Lauke, B., Mai, Y.: Effects of particle size, particle/matrix interface adhesion and particle loading on mechanical properties of particulate – polymer composites. Compos. B **39**, 933–961 (2008)
8. Nonato, R.C., Bonse, B.C.: A study of PP/PET composites: factorial design, mechanical and thermal properties. Polym. Test. **56**, 167–173 (2016)
9. Lei, Y., Wu, Q., Yao, F., Xu, Y.: Preparation and properties of recycled HDPE/natural fiber composites. Compos. Part A Appl. Sci. Manuf. **38**, 1664–1674 (2007)
10. https://www.sigmaaldrich.com/technical-documents/articles/biology/ir-spectrum-table.html. Accessed on 1 July 2019

Rheological Properties of Magnetorheological Polishing Fluid for Micro Mould Polishing

Nurain Abdul Mutalib[1], Izwan Ismail[1,2,3(✉)], Sofarina M. Soffie[1], and Syarifah Nur Aqida Syed Ahmad[1,2]

[1] Faculty of Mechanical and Manufacturing Engineering, Universiti Malaysia Pahang, 26600 Pekan, Malaysia
izwanismail@ump.edu.my

[2] Automotive Engineering Centre, University Malaysia Pahang, 26600 Pekan, Malaysia

[3] Centre of Excellence for Advanced Research in Fluid Flow (CARIFF), University Malaysia Pahang, 26300 Gambang, Pahang, Malaysia

Abstract. Complex surface features such as micro-grooves on micro mould which was produced by the micro-milling process have a significant problem of raised edge or top burr formation. This problem can be solved by using magnetorheological polishing (MRP) technique. MRP capable to produce ultra-fine three-dimensional profile structures. However, a tailored magnetorheological polishing fluid (MRPF) need to be used. This paper presents the effects of MRPF mixture components on their rheological properties. D-optimal mixture design of experiment was used to formulate the MRPF with various composition. The MRPF samples than analyzed using MCR Rheometer for rheological properties. It is found that each MRPF component composition has a significant effect on rheological properties. The result indicates that an optimum composition of MRPF to meet the MRP requirement is by having a volume percentage of carbonyl iron particle (CIP) and abrasive particles at 35% and 10% respectively. The finding of this study produced a significant statistical model of MRPF composition that beneficial for micro mould polishing technology.

Keywords: Magnetorheological polishing · Micro mould · Rheology

1 Introduction

Microfluidic devices were fabricated using many techniques such as micro moulding; photolithography; print and peel; and polydimethylsiloxane (PDMS) processing [1]. Micro milling technique which is used to produce micro mould is low in cost yet produce substantial accuracy of 5 μm. This process can be done without a cleanroom environment and capable to produce complex surface features such as micro-grooves. This technique has been used in microfluidic embossing dies due to its high flexibility, fast and highly precise process. Micro-milling utilizes a high precision computer numerical controlled (CNC) motion system, a high-speed spindle and a miniature cutting tools [2]. However, this process has a significant problem with raised-edge or top-burr-formation. Conventional raised-edge removal using planar polishing method

M. N. Osman Zahid et al. (Eds.): iMEC-APCOMS 2019, LNME, pp. 490–496, 2020.
https://doi.org/10.1007/978-981-15-0950-6_75

is incapable to polish the inside surface of the micro-groove. This problem can be solved by introducing the magnetorheological polishing (MRP) technique. This new process is capable to polished wide range of materials at ultra-fine precision. MRPF is a smart controllable material that can change in viscosity under the application of a magnetic field. General MRPF compositions are magnetisable particle, carrier liquid and non-magnetic polishing abrasive particle.

In MRP, abrasive particles were added in MRPF to increase the material removal rate (MRR) properties. Among the widely used materials in MRPF are carbonyl iron particle (CIP) as a magnetic particle, an abrasive particle such as alumina [3], diamond [4] and ceria [5] and deionised water as a medium carrier. Polishing in MRF is performed by the interaction of MRPF with the sample under the effect of a magnetic field. When the MR fluid is exposed to the magnetic field, it gets stiffened due to the magnetization of CIP. The abrasive particle that gripped by the connected chain of CIP sheared the sample surface where the material removal took place [6]. Jang et al. [4] removed the burr of a different type of machined specimen with the same type of abrasive particle.

Rheological properties such as shear stress and viscosity of the fluid must be considered in the MRPF mixture due to the direct relationship on the stiffening mechanism. This property determined the condition like sedimentation, magnetic field induced yield stress. In MRP, the supreme requirement for the MRPF is having low viscosity at a high magnetic field. The MRPF must be synthesized with optimized composition to meet the requirement for better applicability [7]. Sidpara et al. [8] reported that the different concentration of the CIP and abrasive particle play a significant role in the material removing mechanism. Alumina was selected as abrasive particle because widely used in metal polishing [9].

This paper presents the effects of MRPF mixture components on their rheological properties in micro mould polishing process. The experimental study on rheological properties of shear stress and viscosity were conducted and optimised using D-optimal design of experiment method.

2 Materials and Method

MRPF materials components in this study are 3–5 μm circular shapes carbonyl iron particles (CIP); sharp polygonal shapes alumina powder (500 nm); and deionised water with glyceryl glycol as a medium carrier. Size and morphology of the CIP and alumina are shown in Fig. 1(a) and (b) respectively. The MRPF components are designed in volume fraction where the minimum and maximum values are shown in Table 1 (explain Table 1). The 'D-Optimal Mixture Design of Experiments' was used to investigate the effect of MRPF components on the rheological properties of the MRPF suspension. This method allows investigation and optimization of interdependent mixture variable for desired response specification. Rheological properties of MRPF sample were measured using 'MCR Rheometer 302 Anton Paar' immediately after synthetization process. The rheometer measurement setting was set at 1 mm gap, 25 °C measurement temperature and sandwiched with parallel plate geometry. The rheological measurement included determination of shear stress and viscosity at 0 to 0.4 T magnetic field. The magnetic field was applied perpendicular to the parallel plates of the rheometer, making the obtained particle chains were also perpendicular to the flow direction.

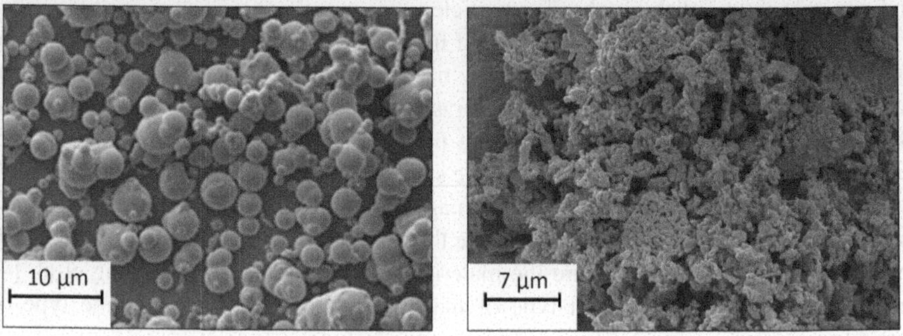

(a) (b)

Fig. 1. Morphology of (a) CIP and (b) Alumina

Table 1. Minimum and maximum range of factors.

Factors	Concentration range (volume %)	
	Minimum	Maximum
DI water	45	70.5
CIP	25	35
Alumina	3.5	10
Glycerol	1	10

3 Results and Discussion

3.1 Effect of CIP Concentration on Shear Stress and Viscosity

The CIP concentration in MRPF shows a significant effect on shear stress and viscosity increment as depicted in Fig. 2(a) and (b) respectively. The MRPF behaved like non-Newtonian liquid where the shear stress build-up as at initial state of shear rate. As shown in Fig. 2(a), the initial shear stress value is increasing as the volume fraction of CIP is increases. The amount the shear stress also increase as the shear rate is in increase. MRPF samples with CIP with the highest CIP concentration of 35 volume % shows the highest value of shear stress. This finding is similar to other MR fluid behaviour where the volume % of the magnetic particle is dependent of shear stress [10]. This is due to higher % of CIP in MRPF causes higher apparent magnetization characteristic of the MRPF. Viscosity behaved similarly as it derived from the shear stress. When the shear rate increase, more energy needed to break the magnetization character in MRPF, thus the viscosity of the MRPF increased as the CIP concentration increases.

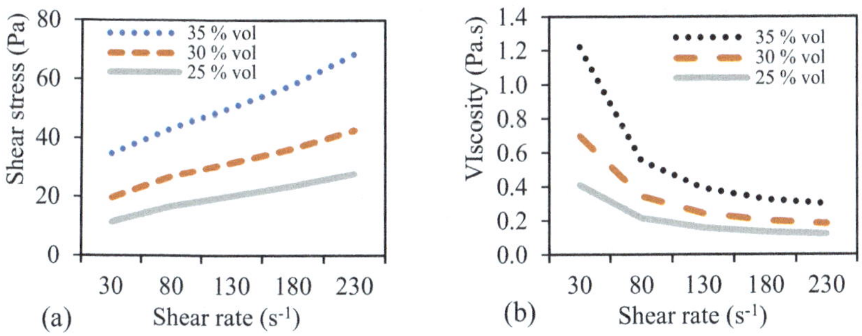

Fig. 2. Effect of (a) shear stress and (b) viscosity on shear rate at different CIP composition.

3.2 Effect of Abrasive Concentration on Shear Stress

For material removal taking place during the MRP process, the concentration of abrasive particle must be in a good range with CIP particle. Figure 3 shows rheological behaviour of different abrasive particle concentration at different shear rate and magnetic field. In Fig. 3(a) significant increasing of shear stress was observed as abrasive particle concentration is increased. Increment of shear stress is more obvious at 10% volume fraction. As depicted in Fig. 3(b), the shear stress is increased as the magnetic field is increases. Figure 3(b) also indicate that the MRPF with lower of abrasive particle concentration (4% volume fraction) shows higher value of shear stress.

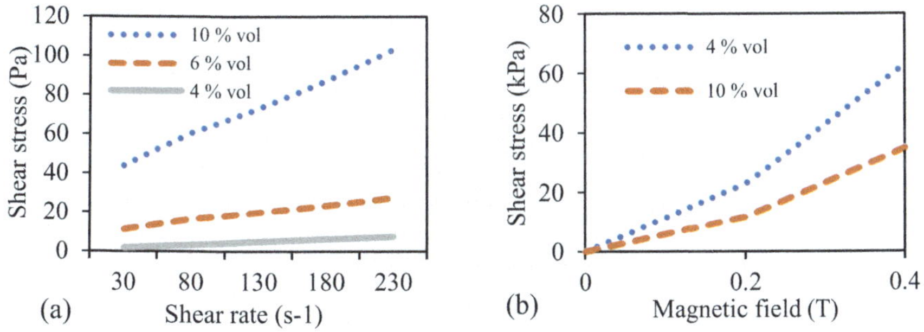

Fig. 3. (a) Effect of shear rate on shear stress at off state (0 T) and (b) effect of magnetic field on shear stress at different abrasive particle concentration.

The increment shear stress as increasing of the abrasive particle was primarily due to the thickening mechanism of abrasive particle where the particle-particle interlocking behaviour happened. The abrasive-particle to CIP size ratio of 10:1 is large enough to let the small abrasive particle to fill the gap between CIP particle. Meanwhile, the High value of shear stress at a lower concentration of abrasive particle is due to stronger CIP chain structure with few interruptions of non-magnetic particle A higher concentration of abrasive particle will deform the chain structure thus lowering the energy needed, to break the chain of CIP [8]. In the polishing process, the surface roughness can be change by having different percentage of abrasive particle concentration. Low concentration of abrasive will lead to low material removal [11].

3.3 D-Optimal Mixture Design of Experiment

Results of the rheological measurements show that the response value of zero magnetic field shear stress ranges from 7.4 to 531.8 pa which has a large ratio of maximum to minimum of 71.6. Due to the large value, results were transformed to Log_{10}. The results then proceed with the analysis of variant (ANOVA) to determine significant design model and lack of fit. The ANOVA result shows the significant design model is 'linear mixture' with the Model F-value of 6.09. Model F value is 0.93%. This indicates a small noise influence in the model. The model Lack-of-Fit F-value is 0.09 which implying the model is fit. The linear model shows that the Log_{10} (shear stress) can be predicted through the model equation as stated in Eq. (1).

$$Log_{10}(Shear\ stress) = -0.002387X_1 + 0.042653X_2 + 0.108367X_3 - 0.006653X_4 \quad (1)$$

where X_1 is the % volume of DI Water, % volume X_2 is CIP, % volume X_3 is Alumina and X_4 is % volume Glycerol. Figure 4 shows the contour plot for three components at a constant value of X_4 of 1%. The most contributed factor is the volume fraction of DI water while the CIP and Alumina shared similar contribution to affect the shear stress.

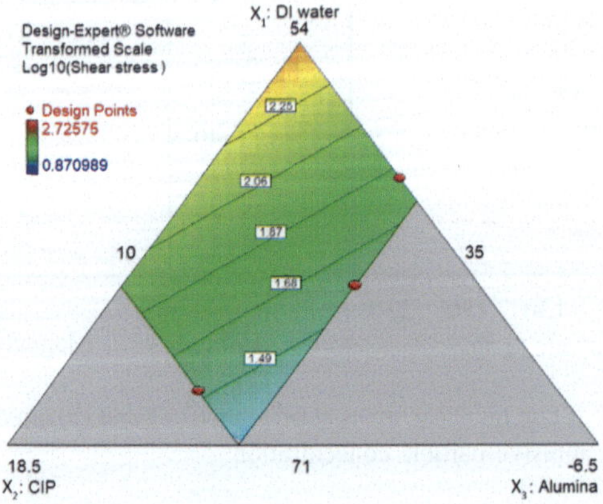

Fig. 4. Contour plot for each component to the effect of shear stress where glycerol is 10 volume percent.

DI water plays important roles to carry the solid particle in absent of magnetic field. This behaviour will be inversed when the magnetic field applied to the MRPF due to higher particle-particle chain bonding strength at a lower volume of carrier fluid. To meet the MRP requirement where the shear stress has to be set at the low value without magnetic field is by having a volume percentage of carbonyl iron particle (CIP) and abrasive particles at 35% and 10% respectively.

4 Conclusions

Experimentation carried out in study shows that the combination of MRPF components can produce a variation of shear stress and viscosity at absent or with the application of the magnetic field. In absent of magnetic field, shear stress is increase with CIP composition. Statistical model investigated in this study can predict the rheological properties of MRPF where the composition of the component in MRPF attribute to changes in the rheological property.

Acknowledgement. This research was supported in part by Fundamental Research Grant Scheme (FRGS), Ministry of Higher Education, Malaysia FRGS/1/2016/TK03/UMP/02/12 (RDU160132).

References

1. Faustino, V., et al.: Biomedical microfluidic devices by using low-cost fabrication techniques: a review. J. Biomech. **49**(11), 2280–2292 (2016)
2. Bordatchev, E., Nikumb, S.: Fabrication of micro moulds and dies using precision laster micromachining and micromilling technologies, vol. 3 (2008)
3. Chen, M., et al.: Design and fabrication of a novel magnetorheological finishing process for small concave surfaces using small ball-end permanent-magnet polishing head. Int. J. Adv. Manuf. Technol. **83**(5), 823–834 (2016)
4. Jang, K.-I., et al.: Deburring microparts using a magnetorheological fluid. Int. J. Mach. Tools Manuf **53**(1), 170–175 (2012)
5. Jain, V.K., Kalia, S., Sidpara, A.M.: Some aspects of fabrication of micro devices by electrochemical micromachining (ECMM) and its finishing by magnetorheological fluid. Int. J. Adv. Manuf. Technol. **59**(9), 987–996 (2012)
6. Sidpara, A., Jain, V.K.: Analysis of forces on the freeform surface in magnetorheological fluid based. Int. J. Mach. Tools Manuf **69**, 10 (2013)
7. Bica, I.: Damper with magnetorheological suspension. J. Magn. Magn. Mater. **241**, 196–200 (2002)
8. Sidpara, A., Das, M., Jain, V.K.: Rheological characterization of magnetorheological finishing fluid. Mater. Manuf. Processes **24**(12), 1478 (2009)
9. Mutalib, N.A., et al.: Magnetorheological finishing on metal surface: a review. In: IOP Conference Series: Materials Science and Engineering, vol. 469, p. 012092 (2019)

10. Rahim, M.S.A., et al.: Thermal conductivity enhancement and sedimentation reduction of magnetorheological fluids with nano-sized Cu and Al additives. Smart Mater. Struct. **26**(11), 115009 (2017)
11. Maan, S., Singh, A.K.: Nano-surface finishing of hardened AISI 52100 steel using magnetorheological solid core rotating tool. Int. J. Adv. Manuf. Technol. **95**(1), 513–526 (2018)

Effect of Aluminum Surface Treatment on the Damping Properties of Aluminum-Rubber Bonding System

Qumrul Ahsan[1(✉)], Adilla Fasha Ahmad Mawardi[1], Sivarao Subramonian[1], Mohd Rizal Alkahari[2], and Azma Putra[2]

[1] Fakulti Kejuruteraan Pembuatan, Universiti Teknikal Malaysia Melaka, Hang Tuah Jaya, Durian Tunggal, 76100 Melaka, Malaysia
qumrul@utem.edu.my
[2] Fakulti Kejuruteraan Mekanikal, Universiti Teknikal Malaysia Melaka, Hang Tuah Jaya, Durian Tunggal, 76100 Melaka, Malaysia

Abstract. Surface treatments on aluminum substrates are introduced to improve the bonding characteristics between rubber and aluminum components that are widely used in automotive industries. This present work involves the modification of surface roughness of aluminum plates either by sand blasting or anodizing or graphene coating followed by primer coating. Finally, rubber was vulcanized on these surfaces and aluminum-rubber interface were characterized. Sand blasted aluminum surface sample provides the best damping performance with a capacity of 50% damping. Although, aluminum treated with graphene has not been shown better damping properties, aluminum treated with sand blasting and graphene shows equivalent damping performance as that of sand blasted sample with similar average surface roughness (Ra) of 3.2 μm.

Keywords: Damping material · Surface roughness · Metal-to-rubber bond

1 Introduction

For many years, rubber to metal bonding has been used extensively to manufacture a variety of components for the automotive industry in order to reduce noise and vibration which occurred in the most of the automotive and engineering applications [1, 2]. Previously, rubber has been continuously bonded to steel [3] which have a density in a range between 7.08 g/cm^3 and 8.05 g/cm^3 depends on its alloying constituents. In recent times, automotive manufacturers have replaced steel with aluminum (Al) in some applications such as engine mountings etc. [4]. Aluminum is getting attention among researcher due to its lightweight (density: 2.70 g/cm^3), ease in processing and has good mechanical properties. However, Al alloys have some limitation which is easy to oxidize [5]. When exposed to air, the surface of Al alloys will form a nano-scale of oxidizing layer. This phenomenon causes poor bonding between Al and rubber. Surface modification of aluminum by graphene nanopletlet impregnation to form the nanocomposite for improved surface hardness [6] and tensile strength [7].

M. N. Osman Zahid et al. (Eds.): iMEC-APCOMS 2019, LNME, pp. 497–502, 2020.
https://doi.org/10.1007/978-981-15-0950-6_76

Over the last couple of decades, the needs in improving bonding strength of metal to rubber bonding were getting higher. Many researchers studied to achieve the good bonding was by improving the surface conditions on a substrate [8, 9]. Preparation of aluminum-rubber components usually follow the conventional sand blasting followed by anodizing and primer coating in order to yield better interface bond between Al and rubber. However entire process is cumbersome and limits the application of volatile organic compounds (VOC) in primer. Therefore, growing attention has been turned to mechanical impregnation of graphene nanopletlets (GNP) onto aluminum surface which could be a solution for improved bonding and limit the use of VOC contents. Thus, surface treatments of metals alloys were introduced to improve the bonding properties between rubber and metal. In this paper, aluminum A6061 alloys with different surface treatments were bonded to rubber by using commercial solvent-based bonding agents to study the damping characteristic of rubber bonded interface with an aim to produce lighter damping material with reduced in noise and vibration for automotive applications. A few types of surface treatments have been performed to attain good bonding strength including sand blasting, phosphate anodizing and graphene coating. Surface roughness measurements were conducted after each treatment and later Oberst Beam Method (OBM) testing was carried out to analyses the damping properties.

2 Methodology

2.1 Surface Treatment of Aluminum

In this study, the Aluminum A6061 alloy samples were first degreased and cleaned with acetone. Later sample surfaces were blasted by a high-speed stream of silica particles impelled by compressed air to eliminate unfavorable oxides and contaminants and to increase surface roughness. The distance of the Al substrate to the nozzle was fixed at ±80 mm. After blasting, the samples were cleaned with compressed air and acetone to obtain dirt free surface before the samples were left dried.

The second type of surface preparation for Al sample was phosphate anodizing where the Al plate as anode was immersed in an electrolytic bath having 100 g of phosphoric per liter of distilled water whereas, the graphite rod was used as the cathode electrode. A current of 1.125 A and a voltage of 20 V were supplied to the bath and the samples were kept for 1 h. After completion sample was washed with water and dried at room temperature.

The third type was nanomaterial coating where graphene nanopletlets (GNP) and polyvinyl alcohol (PVA) solution were swabbed onto the Al surface. Initially, the surface of Al substrate was cleaned with acetone. Drop casting method was used to place the GNP-PVA solution onto the Al substrate in equal amounts and kept in a vacuum oven at a temperature of 60 °C for 8 h so that the water content of the solution evaporates slowly from the Al surface. Finally, the Al plate was taken out from the oven and left to cool at room temperature.

Surface roughness, Ra of Al from different surface conditions was measured by profilometer.

2.2 Rubber Compounding and Bonding onto Aluminum

Rubber used in this study were mixed and prepared at Malaysian Rubber Board, RRIM Research Station based on the compounding formula from previous study [3, 10]. Initially, the primer and topcoat adhesive of bonding agents were applied onto the Al surface at room temperature of 25 °C by brushing method and left to dry for about 30 to 45 min. Then, the rubber strip of 3 mm thick with primer adhesive coated aluminum was placed in mold cavity. Finally, sample was preheated for 2 min and then further hot compressed at an applied stress of 120 kg/cm^2 for 6 min at 150 °C for vulcanization of rubber. Later, rubber bonded sheet materials were cooled at room temperature and withdrawn from the mold. Damping sheets were cut into appropriate dimensions based on ASTM E756-05 [11].

2.3 Oberst Beam Method

Three samples for each surface condition with size of 140 mm × 20 mm × 3 mm were used for OBM test and cantilever beam length was kept constant at 180 mm for all test samples. The sample was held as a cantilever beam and an accelerometer was attached at the free end of samples. Then, vibration frequency in the range of 100–1000 Hz was triggered by a mini shaker. Figure 1 shows the schematic arrangement of OBM test sample and Fig. 2 shows the experimental setup of OBM test system.

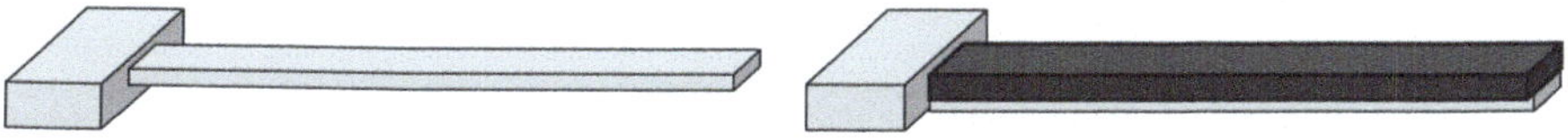

(a) Aluminum beam

(b) Aluminum beam with damping material

Fig. 1. Schematic arrangement of test samples for OBM testing

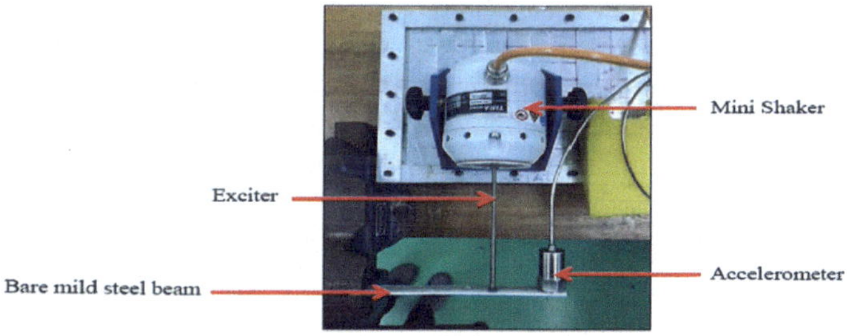

Fig. 2. Experimental set-up for OBM test

Initially, the frequency response function (FRF) measured on the bare aluminum beam was analyzed to determine natural frequencies within the frequency range of 100 to 1000 Hz. Then, measured FRF on the damped beam would be analyzed to determine the natural frequencies and corresponding modal loss factors of the other samples.

3 Results and Discussions

3.1 Surface Roughness Measurements

The average surface roughness, Ra values obtained from the measurement profile on aluminum substrate are shown in Table 1. From Table 1, sand blasting generates almost five times rougher surface of 3.285 μm as compared to untreated aluminum plate sample. Good adhesion occurs only when an adhesive penetrates into the pores, holes and other irregularities of the adhered surface of a substrate and locks mechanically to the substrate. Sample with sand blasted and impregnation of graphene shows the lower surface roughness value because of the smooth coating of graphene layer on blasted surface. However, graphene coating increases the surface roughness when compared with untreated aluminum surface.

Table 1. Average surface roughness before and after treatments

Surface treatment	Average surface roughness, Ra (μm)
Aluminum untreated	0.70 ± 0.1
Sand blasting	3.29 ± 0.1
Blasting + Graphene	3.16 ± 0.1
Phosphate Anodizing	2.47 ± 0.1
Graphene	1.47 ± 0.1

3.2 Oberst Beam Method

Oberst beam method is used to study the effects of surface roughness of Al with different surface conditions on the damping materials. Figure 3 shows the comparison of frequency response functions (FRF) obtained from aluminum beam and aluminum with different surface treatments respectively. The surface treatment plays a positive role on the damping mechanism especially for blasted and blasted/graphene samples with the shift of natural frequency at lower values and reduced wave amplitude as compared to bared Al sample. Obviously improved Al-rubber interface absorbed vibration by efficient transfer of vibrational energy from Al to rubber side. Subsequently, it is also evident in loss factor (η %) comparison graphs (Fig. 4) for different surface treated samples. Loss factor usually represents the combine effect of drift and amplitude reduction of natural frequency. Blasting and blasting/graphene treated samples having higher surface roughness have the loss factor above 4% for 2nd peak whereas the untreated Al shows the lowest loss factor of 1.5% as it has the minimum surface roughness. Similar trend was also observed when considered the resonance at 3rd peak. The more increase in loss factor of damping peaks indicated that more energy was being dissipated during the damping mechanism indicating better bonding.

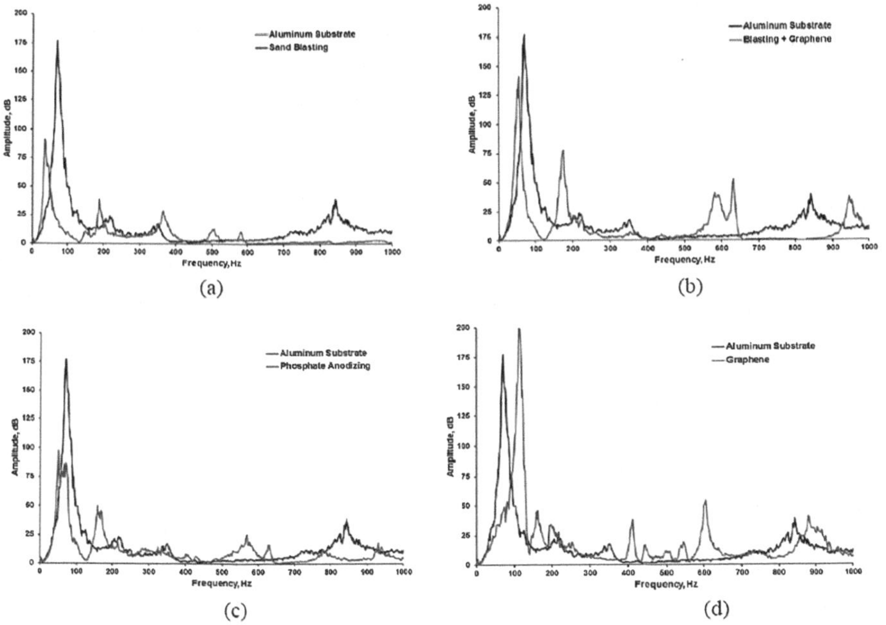

Fig. 3. Comparison of frequency response functions (FRF) obtained from aluminum substrate and aluminum with surface conditions of (a) sand blasting, (b) blasting and graphene (c) phosphate anodizing and (d) graphene

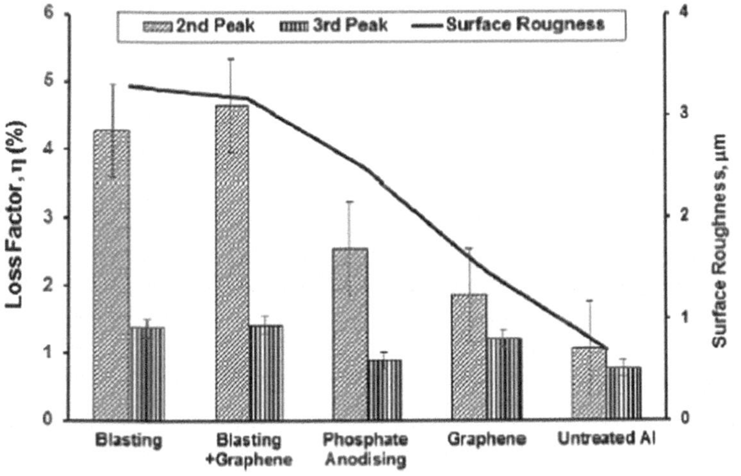

Fig. 4. Effect of surface roughness on the loss factor of untreated and surface treated aluminum samples

4 Conclusion

Damping property prior to adhesive bonding of rubber and aluminum substrate with different surface treatment conditions shows dependence on surface roughness. Sand blasting or blasting/graphene showed highest loss factor as they attained higher surface roughness. However, when aluminum surface is only coated with adhesive and bonded with rubber, the damping properties in terms of loss factor attained the lowest loss factor. Thus, rougher surface condition indicates good damping properties.

Acknowledgement. This research was funded by the grant (PJP/2017/FKM/HI15/201548) of Universiti Teknikal Malaysia (UTeM), Melaka.

References

1. Ansarifar, M.A., Zhang, J., Baker, J., Bell, A., Ellis, R.J.: Bonding properties of rubber to steel, aluminum and nylon 6,6. Int. J. Adhes. Adhes. **21**(5), 369–380 (2011)
2. Shangguan, W.B., Zheng, G., Liu, T.K., Duan, X.C., Rakheja, S.: Prediction of fatigue life of rubber mounts using stress-based damage indexes. Proc. Inst. Mech. Eng., Part L: J. Mater.: Des. Appl. **231**(8), 657–678 (2017)
3. Chander, K.P., Vashista, M., Sabiruddin, K., Paul, S., Bandyopadhyay, P.P.: Effects of grit blasting on surface properties of adhesive bonded adherents. Manuf. Technol. **16**(6), 1371–1375 (2016)
4. Benedyk, J.C., Mallick, P.K.: Aluminum alloys for lightweight automotive structures. In: Materials, Design and Manufacturing for Lightweight Vehicles, pp. 79–113. Woodhead Publishing (2010)
5. Gregory, I.H., Muhr, A.H., Imizan, B.: Prediction of state of cure throughout rubber components. J. Rubber Res. **2**(1), 1–22 (1997)
6. Sahoo, B., Joseph, J., Sharma, A., Paul, J.: Surface modification of aluminum by graphene impregnation. Mater. Des. **116**, 51–64 (2017)
7. Wang, J., Li, Z., Fan, G., Pan, H., Chen, Z., Zhang, D.: Reinforcement with graphene nanosheets in aluminum matrix composites. Scripta Mater. **66**(8), 594–597 (2012)
8. Koruk, H., Sanliturk, K.Y.: On measuring dynamic properties of damping materials using Oberst beam method. In: Proceedings of the ASME 2010 10th Biennial Conference on Engineering Systems Design and Analysis, pp. 1–8 (2010)
9. Institutional Repository: Two advanced styrene-butadiene/polybutadiene rubber blends with a silanized silica nanofiller for potential use in passenger car tire tread compound, pp. 1–53 (2011)
10. Zhang, X.-F., Fan, L., Zhao, X.-H., Zhang, L.-Q., Liu, L., Zuo, Y.: Bonding properties between nitrile-butadiene rubber and aluminum alloy treated by anodizing methods. Appl. Polym. Sci. **112**, 283–289 (2009)
11. Malogi, D., Gupta, A., Kathawate, G.R.: Center impedance method for damping measurement. Adv. Acoust. Vib. **2009**, 1–7 (2009)

Manufacturing Processes

Finite Element Analysis of Baseplate for Failure Estimation in Light Railway Transit Fastening Systems

Noraishah Mohamad Noor, Muhammad Nashrur Faizzi Abdul Razak, and Ahmad Razlan Yusoff(✉)

Faculty of Mechanical and Manufacturing Engineering, Universiti Malaysia Pahang, 26600 Pekan, Pahang, Malaysia
razlan@ump.edu.my

Abstract. Light Railway Transit (LRT) operates daily and connects places in city centre. Hence, it is important for operator company to maintain the operation and maintenance. In railway track maintenance, the fastening systems plays major roles in sustaining the condition of rail, track and the train. This study focuses on the current fastening system failure for product development improvement for maintenance jobs in the railway track. The 3D modelling of the fastening system consists of the base plate, tension clamp, rail pad, washers and the bolt are modelled. Ansys software were analyzed the deformation occurs due to load transmitted by moving train, the amount of stress on the baseplate, the strain occurs and the failure location of the baseplate. Results showed that the failure of baseplate location validates the simulation of the fastening system. The failure location from prediction of the fastening system can identify based on static stress simulation can be used for future baseplate in the fastening systems.

Keywords: Baseplate · Finite element analysis · Strain · Stress · Failure

1 Introduction

The railway track maintenance is one of the important aspects in order to ensure the safety, efficiency and reliability of the train services. Therefore, the major maintenance work is done in every 10 years cycle. In this cycle, fittings, fastenings, switches, ballast, sleepers, baseplate and the rail are inspected, repair or replace if needed [1]. Fastening system used at the viaduct section is known as Direct Fixation Fastening System (DFF). Viaduct section can be defined as the elevated part of the track and the short span bridges that build to cross the valley or obstacle [2]. DFF is installed on the concrete plinth on viaducts, bridges and in some section of the depot. Reason wise, cast concrete plinth are constructed on viaducts or ballast-less track, in order to meet the alignment requirement, provide adjustability of deck construction tolerances, act as a containment feature and as electrical insulation (Figs. 1 and 2).

M. N. Osman Zahid et al. (Eds.): iMEC-APCOMS 2019, LNME, pp. 505–511, 2020.
https://doi.org/10.1007/978-981-15-0950-6_77

Fig. 1. Plinth structure on the viaducts section

Fig. 2. Sandwich tie plate

The forces exerted by the moving train are transmitted into the rail and its fastener and the plinth structure are designed to absorb all the rail forces and distribute it into the viaduct deck. A typical cross section of plinth is in directional longitudinal directions the plinth shall be separated in segments shorter than 8.0 m.

The type of fastening systems used in the elevated track is sandwich tie plate assembly. There are three main function of the system, which is hold the rail in a vertical condition, maintain the correct gauge distance and provide a resilient mounting for noise and impact loading reduction. Sandwich tie plate assembly is the top plate and bottom plate are sandwiched together with rubber made of polyethylene. By standards, the baseplate sandwich stiffness value should be 22kN.

From the company information, the 1st cycle major maintenance was conducted in 2008. However, set of fastening systems are expensive. Therefore, the replacement only covered the baseplate's rubber, new metal bottom plate, new adjustment washer and new plastic rail seat pad. During major maintenance, the technical requirements such as the baseplate spacing, stiffness value and rubber formulation specification are set according to the original OEM specification. Total of 48,000 pieces fleetwide replacement have been done for the 14,214 m of track length for both directions. The installation is completed on 2013 with 50–60 pieces installation per day. 2019 will be the 2nd cycle of major maintenance for DFF. This project will involve the viaduct track area, similar to 1st cycle project and total replacement of the system need to be considered. Therefore, the existing fastening systems need to be considered for replacement with new enhancement fastening systems that more durable with stiffness value able to exceed 25kN based on references from OEM. For initial, the basic information from the existing design is been evaluated via finite element analysis to study the stress distribution and strain.

2 Structural Analysis

Structural analysis is one of the most commonly used in finite element analysis. The structural analysis focuses on finding out the behavior of a physical structure when force or structural loads are provided. There are two types of static structure analysis methods which are non-linear static analysis and linear static analysis. In dynamic analysis, the loads are applied as a function of time or frequency. Loads such as displacements, velocities, accelerations, forces, and stresses make a dynamic analysis more complicated and accurate. The equation of equilibrium for a structure under static loading is as below:

$$[K]\{x\} = \{F\} \tag{1}$$

Where:

{x} = Displacement. [K] = Constant (Linear elastic material behaviour is assumed)

{F} = Force (statically applied and time-varying forces are not considered)

There are two types of structural analysis that can be conducted in Ansys which are implicit and explicit analysis. Both analyses are the way to solve the governing equations of the mechanical problem. Implicit Structural Analysis includes Strength and Vibration Analysis. Table 1 shows the list of materials that involve in material selection for simulation. Meanwhile, the information of the material mechanical and physical properties listed in Table 2.

Table 1. List of fastening system material

Component	Material
Rail	Alloy Steel
Base Plate	Steel + HDPE
Rail pad	HDPE
S clip clamp	Steel
Bolt	Alloy Steel
Washer	Stainless Steel

Table 2. Mechanical and Physical Properties of HDPE, steels and reinforced steel

Element	Density (Kg/m3)	Ultimate tensile strength (MPa)	Yield strength (MPa)	Poisson's ratio	Elastic modulus (GPa)	Thermal expansion (10^{-6} k^{-1})	Impact resistance, IZOD (J/M)
HDPE	958	29.6	25.7	0.46	0.5–1.2	100–200	20–210
Structural steel	7850	0.46	0.25	0.3	207	799.615	0.46
Steel reinforced	8808	0.756	0.522	0.385	322	116.25	10.5

Sources: [3–5]

For boundary conditions, force applied and constraints need to be determined. The type of rail used at LRT Ampang/Sri Petaling line is UIC60 where the weight of the rail is 60 kg/m. Adding up both axle load and rail weight then, the force act on the fastening system and the base plate is 113109.3 N, as shown in Fig. 3a. Figure 3b shows fixed axes for the bottom base plate. Meshing was carried out with Ansys meshing software. Combination of triangular and rectangular mesh was chosen as a quadrilateral mesh of acceptable quality. The rail meshed using hexahedron shaped mesh because it was not the main focus of the analysis area and to reduce analysis time. The other parts used tetrahedron shaped mesh as it was critical to obtain the accurate result for analysis with the help of the refined mesh. In this body sizing the element size used was 5 mm. Figure 3c shows the mesh result of the fastening system and the baseplate.

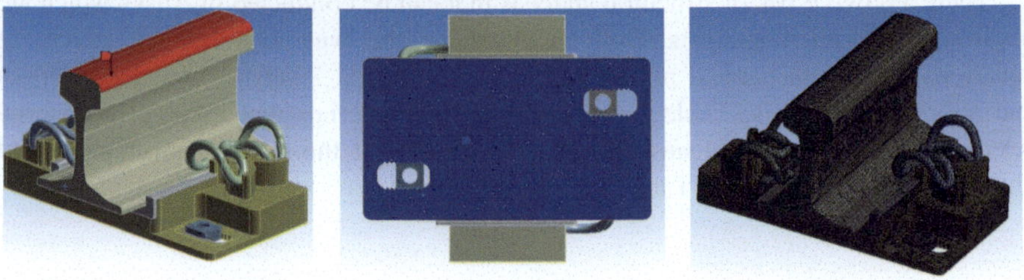

a) Force applied b) Constraints c) Meshing

Fig. 3. Boundary conditions for baseplate simulation

3 Simulation Result

The deformation of the baseplate occurs when loads are given, as shown in Fig. 4. In this research, the base plate is set as steel. Thus, the deformation occurs are expected to be small since the steel stress endurance is around 100 MPa. The stress exerted on the base plate is 21 MPa and the total deformation is 0.002803 mm. Apart from under the rail, the critical area also undergoes deformation due to the vibration and forces occur on the anchor bolt. The critical area on the base plate is the area of connection the anchor bolt to the concrete plinth structure. The thickness of the area is 10 mm. The force that causes vibration and movement around the critical area when a train passed by is 880 N.m. The minimum deformation occurs this area is 0 mm and the maximum is 0.00085 mm.

Stress is defined as the internal forces per unit area exerted at an object subjected to an external force. According to the boundary condition set up in the analysis, the area subjected to force on the base plate is shown as Fig. 5. The area under the rail undergoes most of the forces from the passing trains. The von mises equivalent stress obtain from the simulation at minimum is 22.9 μPa and at maximum is 21.6 MPa. The total stress occurs at the critical area is as follows. The minimum stress is exerted at this area is 500 pPa and the maximum stress is up to 73.344 MPa.

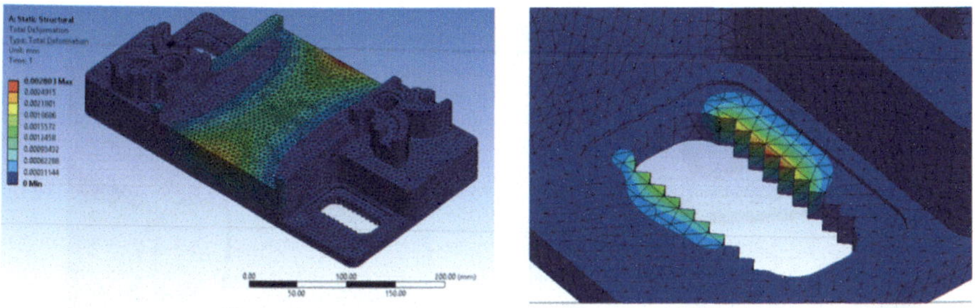

a. Baseplate **b** Critical area

Fig. 4. Total deformation of baseplate

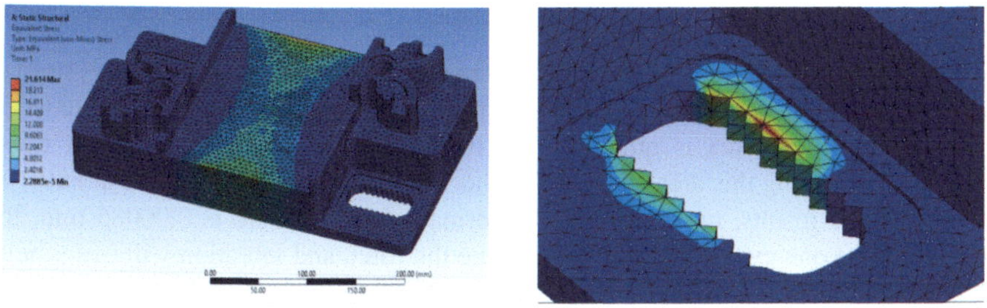

a. Total stress on baseplate **b.**. Highest stress location

Fig. 5. Total stress

A strain is the response of the object when stress is applied. When a force exerted on a material, stress is produced, which can cause the material to undergo deformation. The strain is the amount of deformation in the direction of the applied force divided by the initial length of the material. The strain is also the relative change in the size or shape of an object due to external forces. Figure 6 shows the von mises equivalent strain obtain from the simulation is at minimum 160 p and at maximum 130 μ. The total strain occurs in the critical area is as follows. The minimum strain exerted on this area is 0.003 p and the maximum strain is up to 0.0005 mm/mm.

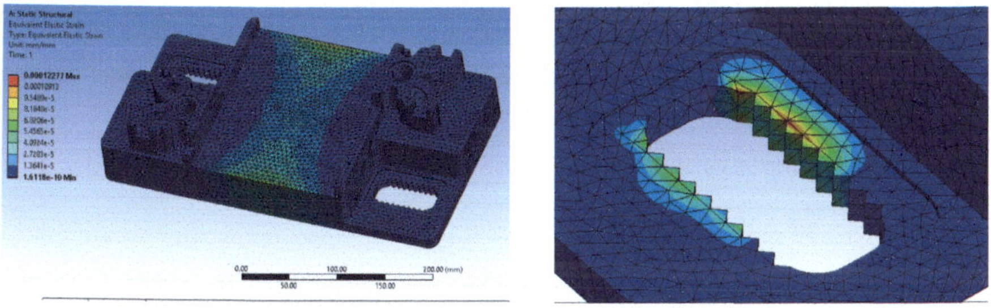

a. Total Strain **b.** Highest strain location

Fig. 6. The total strain on a baseplate

a) Failure in the fastening system b) Failure at baseplate

Fig. 7. Validation with failure location

Figure 7 shows the critical area is the area where the anchor bolt will be inserted into the lock plate and then into the concrete plinth. This area fixed the base plate to the ground. The critical area is quite thin at 10 mm. Loads from passing train cause vibration surrounding this area cause the anchor bolt and lock plate to move at 880 Nm force. From the results, maximum total deformation occurs here is 0.00085 mm. the number is from one cycle of impact. Over time the force and vibration will overwhelm the thin plate and causes it to break or damage.

4 Conclusion

The study conducted by recreating the fastening system assembly and are simulated using ANSYS software. A static structural simulation was executed to study the effect of dynamic force acting on the base plate of the Lord DFF fastening system. Deformation occurs at the center of the base plate. This is expected since the center holds and received directly. However, the deformation is small since the base plate are made of high quality of steel layered with HDPE rubber. The deformation at the critical area however, is more crucial. This is due to the thin plate is expected to sustain the same force received from the center and in addition from visual observation, the force also caused by vibration effect and moving anchor bolt. Total deformation occur are small but not negligible since the deformation occurs on every cycle.

Acknowledgments. The financial supported by Universiti Malaysia Pahang under Internal Research Grant of RDU 190348 are gratefully acknowledged.

References

1. Budai, G., Huisman, D., Dekker, R.J.: Scheduling preventive railway maintenance activities. J. Oper. Res. Soc. **57**(9), 1035–1044 (2006)
2. Mundrey, J.: Railway Track Engineering. Tata McGraw-Hill Education, New York (2009)

3. Khelif, R., Chateauneuf, A., Chaoui, K.J.: Statistical analysis of HDPE fatigue lifetime. Meccanica **43**(6), 567–576 (2008)
4. Ferranti, E., Chapman, L., Lee, S., Jaroszweski, D., Lowe, C., McCulloch, S., Quinn, A.: The hottest July day on the railway network: insights and thoughts for the future. J. Metrol. Appl. **25**(2), 195–208 (2018)
5. Corey, R., Han, J., Khatri, D.K., Parsons, R.L.J.: Laboratory study on geosynthetic protection of buried steel-reinforced HDPE pipes from static loading. J. Geotechn. Geoenviron. Eng. **140** (6), 04014019 (2014)
6. Öchsner, A.: Continuum Damage and Fracture Mechanics. Springer, Berlin (2016)

Enhancement of Surface Integrity in Cryogenic High Speed Ball Nose End Milling Process of Inconel 718

Musfirah Abdul Hadi[1(✉)] and Jaharah A. Ghani[2]

[1] Faculty of Engineering Technology, University Malaysia Pahang, 26300 Gambang, Pahang, Malaysia
musfirah@ump.edu.my

[2] Faculty of Engineering and Built Environment, University Kebangsaan Malaysia, 43600 Bangi, Selangor, Malaysia
jaharahaghani@ukm.edu.my

Abstract. Surface integrity of machined subsurface is an important aspect in any machining process especially in high accuracy demand industries like aerospace. As is known, Inconel 718 is a difficult to machine material and in most cases, Inconel 718 will be resulting an excessive heat generated at the cutting zone which can causes rapid tool wear, damage on machined surface and microstructural defects. Hence, various cooling methods have been made to address these problems and improve the quality of machined surface. In this study, a cryogenic cooling technique using nitrogen liquids (LN_2) was developed for cooling the tool-chip interface during milling Inconel 718. The goal of this paper is to present a comparison study on surface roughness, machined surface microhardness and subsurface microstructure changes between cryogenic cooling and dry techniques. The experiments conducted using a physical vapor deposition (PVD) coated with TiAlN/AlCrN ball nose tungsten carbide for a range of cutting speeds between 140–160 m/min, 0.15–0.20 mm/tooth for a feed rate, and 0.2–0.4 mm for radial depth of cut. The results show that the cryogenic cooling technique is more effective than dry cutting for improving surface roughness and lessening deformation of microstructure changes underneath the machined surface. However, machining in dry technique has produced a high microhardness for machined surface compared to cryogenic cooling technique. Overall, the utilization of the cryogenic technique has improved the surface roughness to a maximum of 88% and reduced the plastic deformation layer, while dry machining can improve the surface microhardness up to 5%.

Keywords: Surface integrity · Cryogenic · End milling · Inconel 718

1 Introduction

Inconel 718 has been widely used in industries that focus on high strength and durability such as aerospace, automotive and marine industries. According to Thakur et al. [1], about 75% of the weight in aerospace use and 50% of the weight in the use of modern jet engines are components made of Inconel 718. Ezugwu and Bonney [2],

M. N. Osman Zahid et al. (Eds.): iMEC-APCOMS 2019, LNME, pp. 512–518, 2020.
https://doi.org/10.1007/978-981-15-0950-6_78

state that Inconel 718 is oxidizing and corrosion resistance which can be used at high temperatures ranging from 217 °C to 700 °C and at the same time maintaining very high strength to its mass weight ratio. Dudzinski et al. [3] categorizes Inconel 718 as an advanced material where it has a good tensile, fatigue, creep and breaking strength.

In metal cutting, high speed machining not only enhances the rate of metal removal, but also gives a positive result in terms of cutting force and surface finish. However, machining Inconel 718 at high speed range presents a serious challenge especially because of its unique combination of properties such as high temperature strength, hardness, and resistance to chemical corrosion which make it difficult to engineered. In a research done by Pawade et al. [4], they found that machining Inconel 718 at high cutting speed will increase the work hardening for machined sub-surfaces at the depth of 30 μm which is up to 68% as compared to its bulk counterpart. The integrity of the Inconel 718 depends on the cutting tool and cutting parameters being used in order to obtain a good result.

The geometry of the cutting tool gives a significant impact on the machining performance. Research done by Subramaniam et al. [5] proofed that tool geometries such as rake angle and nose radius have a strong influence on vibration amplitude during end milling process. Research in ball nose cutting tool has earned attention due to its ability to reduce stress concentration by smooth cutting tip. Saikaew and Baowan [6] found that ball nose cutting tool can give a good surface finished during machining process.

Machining using cryogenic environments is increasingly gaining attention and evolving today due to the use of nitrogen liquids is safe, clean, no coolant toxic, do not require disposal process and most important thing is this method can improve the tool life. The use of cutting fluid represents about 16–20% of manufacturing cost. Therefore, optimum use of cutting fluid is very necessary. However, only a few studies have been carried out to study the effect of application of cryogenic methods using nitrogen liquids to the surface integrity of the newly machined material using ball nose cutting tool.

In this paper, the focus is on surface integrity performance during machining Inconel 718 using nitrogen liquids as a cooling material due to its eco-friendly features. The surface integrity evaluation will have focused on surface roughness, sub-surface microstructure and surface microhardness of the workpiece. The results obtained from cryogenic machining will then comparing to dry machining process.

2 Methodology

The machining process for both dry and cryogenic cutting conditions were conducted on a DMC 635 V eco vertical milling machine. The cryogenic delivery system was applied directly to the cutting zone through a nozzle. The cutting parameters during both machining techniques are cutting speed (*Vc*) at 140–160 m/min, feed rate (*fz*) at 0.15–0.20 mm/tooth, axial depth of cut (*ap*) at 0.3 mm and radial depth of cut (*ae*) at 0.2–0.4 mm.

The workpiece material used was the aged-hardened and solution-treated of Inconel 718 alloy with a dimension of 150 × 100 × 50 mm rectangular block with hardness of

42 ± 2 HRC. The chemical composition of Inconel 718 is shown in Table 1. The machining process was done using a Sumitomo ball nose insert cutting tool with an insert diameter of 10 mm and nominal diameter (insert with holder) of 16 mm. The insert was made from tungsten carbide with multi-layer PVD TiAlN/AlCrN grade ACK 300.

Table 1. Nominal chemical composition of Inconel 718

Elements	Al	B	C	Nb. Ta	Co	Cr	Cu	Fe	Mn	Mo	Ni	Si	Ti
% wt.	0.49	0.004	0.051	5.05	0.3	18.3	0.04	18.7	0.23	3.05	53	0.08	1.05

Surface roughness (*Ra*) was measured using a Mitutoyo Surf Test model SJ-310 series. The measuring force was 0.75 mN with using the stylus tip radius 2 μm and tip angle 60°. The measurements were taken three times at the beginning of the cut in order to eliminate the tool wear effect.

Microhardness testing was done using a Vickers microhardness tester Eseway model EW 442. The force load was set to 50 gf for 15 s each indentation. The sample was 3 × 4 mm and 50 indentations were made on the sample (10 indentation × 5 lines).

For sub-surface microstructure analysis, the prepared sample was mounted, ground, polished and etched by using a Kalling's No. 2 etching solution. After that, the sample was observed under an optical microscope to analyse the revealed microstructure of the machined surface.

3 Results and Discussion

3.1 Surface Roughness Analysis

The difference in surface roughness value resulting from cryogenic and dry machining is shown in Fig. 1. Based on the experimental results, it is clearly shown that surface roughness resulted through cryogenic machining is much lower, in the range of 0.104 μm to 0.161 μm compared to the results for the dry machining, in the range of 0.835 μm to 1.630 μm. The average reduction of surface roughness from dry to cryogenic machining was 88%. The results were found to be in line with the findings of the study conducted by Shokrani et al. [7]. The graph in Fig. 3 also demonstrates that cryogenic machining not only produce better surface roughness values but also produce more consistent results compared to dry machining. The unstable reading during dry machining is believed due to the sticking chips at the tool tip (Fig. 2) which interrupted the contact surface and generated high friction value during machining process. This scenario will let to poor surface finished.

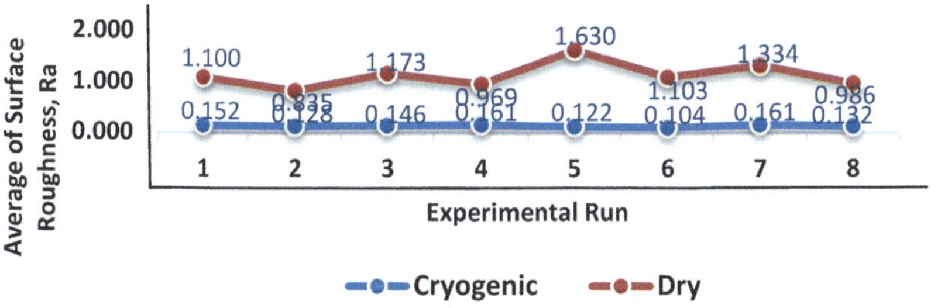

Fig. 1. Comparison of the Ra values between dry and cryogenic machining.

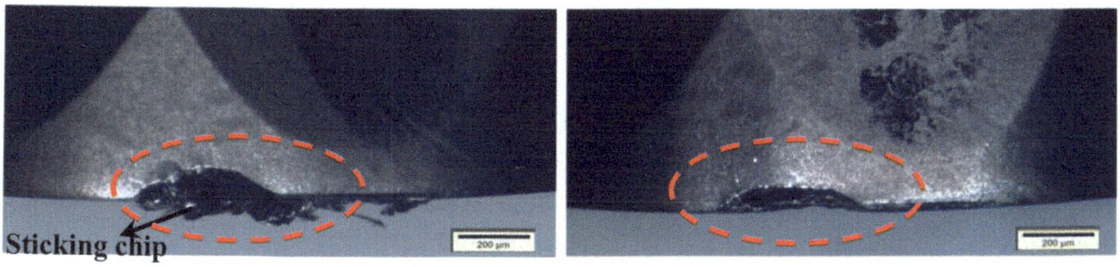

(a) Dry machining (b) Cryogenic machining

Fig. 2. Condition of the cutting tool after (a) dry and (b) cryogenic machining

Generally, Inconel 718 is commonly used to make blades for aircraft engine fans, where conditions for blade surfaces, especially the aerofoil parts, must be smooth to prevent vortices and erosion from any impurities found. In line with the requirements of aircraft manufacturing, the value of surface roughness produced during cryogenic machining experiments is less or equal to 0.2 μm, where this value is less than 0.5 μm corresponding to manual polishing results [8].

3.2 Microhardness Analysis

The analysis of microhardness changes on the sub surface of the workpiece is intended to see the plastic deformation changes in microstructure or work-hardening of the workpiece during dry and cryogenic machining. Overall, the results of the experiments found that both machining techniques had hardened the workpiece on the sub surface after machining process. The most harden part is obtained at the closest to the machined surface. The experimental reading also showed that when the depth of the position of the observation point increased (measured from the machined surface), the value of the microhardness reading would decrease further to the original value of the workpiece which is around 413 ± 22 HV.

Figure 3 shows that dry machining produced harder machined surface as compared to cryogenic machining when machined at *Vc* of 160 m/min, *fz* of 0.15 mm/tooth *ae* of 0.20 mm and *ap* of 0.30 mm. The result shows an increment in the microhardness of nearly 5% as compared for both techniques. Dry machining will increase the compressive stress and cause higher microhardness value in the machined subsurface. This

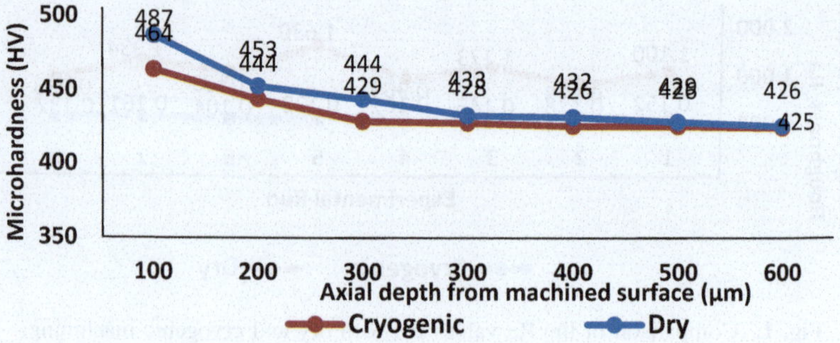

Fig. 3. Measured microhardness during the dry and cryogenic machining at *Vc* = 160 m/min, *fz* = 0.15 mm/tooth, *ap* = 0.30 mm, *ae* = 0.20 mm.

result is in line with the results of the study by Ravi and Kumar [9] on the milling of the hard steel material. This is attributed to the reduction of cutting force and cutting temperature formed during cryogenic machining. The used of nitrogen liquids that are sprayed towards the cutting zone will evaporate and form nitrogen hydraulic cushion between the surface of the cutting tool point and the workpiece points that will carry some of the normal loads. In addition, Umbrello [10] in his study report that the dry machining will attributed to the higher microhardness value due to the higher generation of white layer which normally represent untempered martensite structure.

3.3 Machined Surface Microstructure Analysis

Micro structure changes before and after cryogenic and dry machining are shown in Fig. 4 under 20X magnification. From the results obtained, the plastic deformation near the machined surface of the material was more visible with a depth of 31.79 µm during dry machining compared to cryogenic machining which produced a plastic deformation depth of 12.71 µm. This is due to the fact that dry machining will result in a stronger workpiece adhesion to the cutting tool due to higher cutting temperatures. The same

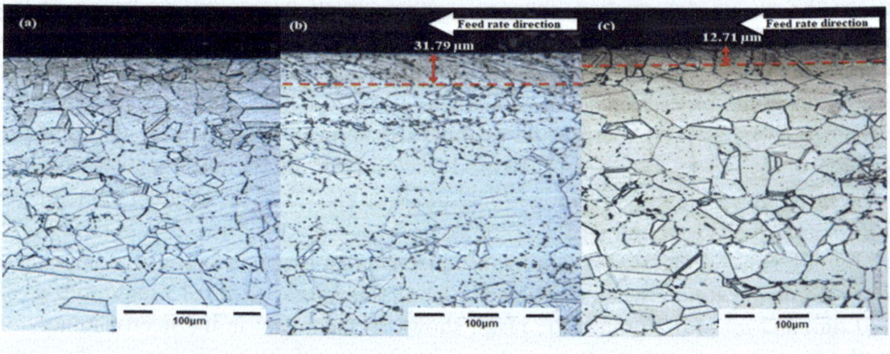

Fig. 4. Comparison of microstructure changes occurring when (a) before machining, (b) dry machining, (c) cryogenic machining.

finding was obtained by Kenda et al. [11] proving that the shear deformation is less likely to occur on the machined surfaces under cryogenic machining compared to dry machining.

The results of this experiment also show that cryogenic cooling allows to reduce the built-in pressure as well as the heat generated during the machining process before the workpiece reaches its yield strength and returns to its original state after machining. In other words, this result indicates that nitrogen liquids can penetrate the tool's tip interface with the workpiece more effectively. This finding further reinforces the evidence that cryogenic cooling can reduce the friction during the machining process.

4 Conclusions

As a conclusion from this study, cryogenic machining produced a better surface roughness as compared to dry machining. The results show that the value of surface roughness that can be obtained during cryogenic machining is equal or less than 0.20 μm, where this value makes it almost like the value for manual grinding (≈0.50 μm). The microhardness value for machined surface has revealed that dry machining produced harder machined surface compared to cryogenic machining by almost 5% increment. The hardest value is located near the machined surface and as it goes deeper toward the bulk surface, the hardness was decreased until match the original hardness value of the material. In analysis for sub-surface microstructure changes, cryogenic machining proved that less plastic deformation occurred compared to dry machining. The used of nitrogen liquids was able to reduce the cutting temperature during machining and allowed the build-in pressure drop before the workpiece reached its yield strength value.

Acknowledgement. This research would not have been conceivable without the specialized help of the Universiti Kebangsaan Malaysia. This research is supported under project RDU1803145. Special thanks to Universiti Malaysia Pahang for providing the facilities for the preparation of this manuscript.

References

1. Thakur, D.G., Ramamoorthy, B., Vijayaraghavan, L.: Machinability investigation of Inconel 718 in high-speed turning. Int. J. Adv. Manuf. Technol. **45**(5–6), 421–429 (2009)
2. Ezugwu, E.O., Bonney, J.: Effect of high-pressure coolant supply when machining nickel-base, Inconel 718, alloy with coated carbide tools. J. Mater. Process. Technol. **153**, 1045–1050 (2004)
3. Dudzinski, D., Devillez, A., Moufki, A., Larrouquère, D., Zerrouki, V., Vigneau, J.: A review of developments towards dry and high speed machining of Inconel 718 alloy. Int. J. Mach. Tools Manuf. **44**(4), 439–456 (2004)
4. Pawade, R.S., Joshi, S.S., Brahmankar, P.K.: Effect of machining parameters and cutting edge geometry on surface integrity of high-speed turned Inconel 718. Int. J. Mach. Tools Manuf. **48**(1), 15–28 (2008)

5. Subramanian, M., Sakthivel, M., Sooryaprakash, K., Sudhakaran, R.: Optimization of end mill tool geometry parameters for Al7075-T6 machining operations based on vibration amplitude by response surface methodology. Measurement **46**(10), 4005–4022 (2013)
6. Saikaew, C., Baowan, P.: Surface finish improvement in ball nose end milling by optimizing operating conditions for different cutting times. Indian J. Eng. Mater. Sci. **22**, 38–50 (2015)
7. Bermingham, M.J., Kirsch, J., Sun, S., Palanisamy, S., Dargusch, M.S.: New observations on tool life, cutting forces and chip morphology in cryogenic machining Ti-6Al-4V. Int. J. Mach. Tools Manuf. **51**(6), 500–511 (2011)
8. Shokrani, A., Dhokia, V., Newman, S.T., Imani-Asrai, R.: An initial study of the effect of using liquid nitrogen coolant on the surface roughness of Inconel 718 nickel-based alloy in CNC milling. Procedia CIRP **3**, 121–125 (2012)
9. Ravi, S., Kumar, M.P.: Experimental investigations on cryogenic cooling by liquid nitrogen in the end milling of hardened steel. Cryogenics **51**(9), 509–515 (2011)
10. Umbrello, D.: Influence of material microstructure changes on surface integrity in hard machining of AISI 52100 steel. Int. J. Adv. Manuf. Technol. **54**(9–12), 887–898 (2011)
11. Kenda, J., Pusavec, F., Kopac, J.: Analysis of residual stresses in sustainable cryogenic machining of nickel based alloy—Inconel 718. J. Manuf. Sci. Eng. **133**(4), 041002–041009 (2011)

Toolpath and Holes Accuracy of Robotic Machining for Drilling Process

Mohd Shahir Kasim(✉), Mohammad Shah All Hafiz, Nurwahida Rosli, W Noor Fatihah Mohamad, Raja Izamshah, Mohd Amran Md Ali, and Abu Abdullah

Advanced Manufacturing Centre, Fakulti Kejuruteraan Pembuatan, Universiti Teknikal Malaysia Melaka, Hang Tuah Jaya, 76100 Melaka, Durian Tunggal, Malaysia
shahir@utem.edu.my

Abstract. Drilling is one of the major machining operations in manufacturing. The application of robots in machining is the alternative technique to produce new products of the future. However, the performance of the robot always been a challenge in production. This paper presents the investigation on the effect of arm robot itinerary, holes orientation and materials type on holes accuracy and toolpath angularity of Aluminum Alloy 6061 (Al 6061) and High-density polyethene (HDPE). A series of drilling experiments by using COMAU robot with 24 runs were conducted at difference combination parameters. The measurement on hole accuracy and toolpath angularity were done by CMM machine. Response surface methodology (RSM) was used as a design of experiment (DOE) for optimization. Evaluation by ANOVA showing that the interaction between arm robot itinerary and material type found to be significant factor for the hole accuracy whilst the orientation of the hole dominant factor affecting the toolpath angularity. Optimization results show that the best accuracy and angularity of the drilling hole when the arm robot itineraries of 971.82 mm (vertical direction) and 1120.65 mm (horizontal direction) for Al 6061 and HDPE respectively.

Keywords: Industrial robot · Precision manufacturing · High-speed drilling

1 Introduction

The application of robots in the manufacturing of engineering part is nothing new to the industry. This alternative technology has been used since the 20th century. The advantage of robot application because of the productivity, repeatability, flexibility, programmability, relatively low cost besides another advantage [1]. The most importance advantage is the ability the robot arm to reach on the specific location which cannot be done by conventional dedicated machine especially in dealing with the complex shape. Hence, the demand of the robot increases yearly. The date from International Federation of Robotics showing the robot sales increased by 16% in 2016 [2].

M. N. Osman Zahid et al. (Eds.): *iMEC-APCOMS 2019*, LNME, pp. 519–524, 2020.
https://doi.org/10.1007/978-981-15-0950-6_79

Machining, drilling, trimming are some of the newest application areas for a robot. The challenge in production as the accuracy of machining has always been a primary interest to product quality. The drawback of robot application such as low stiffness and anisotropic behaviour draws researchers' attention. Yin bu and his clique concluded that stiffness property effects on drilling quality [3]. It was related to robot joints where 70–80% of the 6-DOF serial robot's error comes from the stiffness joints [4]. These problems lead to circularity, position and perpendicularity error [5].

Dimensional tolerance to the dimensions on engineering part is an important aspect in manufacturing. It was specified since the part cannot be produced with exact shape geometry and dimensions. The variation in dimension was reflected by the process capability of removing process, material and product requirement [6]. The permissible deviation is according to the part specification. ISO 2768 and JIS B 0408 specified the different tolerance at different grades (fine, medium, coarse and very coarse). Fine grade products and small dimension require tight permissible tolerance [7, 8]. Hence, the purpose of this experiment is to verify the holes accuracy and toolpath angularity of the robotic machining for the high-speed drilling process.

2 Experimental Setup

In this study three parameters were used as variable input for the DOE namely; arm robot itinerary, materials type, and holes orientation. The responses of this experiment were hole accuracy and tool path angularity. Two different types of materials were used in this experiment that are Aluminium Alloy 6061 (Al 6061) and High-Density Polyethylene (HDPE). Three arm robot itineraries were used are 550 mm, 1000 mm and 1450 mm as shown in Fig. 1.

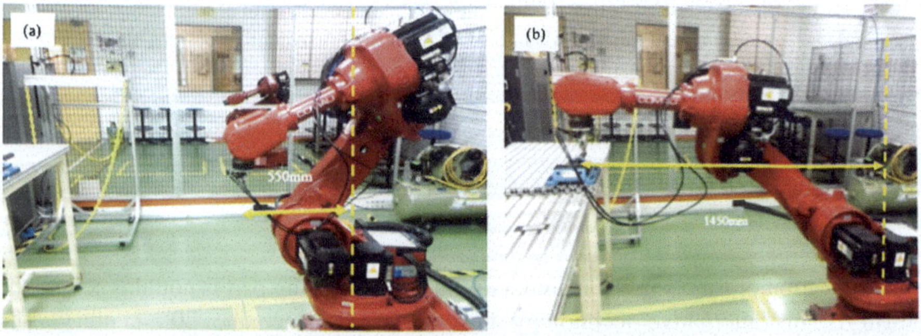

Fig. 1. Arm robot itinerary (a) 550 mm and (b) 1450 mm

The HSS drill bit with the size of 6 mm Ø was used throughout of the experiment. The drill bits were attached on the COMAU robot spindle motor with the speed of 25000 RPM. The numbers of 24 run were tested at different combinations of the cutting parameter which will be evaluated by response surface method (RSM).

The workpiece was drilled in L shape in horizontal and vertical directions. Both holes accuracy and angularity of toolpath were measured by XO Orbit 55 coordinate measuring machine (CMM). Analysis of variance (ANOVA) was performed to identify the dominating factors of the response. It can be identified when the probability of response not changing at different level input where the P-Value is less than 5%.

3 Results and Discussion

3.1 Significant Parameters to Experiment

Variance analysis conducted on the experimental results suggests a quadratic models type for both holes accuracy and toolpath angularity. The purpose of the ANOVA is to investigate which variable inputs significantly affect the responses characteristic. Tables 1 and 2 show the results of the ANOVA analysis.

Table 1. ANOVA analysis for holes accuracy

Source	Sum of squares	df	Mean square	F value	p-value Prob > F	
Model	0.21	5	0.041	3.9	0.0154	Significant
A - Arm itinerary	7.23×10^{-3}	1	7.23×10^{-3}	0.68	0.4198	Not significant
B - Materials type	0.058	1	0.058	5.51	0.0313	Significant
C - Holes orientation	0.015	1	0.015	1.45	0.2446	Not significant
AB	0.057	1	0.057	5.38	0.033	Significant
A^2	0.067	1	0.067	6.31	0.0224	Significant
Residual	0.18	17	0.011			
Lack of fit	0.083	6	0.014	1.59	0.2399	Not significant
Pure error	0.096	11	8.76×10^{-3}			
Cor total	0.39	22				

Table 2. ANOVA analysis for toolpath angularity

Source	Sum of squares	df	Mean square	F value	p-value Prob > F	
Model	15.79	4	3.95	9.91	0.0002	Significant
A - Arm itinerary	0.44	1	0.44	1.09	0.3091	Not significant
B - Materials type	3.24	1	3.24	8.14	0.0102	Significant
C - Holes orientation	9.25	1	9.25	23.21	0.0001	Significant
A^2	2.86	1	2.86	7.18	0.0148	Significant
Residual	7.57	19	0.4			
Lack of fit	2.62	7	0.37	0.91	0.5333	Not significant
Pure error	4.95	12	0.41			
Cor total	23.36	23				

Each material has its own unique stress-strain curve. The ability of the material to withstand at the machine load would affect inaccuracy of the hole produced. Al 6061 tensile strength is 310 MPa whilst HDPE, 29 MPa. This shows that HDPE would easier to break than Al 6061. However, drilling HDPE would be a problematic due to its poor thermal conductivity. HDPE is plastic based materials. Therefore, the heat generated during the material removal process cannot be dissipated by conduction. This causes the material to overheat rapidly, soften and weld itself to the drill thus affecting the diameter accuracy of the hole and toolpath angularity.

3.2 Develop and Validate Prediction Model

The cutting condition that affecting the hole accuracy and toolpath angularity were determined. The developed prediction model as tabulated in Table 3 were used to predict the outcomes response within the variable input range value. For instance, two random combinations for arm robot itinerary and material type are classified during validation are shown in Table 4.

Table 3. Prediction model for holes accuracy and tool path angularity

Materials	Orientation	Hole accuracy	Tool path angularity
Al 6061	Horizontal	$6.87 - 1.03 \times 10^{-3}D + 5.60 \times 10^{-7}D^2$	$86.58 - 7.60 \times 10^{-3}D + 3.62 \times 10^{-6}D^2$
Al 6061	Vertical	$7.04 - 1.31 \times 10^{-3}D + 5.60 \times 10^{-7}D^2$	$86.34 - 7.60 \times 10^{-3}D + 3.62 \times 10^{-6}D^2$
HDPE	Horizontal	$6.82 - 1.03 \times 10^{-3}D + 5.60 \times 10^{-7}D^2$	$85.34 - 7.60 \times 10^{-3}D + 3.62 \times 10^{-6}D^2$
HDPE	Vertical	$6.99\ 1.31 \times 10^{-3}D + 5.60 \times 10^{-7}D^2$	$84.61 - 7.60 \times 10^{-3}D + 3.62 \times 10^{-6}D^2$

Table 4. Results of the validation test for the holes accuracy and tool path angularity

Arm robot itinerary (mm)	Materials	Hole accuracy			Tool path angularity		
		Predicted (mm)	Actual (mm)	% error	Predicted (°)	Actual (°)	% error
651	Al 6061	6.44	6.38	0.93%	90.00	89.98	0.02%
836	HDPE	6.34	6.40	0.95%	89.68	89.92	0.27%

The purpose of the validation test is to confirm conclusions drawn during the analysis and Fig. 2 shows the result of the validation tests. The results reveal that the percentage of error for the predicted values is less than 1%. It was common to justified the acceptable percentage of error for non critical application less than 10% [9]. Therefore, it shows that the quadratic equation for both holes accuracy model and toolpath angularity model can be used as the prediction model.

3.3 Optimization Responses

Table 5 shows the multi-objective optimization results. It shows the suggested arm itinerary which can obtain both minimum errors of hole accuracy and tool angularity simultaneously. The best itineraries are 971.82 mm at vertical direction and 1120.65 mm at the horizontal direction for Al 6061 and HDPE respectively. Based on precision capability, these optimized result of hole diameter falls within IT14 tolerance grade.

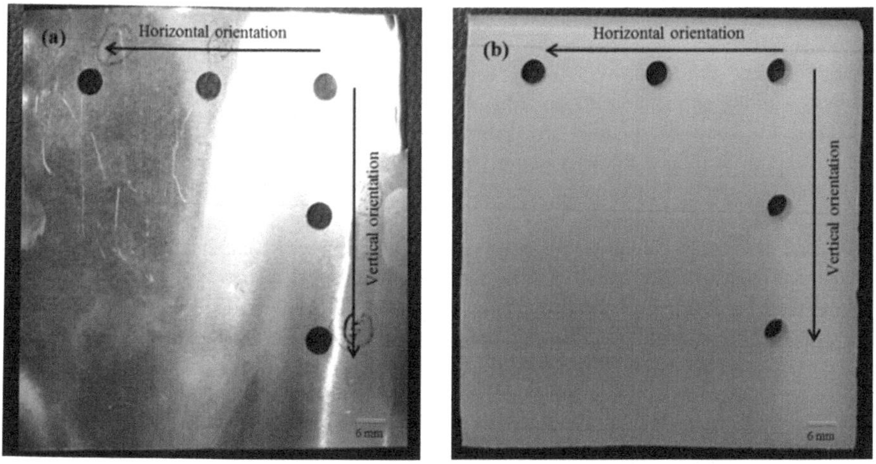

Fig. 2. Results of validation tests (a) Al 6061 at arm robot itinerary, 651 mm; toolpath angularity, 89.98° and (b) HDPE at arm robot itinerary, 836 mm; toolpath angularity, 89.92°.

Table 5. Multi-objective optimization results

Materials	Arm robot itinerary (mm)	Hole diameter (mm)	Toolpath angularity (°)
Al 6061	971.82	6.34	89.31
HDPE	1120.65	6.28	89.82

4 Conclusion

There are several conclusions can be rendered based on the study conducted:

1. Interaction between arm robot itinerary and material type found to be a significant factor for the hole accuracy whilst the orientation of the hole is a dominant factor affecting the toolpath angularity.
2. The error for the prediction models is less than 1%.
3. The best arm robot itinerary for COMAU robot based on the machining parameters is 971.82 mm and 1120.65 mm with the horizontal direction.

Acknowledgement. The authors are grateful to the Advanced Manufacturing Center, Universiti Teknikal Malaysia Melaka and the Ministry of Education Malaysia for funding this research project through grants no: FRGS/2018/FKP-AMC/F00378.

References

1. Taiwantrade: How Industrial Robot Application in Manufacturing Is Changing The Way We Manufacture? (2018). https://www.taiwantrade.com/seo/introduction-to-internet-of-things/application-of-robots-in-manufacturing-industries.html. Accessed 12 Apr 2019

2. International Federation of Robotics: Executive Summary World Robotic 2017 Industrial Robots (2017)
3. Bu, Y., Liao, W., Tian, W., Zhang, J., Zhang, L.: Stiffness analysis and optimization in robotic drilling application. Precis. Eng. **49**, 388–400 (2017)
4. Dumas, C., Caro, S., Garnier, S., Furet, B.: Joint stiffness identification of six-revolute industrial serial robots. Robot. Comput. Integr. Manuf. **27**(4), 881–888 (2011)
5. Garnier, S., Subrin, K., Waiyagan, K.: Modelling of robotic drilling. Procedia CIRP **58**, 416–421 (2017)
6. Day, D., Raines, M., Swift, K.: Process capable tolerancing. Mach. Des. (2005)
7. Japanese Standard: JIS B 0408:1991 - General dimensional tolerances for parts formed by press working from sheet metal. Japanese Standards Association (2015)
8. International Standard: ISO 286 - Part 2: Geometrical product specifications (GPS) - ISO code system for tolerances on linear sizes. (2010)
9. Hills, R.G., Trucano, T.G.: Statistical Validation of Engineering and Scientific Models: Background, New Mexico (1999)

Optimization of Speed Cylinder and Distance Speed Cylinder Hydraulic Movement of Kobelco Tire Curing Machine

Deri Teguh Santoso[1](✉) and Pajar Barokah[2]

[1] Universitas Singaperbangsa Karawang, Karawang, West Java 41361, Indonesia
deri.teguh@ft.unsika.ac.id
[2] PT. Multistrada Arah Sarana, Tbk., Bekasi, West Java 17530, Indonesia

Abstract. The process of making tires is using the curing process, this process using the Kobelco curing machine. Press Unit Problem contribute 51.5% of the total problem in curing process. Press unit problems caused by dry time in press cylinder unit to open or close movement still takes long time which is not productive. Improvements made by modifying the speed cylinder and distance speed cylinder movement. Optimization of Speed Cylinder by modifying the setting of speed parameters on the cylinder selenoid valve, while optimizing the distance speed cylinder movement by modifying the PLC program to obtain optimal speed of movement distance. The results obtained after optimization obtained 100 mm/s of hydraulic speed cylinder, and 45° mechanical flow control. The problem of the press unit decreases, while the dry time reduced by 37% so that the production of tires increases 12 pcs tires/day/machine, so that total saving cost as much as IDR. 84,000,000.

Keywords: Tire · Curing process · Cylinder speed · Dry time

1 Introduction

The manufacturing industry in Indonesia is increasing, and as shown from Indonesia's statistical export in January 2019, it was dominated by the processing industry exports, 73.24% of which are Indonesia non-oil and gas commodities [1]. The manufacturing industry is rapidly developing, one of which is the tire industry. This industry is growing, it is influenced by the rapid development of the manufacturing industry. Tires generally use composite materials that are non-isotropic, consisting of several rubber components combined with textile materials and steel material to strengthen the tire structure [2].

The tire production process consists of several process stages as seen in Fig. 2. The last process in the tire production process is the vulcanization process or better known as the curing process, because the machine used is the Kobelco curing machine. The main problem of the curing process is its long dry time, which often causes failure in the press unit of the curing machine, resulting in productivity losses.

M. N. Osman Zahid et al. (Eds.): iMEC-APCOMS 2019, LNME, pp. 525–530, 2020.
https://doi.org/10.1007/978-981-15-0950-6_80

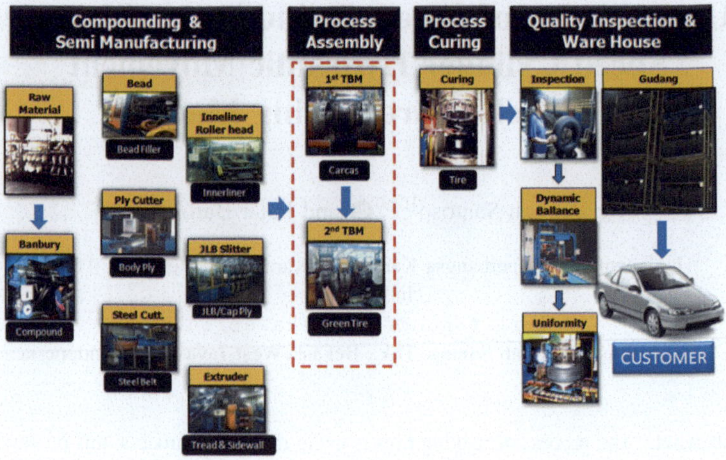

Fig. 1. Tire manufacturing processes (Source: PT. Multistrada Arah Sarana, Tbk., STRADA)

Several research studies related to the curing process have been carried out. a research study conducted by Pandya et al. [3] proved that reducing the delay time during the curing process can improve product quality and reduce utility costs by causing the temperature of the delay time to be lowered at from 38 °C to 4 °C, this research study conducted an experimental study and CFD simulation. Other research studies related to optimizing the curing process by reducing curing time by 35 min, thus increasing productivity using the Finite Element Analysis (FEA) method [4], regarding the effect of vulcanization on mechanical properties [5]. Another study is conducted to measure the effect of temperature vulcanization using mixed methods [6].

2 Research Methodology

2.1 Curing Process

The curing process (vulcanization process) for this study is conducted using Kobelco curing machine. This process is one of the processes of a series of tire manufacturing processes carried out in the last production before quality check, as seen in Fig. 1. For the outside of the tire, the process is conducted with temperature of 185 °C and pressure of 10.5 bar, while for the inside the tire uses a combination of steam and N_2 gas works alternately according to the specified product specifications, the pressure used is 15 bar for steam and 21 bar for N2 gas. The curing process takes between 10–30 min, depending on the type of tire used. In Fig. 2 it can be seen a total of 13 min for the curing process is used for the A45 Eco type. From the latest data obtained, this type of A45 Eco produces 31 pcs tire/mold/work shift.

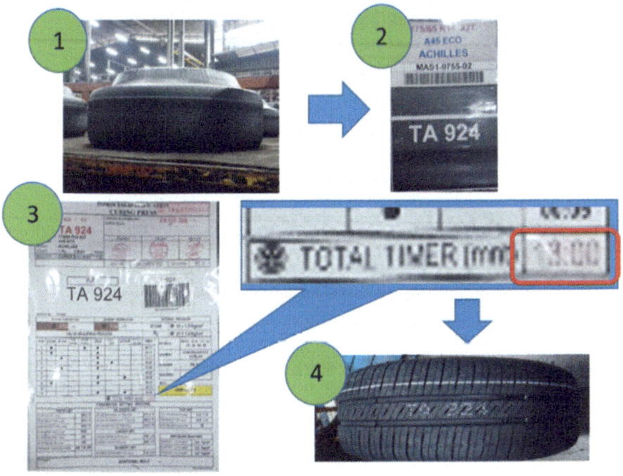

Fig. 2. Curing process A45 Eco type

2.2 Dry Time

Dry time is the time needed by the curing machine to move from the previous curing process to the next curing process, as seen in Fig. 3. Dry time can be interpreted as the preparation and ending time of the curing machine before the next curing process productivity will increase, and the longer the dry time the lower productivity is, which will increase the loss of production targets. Delay time [3] is not the same as dry time, because delay time is idle time, while dry time is part of curing process. Dry time has 10 different processes with a total dry time of 155 s, then 10 kinds of processes that occur during dry time will be optimized so as to accelerate the motion of the components that work so that the dry time is shorter. Figure 4 marked red which is the process in dry time will be optimized for shorter time.

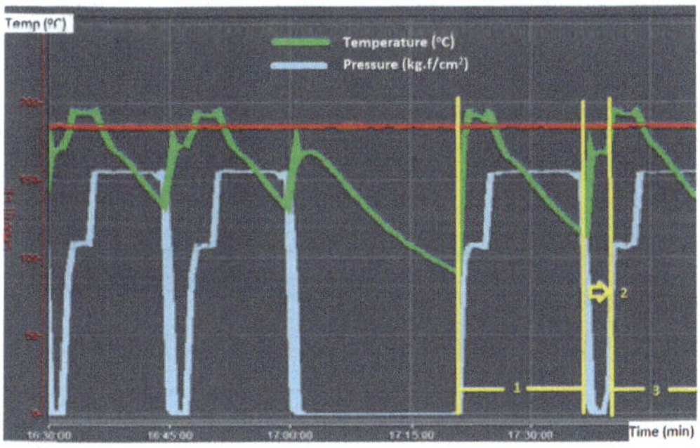

Fig. 3. Dry time (2) is located between previous curing time (1) and next curing time (3)

Working Step	Actual
Mold open	41
Unloader in	6.06
Top ring up	8.49
Unloader down	11.99
Chuck unload open	4.41
Unloader up out	11.59
Loader in down	6.4
Shapping	18.46
Loader up out	9.18
Mold close	37.48
Total Time	**155.1**

Fig. 4. Dry time breakdown

2.3 Research Framework

This research study, aims to shorten dry time by accelerating the speed of TCR hydraulic cylinder movement (Fig. 5, part A), which current cylinder speed is 40.5 mm/s (top ring up; Fig. 4). Optimization is done by modifying the parameter settings of the speed cylinder on the selenoid valve cylinder using transducer (Fig. 5, part B) and mechanical flow control (Fig. 5, part C). Optimization of the speed cylinder movement is also done by modifying the PLC program to adjust the suitability of the actual position with TCR hydraulic cylinder speed changes to obtain the optimal of dry time.

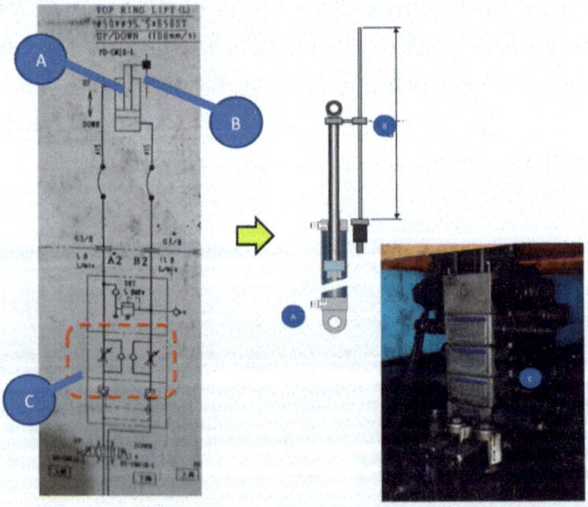

Fig. 5. (A) TCR hydraulic cylinder; (B) Transducer; (C) Mechanical flow control.

Optimization using two independent factors are TCR hydraulic speed cylinder (X_1) with range level of 40–100 mm/s, and mechanical flow control (X_2) range level of 15°–90° (Table 1). There are two independent factors with 6 and 7 levels respectively, so that the full factorial is 42 possibilities. The dependent factor is Dry Time (Y), the lower the dry time value, the better the result.

Table 1. Design of Experiment (DoE) of TCR up (top ring up)

Independent factor	Level						
	1	2	3	4	5	6	7
TCR hydraulic speed cylinder (X_1); (mm/s)	40	50	60	70	80	90	100
Mechanical flow control (X_2)	15°	30°	45°	60°	75°	90°	

3 Result

3.1 Dry Time Reduction

The optimization obtained after performing the Design of Experiment (Table 1), hydraulic cylinder is 100 mm/s and mechanical flow control 45°. The overall actual data speed after optimization can be seen in Fig. 6. The dry time before optimization is 119.7 s, which is the total time to be optimized. After optimization is implemented, get a dry time reduction result of 37%, producing a total dry time of 74.9 s.

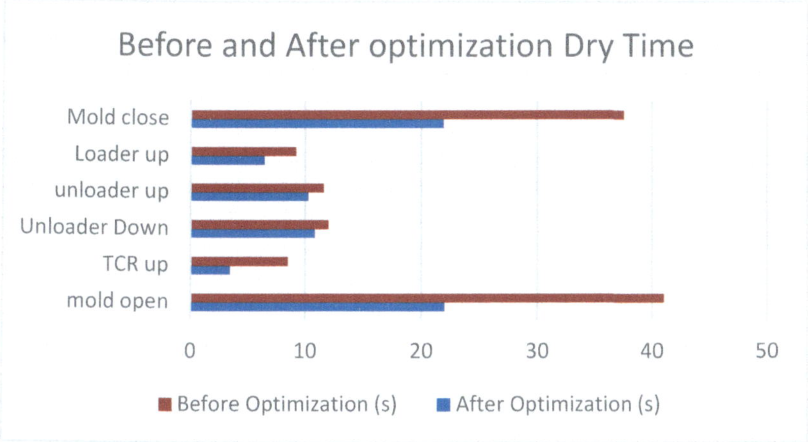

Fig. 6. Comparison after dry time optimization

3.2 Increase Productivity

The dry time decrease was 37% or equivalent to 44.78 s. This shows that the total time saved is 1388 s or equivalent to 2 pcs/mold/workshift. The condition of the machine uses 2 pcs molds, and the work system in 3 workshifts. If the price is IDR 350,000/pc tire, the total productivity increased is IDR 84,000,000/month.

4 Conclusions

The curing process that has been optimized is related to the dry time. Dry time has a number of processes that take too long to cause productivity to decline. Optimization of the curing process by increasing the speed cylinder, increasing the cylinder speed by modifying the selenoid valve cylinder, and modifying the PLC program is able to shorten the dry time. The results obtained after optimization are a decrease of 37% dry time, and an increase productivity worth IDR 84,000,000/month.

Acknowledgment. PT. Multistrada Arah Sarana, Tbk have facilitated the research, and proceeding funded by Kementerian Riset dan Pendidikan Tinggi Republik Indonesia.

References

1. Badan Pusat Statistik (BPS-Statistics Indonesia): https://www.bps.go.id/publication/2019/03/29/c37da964398b5f32c85238 1f/buletin-statistik-perdagangan-luar-negeri-ekspor-menurut-kelompok-komoditi-dan-negara-januari-2019. Access 24 Apr 2019
2. Leister, G.: Passenger Car Tires and Wheels: Development-Manufacturing-Application, 1st edn. Springer, Switzerland (2018)
3. Pandya, M., Patel, R.N., Amarnath, S.K.P.: Determination of time delay and rate of temperature change during tyre curing (Vulcanizing) cycle. In: Chemical, Civil and Mechanical Engineering Tracks of 3rd Nirma University International Conference, Procedia Engineering, vol. 51, pp. 828–833. Elsevier, New York (2013)
4. Wang, Y., Su, B., Wu, J.: Simulation and optimization of giant radial tire vulcanization process. In: International Conference on Advances in Computational Modelling and Simulation, Procedia Engineering, vol. 31, pp. 723–726. Elsevier, New York (2012)
5. Magioli, M., Sirqueira, A.S., Soares, B.G.: The effect of dynamic vulcanization on the mechanical, dynamics mechanical and fatigue properties of TPV based on polypropylene and ground tire rubber. Polym. Test. **29**, 840–848 (2010)
6. Mansilla, M.A., Marzocca, A.J., Macchi, C., Somoza, A.: Natural rubber/styrene-butadiene rubber blends prepared by solution mixing: influence of vulcanization temperature using a Semi-EV sulfur curing system on the microstructural properties. Polym. Test. **63**, 150–157 (2017)

Eco Design for Rooftop in Urban Housing

Norasmiza Mohd and Zubair Khalil(✉)

Faculty of Mechanical and Manufacturing Engineering,
Universiti Malaysia Pahang, 26600 Pekan, Pahang, Malaysia
zubair@ump.edu.my

Abstract. Malaysia is hot tropical climate country. This effects the residential environment over thermal comfort acceptable limit throughout the year. The main aim of this study is to study important parameters to house design on reducing temperature without air conditioning system. Four parameters were considered in this study. They were air flow types, gap holes, radiant barrier, and double skin façade (DSF) roof. An experiment was conducted and it has been revealed that air flow inward, the roof has a gap hole, radiant barrier at bottom of roof panel and thickness of DSF 1 cm shown the most significant results in reducing the temperature. The maximum temperature difference between indoor and outdoor is 7 °C. The eco design of roof in the house is considered a better option for this climate not only for its ability to provide natural air circulation, but also in reducing energy consumption up to 30% per year. Implementing this design in a roofing system is to enhance the heat dissipation as well as contributing the thermal comfort for human in residential building.

Keywords: Energy saving · Eco design · Thermal comfort

1 Introduction

Located closed to the Equator, Malaysia is experiencing the hot and humid climate. Additionally, global warming effects increase the temperature to even higher level [1]. In the projection done by Tang shown that the temperature will increase up to 4 °C by the end of 21st century. The increase rate is about 0.25 °C per decade in Peninsular Malaysia [2]. As a result, overheating are becoming significant problem in buildings especially residential type buildings in Malaysia [3]. Many types of research have been undertaken to find solution to the problem and subsequently study the performance of a ventilation system in house configuration. Common and conventional methods used to solve overheating is the usage of air conditioners. It could realize comfortable indoor environment even within warm city climate. However, continuous usage of air conditioners leads to high energy consumption and heat release from air conditioners external units are the root problem that causes global warming. In order to solve this problem conclusively, alternative cooling method to air conditioners is a must. This paper reveal natural's application of biomimicry technique based from termite mold [4]. This technique enables control of temperature and humidity to thermal comfort using air flow channel within structure. The acceptable temperature range of thermal comfort was from 25.5 °C to 28 °C and ideal comfort for humidity is within 30% to

M. N. Osman Zahid et al. (Eds.): iMEC-APCOMS 2019, LNME, pp. 531–536, 2020.
https://doi.org/10.1007/978-981-15-0950-6_81

60% [5]. Objective of this study is to design a eco house that apply the passive cooling technique which comfortable to human body at temperature between 25 °C to 28 °C and investigate the parameter effectiveness for cooling process of house.

2 Methodology

This research was focused on the investigation of four parameters which is air flow types, gap holes and radiant barrier availability, double skin façade (DSF) on their significant contribution to reduce temperature inside the model of the house using passive to semi active cooling technique. Figure 1 shows the design of the house and parameters employed in the experiment. The material used to fabricate the model house is acrylic. The thermal conductivity of acrylic is 0.2 W/m·K.

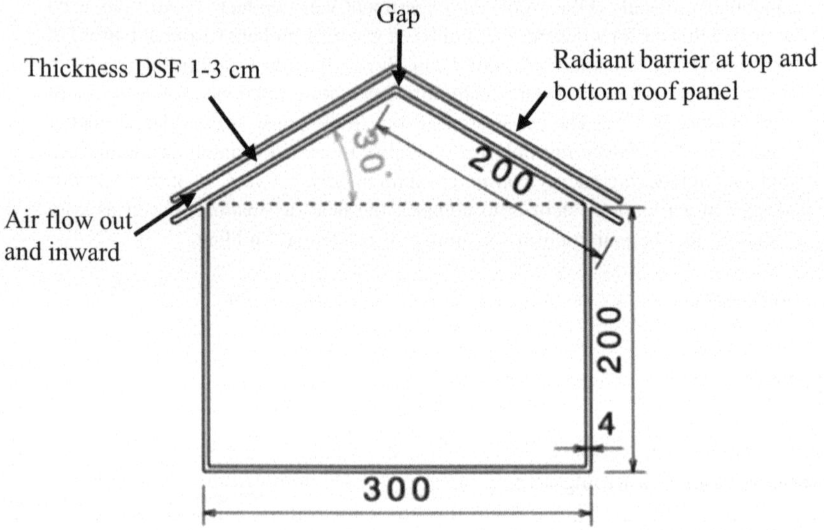

Fig. 1. House dimension in mm and the parameter indicator.

Natural ventilation is governed by volume and distribution of openings in the building's envelope, the internal pattern of air flow and pressure generated by an external force. This section outlines the importance of the flow of air in closed enclosure and opening at the roof. Hot air will rise upward due to buoyancy effect thus the opening at the highest point of the roof will serve as an outlet for releasing the hot air to the environment.

Double skin façade (DSF) is a combination of multiple layers of roof into one system. The roof is separated by a layer of air or gas in-between called air cavity that acts as an added layer of insulation. This will prevent unwanted heat from entering the house and maintaining room temperature in no-flow state (conduction). DSF able to serve as dual function of convection and conduction heat transfer to optimize its functions. In this study, DSF thickness of 1 cm and 3 cm was investigated to determine which significantly reduce temperature.

The space between the DSF and gap holes can be positioned to set up a convection flow induced by the density difference between cold and warm air. This is known as natural or free convection flow type. Then, force convection using mechanical fan controlled the air flow direction of either inward or outward were also investigated.

A radiant barrier is a highly reflective material able to reflect direct radiant heat. Material used in this study as the radiant barrier was aluminium. Thermal conductivity and emissivity of aluminium is 235 W/m·K and 0.04 respectively. In this experiment, the location of the radiant barrier is to determine whether the barrier at the bottom or top roof will give a better reflective performance.

3 Results and Discussion

3.1 Double Skin Façade (DSF)

From Fig. 2, the temperature indoor for thickness 1 cm is lower than thickness 3 cm and outdoor temperature. At 2.00 pm, thickness 1 cm produce bigger temperature difference between indoor and outdoor temperature. From this result, thickness 1 cm can reduce temperature up to 7 °C while 3 cm is 2 °C. Convection heat transfer occurs when there is air flow in the air cavity of DSF. The thickness of DSF must be smaller as possible so that the velocity of air that flow will be higher. Incompressible fluids have to speed up when they reach a narrow section in order to maintain a constant volume flow rate. Using this method, the rate of fluid flow increase during convection heat transfer which contributes to increasing the rate of heat removal.

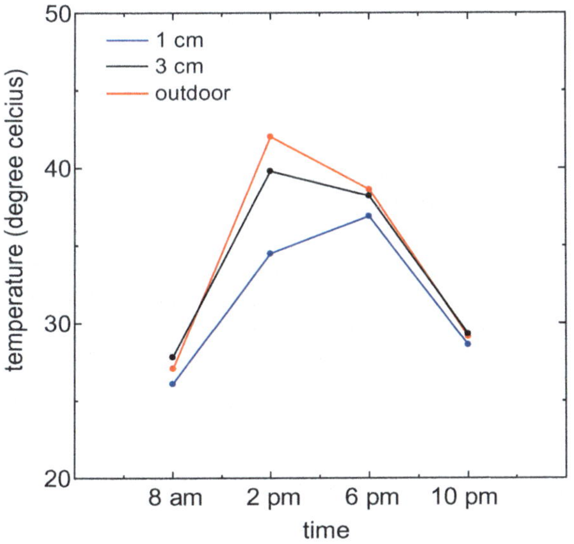

Fig. 2. Average temperature (°C) against time for thickness of DSF.

3.2 Air Flow

Inward air flow into the DSF exhibit in Fig. 3, shown significant difference temperature for both time, 2.00 pm and 6.00 pm compare to outdoor and outward air flow where air were sucked from within DSF. Temperature difference of inward and outward air flow to outdoor were 7 °C and 5 °C respectively. The difference seems to caused from the air mass flow rate and pressure where inward air flow have higher rating of both while outward might need higher fan power to produce similar results to inward flow. The reason is because the intended low pressure at outlet by the fan was inadequate due to lost of pressure in the opening space between fan and DSF.

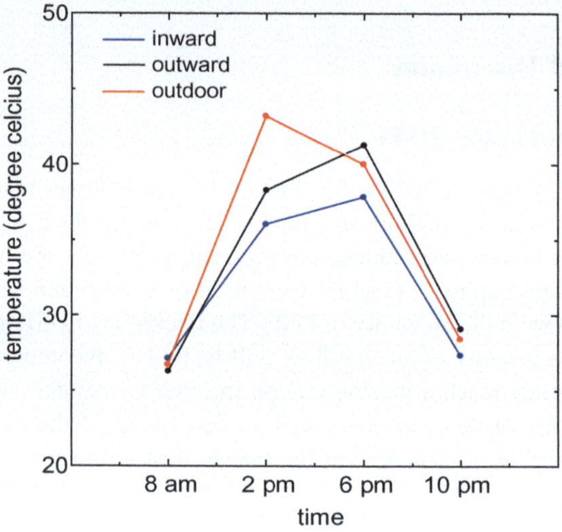

Fig. 3. Average temperature (°C) against time for air flow directions.

3.3 Gap

Based on Fig. 4, the average temperature for the roof with gap exhibit higher temperature difference between indoor and outdoor than roof without gap. At 2.00 pm, temperature difference for roof without gap was about 2 °C. On the other, for roof with gap, temperature difference was 7 °C. From these results, it is deduced that gap hole in roof is significance in reducing temperature. The reason is it offers a good outlet for ventilation because hot air naturally accumulates at the highest point of the roof. While in the case for roof without gap, it will be an enclosed design that does not allow any heat dissipate and air flow, thus resulting to heat gain and leads to a very high internal temperature.

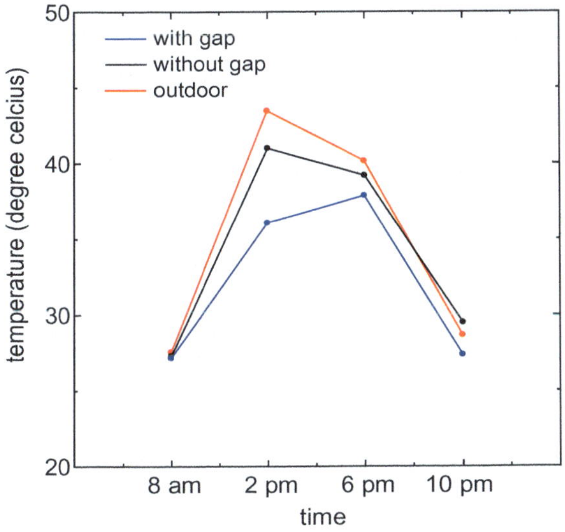

Fig. 4. Average temperature (°C) against time for gap opening availability

3.4 Radiant Barrier

From Fig. 5, temperature indoor for location radiant barrier at the bottom of the roof is slightly lower than the radiant barrier at the top of the roof. Temperature difference for radiant barrier at bottom and top of the roof about 6 °C and 5 °C respectively. So the indoor temperature is lower than the outdoor temperature when the radiant barrier is installed at the roof of the house. However, the differences were too small. We can

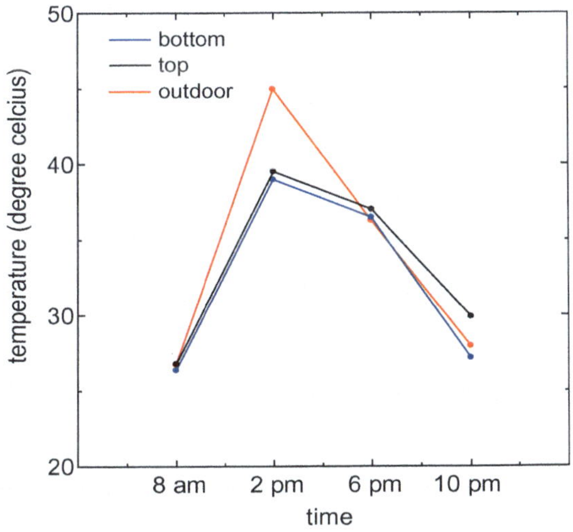

Fig. 5. Average temperature (°C) against time for radiant barrier position

conclude that while availability of radiant barrier is necessary in reducing temperature, their position is not significant. This conclusion is agreed by Lee stating that the peak of ceiling heat flux and house with radiant barrier system show a decrease of about 21% to 25% due to the application of the radiant barrier [6].

4 Conclusion

From the experiment, it has been revealed that inward air flow, roof with gap, radiant barrier at bottom of roof panel and DSF with 1 cm thickness exhibited significant performance on temperature reduce. The maximum temperature difference between indoor and outdoor is 7 °C. Implementing the eco design in roofing system will enhance the heat dissipation as well as contributing the thermal comfort for human in residential building. However, the efficiency of heat reduction depends on roof design and climatic boundaries including air temperature, relative humidity, direction and velocity of wind and sun radiation which differs from day to day. Therefore, to effectively improve heat rejection from buildings by natural means, the physical characteristics of the building should be sufficiently understood. Also, it should be noted that general single roof system adapted in Malaysia shown a higher temperature inside the residential area compare to outdoor. When compare to DSF systems where temperature is reduced, it is obvious that the single roof system is one the root of problems that leads to resident inconvenient, air conditioners with high energy consumption usage and others.

Acknowledgement. The authors are grateful to Faculty of Mechanical & Manufacturing Engineering and Innovative Manufacturing, Mechatronics & Sports Laboratory for providing the facilities to carry out this research work. The publication of the manuscript is funded by Universiti Malaysia Pahang (RDU 1703251).

References

1. Marasso, E., et al.: Global and local environmental and energy advantages of a geothermal heat pump interacting with a low temperature thermal micro grid. Energy Convers. Manag. **172**, 540–553 (2018)
2. Ho, D.T.K.: Climate change in Malaysia: Trends, contributors, impacts, mitigation and adaptations. Sci. Total Environ. **650**(Pt 2), 1858–1871 (2019)
3. Kazanci, O.B., Olesen, B.W.: Thermal indoor environment and energy consumption in a plus-energy house: cooling season measurements. Energy Procedia **78**, 2965–2970 (2015)
4. Radwan, G.A.N., Osama, N.: Biomimicry, an approach, for energy efficient building skin design. Procedia Environ. Sci. **34**, 178–189 (2016)
5. Lu, S., et al.: Field study of thermal comfort in non-air-conditioned buildings in a tropical island climate. Appl. Ergon. **66**, 89–97 (2018)
6. Lee, S.W., Lim, C.H., Salleh, E.I.B.: Reflective thermal insulation systems in building: a review on radiant barrier and reflective insulation. Renew. Sustain. Energy Rev. **65**, 643–661 (2016)

Investigation on the Effect of Multiple Passes in Plain Waterjet Cleaning of Paint

Mohd Nazir Mat Nawi[1,2(✉)], Hafiz Husin[1,2], M. A. Gebremariam[1], and Azmir Azhari[1]

[1] Faculty of Manufacturing Engineering, Universiti Malaysia Pahang, 26600 Pekan, Malaysia
mhd_nazir@yahoo.com
[2] Centre for Foundation Studies, International Islamic University Malaysia, 26300 Gambang, Malaysia

Abstract. Paint removal process in automotive coating is widely used in vehicle component recycling industry. The need of utilization and recycling the automotive component without producing secondary pollution from the paint removal process is recently become a major concern globally. Water jet cleaning is a new method for paint removal and getting recognition because of environmental friendly and it is better than mechanical cleaning such as sand blasting, brushing with water, hydropneumatic cleaning, controlled dry sanding, low pressure water projection and low pressure water spray. The present study focuses on the investigation of effect of multiple passes in plain water jet cleaning on paint removal process. A new method of multiple passes treatment is applied in plain water jet cleaning to access its effect on surface roughness and paint removal rate. It was found that, with increasing of number of passes, the surface roughness and paint removal rate is slightly increase. It is also found that the increase in water jet pressure will increase the surface roughness and paint removal rate. This is probably because increasing pressure will leads to more energy to remove the paint. It is found also that the increase in traverse rate increase the surface roughness and decrease paint removal rate. Based on the present study, it is a high prospect to apply multiple passes of paint removal using plain water jet in automotive industry.

Keywords: Paint removal · Waterjet cleaning · Multiple jet passes

1 Introduction

A removal process of deposits, contaminates and coating materials from the substrates of parts or manufacturing product is defined as a fundamental in industrial technology. Applications of this technique involve in various fields including automotive paint removal industry. Paints are resulting from pigments (such as iron oxide, carbon black or aluminium among others) combined with synthetic resins (thermoplastics or thermostable) which permit the bonding between the paint and the substrates [1].

In paint removal, the main objective is high cleaning efficiency and paint erosion with minimum substrate damage. The current study focuses on cleaning method mostly using laser. Cleaning by laser is found to cause damage to the substrate, alteration of

M. N. Osman Zahid et al. (Eds.): iMEC-APCOMS 2019, LNME, pp. 537–543, 2020.
https://doi.org/10.1007/978-981-15-0950-6_82

substrate colour and high cost. It was observed that the laser cleaning is almost 20 times more expensive than conventional method leads to its application limited to historical and ornamental usage [1]. Laser ablation create environmental health hazard with the rapid improvement of this technology fail to follow the security and health assessment and this is proven by the lack of study on stone emission using laser ablation [2]. There are numerous aspects to be addressed in paint removal where water jet cleaning method is prove to produce less environmental health hazard as compared to laser since water is used for cleaning. High interest of research on paint removal using laser leads to scarce data on water jet application on paint removal. Considerable works using water jet for paint removal was done merely to study on parameters such as stand off distance, water jet pressure, kinetic energy, traverse rate and attack angle. Theoretical and experimental analysis done on coating removal from passenger-vehicle plastic for recycling by using water jet demonstrates that regardless the changes in stand off distance, traverse rate and pressure, the paint removal decrease when water jet angle increases due to decrease in force of the jet impact [3]. Parameter optimization in research on de-painting of steel substrate results a linear relationship between water jet kinetic energy (loading intensity), stand off distance (loading frequency) and paint removal [4].

All of these studies were made with single pass cleaning. The recent research that study on this multiple passes cleaning using laser cleaning results on two passes shows more efficiency and less damage to the stone as compared to single pass [5]. This is due to the removal was done by lower irradiance level (lower than damage threshold) than single pass. The present work attempts to investigate the effect of multiple passes in plain waterjet cleaning of paint.

2 Experimental Work

2.1 Material

Materials from automotive parts were cut away from standard plate by mechanical sawing. Substrates were ready to be machined after primary cleaning with dry compressed air to remove dust. The mass of the cleaned substrate was measured before and after the paint removal process to study the paint removal by weight loss method. The tensile mechanical properties of the paint on the surface of automotive plastics component are shown in Table 1 [3].

Table 1. Mechanical properties of the paint on the surface of automotive plastics component.

Paint	Elastic modulus, E (Gpa)	Yield strength, $\sigma_{0.2}$ (Mpa)	Ultimate strength, σ_b (Mpa)
Top coating	1.59–1.76	11.34–15.85	17.43–21.22
Intermediate coating	1.54–1.88	8.99–13.10	10.71–13.75
Primer	0.52–0.77	4.93–7.11	7.83–10.60

2.2 Equipment

The whole experiments were conducted using a self-developed waterjet cutting machine. The currently developed machine has a pump which can generate water pressure of up to about 100 MPa due to its low cost. The type of the pump used is an air driven liquid pump. The machine is controlled by a computer numerical control (CNC) system. Table motion is controlled along multiple axes (*X* and *Y*) as well as the nozzle that moves in the *Z*-axis (depth). The cutting head consists of a ruby orifice of 0.127 mm in diameter and a tungsten carbide focusing tube of 0.76 mm and 76.2 mm in diameter and length respectively. The developed CNC waterjet machine is shown in Fig. 1.

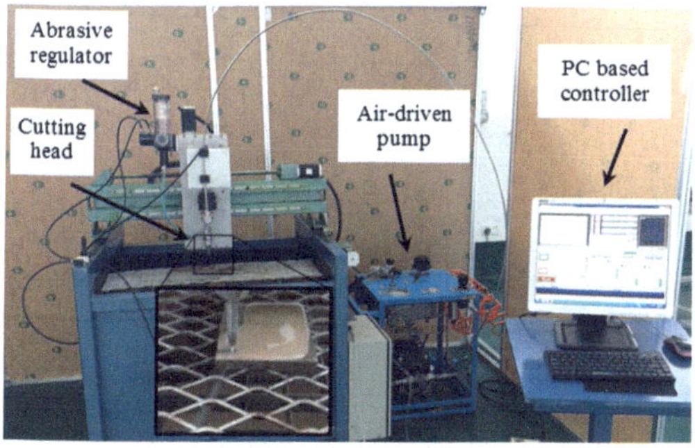

Fig. 1. The self-developed CNC waterjet machine [6].

2.3 Experimental Design

The paint on automotive component was removed by waterjet cleaning using multiple cleaning passes with various water jet parameters. Stand off distance and impact angle are kept constant at 10 mm and 90 respectively. The machining parameters and their minimum and maximum levels used in the present study are shown in Table 2.

Table 2. Machining parameters and their levels.

No	Machining parameters	Value
1	Number of passes, *n*	1, 3
2	Traverse rate, *u* (mm/min)	500, 1000
3	Pressure, *P* (MPa)	34, 69

3 Result and Discussion

3.1 Effect of Parameters on Paint Loss

Figure 2 illustrates the effect of traverse rate on paint loss on the painted surface at different number of passes and pressures. It can be observed that a higher paint loss occurs during cleaning at a higher pressure. Also, it is to note that with higher number of passes produces more paint loss. Furthermore, it can be noticed that increasing traverse rate produces less paint loss due to less interaction between the jet and workpiece thus leads to ineffective erosion process [9]. It is interesting to notice that with decreasing traverse rate from 1000 to 500 mm/min and increasing pressure, more paint loss was acquired with more number of passes. This is due to more amount of paint that is lost at every cleaning pass.

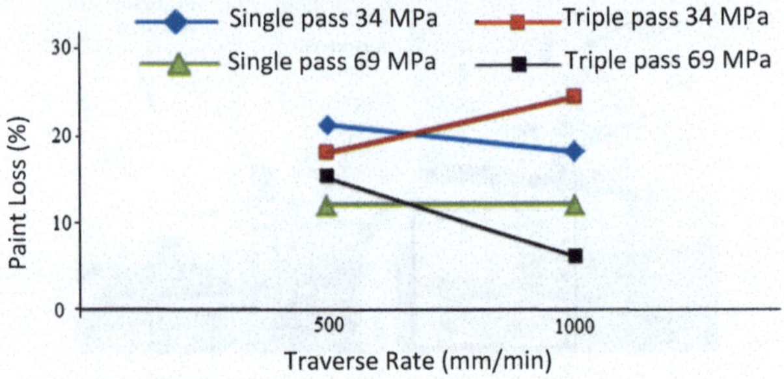

Fig. 2. Effect of traverse rate on paint loss for different cleaning passes and pressures.

3.2 Effect of Parameters on Surface Roughness

Figure 3 shows the effect of traverse rate on surface roughness on the painted surface at pressure 34 MPa. It can be noticed that increasing the number of passes results lower surface roughness. This indicates the smoothening action on the same surface by the second and third pass to remove uneven surface left by previous cleaning pass similar to that cutting with multipass [7]. Furthermore, increasing the number of passes and traverse rate result a slightly lower surface roughness. Lower traverse rate produces a cleaner removal process with a smoother surface finish [8]. Moreover, a higher traverse rate leads to less interaction between the water particle and paint surface [9]. It can be seen that increasing the pressure results in a slightly lower surface roughness for triple pass as compared to a single pass. This is due to increasing kinetic energy of the particles at a higher pressure thus removing more materials similar to cutting ceramic [9].

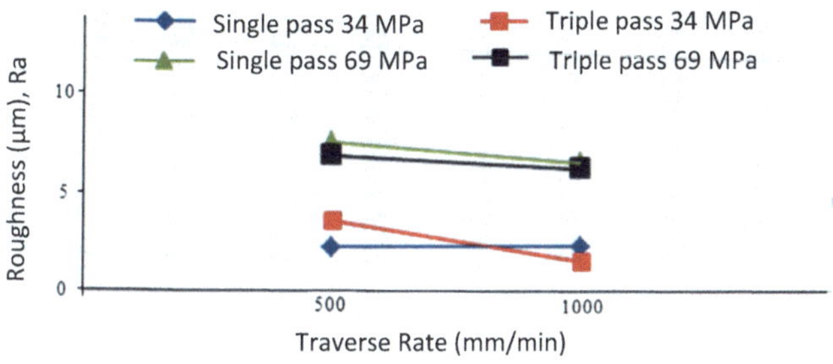

Fig. 3. Effect of traverse rate on surface roughness for different cleaning passes and pressures.

3.3 Visual Inspection

Figure 4 show visual effect of paint removal using single pass at different traverse rate and pressures. In all Fig. 4(a), (b), (c), and (d) it is note that the existence of black area/lines in the picture is the paint area which has not yet been removed. Cleaning at lower pressure at 34 Mpa, smoother surfaces are produced as compared to cleaning at higher pressure of 69 MPa as indicated in Fig. 5. It seems that the pressure is too high at 69 MPa to not only removing the paint but some portions of the substrate as shown in Fig. 5(c) and (d). Furthermore, a higher traverse rate yields less paint erosion and lower surface roughness since less water particle to clean the surface area as nozzle speed increases.

Fig. 4. Visual effect of paint removal using single pass at different traverse rates and pressures, (a) traverse rate 500 mm/min, pressure 34 MPa (b) traverse rate 1000 mm/min, pressure 34 MPa (c) traverse rate 500 mm/min, pressure 69 MPa (d) traverse rate 1000 mm/min, pressure 69 MPa.

Figure 5 shows effect of paint removal using three passes at different traverse rates and pressures. Similar conclusions can be made regarding the effect of traverse rate and pressures as cleaning using a single pass as shown in Fig. 4. Furthermore, comparing both Figs. 4 and 5 for single and three passes respectively, it is found that three passes remove more paint than single pass as indicated by the lacking of black area/lines. This is due to continuous removal of paints at every cleaning pass. Increasing number of passes decreasing the surface roughness as seen in Figs. 3 and visually proven by smoother surface in Fig. 5 as compared to Fig. 4.

Fig. 5. Visual effect of paint removal using three passes at different traverse rates and pressures, (a) traverse rate 500 mm/min, pressure 34 Mpa (b) traverse rate 1000 mm/min, pressure 34 Mpa (c) traverse rate 500 mm/min, pressure 69 Mpa (d) traverse rate 1000 mm/min, pressure 69 Mpa.

4 Conclusion

It was shown that automotive part was successfully cleaned using waterjet machine with multiple number of cleaning passes. The cleaning performance of the waterjet machine is satisfactory with acceptable surface removal qualities. The paint loss decreases with an increase in the traverse rate. Meanwhile, the surface roughness increase with an increase in the traverse rate. It can be concluded that increasing waterjet cleaning passes can be successfully used for recycling automotive component with acceptable cleaning qualities.

Acknowledgement. The Authors would like to thankfully acknowledge the financial support from the Universiti Malaysia Pahang through RDU1703313.

References

1. Sanmartín, P., Cappitelli, F., Mitchell, R.: Current methods of graffiti removal: a review. Constr. Build. Mater. **71**, 363–374 (2014)
2. Vergès-Belmin, V., Wiedemann, G., Weber, L., Cooper, M., Crump, D., Gouerne, R.: A review of health hazards linked to the use of lasers for stone cleaning. J. Cult. Herit. **4**, 33–37 (2003)
3. Zhang, H., Chen, M.: Theoretical analysis and experimental study on the coating removal from passenger-vehicle plastics for recycling by using water jet technology **67**(11), 2714–2726 (2015)
4. Teimourian, H., Shabgard, M.R., Momber, A.W.: De-painting with high-speed water jets: paint removal process and substrate surface roughness. Prog. Org. Coat. **69**(4), 455–462 (2010)
5. Penide, J., et al.: Removal of graffiti from quarry stone by high power diode laser. Opt. Lasers Eng. **51**(4), 364–370 (2013)
6. Murugan, M., Gebremariam, M.A., Hamedon, Z., Azhari, A.: Performance analysis of abrasive waterjet machining process at low pressure performance analysis of abrasive waterjet machining process at low pressure, pp. 0–6 (2018)
7. Wang, J., Guo, D.M.: The cutting performance in multipass abrasive waterjet machining of industrial ceramics **133**, 1–7 (2003)

8. Folkes, J.: Waterjet-an innovative tool for manufacturing. J. Mater. Process. Technol. **209** (20), 6181–6189 (2009)
9. Tanmay Tiwari, A.M., Sourabh, S., Nag, A., Dixit, A.R., Das, A.K.S.A.K., Mandal, N.: Parametric investigation on abrasive waterjet machining of alumina ceramic using response surface methodology. In: Parametric Investigation on Abrasive Waterjet Machining of Alumina Ceramic Using Response Surface Methodology, pp. 0–7 (2018)

Investigation on the Effect of Abrasive Waterjet Parameter on Machining Stainless Steel

Hafiz Husin[1,2(✉)], Mohd Nazir Mat Nawi[1,2], M. A. Gebremariam[1], and Azmir Azhari[1]

[1] Faculty of Manufacturing Engineering, Universiti Malaysia Pahang, 26600 Pekan, Malaysia
hafidzuka@yahoo.com
[2] Centre for Foundation Studies, International Islamic University Malaysia, 26300 Gambang, Malaysia

Abstract. Abrasive waterjet machining (AWJM) can perform various machining operations on almost any material including hard to machine materials. However, performance of a process depends on the chosen set of parameters. AWJM has major parameters such as traverse rate, water pressure, standoff distance and abrasive flowrate that will affect its performance. The present study investigates the effects of AWJM parameters (i.e. traverse rate, water pressure and standoff distance) on the width, depth and roughness of a channel produced at the surface of stainless steel. The results showed that by reducing the standoff distance and water pressure, the width of the channel was also reduced. For channel depth, increasing traverse rate produced a shallower channel depth. In contrast, increasing the water pressure produced a deeper channel depth. The surface roughness of the channel showed significant improvement by reducing the water pressure at a lower traverse speed and standoff distance. A proper selection of parameters is required in order to produce a suitable channel during AWJM surface texturing process.

Keywords: Abrasive waterjet machining · Waterjet channeling

1 Introduction

This paper discusses the effects of abrasive waterjet machining of parameter on surface texturing on stainless steel. Waterjet either in plain form or mixed with abrasive can perform various machining operation on almost any material including hard to machine material. The application is endless and it is nontraditional machining method that has been chosen by manufacturers due to its flexibility and reliability [1]. Performance of any process depends on the parameter chosen. AWJM has major parameter such as traverse rate, water pressure, standoff distance, abrasive flowrate and abrasive particle size that affects the performance of the AWJM [2]. AWJM has no machine vibration or

M. N. Osman Zahid et al. (Eds.): iMEC-APCOMS 2019, LNME, pp. 544–549, 2020.
https://doi.org/10.1007/978-981-15-0950-6_83

tool chatter and conventional tools are prone to edge chipping and tool wear [3]. The AWJM process is clean, do not create dust, chips and chemical pollution because the eroded material is carried by the waterjet thus eliminating dust, air pollution and as a result, it is environmentally friendly [4].

2 Experimental Work

2.1 Material

The samples used in the experiment was 25 mm × 25 mm × 4 mm Stainless Steel SS400 cut by mechanical saw. Stainless Steel was chosen because many studies have demonstrated the advantages of using Stainless Steels in the field of automotive and other industrial applications [3] (Table 1).

Table 1. Workpiece characteristic.

Material	Density (kg/m^3)	Tensile strength (MPa)	Young's Modulus (GPa)
SS400	7860	400	190–210

2.2 Equipment

The machining setup used Haskel air driven liquid pump with maximum air pressure of 150 Psi and the pump can supply a maximum output of 33,000 Psi. The machine is equipped with a gravity feed type of abrasive hopper, an abrasive feeder system with a pneumatically controlled valve. This pump is reliable, compact, robust, and easy to maintain while the cost of the pump is way cheaper than the standard electrical or hydraulic type pumps which are used in a commercial waterjet machine. The machine is controlled by a computer numerical control (CNC) system. The developed machine is shown in Fig. 1. The nozzle and abrasive characteristic are mentioned in the table below. The abrasive with mesh size of 80 ($\approx$177 μm). This size was selected due to its most applications in industrial operations of abrasive water jet machining (Table 2).

Table 2. Nozzle and abrasive characteristic.

Abrasive material	Abrasive size particle	Orifice diameter	Nozzle diameter
Garnet	75 to 106 μm	0.05 mm	1.22 mm

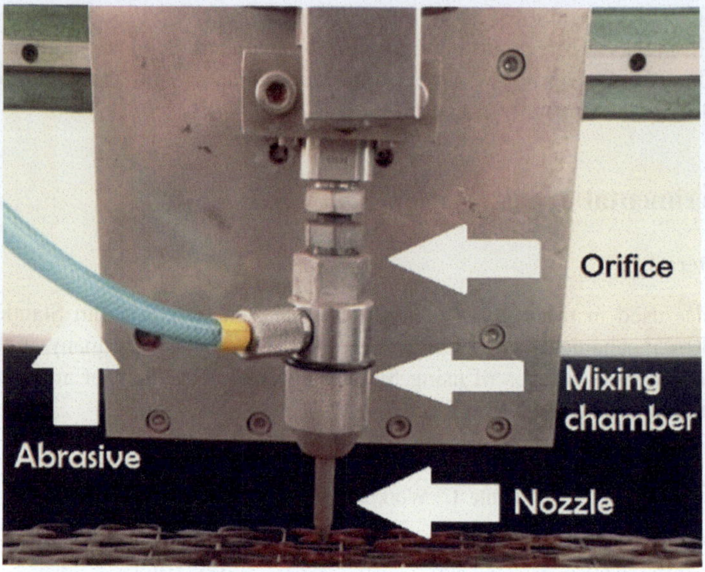

Fig. 1. The self-developed waterjet machine.

2.3 Experimental Design

In the current study, three control parameter were selected as a factor shown in the table below. The parameters and level were selected by primarily based on preliminary experiment and previous literature review for on cutting process. In the experiment a channel was produce by abrasive waterjet. In the experiment eight runs of channeling were conducted on the workpiece surface. The machining performance of all channel samples was assessed in terms of width, depth and the roughness of the channel. The width and the depth of the channel was measured using optical video measuring system. The measurement was taken three times and average value is taken to minimize errors. A surface roughness measuring device, MarSurfPs1 was used in this study. It features a diamond stylus which travels along a straight line over the surface. All of measurement of surface roughness were recorded in micrometer (μm). The surface roughness was measured inside the channel depth. Three different measurements were taken due to variation of roughness data for each surface so that the average could be calculated (Table 3).

Table 3. List of machining parameters.

No	Machining parameters	Value
1	Traverse rate (mm/min)	50, 150
2	Pressure (MPa)	40, 80
3	Standoff distance (mm)	2, 10

3 Result and Discussion

3.1 Effect of Parameters on Response

Figure 2(a) shows the effect of standoff distance over channel width. Result shows that standoff distance is proportionate with channel width. An increase of the channel width was found with the increase of standoff distance. The characteristic of waterjet that widen after exiting the nozzle confirms the result since the width of channel become bigger as the standoff distance got higher. The finding aligns with the experimental observation done by [5] which stated that fully-developed lateral outflow jetting causing the contact zone to expand. Researcher found that higher standoff distance allows the jet to expand before impingement [6]. Figure 2(b) displays the effect of standoff distance on channel depth. Result show that standoff distance is inversely proportional with channel depth. An increase of standoff distance will cause the depth of the channel to reduce. Higher standoff distance allows the jet to enlarge before making contact with the workpiece substrate. The jet enlargement lowering the densities of abrasive particles. This situation lowers the penetration depth of the channel [6].

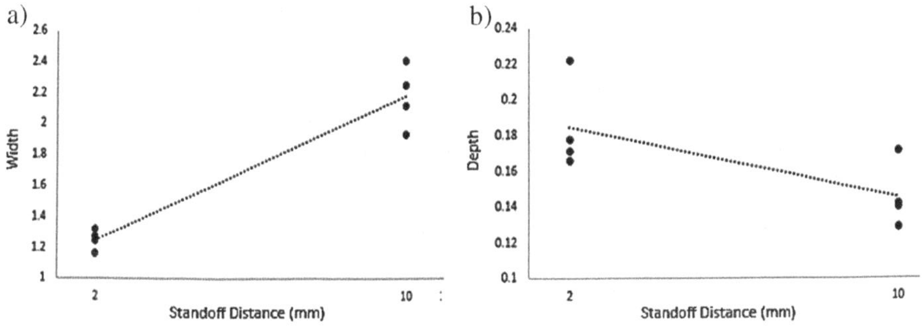

Fig. 2. Effect of standoff distance for channel width and depth.

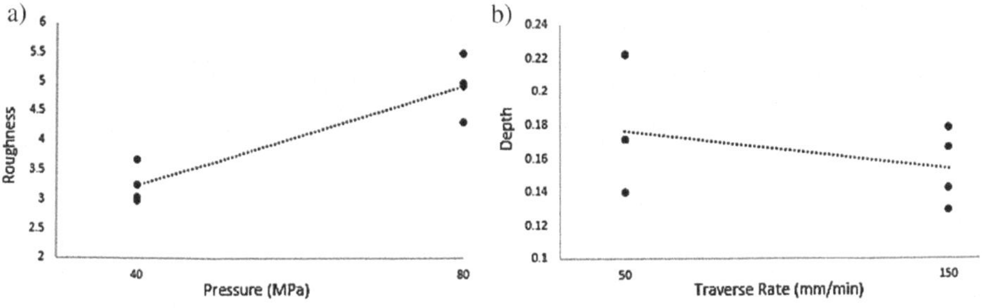

Fig. 3. (a) Effect of pressure on roughness (b) Effect of traverse rate on channel depth.

Figure 3(a) displays the effect of pressure on roughness of the channel surface. The result shows that pressure is directly proportionate with roughness of the channel surface. The roughness of the channel to increase with the increase of pressure. This finding approves the fact that a higher air pressure results in higher particle velocities and consequently higher particle kinetic energies which increases the material removal rate. Higher particle kinetic energies cause larger dents on the surface and consequently a rougher surface [7]. Figure 3(b) displays the relation between traverse rates and channel depth. Result shows that traverse rate inversely proportional with channel depth. An increase in traverse rate decreases the channel depth. As nozzle traverse speed increases, depth of cut decreases [3]. When traverse rate is high, fewer particles impact a given area of workpiece material. This has resulted in less material removal hence a shallower channel depth [7]. Similar finding which concludes that higher nozzle speed results in a shorter time of the jet exposing on to the material which, leading to fewer particles contributing to material removal, hence resulting to shallower channel depth [8].

3.2 Visual Inspection

This phase of the experiment is to study on visual inspection of the samples using optical video measuring system. Figure 4(a) shows the top view of the channel produce by AWJM. The figure shows the width of the channel. Figure 4(b) shows the depth of a

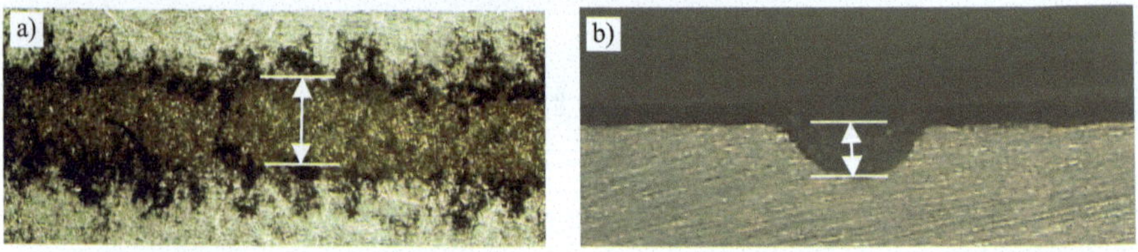

Fig. 4. (a) Channel width top view (b) Channel depth side view.

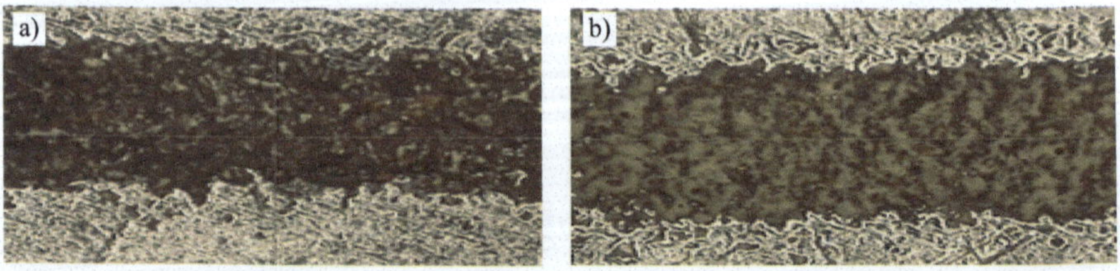

Fig. 5. (a) Channel inner surface with high pressure (b) Channel inner surface with low pressure.

channel after polishing the side edge of the workpiece. Figure 5(a) show inner surface of the channel that has higher roughness with high pressure compare to Fig. 5(b) which has lower roughness due to lower pressure.

4 Conclusion

An experimental investigation has been carried out to understand the effects of abrasive waterjet parameter on machining stainless steel. It has been found that by reducing the standoff distance and water pressure, the width of the channel was reduced. For channel depth, an increase in traverse rate produced a shallower channel depth while an increase in the water pressure produced a deeper channel depth. Finally, it was found that the surface roughness of the channel surface showed significant improvement by reducing the water pressure at a lower traverse rate and standoff distance.

Acknowledgement. The Authors would like to thankfully acknowledge the financial support from the Universiti Malaysia Pahang through RDU1703313.

References

1. Folkes, J.: Waterjet—an innovative tool for manufacturing. J. Mater. Process. Technol. **209**, 6181–6189 (2009)
2. Kumar, R.S., Gajendran, S., Kesavan, R.: Estimation of optimal process parameters for abrasive water jet machining of marble using multi response techniques. Mater. Today Proc. **5** (5), 11208–11218 (2018)
3. Supriya, S.B., Srinivas, S.: Machinability studies on stainless steel by abrasive water jet - review. Mater. Today Proc. **5**(1), 2871–2876 (2018)
4. Kulekci, M.K.: Process and apparatus developments in industrial waterjet applications **42**, 1297–1306 (2002)
5. Chillman, A., Ramulu, M.: A general overview of waterjet surface treatment modeling (2009)
6. Azmir, M.A., Ahsan, A.K.: A study of abrasive water jet machining process on glass/epoxy composite laminate. J. Mater. Process. Technol. **209**(20), 6168–6173 (2009)
7. Wang, J., Moridi, A., Mathew, P.: Micro-grooving on quartz crystals by an abrasive air jet **225**, 2161–2173 (2011)
8. Dadkhahipour, K., Nguyen, T., Wang, J.: Mechanisms of channel formation on glasses by abrasive waterjet milling. Wear **292–293**, 1–10 (2012)

Extension of an Analytical Model for a Contour-Parallel Strategy in the Triangular Pocket Machining

Mochammad Chaeron[1], Budi Saputra Wahyuaji[2], and Apriani Soepardi[1(✉)]

[1] Industrial Engineering Department, Universitas Pembangunan Nasional Veteran Yogyakarta, Depok, Indonesia
apriani.soepardi@upnyk.ac.id
[2] PT. Honda Prospect Motor, Karawang, Jawa Barat 41363, Indonesia

Abstract. The main goal of machining methods is to achieve a given final geometric shape within the shortest time. In this study, we propose a new model of the contour-parallel machining strategy for triangular pockets in order to minimize tool path length. We have extended an analytical model by appending additional tool path parts to the existing tool path model for removing scallops, which remained an area along the boundary. The result shows that the proposed analytical model can give the shortest tool path among other models, so can reduce the amount of pocket machining time.

Keywords: Contour-parallel · Milling machining · Tool path · Triangular pocket

1 Introduction

Milling is the mechanical operation of removing material from a piece of stock through the use of a rapidly spinning circular milling tool in order to form a given final geometric shape [1]. Two basic topologies of tool paths, strategies for pocket milling are directional-parallel machining strategy and contour-parallel machining strategy. A pocket milling is one of the most common mechanical operation in machining of metal parts. Almost 80% of the milling process to produce mechanical parts are machined by numerical control (NC) pocket milling [2]. The pocket machining is commonly met in the roughest and the finishing phase of molds and dies manufacturing.

Contour-parallel tool paths are among the most widely used tool paths for pocket machining. This method commonly used as cutting tool paths, especially for the large-scale material removal in 2.5D end milling. If such features have two or more hard faces, then the possibility of cusps arises with direction-parallel tool paths [3]. Therefore, contour parallel tool paths have acquired attention from many researchers [4–6]. Although the contour-parallel tool path has over the years been researched with the pairwise offset approach and the Voronoi diagram, but these are computationally expensive [5]. The advantage of this method is that most of the contour parallel

M. N. Osman Zahid et al. (Eds.): iMEC-APCOMS 2019, LNME, pp. 550–557, 2020.
https://doi.org/10.1007/978-981-15-0950-6_84

machining time, the tool is always in contact with the material, which reduces idle time spent in lifting, positioning, and plunging of the tool.

The purpose of pocket machining process is to eliminate all the material inside a pre-defined contour between two horizontal planes in a minimum machining time [7]. Therefore, we propose the new analytical model for extending the analytic model of contour-parallel strategy [8] in order to minimize tool path length. We only discuss in case of triangular pocket machining in this paper.

2 Methods

2.1 Previous Work

The contour-parallel strategy gives a possibility for generating tool paths, which not leaves the scallops. In this condition, there must exist the overlap, which depends on the smallest angle of the triangular pocket [8]. Its mechanism is composed of two steps that uses the diameter of tool $2r$. *First*, move inward offset with half of tool diameter r from pocket boundary. The result of first inward offset is the triangle ABC. *Second*, move inward offsets to the triangle ABC by a distance of $2(r - \rho)$ and so on. The triangular pocket machining using this strategy is shown in Fig. 1.

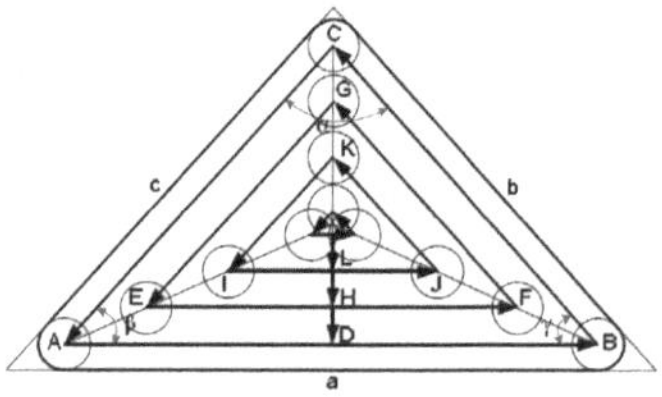

Fig. 1. Illustration of the basic contour-parallel machining strategy with overlap [10].

From this figure, there are two main segments in this strategy: (1) The boundary-parallel paths (ABC, EFG, IJK, and so on) correspond to the parts which surround the pocket boundary; (2) The connector paths (DH, HL, and so on) correspond to the parts which connect to the pocket boundary-parallel segments.

The Length of Boundary-Parallel Paths

The closest offset to the pocket boundary (outer tool path) is a line of distance of tool radius r from the pocket contour line. So, for the triangle ABC is given by

$$\overline{AB} + \overline{BC} + \overline{CA} = (a + b + c) - 2 \cdot r \cdot \left(Cot(\alpha/2) + Cot\left(\beta/2\right) + Cot(\gamma/2)\right) \quad (1)$$

Furthermore, all of the inner offset lines must be overlapped to avoid un-machined area (the shaded section in Fig. 2). The β is the smallest angle between two straight lines corresponding to the contour line in the triangle pocket. From Fig. 2 is obtained

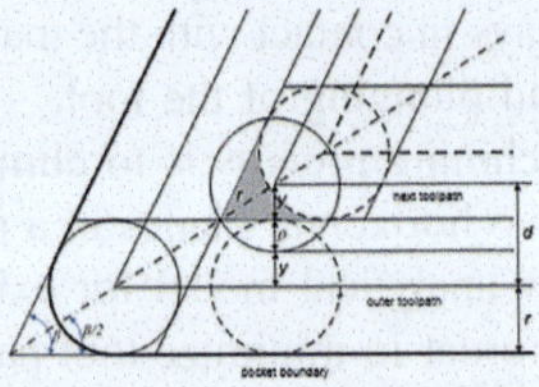

Fig. 2. Illustration of the overlap for avoiding remained area.

$$Sin\left({}^{\beta}\!/_{2}\right) = {}^{y}\!/_{r} \tag{2}$$

The overlap width of the closest two offset lines is

$$\rho = r - y = r - r \cdot \left(Sin\left({}^{Min\{\alpha,\beta,\gamma\}}\!/_{2}\right)\right) = r \cdot \left(1 - Sin\left({}^{Min\{\alpha,\beta,\gamma\}}\!/_{2}\right)\right) \tag{3}$$

Consequently, the offset distance of tool path is given by

$$d = 2 \cdot y + \rho = 2 \cdot (r - \rho) + \rho = 2 \cdot r - \rho \tag{4}$$

By distance of $2 \cdot r - \rho$, for the triangle EFG (see Fig. 1) can be computed

$$= (a + b + c) - ((2 \cdot r) + 2 \cdot (2 \cdot r - \rho)) \cdot \left(Cot({}^{\alpha}\!/_{2}) + Cot\left({}^{\beta}\!/_{2}\right) + Cot({}^{\gamma}\!/_{2})\right) \tag{5}$$

Determining the number of triangles, which parallel to pocket boundary, is firstly needed the value of *h* (Fig. 3).

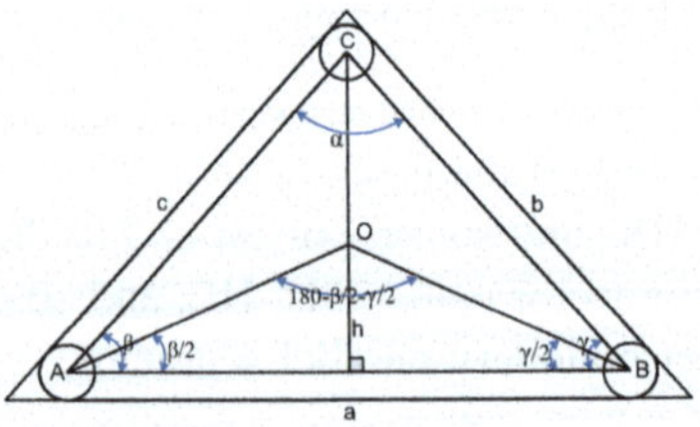

Fig. 3. The *h* value for determining the number of triangles.

From the figure above, the *h* value can be obtained based on sinus law

$$\frac{\overline{AB}}{Sin\left(180 - {}^{\beta}\!/_{2} - {}^{\gamma}\!/_{2}\right)} = \frac{\overline{OA}}{Sin({}^{\gamma}\!/_{2})} = \frac{\overline{OB}}{Sin\left({}^{\beta}\!/_{2}\right)} \tag{6}$$

$$h = \frac{\overline{AB} \cdot Sin\left(\beta/2\right) \cdot Sin\left(\gamma/2\right)}{Cos\left(\alpha/2\right)}$$
$$= \frac{\left(a - r \cdot \left(Cot\left(\beta/2\right) + Cot\left(\gamma/2\right)\right)\right) . Sin\left(\beta/2\right) \cdot Sin\left(\gamma/2\right)}{Cos\left(\alpha/2\right)} \quad (7)$$

The number of triangles, which parallel to pocket boundary, is given by

$$n = \frac{h}{2 \cdot r - \rho} = \frac{\left(a - r \cdot \left(Cot\left(\beta/2\right) + Cot\left(\gamma/2\right)\right)\right) . Sin\left(\beta/2\right) \cdot Sin\left(\gamma/2\right)}{(2 \cdot r - \rho) \cdot Cos\left(\alpha/2\right)} \quad (8)$$

The total length of tool paths, which parallel to pocket boundary, is given by

$$\sum_{i=1}^{n} (a + b + c) - ((2 \cdot r) + 2 \cdot (2 \cdot r - \rho) \cdot (1 - i)) \cdot \left(Cot\left(\alpha/2\right) + Cot\left(\beta/2\right) + Cot\left(\gamma/2\right)\right) \quad (9)$$

where the value of n is rounded up.

The Length of Connecting Paths
The total distance of connector segments is given by

$$(n - 1)(2r \cdot \rho) \quad (10)$$

Thus the function $f(\alpha, \beta, \gamma, a, b, c, \rho, r)$, i.e. the total length of the tool paths in contour-parallel machining strategy, is the sum of (9) and (10).

2.2 Previous Development of the New Analytical Model in the Contour-Parallel Strategy Work

Given α, β, and γ are the inner angles of the triangle corresponding to the three sides a, b, and c, respectively, ρ is the cutter overlap coefficient (between 0 and 1) denoting the degree of overlap between two adjacent cutting paths, and r is the radius of the cutting-tool. Related to the proposed model, the value of ρ is zero because the proposed model does not need the overlap. The triangular pocket machining using contour-parallel strategy without overlap is illustrated in Fig. 4. As shown in Fig. 4, machining is started from point B and ended at B again, so the first tool path is B-C-A-B. Then, the machining is forwarded to ascending path BE with an angle of $\frac{1}{2}\gamma$. This path is used to remove the scallops also. A triangular pocket with all three different angles ($\alpha \neq \beta \neq \gamma$) has an ascending path that equals a tool path with the smallest angle. It is due to the fact that an ascending path with the smallest angle is an ascending path with the longest path.

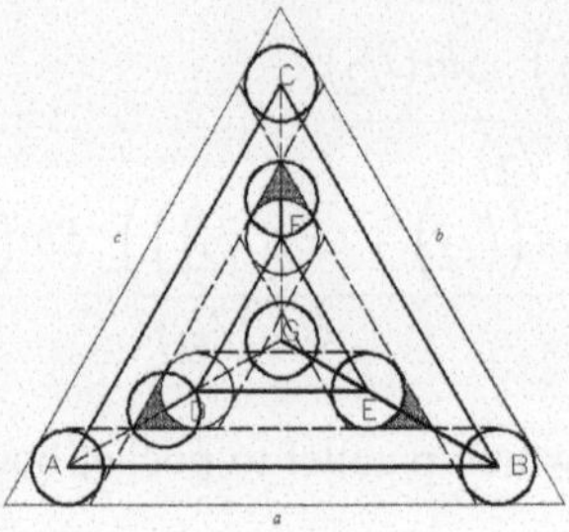

Fig. 4. Illustration of the proposed tool path of contour-parallel strategy without overlap.

Furthermore, the tool path is continued by starting from point E, thus the second tool path is E-F-D-E. This path has a track for removing unmachined area from point F and D. For eliminating the scallops, cutting-tool at the point D and F performs re-machining by distance of *r* with ½β and ½γ respectively. The machining process is repeated until the whole of triangular area has milled.

3 Results and Discussion

3.1 Derivation of the Proposed Analytical Model

From Fig. 4, there are three main segments in the proposed model: (1) The boundary-parallel paths (ABC, DEF) correspond to the parts which surround the pocket boundary; (2) The ascending paths connected to tool path which parallel to AB; (3) The scallop removal paths.

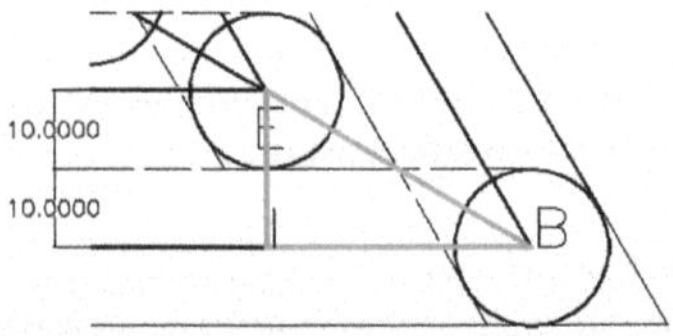

Fig. 5. Illustration of the ascending paths using the proposed model.

In the proposed model, the length of boundary-parallel paths is the same with Eq. (1) above. From Fig. 5 can be seen that the length of ascending paths in the triangle BEI is given by

$$BE = \frac{2 \cdot r}{Sin\left({}^{1}\!/_{2}\gamma\right)} \tag{11}$$

So, the total length of ascending paths is given by

$$(n)\frac{2\cdot r}{Sin^{1}/_{2}(\min(\beta,\gamma))} \quad \text{for n} \leq 1 \tag{12a}$$

$$(n-1)\frac{2\cdot r}{Sin^{1}/_{2}(\min(\beta,\gamma))} \quad \text{for n} > 1 \tag{12b}$$

While the total length of scallop removal paths is given by

$$(n)\cdot 2\cdot 2\left(\frac{r}{Sin^{1}/_{2}(\max(\beta,\gamma))}-r\right) \quad \text{for n} \leq 1 \tag{13a}$$

$$(n-1)\cdot 2\cdot 2\left(\frac{r}{Sin^{1}/_{2}(\max(\beta,\gamma))}-r\right) \quad \text{for n} > 1 \tag{13b}$$

Then, the total length of tool paths is the sum of Eqs. (9), (12a), (12b) and (13a), (13b).

3.2 Numerical Example

To illustrate the computation of the proposed model, we include the following numerical example. *Example 1*, an equilateral triangular pocket has three equal angles of 60° ($\alpha = \beta = \gamma$) and sides of 100 mm ($a = b = c$), curvature radius of corners and cutting tool radius are 10 mm (Fig. 6a). In Fig. 6b, the illustrative toolpaths using the proposed machining model is presented for numerical example 1.

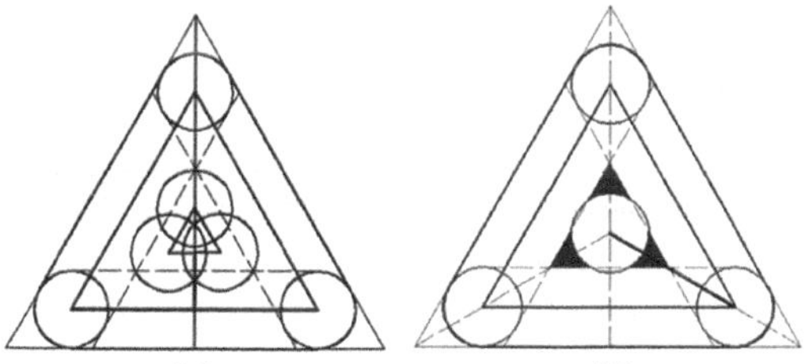

(a) **(b)**

Fig. 6. Illustration of triangular pocket for example 1.

For *Example 2*, given a triangular pocket, which has three different angles α, β, and γ of 90°, 53°, and 37° respectively, with sides of a, b, and c of 100 mm, 80 mm, and 60 mm respectively. Its curvature radius of corners and radius of cutting tool is 5 mm.

The comparison result of the machining models for two numerical examples is shown in Table 1. From this table is known, that the proposed analytical model of the contour-parallel machining strategy cannot reduce the length of tool path of an equilateral triangular pocket with all three equal angles and sides (Example 1). However, this model can shorten the tool path length for a triangular pocket, which has all three different sides and angles (Example 2).

Table 1. Comparison of machining length among the analytical models.

Machining strategy	Model	Total length of machining (mm)	
		Example 1	Example 2
Direction-parallel	[9]	388.231	407.846
	[10]	297.175	318.347
Contour-parallel	[8]	275.309	315.660
	Proposed new model	276.080	296.578

4 Conclusion

The proposed model cannot decrease the tool path length for all types of triangular pocket. However, the proposed model can give shorter total length of machining for all types of triangular pocket than the direction-parallel milling strategy, with average 17%. This result shows that there is the improvement of the pockets machining time using the contour-parallel milling compared with the zigzag milling strategy.

References

1. Arya, S., Cheng, S.W., Mount, D.M.: Approximation algorithm for multiple-tool milling. Int. J. Comput. Geom. Appl. **11**(3), 339–372 (2001)
2. Held, M.: VRONI: an engineering approach to the reliable and efficient computation of Voronoi diagrams of points and line segments. Comput. Geom. Theory Appl. **18**(2), 95–123 (2001)
3. Mansor, M.S., Hinduja, S., Owodunni, O.O.: Voronoi diagram-based tool path compensations for removing uncut material in 2½D pocket machining. Comput. Aided Des. **38**(3), 194–209 (2006)
4. Choy, H.S., Chan, K.W.: A corner-looping based tool path for pocket milling. Comput. Aided Des. **35**(2), 155–166 (2003)
5. Park, S.C., Choi, B.K.: Uncut free pocketing tool-paths generation using pair-wise offset algorithm. Comput. Aided Des. **33**(10), 739–746 (2001)
6. Kim, B.H., Choi, B.K.: Machining efficiency comparison direction-parallel tool path with contour-parallel tool path. Comput. Aided Des. **34**(2), 89–95 (2002)
7. Bouard, M., Pateloup, V., Armand, P.: Pocketing toolpath computation using an optimization method. Comput. Aided Des. **43**(9), 1099–1109 (2011)

8. Chaeron, M.: Model Analitis Panjang Lintasan Pahat Untuk Pemesinan bentuk rongga (pocket) segitiga. Jurnal Teknologi Industri, X(3) (2006). (in Indonesian)
9. Veeramani, D., Gau, Y.S.: Selection of an optimal set of cutting-tools for a general triangular pocket. Int. J. Prod. Res. **35**(9), 2621–2638 (1997)
10. Chaeron, M., Wahyuaji, B.S., Soepardi, A.: Development of a direction-parallel strategy for shorting a tool path in the triangular pocket machining. Jurnal Ilmiah Teknik Industri **18**(1), 1–7 (2019). https://doi.org/10.23917/jiti.v18i1.7151

Study on Operational Characteristic of Microwave Oven Driven Plasma Spray Device

Muhammad Fahmi Izuwan(✉), Ahmad Redza(✉), and Mohd Nizar(✉)

Faculty of Mechanical and Manufacturing Engineering,
Universiti Malaysia Pahang, 26600 Pekan, Malaysia
mfl8004@stdmail.ump.edu.my,
{ahmadredza,mnizar}@ump.edu.my

Abstract. Microwave oven induced plasma method is a novel application of microwave oven to generate plasma for coating process. By using integrated microwave system with low cost 2.45 GHz magnetrons, low initial setup cost is able to be achieved in compare to conventional plasma spray methods. The innovation of microwave oven induced plasma spray is seen as a good candidate for future research study due to its easily operable at low power. It requires relatively low power compared to conventional plasma spraying method and able to be generated at atmospheric pressure conditions. However, the research regarding this microwave oven induced plasma spray is very less and the mechanisms are still more to be discovered. Therefore, our research is focusing on the understanding of the operational characteristics of microwave oven driven plasma spray device. 0.8 kW output power is used in this research, and by controlling the working gas flow rate and antenna outlet diameter, the plasma had been able to be generated at 2, 3, and 4 mm antenna outlet diameter with 10, 15, and 20 lpm flow rates of argon gas. The widest plasma plume had been able to be generated was at 20 lpm for the each of antenna diameter, and at 3 mm of outlet antenna diameter with 15 lpm of argon (Ar) gas flow rates, it is the acceptable condition for producing plasma plume.

Keywords: Microwave plasma · Plasma spray · Plasma plume

1 Introduction

Coating technology is one of surface engineering method which can be used to enhance the surface properties for corrosion and wear protection. It also a modification treatment of surface by adding new materials on the surface of substrates [1]. Plasma spraying method is the most versatile of the thermal spray processes which capable of spraying all materials that are considered spray-able. It uses electrical energy to produce heat in form of thermal energy which caused by ignition of gas plume and other type of way. Plasma spraying is one of the processes to fabricate thick coating which has costly set up due to some of the process of thermal spray coatings require very expensive apparatus and lead to a high initial set up cost [2]. There are three types of plasma spray power source were used which is direct current (DC) plasma, radio

M. N. Osman Zahid et al. (Eds.): iMEC-APCOMS 2019, LNME, pp. 558–563, 2020.
https://doi.org/10.1007/978-981-15-0950-6_85

frequency (RF) plasma, and microwave (MW) plasma. Microwave plasma is easy to be generated at low power less than 1 kW [3] and has stability in wide range of pressure [4]. Moreover, microwave plasma does not require electrode for electric discharge as compared with DC plasma. It also can be operated both with and without external magnetics field if compared to RF plasma. Furthermore, microwave plasma was able to be generated at atmospheric condition and able to coat at much narrow area [4]. On the other hand, conventional plasma spray method has extreme temperature plasma (above 10000 K) [5] and only operable at high power (above 40 kW). This extreme temperature plasma is very useful to fabricate high melting point material in coating process [6]. However, the excessive heat input from this plasma may effect in deterioration of material phase structure and degraded some function as well in some materials. Thus, it will cause the difficulty to fabricate coating onto low melting point substrates such as plastic, resin and polymers.

Here, the innovation of microwave oven induced plasma spray is seen as a good candidate due to its stability in various pressure ranges and easy to be generated at low power. In this study, commercially available microwave oven is use to generate plasma. By using integrated microwave generators, this innovation of microwave oven induced plasma spray will reduce the cost and easier to handle. The magnetron and the waveguide that produce the microwave at the center of resonant cavity leads to low initial setup cost compare to other plasma spray methods [6]. However, the research concerning this microwave oven induced plasma spray is very less and the mechanisms are still more to be ascertained. Therefore, our investigation is centering on the comprehension of the operational characteristics of microwave oven driven plasma spray device at different antenna outlet diameter and gas working flow rate by using same output power which is 0.8 kW. This different in antenna outlet diameter and gas flow rates will produce different length and width of plasma plume.

2 Experimental Procedure

2.1 Process

Figure 1 shows the experimental schematic diagram of the microwave oven plasma spray device that is used in this study. Microwave with the frequency of 2.45 GHz is transmitted from generator through a waveguide to the resonant cavity. The antenna which is made of Cu-Zn material is positioned at the center of the resonant cavity to utilize the concentration of electric field on the tip of the antenna for plasma generation. The working gas is supplied through the antenna axially. To avoid microwave reflection to microwave generator, dummy load is placed at the position shown in the diagram. The possibility of arcing inside microwave oven cavity is also able to be reduced [7]. Thus, all microwave that oscillated by generator are absorbed if the plasma plume fails to be generated. At the tip of the antenna, high-intensity electric field is generated from the induced electrical breakdown as a form of plasma plume at the downstream.

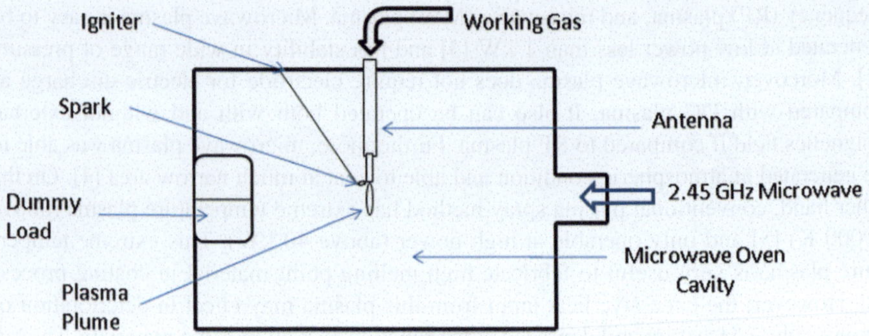

Fig. 1. Schematic diagram of the microwave oven induced plasma spray.

2.2 Experimental Setup

The investigation of the plasma ignition condition for the microwave oven induced plasma is conducted in this experiment. The experimental condition of plasma ignition of induced plasma is shown in Table 1. The operations time is limited until 210 s to minimize the damage taken by the antenna. The images of the plasma plume produce were taken by camera and will be analyzing using image processing method. On the other hand, the effect of the gas flow velocity in stable plasma production can be interpreted by Reynolds number. Reynolds number is calculated by applying the antenna outlet diameter (OD) of which is 2, 3 and 4 mm respectively. Reynolds number (Re) will indicate whether the flow is laminar and turbulence using equation below.

$$Re = \frac{p_2 V_2 D_2}{\mu_2} \tag{1}$$

Table 1. Experimental condition for plasma ignition.

Antenna outlet diameter (mm)	2, 3, 4
Working gas flow rate (l/min)	10, 15, 20
Output power (kW)	0.8
Working gas	Argon

3 Result and Discussion

The plasma shows more stable discharge when laminar flow take place rather than turbulence flow [8]. Based on table below, Reynolds number from all outlet diameter of antenna are below 2000 which is laminar. In addition, due to high plasma temperature in the central region, plasma will have high molecular velocity as a results of small Reynolds number [9]. Therefore, it can be concluded that 2 mm, 3 mm and 4 mm is applicable for plasma ignition (Table 2).

Table 2. Calculated Reynolds number by outlet diameter of antenna.

Antenna OD (mm)	Calculated Reynolds number		
	10 lpm	15 lpm	20 lpm
2	121.53	131.10	192.59
3	65.63	107.01	89.01
4	747.43	682.60	563.09

The experimental result for plasma ignition condition is shown in Table 3. All working gas flow rate are able to generate plasma and the width of the plasma plume increase when the increasing of the flow rate of the Ar is shown in Fig. 2. Thermal pinching effect which play important role for the plasma shape of the velocity profile in the nozzle antenna can be observe start at 15 lpm of flow rates [10]. In addition, the length of plume only is affected by changing the gas flow rate of Ar [6]. Moreover, the applied value of voltage, composition of gas and gas flow rate is affected by the length of the plasma [11]. On the other hand, the increasing of gas flow rate will cause shape of plasma becomes narrower.

Table 3. Experimental result for plasma ignition at 2, 3 and 4 mm antenna OD.

Antenna Outlet Diameter	Gas Flow Rates		
	10 lpm	15 lpm	20 lpm
2 mm			
3 mm			
4 mm			

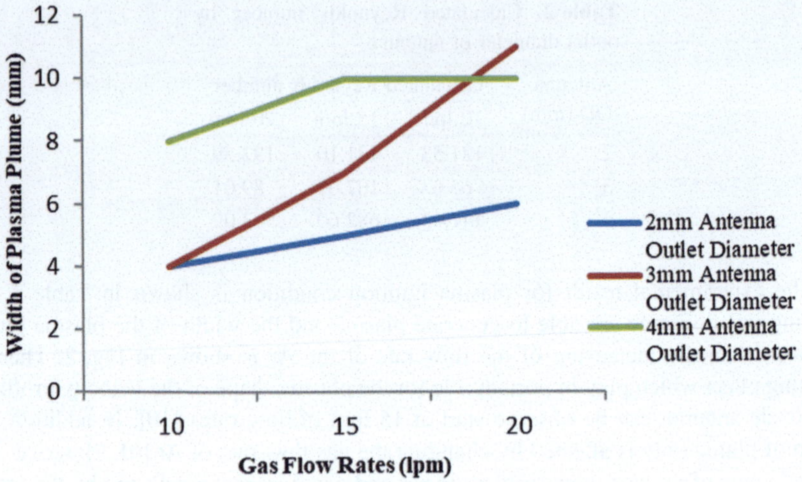

Fig. 2. Results of width of plasma plume versus gas flow rates.

4 Conclusion

The summary of the things that had been clarified from the study on operational characteristic of microwave oven driven plasma spray device and the evaluation are listed below.

1. Microwave oven induced plasma spray able to be developing at low power which is 0.8 kW.
2. All our nozzle antenna design has acceptable calculated Reynolds number which is below 2000 which indicates laminar flow.
3. From the plasma ignition condition experiment, thermal pinching effect play important role for the plasma shape of the velocity profile in the nozzle antenna.

Acknowledgement. The authors gratefully appreciated University Malaysia Pahang for the financial support through Internal Research Grant RDU1703255 and RDU1703253.

References

1. Davis, J.R.: Surface engineering for corrosion and wear resistance. ASM Int. **51**(1) (2001)
2. IRS Surfacing Technologies Ltd.: Advantages and Disadvantages of Thermal Spray Coatings (2015). https://www.irsltd.co.uk/advantages-and-disadvantages-of-thermal-spray-coatings/. Accessed 17 Jan 2019
3. Yasui, T., Ahmad Mokhtar, A.R., Fukumoto, M.: Development of low power plasma spray process using atmospheric pressure microwave plasma (2013)
4. Redza, A.: Study on operational characteristics of low power atmospheric pressure microwave plasma spray method. Doctor of Engineering. Toyohashi University of Technology (2016)

5. Liu, Y., Wang, P.J., Tzeng, C.C.: Thermal temperature measurements of plasma torch by alexandrite effect spectropyrometer. Adv. Opt. Technol. **2010**, 7 (2010)
6. Rosario, L.M.D., et al.: characterization of a microwave-induced atmospheric-pressure Ar-N2 plasma pencil. IEEE Trans. Plasma Sci. **45**(2), 301–309 (2017)
7. Brooks, D.J., Douthwaite, R.E.: Microwave-induced plasma reactor based on a domestic microwave oven for bulk solid state chemistry. Rev. Sci. Instrum. **75**(12), 5277–5279 (2004)
8. Jin, D.J., Uhm, H.S., Cho, G.: Influence of the gas-flow Reynolds number on a plasma column in a glass tube. Phys. Plasmas **20**(8), 083513 (2013)
9. Morgan, R.E., Prince, A.S., Selvidge, S.A., Phelps, J., Martin, C.L., Lawrence, T.W.: Plasma torch, pp. 474–484 (2000)
10. Redza, A., Yasui, T., Fukumoto, M.: Deposition of hard chrome coating onto heat susceptible substrates by low power microwave plasma spray. In: IOP Conference Series: Materials Science and Engineering, vol. 114, no. 1 (2016)
11. Sarani, A., Nikiforov, A.Y., Leys, C.: Atmospheric-pressure plasma jet in He/H_2O mixture. **39**(11), 2358–2359 (2011)

Comparative Study of Tool Path Strategies in CNC Machining for Part with B-spline Surfaces

Zainal Fahmi Zainol Abidin and Muhammed Nafis Osman Zahid(✉)

Faculty of Mechanical and Manufacturing Engineering, Universiti Malaysia Pahang, 26600 Pekan, Pahang, Malaysia
nafis@ump.edu.my

Abstract. CNC machines are widely used in production of various machinery components including turbine blades, impellers, rotors, propellers etc. Most of these components are built-up from free form surfaces which considered complex shapes and required proper set up for machining. This paper presents optimization of toolpath pattern for cutting parts with B spline surfaces in 4 axis machining. Generally the operation is carried out by using 4 axis machining methods which employs variable streamline operations in the finishing process. The appropriate selection of a toolpath pattern can significantly improve productivity and lead to lower production times. Different toolpath scenarios are simulated in CADCAM prior to real cutting process. In order to execute the comparative study of tool path strategies, all common cutting parameters (spindle speed, feed rate, tool diameter, plunge-rate, and depth of cut) are set to be constant. The toolpath strategies employed in this study includes helical or spiral, zig, zigzag and zigzag with lift. Cutting operation built-up and validation are performed through NX10, VERICUT and CNC machining. The objective is to optimize the machining process for B-spline model by selecting the shortest toolpath with maximum volume removal based on using variable streamline operation. The result indicates different tool path strategies based on the level of B spline curvature exhibit in the component.

Keywords: Toolpath pattern · B-spline · Free-form surface

1 Introduction

B-spline surfaces can be categorized as free form surfaces that require complex setup in machining process. Complex surfaces are being used in various areas of work such as to create medical replicas and in automotive and aerospace industries. Over the years, many researchers study the best formulae to optimize the machining process of free-form surface in terms of machining efficiency and surface quality. Lasemi et al. [1] through their review paper of the fundamental issues and new developments in CNC machining of freeform surfaces emphasize three major issues need to be considered in freeform surface machining which is tool path, tool orientation, and tool geometry. Zhang et al. [2] in their research has apply various toolpath in computer numerical control milling of a complex freeform surface machining. The objective of the study is

M. N. Osman Zahid et al. (Eds.): iMEC-APCOMS 2019, LNME, pp. 564–569, 2020.
https://doi.org/10.1007/978-981-15-0950-6_86

to understand how 3D tool paths influence the cutting response in term of machining efficiency, surface quality, and form accuracy. In terms of machining efficiency, Li et al. [3] in their research considered a basic criterion to choose a pattern which results in less machining time or the short path length. It has been found that toolpath pattern has significant influence on toolpath length or machining time. The influence is dependent on the shape and size of the stock and island contours, and the size of cutter used. He considered a basic criterion to choose a pattern which results in less machining time or the short path length. They found that the tool travelling time is usually a very small part of the total machining time (<2%), whereas the rest of the total machining spends on cutting material. Hence, they proposed an approach to select proper toolpath pattern. Currently, the development of cutting pattern optimization is involving with 3-axis milling machine [4–6] and too little finding involving 4-axis machining. Therefore, experiment has been conducted to find the optimization of cutting pattern in 4-axis machining. In this paper, a comparative study toward the different types of toolpath which lead to the effect of machining time and volume excess is proposed. The goal of these studies is to optimize the machining process for B-spline surfaces model in finishing operation by using 4- axis CNC machine.

2 Method of Machining Operation

The removal of excess material in finishing operation for B-spline surface can be done in a number of ways. Most of the CAM systems would provide for different type of finishing operation options which the user can choose considering the type of surface to be machined. In complex surface machining, the common method use is variable streamline operation. This method builds an implied drive surface from the selected geometry. Streamline drive method enables completely flexible tool path creation that suit for complex surface machining. A well-ordered grid of regular faces also not required in this operation [8]. Since the B-spline surface is categorized as the complex surface in CNC machining, this method considered as the most suitable operation to be used in this experiment. In order to study the cutting pattern effects, other machining parameters for example feed rate, spindle speed, cutting tool diameter and depth of cut are set to be constant while the cutting pattern is set to be varies. There are four cutting path strategies being used for variable streamline operation in NX10 and each cutting method has advantages and disadvantages. In this study, 4 methods have been compared to select the proper toolpath that satisfying minimum machining time (Fig. 1):

a. Helical or Spiral: Tool progressively deforms the blank with a Spiral movement from the top going towards the maximum depth [7].
b. Zig: This takes a linear path in only one direction of flow.
c. ZigZag: This tool takes a zigzag path at every level of depth
d. ZigZag with Lift: This implement a mixed cut direction by default.

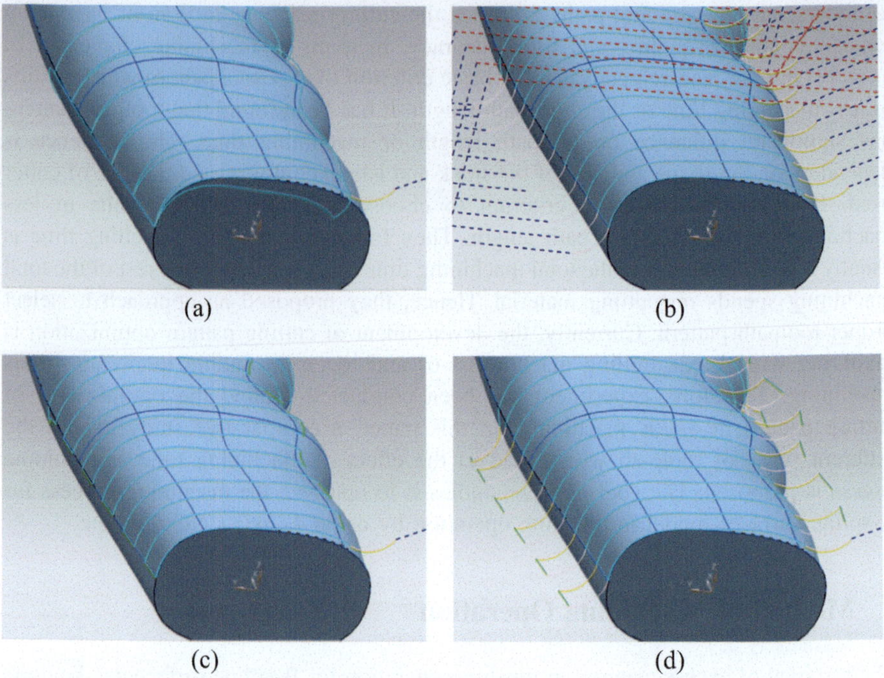

Fig. 1. (a) Helical or spiral (b) Zig, (c) Zigzag (d) Zigzag with lift

2.1 Vericut Software

Vericut software is a computer aided machining (CAM) software that capable to simulate machining process and provide machining data through the simulation process in example cutting time, cutting distance and volume excess information. Vericut has been used in these studies to obtain the volume excess information since NX10 is not providing the volume excess information. Hence, after simulation in NX10 is done, every model is simulated again in Vericut software to find the volume excess information (Fig. 2).

3 Result and Analysis

Simulation have been conducted towards three B-spline model that consisting B-spline free form surface. Based on the simulation in NX10 (to find cutting time and cutting length information) and Vericut (to find volume excess information), the results obtained as per table below (Fig. 3 and Table 1).

Optimal cutting pattern is measured by the shortest cutting time and cutting length obtained with minimum volume access. Based on the result obtained from simulation, the optimal cutting pattern selected for wind turbine blade is helical or spiral with the shortest cutting time and cutting length among other cutting pattern tested. Even the volume excess is not at the maximum removal amongst of all cutting pattern but the

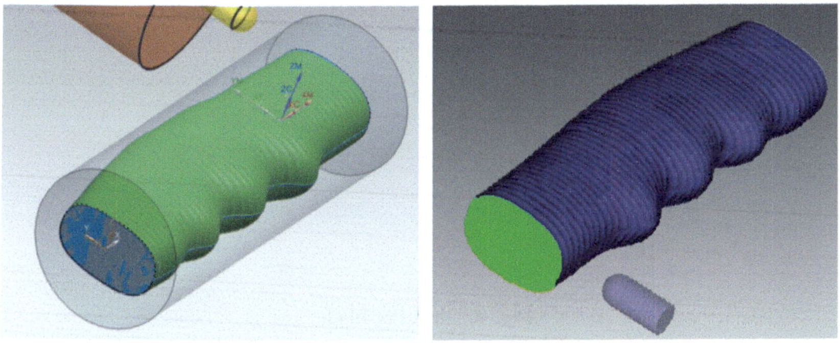

(a) (b)

Fig. 2. (a) Simulation in NX10 (b) Simulation in Vericut

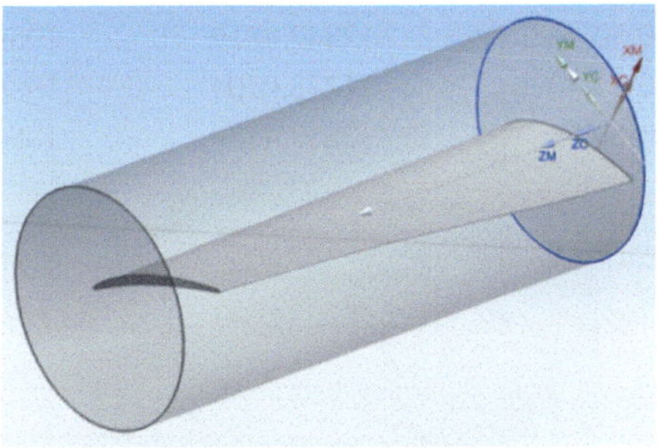

Fig. 3. Turbine blade.

Table 1. Result for wind turbine blade.

Cutting pattern	Cutting time (sec)	Cutting length (mm)	Volume access (m^3)
Helical or spiral	0:44:12	11212.3703	8971.1387
Zig	0:50:00	15447.6135	8894.3072
Zig zag	0:44:11	11205.0734	8895.5309
Zig zag with lift	0:47:36	12074.3683	8894.2494

value is still acceptable with only slight different compared to other cutting pattern value tested (Fig. 4 and Table 2).

For the impeller blade, the optimal cutting pattern selected is Zigzag since it has the shortest cutting time and cutting length. Volume excess value only has slight different amongst other cutting pattern and the value is still acceptable (Fig. 5).

While for the handgrip shape, the optimal cutting pattern selected is Helical or spiral since it has the shortest cutting time and cutting length. Volume excess value only has slight different amongst other cutting pattern and the value is still acceptable (Table 3).

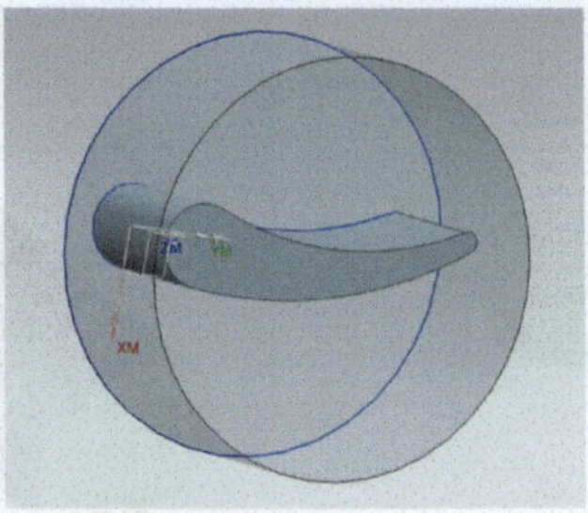

Fig. 4. Impeller blade.

Table 2. Result for impeller blade.

Cutting pattern	Cutting time (sec)	Cutting length (mm)	Volume access (m^3)
Helical or spiral	0:45:23	11456.0492	1086.5576
Zig	0:50:46	12842.4819	1086.5242
Zig zag	0:44:40	11278.0704	1086.5331
Zig zag with lift	0:50:55	12842.4139	1086.5244

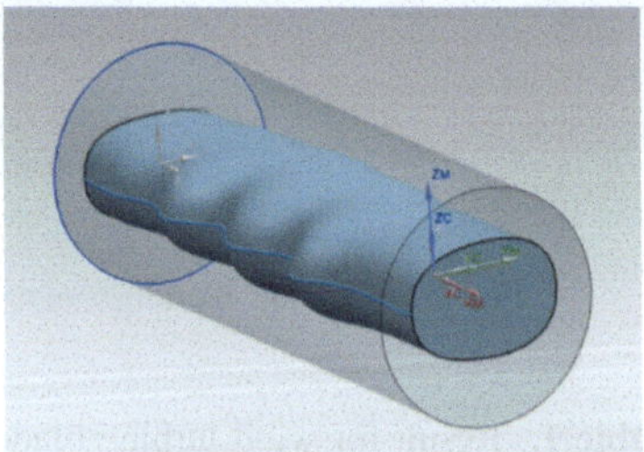

Fig. 5. Handgrip.

Table 3. Result for handgrip.

Cutting pattern	Cutting time (sec)	Cutting length (mm)	Volume excess (m^3)
Helical or spiral	1:39:49	25154.5698	12317.5911
Zig	1:51:41	30246.3468	12312.9727
Zig zag	1:39:52	25170.5534	12316.7004
Zig zag with lift	1:51:50	28158.6369	12312.9549

4 Conclusion

The results of the comparison show that the optimal toolpath is dependent on geometry of the part and the type of the used toolpath. In general there is no specific toolpath type that can be used specifically for every B-spline model since every model from simulation give different type of optimal cutting pattern.

Acknowledgement. We acknowledge with gratitude to Ministry of Higher Education Malaysia for providing a financial support under Fundamental Research Grant Scheme (FRGS) (RDU160130)(FRGS/1/2016/TK03/UMP/02/1) and Faculty of Manufacturing Engineering, Universiti Malaysia Pahang which realize this research project.

References

1. Lasemi, A., Xue, D., Gu, P.: Computer-aided design recent development in CNC machining of freeform surfaces: a state-of-the-art review. Comput. Des. **7**(42), 641–654 (2010)
2. Zhang, X.F., Xie, J., Xie, H.F., Li, L.H.: Experimental investigation on various tool path strategies influencing surface quality and form accuracy of CNC milled complex freeform surface. Int. J. Adv. Manuf. Technol. **59**, 647–654 (2012)
3. Li, H., Dong, Z., Vickers, G.W.: Optimal tool path generation for 2½D milling of dies and molds. In: International IFIP TC5/WG5.3 Conference SSM 1998 Sculptured Surface Machining Conference, pp. 9–11 (1998)
4. EL-Midany, T.T., Elkeran, A., Tawfik, H.: Toolpath pattern comparison: contour-parallel with direction-parallel. In: Geometric Modeling and Imaging - New Trends (GMAI 2006), pp. 77–82 (2006)
5. Prajapati, R., Rajurkar, A., Chaudhary, V.: Tool path optimization of contouring operation and machining strategies for turbo machinery blades. Int. J. Eng. Trends Technol. (IJETT) **4** (5), 1731–1737 (2013)
6. Panchal, P.R., Makwana, S.J., Makwana, S.J.: Comparative study of tool path strategies in NX for pocket machining. Int. J. Eng. Technol. Sci. Res. (IJETSR) **4**(10), 41–48 (2017)
7. Yanamundra, K.K., Karthikeyan, R., Naranje, V.: Finite element simulation and experimental verification of incremental sheet metal forming. In: IOP Conference Series: Materials Science and Engineering, vol. 346, no. 1, pp. 1–12 (2018)
8. http://www2.me.rochester.edu/courses/ME204/nx_help/index.html#uid:drive_area_milling_dialog. Accessed 21 May 2019

Taguchi Multi Respond for Eri Silk/Cotton Yarn Process Parameters Optimization Using Rieter R35 Rotor Open End Machine

Ridya Amerani Pra Lovian[1,2(✉)], Cucuk Nur Rosyidi[1,2], and Eko Pujiyanto[1,2]

[1] Master Program of Industrial Engineering, Universitas Sebelas Maret, JL. Ir. Sutami 36A, Surakarta 57126, Indonesia
ralovian@gmail.com
[2] AK Tekstil Solo, JL. Ki Hajar Dewantara, Surakarta 57126, Indonesia

Abstract. Eri worm (*Samia Cynthia ricini*) is a polyphagous animal whose able to be fed by 29 species of plants such as tapioca and castor. Eri worm produces cocoon which will be furthered process into fiber in the kind of staple or short fiber. The fiber then usually blends with other fiber such as cotton or polyester. In Indonesia, there have been Eri worm cultivation from livestock, yarn making to fabric making, with poor quality, because the process is manually and traditionally. In this paper, the eri silk fiber will be blended with cotton fiber to develop a new blend fiber yarn using modern short spinning process for quality improvement. The blended materials will be processed on a Drawing Machine and then the results from that machine will be further processed using Rotor Open End RIETER R35. Taguchi multi respond is used in this paper to determine the optimal process parameters of eri silk/cotton yarn production. The results of this research can be used to increase the quality of the yarn production which currently processed using manual machine. Shows that the parameters are speed rotor with two levels are 60,000 rpm and 80,000 rpm, and speed open roll with two levels are 6,000 rpm, and 8,000 rpm, and twist multiply with two levels are 4.3 and 4.6. The optimal process parameters for eri silk/cotton yarn production were found at 60,000 rpm of rotor speed, 8,000 rpm of opening roll speed, and 4.6 of twist multiply.

Keywords: Eri silk · Yarn spinning · Taguchi

1 Introduction

Eri worm (*Samia Cynthia ricini*) is a polyphagous animal whose able to be fed by 29 species of plants such as tapioca and castor [1]. Different with other silk worm (*Bombyx Mori L*) which fed with mulberry leaf, eri silk produces staple fiber from the cocoon which left broken to let the pupa coming out by itself. Staple fiber have some advantages, one of them is on the process of staple fiber which can be mixed or blended with other natural fiber like cotton or synthetic fiber like polyester. Fiber blending is very common practice in textile industry to produce yarn to achieve certain characteristics which cannot be resulted from the use of single fiber type [2]. In Indonesia, eri

M. N. Osman Zahid et al. (Eds.): iMEC-APCOMS 2019, LNME, pp. 570–576, 2020.
https://doi.org/10.1007/978-981-15-0950-6_87

silk yarn has been produced in traditional way. The staple fiber must be twisted and pulled out manually by hand and rolled up on traditional wheel. This method will produce yarn with high unevenness, hence the yarn has poor quality on tenacity. A modern machine can improve the quality of eri silk yarn through a spinning process. Several machines are used in the yarn process, such as Blowing, Carding, Drawing frame and Open End machine.

Several research has been conducted in eri silk yarn process. For example, there is a study using Design of Experiment approach to determine and known how the eri silk works, and find the best formula for blended fiber silk with waste staple rayon with 50:50 compositions [3]. Another research has been conducted to compare the mulberry feeding silk and eri silk. The results showed that eri silk has a higher elasticity and stronger against pests and insects. With this characteristics, eri silk fiber can increase the characteristics of the resulted yarn [4]. In Indonesia, there have been Eri worm cultivation from livestock, yarn making to fabric making, with poor quality, because the process is manually and traditionally. The Quality of yarn can be determined the seven quality tools, and for reducing nonconformance of yarn quality, we can use fishbone diagram, why analysis and pareto to increase the quality of yarn [5]. In this paper, the eri silk fiber will be blended with cotton fiber to develop a new blend fiber yarn using modern short spinning process for quality improvement. In this paper, the Order Preference Technique Similarity with Ideal Solutions (TOPSIS) is used to determine the parameter of setting machine. TOPSIS has several advantages: simple, taking into account all kinds of criteria, rational and easy to understand, and easy calculations [6].

In the present work a study is conducted to blend eri silk and cotton in draw frame process with different composition as an attempt and then the results from that machine will be further processed using Rotor Open End RIETER R35 to determine good quality eri silk/cotton yarn especially on tenacity characteristics.

2 Materials and Method

The eri silk fibers were prepared from eri silk cocoon (*Samia Cynthia ricini*) which have been boiled and dried under the sun in two days. After that, cocoon is put into a blowing machine and carding machine to make sliver as the feeding material of draw frame. Blending process is carried out in the draw frame due to easy control of weight ratios between eri silk and cotton fibers with blending six doubling of sliver [4]. By feeding two eri silk sliver with four cotton slivers in the blend ratio of 33/67, then the slivers will be drawn again through draw frame with eight doubling of blending eri/cotton sliver to get final sliver product. The sliver from draw frame are then fed as material in open end machine to produce open end yarn of 10^s.

In order to prepare the yarn samples, Orthogonal array of $L_4(2^3)$ Taguchi method is used. We use four number of experiment with three factors namely rotor speed, open roll speed and twist multiplier. Each factor consists of two level as shown in Table 1. The table shows this method is fractional factorial experimental matrix that is orthogonal and balanced. The orthogonal array $L_4(2^3)$ is shown in Table 2. The workshop condition have temperature of 32 °C and RH of 65%.

Table 1. Factors and level of experiment

Factors (unit)	Factor designation	Level	
		Level-1	Level-2
Speed rotor	A	60,000	80,000
Speed open roll	B	6,000	8,000
Twist multiply	C	4.3	4.6

Table 2. $L_4(2^3)$ orthogonal array

Experiment no.	Column			Column		
	A	B	C	A	B	C
1	1	1	1	60,000	6,000	4.3
2	1	2	2	60,000	8,000	4.6
3	2	1	2	80,000	6,000	4.6
4	2	2	1	80,000	8,000	4.3

Two responses are considered in this research, namely tenacity and elongation. The specimens will be measured by following standard test method of SNI 7650:2010 using Statigraph L tensile tester. The gauge length was kept at 50 mm and force 100 N, at strain speed 500 mm/min. A total of 20 tests were conducted for each sample.

3 TOPSIS Method

Decision making problem is the process of finding the best choice among a set of alternatives. Decision problems such as ranking, choice and sorting problem are difficult since they involve several criteria. Multi Criteria Decision Analysis method (MCDA) has been developed to support decision makers in determining choices in decision making. MCDA problems can be expressed in a matrix format (decision matrix) as shown in Table 3.

Table 3. Decision matrix

	C_1	C_2	....	C_n
A_1	x_{11}	x_{12}	....	x_{1n}
A_2	x_{21}	x_{22}	....	x_{2n}
A_m	x_{m1}	x_{m2}	....	x_{mn}

In Table 4, A_1, A_2, ..., A_m denote all possible alternatives among which decision makers have to choose, C_1, C_2,..., C_n denote the criteria with which alternative performance are measured, x_{ij} denotes the rating of alternative A_1 with respect to the criterion C_j.

The main principle of TOPSIS is that the chosen alternative must have the closest distance from ideal solution and have the farthest distance from non-ideal solution in term of Euclidean distance. According to [7] TOPSIS has the following calculation steps:

1. Perform Normalization
   (a) Perform normalization of distributive data.

$$n_{ij} = \frac{a_{ij}}{\sqrt{\sum x_{ij}^2}} \tag{1}$$

   The distributive normalization is the decision matrix (a_{ij}) divided by the square root of the sum of each element (a) the square in the column.
   (b) Perform normalized ideal data

$$r_{ai} = \frac{X_{ai}}{U_a^+} \tag{2}$$

$$r_{ai} = \frac{X_{ai}}{U_a^-} \tag{3}$$

   Ideal idealization (r_{ai}) divides each matrix (x_{ai}) with the highest value in each U_a^+ column, if the criteria should be maximized. If the criteria is to be minimized, then each x_{ai} must be divided by the lowest value in each column.
2. Calculate the normalization of the weight of the decision matrix by Eq. (4).

$$v = w_i \cdot n_{ai} \tag{4}$$

3. Determine the value of positive ideal solution and negative ideal solution using the Eqs. (5) and (6) respectively:

$$A^+ = \left(v_i^+, \ldots, v_m^+ \right) \tag{5}$$

$$A^- = \left(v_i^-, \ldots, v_m^- \right) \tag{6}$$

   Where $V_i^+ = \max_a (v_{ai})$ if criteria *i* is maximized and $V_i^- = \min_a (v_{ai})$ if criterion *i* is minimized
4. Calculate the distance of each alternative from the ideal solution using the Eqs. (7) and (8).

$$d_a^+ = \sqrt{\sum_i (v_i^* - v_{ai})^2} a = 1, \ldots, m \tag{7}$$

$$d_a^- = \sqrt{\sum_i (v_i^- - v_{ai})^2}, a = 1, \ldots, m \tag{8}$$

5. Calculate the nearest relative value to the ideal solution using Eq. (9)

$$K = \frac{da^-}{da^+ + da^-} \quad (9)$$

6. Specifies the order of preference

4 Results and Discussion

TOPSIS calculation steps in this paper are as follow, the data as the results of experiments are shown in Table 4.

Table 4. Ideal normalization

Experiment	Tenacity (cN/Tex)	Elongation (%)
1	7.88	6.97
2	7.63	6.53
3	10.21	7.26
4	10.59	7.2

Calculates the normalization of the weight of the decision matrix by multiplying the weight of each criterion with the normalized results by ideal method. The calculation results can be seen in Table 5. Determine the value of a positive ideal solution and a negative ideal solution using ideal normalization. The calculation results at Table 6.

Table 5. Normalization of decision matrix weight

Experiment	Tenacity (cN/Tex)	Elongation (%)
1	0.5	0.5
2	0.52	0.53
3	0.39	0.48
4	0.37	0.48

Tabel 6. Ideal positive and negative

A+	A−
0.52	0.53
0.53	0.48

Calculating the distance of each alternative from the ideal solution using ideal normalization can be seen in Table 7. Calculate the nearest relative value to the ideal solution using ideal normalization. After that, determining the order of preference using ideal normalization can be seen in Table 8. Based on the results of distributive calculations, the sequence of ranking is experiment 2, 1, 3 and 4.

Table 7. The distance of each alternative uses the distributive and ideal normalization

Experiment	di+	di−
1	0.04	0.13
2	0	0.15
3	0.14	0.01
4	0.15	0

Table 8. Final ranking

Experiment	K
1	0.78
2	1
3	0.09
4	0.03

In this research, the weight of criteria is assumed to be the same. According to the TOPSIS final ranking, the best tenacity and elongation quality is experiment number 2, which parameter setting are rotor speed 60,000 rpm, opening roll speed 8,000 rpm and twist multiply 4.6 and the worst performance is experiment number 4, which have rotor speed 80,000 rpm, opening roll speed 8,000 rpm and twist multiply 4.3 based on the TOPSIS ranking.

5 Conclusion

In this research, Taguchi Multi Respond was used to determine optimal process parameters of eri silk/cotton yarn. The yarn was processed on an open-end rotor spinning machine to produce mixed yarns that has the better quality characteristics in term of tenacity and elongation. TOPSIS was used to solve the multi respond problem due to its simple calculations. From the data analysis, the optimal process parameters for eri silk/cotton yarn production were found at 60,000 rpm of rotor speed, 8,000 rpm of opening roll speed, and 4.6 of twist multiply. The result of this study provided a useful data and information for eri silk/cotton yarn production. Further research may be conducted by involving more factors and responds such as ratio of blending eri/cotton fiber on factors and unevenness of yarn on responds.

References

1. Garuraj, Kumar, K.P., Naika, R.: Rearing performance of eri silkworm, samia cyntia ricini (boiduval) (leooidoptera: saturniidae) on cultivars of castor. J. Entomol. Zool. Stud. **5**(5), 816–821 (2017)
2. Choudhuri, P.K., Majumdar, P.K., Sarkar, B.: Studies on tensile properties of eri silk/polyester blended yarn using design of experiment methodology. Inst. Eng. (India), Ser. E **94**(1), 37–46 (2013)
3. Dharma, F.P., Hardiman, H.D., Ikatrinasari, Z.F., Purba, H.H.: New development fiber material: use DoE approach to determine the best formula for blended fiber silk (Samiya Cynthia Riccini and Semi-Natural Fiber). In: IOP Conference on Series: Materials Science and Engineering, vol. 508, p. 012105 (2019)
4. Chollakup, R., Suesat, J., Ujjin, S.: Effect of blending factorson eri silk and cotton blended yarn and fabric characteristics. Macromol. Symp. **264**, 44–49 (2008)
5. Dharma, F.P., Ikatrinasari, Z.F., Purba, H.H., Ayu, W.: Reducing nonconformance quality of yarn using pareto principles and fishbone diagram in textile industry. In: IOP Conference on Series: Materials Science and Engineering, vol. 508, p. 012092 (2019)
6. Nazar, Y., Lovian, R.A.P., Raharjo, D.C., Rosyidi, C.N.: Supplier selection and order allocation using TOPSIS and linear programming method at Pt. Sekarlima Surakarta. In: AIP Conference Proceedings, vol. 2097, p. 030050 (2019)
7. Ishizaka, A., Nemery, P.: Multi Criteria Decision Analysis Method and Software. Wiley, Hoboken (2013)

Effect of Machining Process Parameters on Acceleration Signal in Determining Surface Quality of Milling Process at Ductile Iron

Norlida Jamil and Ahmad Razlan Yusoff(✉)

Faculty of Mechanical and Manufacturing Engineering, Universiti Malaysia Pahang, 26600 Pekan, Pahang, Malaysia
norlidajamil@gmail.com, razlan@ump.edu.my

Abstract. The present work studies the effect of machining parameters for determining surface quality during milling process of ductile iron under different lubrication conditions. It was conducted by adopting direct and indirect measurement using two accelerometers with coated and uncoated tool. The experiment data was collected by inputs of spindle speed, feed rate, axial-radial depth of cut, and MQL flow rate. The response in term of vibration signal was measured and extracted from time series data into kurtosis and skewness analyses. For direct measurement, surface qualities from experimental configuration of workpiece were measured using roughness tester for verification and comparison. Alternatively, the result verified the different effect of machining process parameters contributes to different surface quality based on kurtosis and skewness analyses for the indirect measurement.

Keywords: Machining parameter selection · Vibration signal · Coated tool · Uncoated tool · Kurtosis · Regression · Minimal Quantity Lubrication (MQL) · Surface roughness

1 Introduction

Milling is one of common and efficient cutting process and the application has contributed into various engineering and manufacturing industries. The efficiency of milling is highly dependent on quality performance like surface finish, tool life and wear, cutting force, temperature and coolant consumption [1]. The influential factors like spindle speed, feed rate, axial-radial depth of cut, and materials used are considered as parameter input [2]. This is to reduce the complication that leads to the milling tool-workpiece failure [3] and increase the processing cost [4] subsequently. In the recent decades, industries have been moving towards sustainable production besides environmentally friendly and therefore MQL was adapted as potential substitute for traditional cutting coolant. Khan and Dhar outlined advantages using vegetable-based oil besides of biodegradability, high lubrication and stability [5]. Boswell reviewed that over the years, MQL are significantly comparable to traditional flood coolant [6]. For instance, the range of cutting fluid consumption is between 50 ml/h to 2000 ml/h [7] whereas other studies stated even lower flow rate of 10–100 ml/h [8], which is

M. N. Osman Zahid et al. (Eds.): iMEC-APCOMS 2019, LNME, pp. 577–583, 2020.
https://doi.org/10.1007/978-981-15-0950-6_88

extremely low compared to flood cooling; approximately 1200000 ml/h. Moreover, studies found that MQL can be applied to eliminate/lessened obstacle in dry machining during end milling hardened steel, such as generation of temperature and force, material softening, poor surface, tool wear and premature tool failure [1]. There are two types of monitoring or measurement method in cutting processes: direct measurement and indirect measurement. Direct measurement method estimates cutting condition based on signal analysis captured by one or more sensors in order to monitor and predict current cutting process. The signal can be representing various features, such as vibration, cutting force, acoustic emission and sound [9]. For example, Madhusadana et al. [10] installed an accelerometer whereas Jamil et al. [11] mounted multiple accelerometers during milling process to enhance information derived from sensors. These signals next were extracted into specific features such as time domain, frequency domain, wavelet transform and etc. before constructed into monitoring model using support vector machine (SVM) based, regression and others. A detection system by [12] extracted feature information from time domain using statistical parameters of central moment, include the root mean square, standard deviation, kurtosis and skewness of time series data collected. In contrast, indirect measurement was conducted using optical equipment such as optical microscope and portable measuring device like potable roughness tester [11] to measure tool-workpiece surface condition.

In order to evaluate the correlation between direct and indirect measurements besides selecting the machining parameter for end milling ductile iron, experiments were carried out in terms of spindle speed, feed rate, axial-radial depth of cut, and MQL flow rate in end milling ductile iron. The results from cutting process vibration signal were measured using time domain analysis before extracted into quantification analysis. On the other hand, surface roughness of milling ductile iron was measured for direct measurement. Coated/uncoated tool material and MQL effects also were observed and discussed further in this article.

2 Experimental Method

The experiments were performed on Makino KE-55 vertical computer numerical control (CNC) knee-type milling machine. The cutting tools were eight-flute solid carbide end mill cutter with 20 mm diameter. An uncoated tool and a titanium aluminium nitride (TiAlN)-coated tool were used to observe and study the influence of coated on cutting measurement. Ductile iron FCD450 used as the workpiece with geometry of 150 mm × 150 mm × 50 mm.

The experimental set-up also includes of two PCB piezotronics accelerometers as illustrated in Fig. 1. The single and tri-axial axes accelerometers were mounted on spindle head and the ductile iron respectively. Both accelerometers at 100 mV/g of sensitivity values were used to measure vibration signal during cutting process. Meanwhile, the minimal quantity lubricant was used instead of presence cutting fluid or flood coolant. The MQL was supplied by Unist Coolubricator model using flexible plastic dual use nozzle during the experiment at 150 psi (1.0342 Mpa), whereby both lubricant output and air blow off to clear chip. The model is capable of supplying particles of oil-air mixture from ejector nozzles, while discharging MQL using 0.03 ml per stroke 1 drop pump. The oil used up in the experiment was the Unist Coolube 2210 lubricant. It is a hundred percent natural biodegradable lubricant, non-toxic and made from renewable vegetable product that is friendly to the environment and machinist. The experiments were carried out in a single path single pass system using coated and uncoated tool with overhang length of 40 mm. Spindle speed (n), feed rate (v_f), axial depth of cut (a_p), radial depth of cut (a_e), and MQL volume flow rate ($\dot{Q}$) as the input parameters (Table 1). The signals during cutting process were measured by accelerometers and converted by NI USB-4431 data acquisition via Dasylab software under 10000 Hz of sampling rate and 20000 units block size. The signal was further analysed using Matlab for time domain analysis as indirect method measurement. On the other hand, as for direct measurement, a total of nine reading of surface roughness, arithmetic average (Ra) were taken averagely at three different point for one single pass using Mitutoyo roughness tester.

Table 1. Machining conditions

Machining parameters	Values
Spindle speed, n (rev/min)	1000, 1487, 2015, 2495, 3026
Feed rate, v_f (mm/min)	120, 165, 375, 520, 720
Axial depth of cut, a_p (mm)	0.75, 1.25, 1.75, 2.5, 3.5
Radial width of cut, a_e (%)	25, 50, 75, 100
MQL nozzle spraying angle (°)	135
MQL volume flow rate, $\dot{Q}$ (ml/h)	9, 18, 27, 36, 45

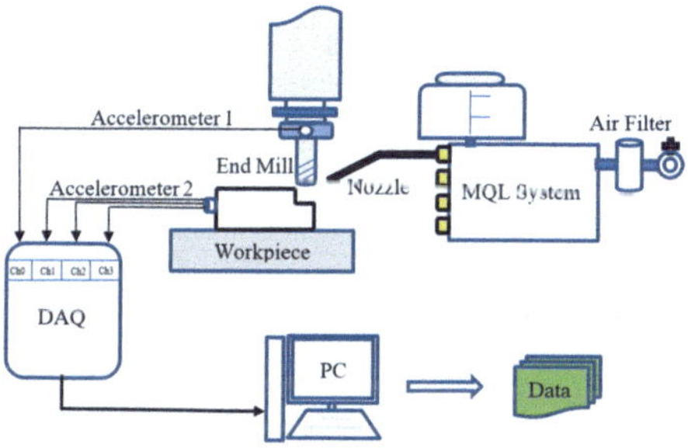

Fig. 1. Schematic of experimental set-up.

3 Results and Discussion

Figure 2 shows the time domain computed from accelerations for a spindle speed of 3026 rev/min, feed rate of 720 mm/min, axial depth of cut 3.5 mm and 100% radial depth of cut at 18 ml/h of MQL. During the 10 s of cutting, it is visible that the amplitudes were approximately 3000 mm/s^2. A similar trend was reported for other parameters and environments. Generally, cutting process at high speed and feed rate increase the vibrations and can be reduced by applying coating and MQL. Since there is no significant difference were observed in cutting process using time domain analysis, thus, kurtosis and skewness analyses were selected for next indirect measurement.

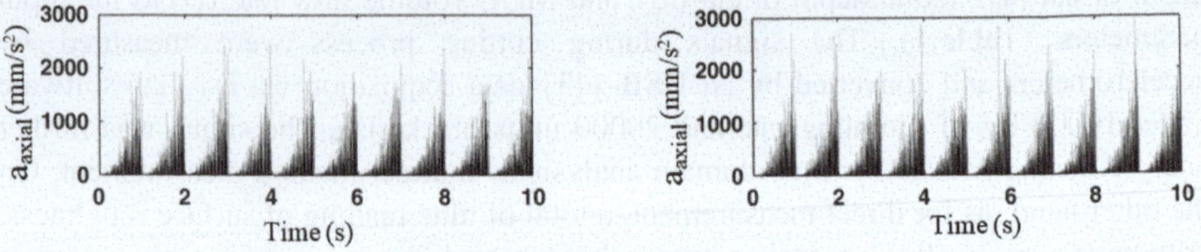

Fig. 2. Time domain for coated (left) and uncoated (right) tool of axial 18 ml/h of MQL.

Figure 3 shows the variation in normalized kurtosis and skewness data for coated tool and uncoated under dry cutting besides at 9, 18, 27, 36, 45 ml/h of MQL flow rate. Since the signals by coated and uncoated tool were extracted from accelerometer 1 and 2, which mounted at cutting tool and workpiece respectively. Thus, there were four sets of data to be examined thoroughly along the study. It is visible that all set data were measured at its highest reading (1.00) during dry cutting and the best flow rate applied for all is at 45 ml/h. It is also noticeable that difference between point is more visible at skewness graph, such as at 36 ml/h of flow rate for both graphs. Since 75% of uncoated tool data measured is higher than coated tool. Thus, it is verified that coated tool performs better result along the MQL experiment. Furthermore, acceptable range for MQL flow rate is 36 and 45 ml/h.

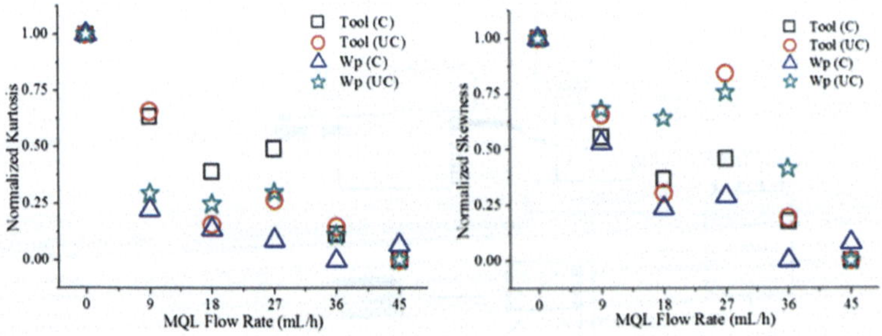

Fig. 3. Kurtosis (left) and skewness (right) schematic for MQL.

There are five different axial depth of cut (0.75 mm, 1.25 mm, 1.75 mm, 2.5 mm, 3.5 mm) were conducted. As illustrated in Fig. 4, only set data by accelerometer 2 using coated tool displays the maximum reading during 1.75 mm meanwhile others at 3.5 mm depth of cut. In contrast, the lowest reading obtained which considered as best parameter for all four groups data were 0.75 mm and 1.75 mm. The result also surprisingly shows that uncoated performs better in these range of axial cut. When the graph is divided into two part at 2.0 mm axial depth of cut, it is visible that all coated reading is higher than uncoated and alternately change at 2.5 mm and 3.5 mm cut. Meanwhile, radial width of cut as in Fig. 3(c) used were 25, 50, 75, and 100% radial as 5 mm, 10 mm, 15 mm, and 20 mm width of cut respectively. Experimental conducted using uncoated tool demonstrates best result at lower radial cut, which is 25 and 50%. Meanwhile, coated tool performs well under larger radial which is at 15 mm and 20 mm width of cut. Besides, it is visualising that data plotted are collectively under 0.50 of kurtosis and skewness at 75% and full radial width of cut. For the direct measurement, surface roughness was applied to observe the cutting process performance index. Figure 5 shows a variation of Ra values during experiment at 0, 9, 18, and 36 ml/h of MQL using coated and uncoated tool. It is visible that highest Ra value was measured at 1.1716 μm during dry cutting using uncoated tool and decreased to 0.9552 μm when MQL is applied at 9 ml/h. Moreover, all Ra values under dry milling were measured higher than when MQL was applied. In terms of cutting tool condition, the Ra values obtained using coated tool were measured slightly inferior than uncoated one. For instance, average surface roughness as feed rate at 9 ml/h and speed at 18 ml/h using coated tool were 0.785 μm and 0.95 μm, which lesser than while using coated tool; 0.874 μm and 1.063 μm respectively.

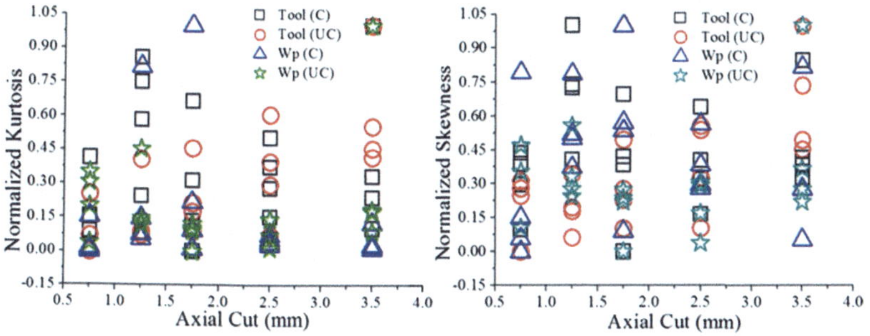

Fig. 4. Kurtosis and skewness schematic for axial depth of cut.

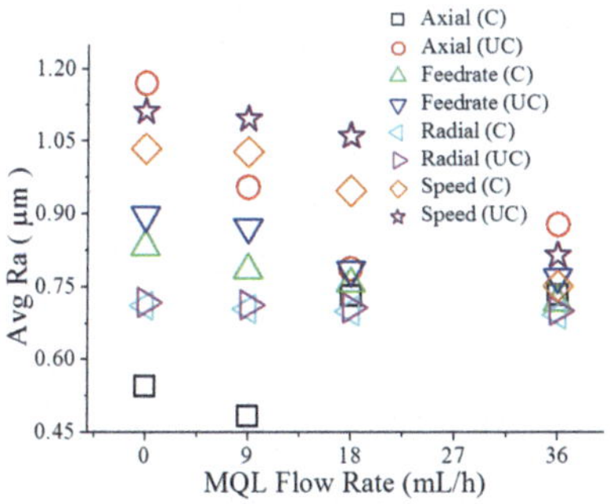

Fig. 5. Correlation of parameters and surface roughness

4 Conclusions

The current experiment with direct and indirect measurement and subsequent analysis to select machining parameter illustrates both measurements shows enhancement in acquiring and analyse data. Besides, increase of MQL flow rate up to 36 ml/h during milling decreases the cutting vibration amplitude and surface roughness. It is recommended to apply MQL system during cutting process over dry cutting. The accelerations and average surface roughness measured were reduced up to 70% when using coated tool over uncoated tool even though result surprisingly shows that uncoated performs better in smaller depth of cut. It is recommended to extend the experiment using variety types of coated tool to study the correlations between them. Graph plotting using kurtosis and skewness of central moment is preferred. However, from present study, there is no significant difference interpretation between kurtosis and skewness. Thus, it was decided to use kurtosis plot in further experimental work.

Acknowledgement. The authors would like to thank the Faculty of Manufacturing Engineering in University Malaysia Pahang (UMP) and University Malaysia Pahang for financial support under PGRS180321.

References

1. Mia, M., et al.: Optimization of MQL flow rate for minimum cutting force and surface roughness in end milling of hardened steel (HRC 40). Int. J. Adv. Manuf. Technol. **89**(1–4), 675–690 (2017)
2. Kumar, A.S., Deb, S., Paul, S.: A study on micro-milling of aluminium 6061 and copper with respect to cutting forces, surface roughness and burr formation. In: ASME 2018 13th International Manufacturing Science and Engineering Conference. American Society of Mechanical Engineers (2018)
3. Vetrichelvan, G., et al.: An investigation of tool wear using acoustic emission and genetic algorithm. J. Vib. Control **21**(15), 3061–3066 (2015)
4. Liu, C., Wang, G., Li, Z.: Incremental learning for online tool condition monitoring using Ellipsoid ARTMAP network model. Appl. Soft Comput. **35**, 186–198 (2015)
5. Khan, M., Dhar, N.: Performance evaluation of minimum quantity lubrication by vegetable oil in terms of cutting force, cutting zone temperature, tool wear, job dimension and surface finish in turning AISI-1060 steel. J. Zhejiang Univ.-Sci. A **7**(11), 1790–1799 (2006)
6. Boswell, B., et al.: A review identifying the effectiveness of minimum quantity lubrication (MQL) during conventional machining. Int. J. Adv. Manuf. Technol. **92**(1-4), 321–340 (2017)
7. Tschätsch, H., Reichelt, A.: Cutting fluids (coolants and lubricants). In: Applied Machining Technology, pp. 349–352. Springer (2009)
8. Tai, B.L., et al.: Minimum quantity lubrication (MQL) in automotive powertrain machining. Proc. CIRP **14**, 523–528 (2014)
9. Zhou, Y., Xue, W.: Review of tool condition monitoring methods in milling processes. Int. J. Adv. Manuf. Technol. 1–15 (2018)
10. Madhusudana, C., Kumar, H., Narendranath, S.: Condition monitoring of face milling tool using K-star algorithm and histogram features of vibration signal. Eng. Sci. Technol. Int. J. **19**(3), 1543–1551 (2016)

11. Jamil, N., Yusoff, A.: Kurtosis quantification of different minimal quantity lubrication effects in machining cast iron with coated and uncoated tool. In: Proceedings of Asia International Conference on Tribology. Malaysian Tribology Society (2018)
12. Huang, P.B., Ma, C.-C., Kuo, C.-H.: A PNN self-learning tool breakage detection system in end milling operations. Appl. Soft Comput. **37**, 114–124 (2015)

Study of Cutting Speed Effects on Lubricant Oil Film Thickness Under Minimum Quantity Lubrication

Nur Izzati Khoirunnisa Ismail[1], Nurrina Rosli[1(✉)], and Kenji Amagai[2]

[1] Faculty of Mechanical and Manufacturing Engineering, Universiti Malaysia Pahang, 26600 Pekan, Pahang, Malaysia
nurrinarosli@ump.edu.my
[2] Department of Mechanical Science and Technology, Gunma University, Kiryu, Gunma 376-8515, Japan

Abstract. In recent decades, there has been increasing interest in the study of Minimum Quantity Lubrication (MQL) due to its outstanding performance despite the minimal usage of cutting oil. However, study focusing on the behavior of oil mist during the MQL machining process is still scarcely reported. It is important to clarify this matter in detail as to explain how the lubricant oil mist can successfully reach the narrow cutting zone. The aim of this study was to investigate the cutting speed effects on the behavior of lubricant oil film by measuring its thickness accumulation on the workpiece after the MQL milling process. Measurement was conducted by using Laser Induced Fluorescence (LIF) method. Results showed that the average thickness of oil film generated at the center of milling path was approximately at 0.37 mm. Penetration ability of lubricant oil to reach the narrow cutting zone dropped with increasing cutting speed and subsequently leading to accumulation of thicker oil film at the cliffs of milling path. Further investigation is needed to clarify whether the nozzle position or the cutter flute may be the attributor of this phenomena. Moreover, it was found that the MQL machining must be conducted appropriately to ensure the oil mist can successfully lubricating the cutting zone on the entire workpiece.

Keywords: Minimum Quantity Lubrication · Lubricant oil behavior · Laser Induced Fluorescence

1 Introduction

Minimum Quantity Lubrication (MQL) has emerged as the most versatile lubrication method for its adaptability in various machining process such as turning, milling, drilling and grinding [1]. The lubricant oil delivered by MQL is generally in oil mist form, resulted from the atomization process between the oil and high pressurized air [2]. This makes the total oil consumption in MQL become much less than that of conventional lubricating method. Hence, the manufacturing cost as well as environmental hazards also can be cut down.

M. N. Osman Zahid et al. (Eds.): iMEC-APCOMS 2019, LNME, pp. 584–589, 2020.
https://doi.org/10.1007/978-981-15-0950-6_89

Apart from being the main source to help reducing the oil consumption, the pressurized air in MQL also plays an important role to drive the lubricant oil successfully penetrating the narrow cutting zone in the tool-workpiece interfaces. This penetration ability of lubricant oil has been well reported to effectively enhance the performance of MQL machining process [3], despite the minimal volume of oil being used. Specifically, the enhancement can be attributed to the lubricating effects sourced from the thin lubricant oil film generated in the contact area between the tool and workpiece. This oil film is crucial in order to combat wear and corrosion of cutting tool as well as reduce the surface roughness of workpiece [4]. The oil film is also able to shield the finished surface of workpiece from abrasions [5].

Although there are many investigations involving the MQL machining process have been conducted in the past, the study that is solely focusing on the behavior of oil mist during the machining process is still scarcely reported. It is important to clarify this matter in detail as to explain how the lubricant oil mist can successfully reach the narrow cutting zone. However, limitation to set up the measurement apparatus during the machining process is on-going has restraint such important study. Here, this paper is aimed to study the cutting speed effects on the behavior of lubricant oil film by measuring its thickness accumulation on the workpiece after the MQL milling process. Although the measurement was not conducted instantaneously during the milling process, the mechanism of lubricant oil successfully reaching the cutting zone can still be predicted through this analysis.

2 Experimental Setup and Procedure

2.1 Preparation of Lubricant Oil and Machining Process

MQL generator system from Unist Coolube® was installed on an CNC milling machine to supply the lubricant oil mist to the workpiece. Aluminum alloy of AA6061 in 45 mm-thickness and 100 mm^2-top surface area was utilized as the workpiece. The cutting tool was a coated carbide end mill in 8 mm-diameter. The cutting fluid is made of natural based oil with 18.5 cSt-viscosity. The oil was initially dissolved with fluorescent dye, i.e. coumarin153 in 0.06 wt%-concentration to allow its thickness measurement after the milling process.

The oil solution was sprayed to the workpiece through the nozzle attached in the MQL generator. The nozzle orifice is in 2 mm-inner diameter. The illustration of milling setup is shown in Fig. 1. The workpiece was milled from the starting point, *S* to the other tip, side by side on the entire surface of workpiece. The MQL spray nozzle was directed to the cutting zone, positioned parallel in the feed direction with orifice situated 5 mm from the tip of cutting tool. The Experiments were conducted in various cutting speed, i.e. 30, 35 and 40 m/min at constant feed rate of 270 mm/min.

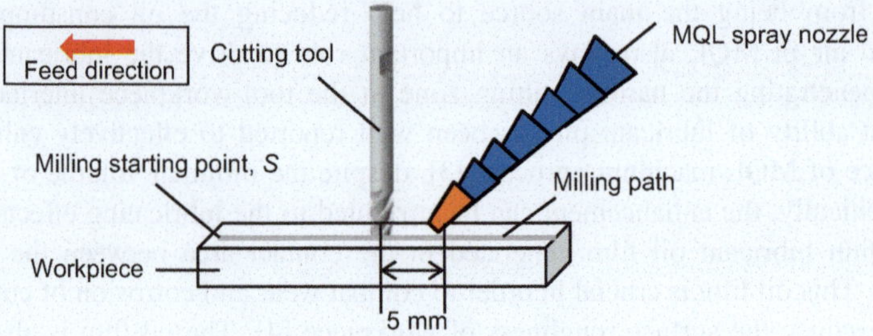

Fig. 1. Illustrations of machining setup.

2.2 Measurement of Lubricant Oil Film Thickness

After the milling process, the thickness of lubricant oil film accumulated on the workpiece was measured by using a non-intrusive technique called as Laser Induced Fluorescence (LIF) method. As shown in Fig. 2, the milling path was irradiated by a line beam shape of diode laser at the center of milling path *C*, cliffs of milling path in left, *L* and right, *R* sides, respectively. The diode laser is in 200 mW-power and 405 nm-wavelength. Upon the laser irradiation, coumarin153 contained in the oil solution excited and consequently caused the oil solution to emit a fluorescence light in 532 nm-wavelength. A video camera was then used to record the emitted light intensity at 30 frames per second in 1080 × 1920 pixels. A green filter was placed on the emitted light path to filter undesirable wavelength.

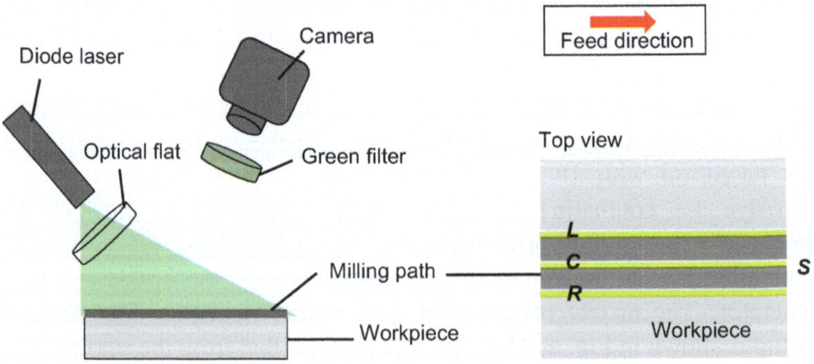

Fig. 2. Illustrations of LIF method in the measurement of lubricant oil film thickness.

Using this LIF method, a calibration procedure is essential to acquire relationship between the lubricant oil film thickness and fluorescence light intensity. The relationship is theoretically based on the Eq. 1 [6] as follows.

$$I_e = \phi I_0 \cdot \exp(-c_{dye} \cdot \delta \cdot \in_\lambda) \quad (1)$$

where I_e as emitted light intensity, I_0 as incident light intensity, δ as lubricant oil film thickness, c_{dye} as dye concentration in lubricant oil solution and $\in_\lambda$ as molar absorption of the fluorescent dye. In this study, the calibration gave a linear model of Eq. 2 as follows.

$$I_e = 49.401\delta + 18.356 \quad (2)$$

3 Results and Discussion

The results of lubricant oil film thickness at feed rate of 270 mm/min and cutting speed of (a) 30 m/min, (b) 35 m/min and (c) 40 m/min are shown in Fig. 3. The graphs were constructed by plotting the distance from milling starting point versus to the oil film thickness at left, center and right sides. Overall, it is depicted from the results that lubricant oil mist tends to spread on the cliffs of milling path. This incident was obviously seen at the lowest cutting speed of 30 m/min, where the oil film constantly fluctuated along the milled area but with larger thickness at left and right cliffs. The nozzle position being set parallel to the feed direction probably cause this incident, where study by Rahim and Dorairaju [7] stated that nozzle positioned at 45° from the feed direction will enhance the machining operation to enhance the penetration ability of oil mist in the cutting zone.

As the cutting speed increased, the oil film barely appeared at the center the oil mist. This specifically implies that the penetration ability of lubricant oil to reach the cutting zone dropped with increasing cutting speed as the oil mist might be intensely sent off to the air during the milling process. Only when the lowest cutting speed was set at 30 m/min, the oil film accumulated at the center of milling path with average thickness of 0.37 mm. According to Hadad and Sharbati [2], increasing cutting speed may add the power of barrier led by the flute of cutting tool. Hence, the ability of oil mist to slide off between the flute spaces decreased.

Moreover, lubricant oil film at the cliffs of milling path under cutting speed of 35 m/min was clearly seen to accumulate at the milling end point, rather than at the milling starting point as can be seen under higher cutting speed of 40 m/min. This suggests that the lubricant oil spraying was slightly delayed from the milling process during the cutting speed was operated at 35 m/min. On the other hand, the oil spraying under cutting speed of 40 m/min may has already begun even before the milling process was started.

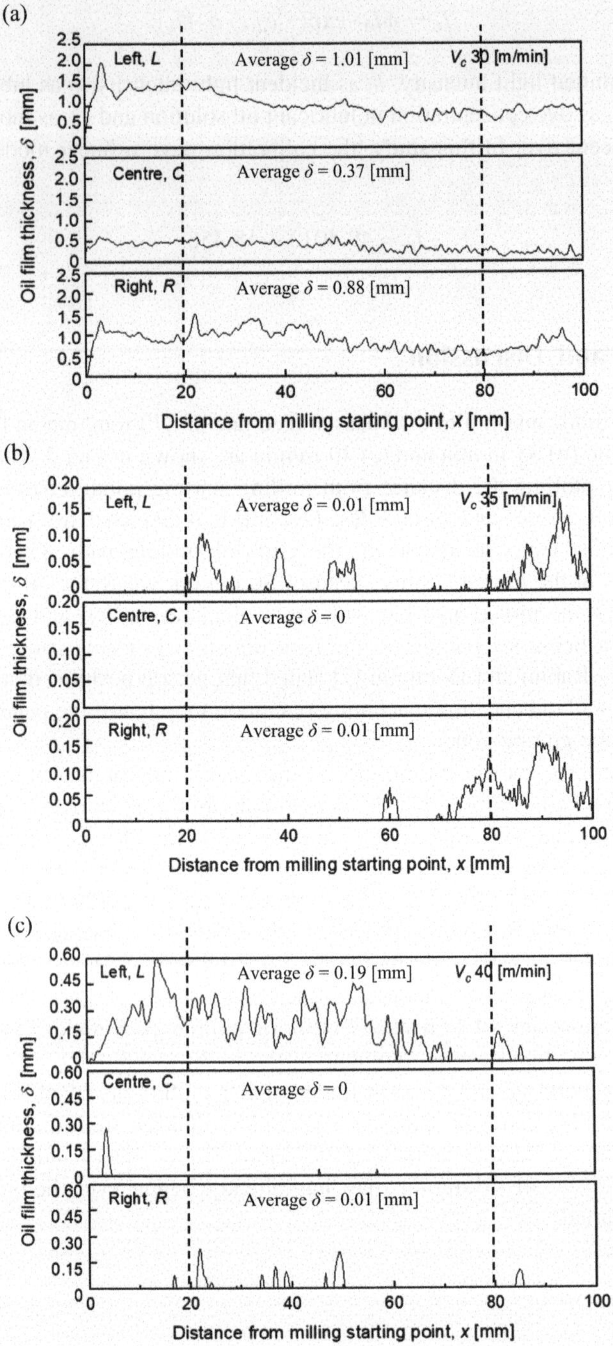

Fig. 3. Oil film thickness under feed rate of 270 mm/min and cutting speed (a) 30 m/min, (b) 35 m/min and (c) 40 m/min.

4 Conclusions

The average thickness of oil film generated at the center of milling path was found approximately at 0.37 mm. Penetration ability of lubricant oil to reach the narrow cutting zone dropped with increasing cutting speed and subsequently leading to accumulation of thicker oil film at the cliffs of milling path. Further investigation is needed to clarify whether the nozzle position or the cutter flute may be the attributor of this phenomena. Moreover, the MQL machining must be conducted appropriately to ensure the oil mist can successfully lubricating the cutting zone on the entire workpiece. Although the oil film being measured on the machined surface after the cutting process are considered as leftovers oil that remained on the surface, this study still give a bit of an overview of lubricant oil mist capability to penetrate the narrow cutting zone. However, a further study with correlation of the effects of lubricant oil film thickness on the surface roughness of workpiece is therefore need to be undertaken to support the explanation.

Acknowledgement. This project was supported by UMP Research & Innovation Department (Grant no. RDU1703158) and Ministry of Higher Education Malaysia (Grant no. RDU190124).

References

1. Maruda, R.W., et al.: A study on droplets sizes, their distribution and heat exchange for minimum quantity cooling lubrication (MQCL). Int. J. Mach. Tools Manuf **100**, 81–92 (2016)
2. Hadad, M., Sharbati, A.: Thermal aspects of environmentally friendly-MQL grinding process. Proc. CIRP **40**, 509–515 (2016). CONFERENCE 2016
3. Sun, H., Feng, Y.H.: The research of minimum quantity lubrication mechanism in grinding with mechanical mechanics. Appl. Mech. Mater. **252**, 129–133 (2012)
4. Davim, J.P.: Sustainable Manufacturing, 1st edn. ISTE Ltd., London (2009)
5. Hagemeier, T., Hartmann, M., Kühle, M., Thévenin, D., Zähringer, K.: Experimental characterization of thin films, droplets and rivulets using LED fluorescence. Exp. Fluids **52**, 361–374 (2012)
6. Li, S., Yu, A., Lu, R.: Fluorescence quenching of coumarin 153 by hydroxyl-functionalized room temperature ionic liquids. Spectrochim. Acta - Part A Mol. Biomol. Spectrosc. **165**, 161–166 (2016)
7. Rahim, E.A., Dorairaju, H.: Evaluation of mist flow characteristic and performance in minimum quantity lubrication (MQL) machining. Meas. J. Int. Meas. **123**, 213–225 (2018)

Tool Deterioration of Stainless Steel 316 in Wet Milling Operation Using Carbide Tool

Amirul Ashraf Makhtar[1], Nurul Hidayah Razak[1(✉)], and Zhan Chen[2]

[1] Faculty of Mechanical and Manufacturing Engineering, Universiti Malaysia Pahang, 26600 Pekan, Pahang, Malaysia
hidayahrazak@ump.edu.my

[2] Department of Mechanical Engineering, Auckland University of Technology (AUT), 55 Wellesly St East, 1010 Auckland, New Zealand

Abstract. Tool deterioration of cutting insert is one of the crucial problem in machining stainless steel due to its excellent mechanical properties. This current work presents a study of tool deterioration of stainless steel 316 in milling using carbide tools. Experiments were conducted in wet condition where tool progression was monitored carefully in every milling passes and the wear criterion was measured. Result shows that the dominant tool deterioration mechanism of carbide tool is tool wear at the flank area where the tool gradually wear from the first milling passes to the final pass.

Keywords: Tool deterioration · Stainless steel 316 · Carbide tool · Milling

1 Introduction

Nowadays, stainless steel 316 a widely used steel in the industry and also for future development planning to be used in desalination industry, phosphoric acid industry and portable water industry [1]. This strongly shows that stainless steel has a high demand and usage in the industry. Stainless steel has a strong corrosion resistance and a great resistance to stress corrosion cracking while having a yield and tensile strength greater than ferritic grades and austenitic [2]. It is also chosen for its great weld ability and formability properties.

Due to high mechanical properties of stainless steel, tool deterioration has become of the major concern. As the machining process occur gradually, this will lead to tool wear and what this will do is that it will affect the tool life and resulting in an imperfect finish product [3]. Currently there are limited source as only a small amount of studies that have been conducted on tool wear for stainless steel 316 in up milling operation using carbide tool in dry cutting condition. Thus, this study aims to monitor the progression tool deterioration and understand how it fails during machining.

M. N. Osman Zahid et al. (Eds.): iMEC-APCOMS 2019, LNME, pp. 590–595, 2020.
https://doi.org/10.1007/978-981-15-0950-6_90

2 Experimental Method

All experiments were conducted using a Makino KE55 milling machine. The workpiece was clamped on the machine as shown in Fig. 1. Stainless steel 316 was used in the experiments and the composition specification is provided in Table 1. The dimension of the work piece was 75 × 100 × 30 in millimeter in rectangular form. Tungsten carbide tools of SUMITOMO EAX2016EL-145 with TiAlN coating were used in the experiments as shown in Fig. 2. A single flute tool holder was used, enabling unambiguous tool deterioration monitoring of each tool. The holder was 16 mm diameter and the base circle radius from the cutting edge to the tool insert was 9 mm.

Fig. 1. Up milling set up and workpiece clamping

Table 1. Stainless steel 316 compositional specification

Grade		C	Mn	Si	P	S	Cr	Mo	Ni	N
316	Min	–	–	–	0	–	16.0	2.0	10.0	–
	Max	0.08	2.00	0.75	0.045	0.03	18.0	3.00	14.00	0.10

Fig. 2. Tungsten carbide cutting insert.

Milling experiments has been conducted by using machining parameters in Table 2 and a repetition experiment was performed for each parameters. These parameters are commonly used when machining stainless steel. After each milling pass (1 pass = 100 mm), the tool insert was removed from the tool holder and the condition of tool insert was examined using an Optical video measuring system (OVMS). This experiment was continued until it reaches the wear criterion, $V_\mathbf{B} \sim 0.3$ mm as stated in the ISO Standard 8688-2 [4].

Table 2. Machining parameters

Machining parameters	Values
Feed rate (mm/tooth), f	0.05
Cutting speed, V_c	100
Spindle speed (rpm)	1000
Depth of cut (mm), DoC	1
Cutting condition	Wet

3 Result and Discussion

The progression of the tool deterioration of tungsten carbide insert (insert 1) is presented in Fig. 3(a–d). Based on the Fig. 3b, after the second pass the VB value was 0.20 mm and progressively wear to VB $\sim$ 0.206 mm after the fourth pass as shown in Fig. 3c. The cutting insert has reached the wear criterion, flank wear of $V_B > 0.3$ mm after the eighth passes, where the VB was 0.3243.

Similar observation can be seen on the second insert where the VB $\sim$ 0.1956 mm as shown in Fig. 4b. Then the cutting insert continued experiencing the slow wearing process and reached the wear criterion after the seventh pass, VB $\sim$ 0.1956 mm. This slow rate of tool deterioration has suggested that flank face of the cutting insert experienced some small "pieces" lost along the flank area in the beginning of the milling pass and started to be visible after the final pass.

Figure 5 shows the trend of progression wear for the both cutting inserts. Based on this figure, it can be seen that both of the insert experience an identical trend which is a gradual loss. According to the ISO Standard [4], this phenomenon can be classified as tool wear. This observation also can be found in Razak et al. [5].

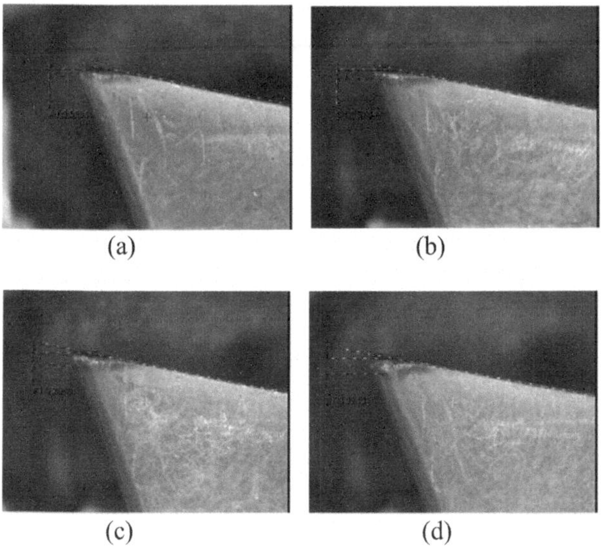

Fig. 3. Optical video measuring system images of side flank face, (a) before machining, (b) second pass, V_B = 0.20 mm, (c) fourth pass, V_B = 0.206 mm and (d) eighth pass, V_B = 0.3243 mm for V_c = 100 m/min, f = 0.05 mm/tooth and DoC = 1.0 mm

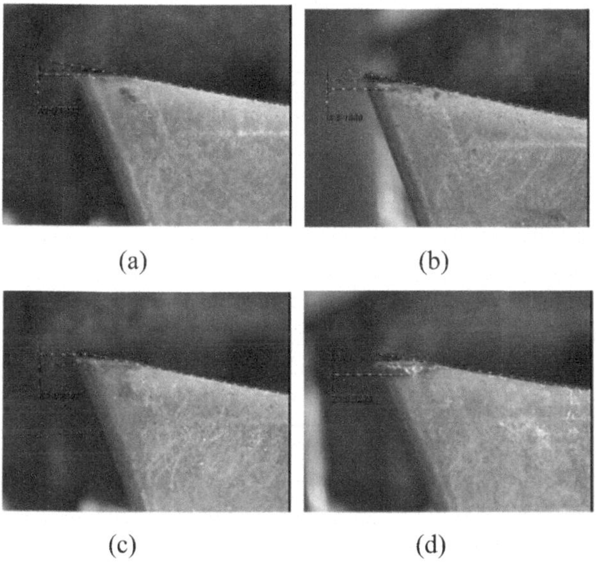

Fig. 4. Optical video measuring system images of side flank face, (a) before machining, (b) second pass, V_B = 0.1956 mm, (c) fourth pass, V_B = 0.2162 mm and (d) seventh pass, V_B = 0.3243 mm for V_c = 100 m/min, f = 0.05 mm/tooth and DoC = 1.0 mm

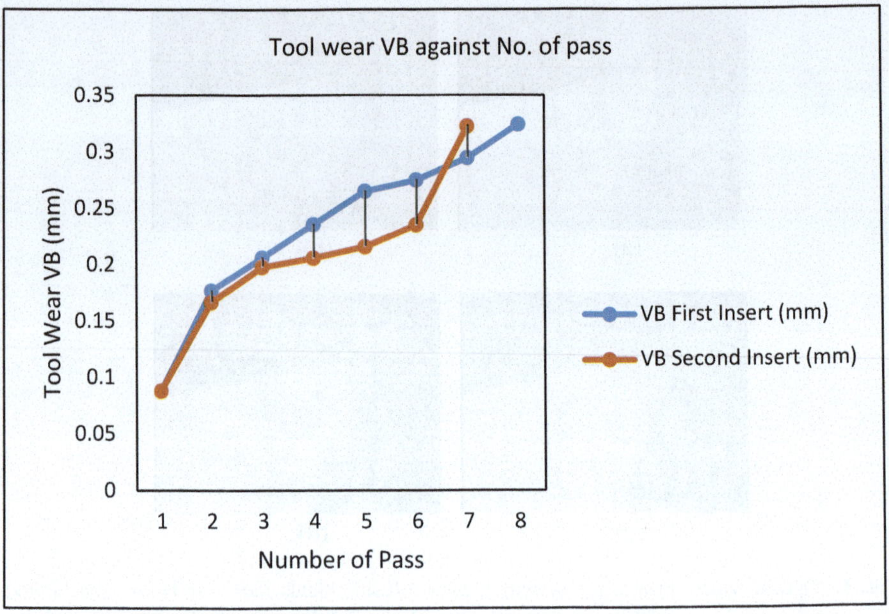

Fig. 5. Tool wear V_B against No. of pass

4 Conclusions

From the observation from the duplicate experiments, it can be concluded that the tool deterioration of stainless steel 316 using carbide insert is tool wear. It was observed that the cutting tools experienced an identical slow rate of tool deterioration from the beginning to the final pass. Small "pieces" lost was found at the flank face of the cutting insert in the very beginning of milling passes and continue to apparent in the final pass.

Acknowledgment. The author gratefully acknowledges Universiti Malaysia Pahang (UMP) for funding this project under internal research grant (RDU1703132).

References

1. Uysal, A., Demiren, F., Altan, E.: Applying minimum quantity lubrication (MQL) method on milling of martensitic stainless steel by using nano Mos2 reinforced vegetable cutting fluid. Proc. - Soc. Behav. Sci. **195**, 2742–2747 (2015)
2. Habak, M., Lebrun, J.L.: An experiment study of the effect of high-pressure water jet assisted turning (HPWJAT) on the surface integrity. Int. J. Mach. Tools Manuf. **51**, 661–669 (2011)

3. Hadi, M.A., Ghani, J.A., Haron, C.H.C., Kasim, M.S.: Comparison between up-milling and down-milling operations on tool wear in milling Inconel 718. Proc. Eng. **68**, 647–653 (2013)
4. ISO 8688–2: International Organisation for Stardardization, Geneva (1989)
5. Razak, N.H., Chen, Z., Pasang, T.: Progression of tool deterioration and related cutting force during milling of 718Plus superalloy using cemented tungsten carbide tools. Int. J. Adv. Manuf. Technol. **86**, 9–12 (2016)

Effect of Lubricating Oil on Sliding Loss and Power Loss of Nylon Gear

Mohammad Asaduzzaman Chowdhury[1], Md. Azizul Islam[1], Dewan Muhammad Nuruzzaman[2(✉)], Bengir Ahmed Shuvo[1], Rajib Nandee[1], and Uttam Kumar Debnath[1]

[1] Department of Mechanical Engineering, Dhaka University of Engineering and Technology, Gazipur, Gazipur 1700, Bangladesh

[2] Faculty of Mechanical and Manufacturing Engineering, Universiti Malaysia Pahang, 26600 Pekan, Pahang Darul Makmur, Malaysia

dmnuruzzaman@ump.edu.my

Abstract. Sliding loss and power loss are the most dominating concern to mechanical power transmission system by any type of gear. Power loss is profoundly influenced by the friction force and friction loss between two gears in contact. In this study, four types of lubricating gear oils 5W30, 10W30, 10W40 and 15W50 were tested for different properties using nylon gears test set up. Test results revealed that 5W30 oil has highest stabilization temperatures among all tested oils. On the other hand, 15W50 oil exhibited lowest stabilization temperatures. It was also observed that gear oil 10W30 has higher stabilization temperatures than gear oil 10W40. The obtained results show that friction coefficient gradually decreases with the increase in rotational speed ranging from 500 to 1000 RPM for all tested gear oils. Experimental data also show that power loss of nylon gear increases with the increase in sliding speed.

Keywords: Sliding loss · Power loss · Nylon gear · Lubricating oil

1 Introduction

The gearbox consists of different parts such as gears, shaft, bearings, seals etc. The effectiveness of a gearbox relies on high losses of the energy. Gear lubricant is an important substance to reduce friction force, temperature, corrosion, and wear. For transmitting rotational motion, characteristics of two meshing gears are very important. During running-in operation and through the gear ratio, gears always produce a change in torque in order to create a mechanical advantage. Frictional loss or power loss is very important for gearing application. Factors influencing the gear box power loss were investigated and it was reported that no-load losses can be minimized at low temperatures [1]. The research findings suggested that low-loss gears can significantly contribute to load-dependent power loss reduction in gearing application. Using a FZG back-to-back test rig, different types of durability tests such as micropitting and pitting were carried out [2]. Moreover, scuffing and wear tests were also performed in this investigation. The obtained results showed that in gear transmission, there is a limitation for lowering the quantity of oil without any deleterious effect on load carrying

M. N. Osman Zahid et al. (Eds.): iMEC-APCOMS 2019, LNME, pp. 596–602, 2020.
https://doi.org/10.1007/978-981-15-0950-6_91

capacity. The influence of oil temperature on gear failures was investigated and test results revealed that gear failures such as micropitting, pitting, scuffing and wear were influenced by oil temperature [3]. Characteristics of different types of polymer based spur gears under different loading conditions were investigated in detail [4, 5]. The effects of gear parameters, gear material and service conditions on the durability of polypropylene spur gears were investigated extensively [6]. The obtained results revealed that durability of test gear was significantly influenced by different gear parameters. Recently, lubricated PEEK gears were experimentally investigated under various loading conditions to examine the failure modes of the contact surfaces [7].

In this study, different characteristics of lubricating gear oils were investigated using nylon gear test set up. Four types of lubricating gear oils 5W30, 10W30, 10W40 and 15W50 were tested for stabilization temperature, friction coefficient, sliding loss and power loss.

2 Experimental Data and Methodology

Tables 1, 2 and 3 show the parameters of torque transducers, geometric properties of the pinions used in the test gearbox and specifications of rolling bearings:

Table 1. Parameters of torque transducers.

Transducer	Capacity (Nm)	Non-linearity (%)^x	Hysteresis (%)^x	Repeatability (%)^x	Temperature range^y (1C)
Input torque	5645	±.0025	±0.030	±0.05	+20 to +75
Output torque	2245				

x is% of rated output and y is Compensated.

Table 2. Geometrical factors of the pinions.

Parameter	Pinion number	
	1	2
Modulus (m) in mm	10	10
Number of teeth (z)	38	50
Addendum modification (x)	0.415	0.3814
Face width (b) in mm	32	32
Pressure angle (α)	20	20
Gear ratio ($I = z_2/z_1$)	1.32	

Table 3. Specification of roller bearings.

Quantity	Type	Reference	d (mm)	D (mm)	r (mm)	Basic load dynamic (lbs)	Basic load static (lbs)
1	Roller bearing	6207	35	72	2.0	5790	3450
1	Roller bearing	6207	35	72	2.0	5790	3450

The physical and chemical characterizations of the lubricants are displayed in Table 4. Experimental test sequence data are shown in Table 5.

Table 4. Chemical and physical properties of the gear oil (synthetic, Mobil Co).

Parameter	Unite	5W30	10W30	10W40	15W50	ASTM method
Viscosity @40 °C	mm^2/sec (cSt)	61.7	78.1	95.9	114	D455
Pour Point	°C	−39	−37	−33	−30	D97
Flash Point	°C	230	232	232	236	D92
Density@15.6 °C	g/ml,	0.856	0.86	0.860	0.876	D4052
Sulfated ash	mass%	0.80	0.96	0.91	0.98	D874

Table 5. Experimental test sequence data.

Experimental sequence	Input speed (rpm)	Input torque (N-m)	Test duration (hours)
1	500	1, 2, 4, 6	10 h
2	750	1, 2, 4, 6	10 h
3	1000	1, 2, 4, 6	10 h

In this investigation, the sliding loss was calculated by using equations of previous investigation [8]. Previous investigations also recommended the experimental formulas for the estimation of the friction coefficient [9, 10]. In present study, we considered the formula [10] to evaluate friction coefficient of gear power transmission system.

3 Results and Discussion

Figure 1 shows the oil sump stabilization temperatures for different types of tested gear oils and for different rotational speeds ranging from 500 to 1000 RPM. The obtained results revealed that 15W50 oil has lower stabilization temperatures compared to any other selected gear oil, while 5W30 oil indicates the utmost stabilization temperatures. Gear oils 10W30 and 10W40 also show higher stabilization temperatures than gear oil 15W50 for all experimental conditions.

Figure 2 demonstrates the test results in such a way that, on x-axis, the oil sump adjustment temperatures $T_{or} = T_{Oil} - T_{Room}$ are plotted. The variances between the sump oil and outer gearbox wall temperatures were determined by $T_{ow} = T_{Oil} - T_{wall}$. Moreover, the rotational speed is directly proportional to sliding velocity which enhances the shear action of film within nominal contact zone and the increase in μ causes sliding loss and power loss according to Figs. 3 and 4 respectively.

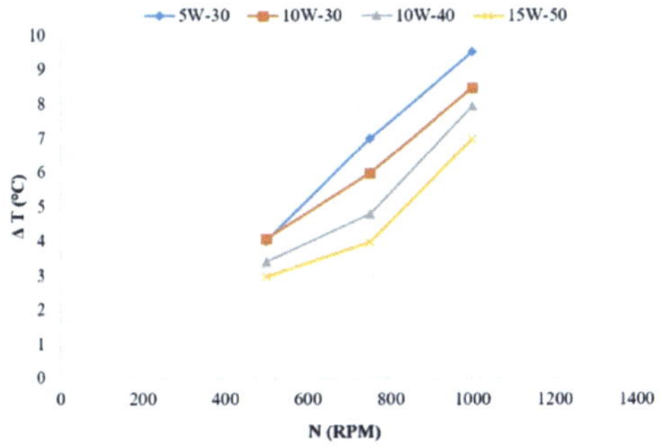

Fig. 1. Relation between oil sump stabilization temperature difference (ΔT) and shaft speed (RPM) at 1 N-m for nylon gear.

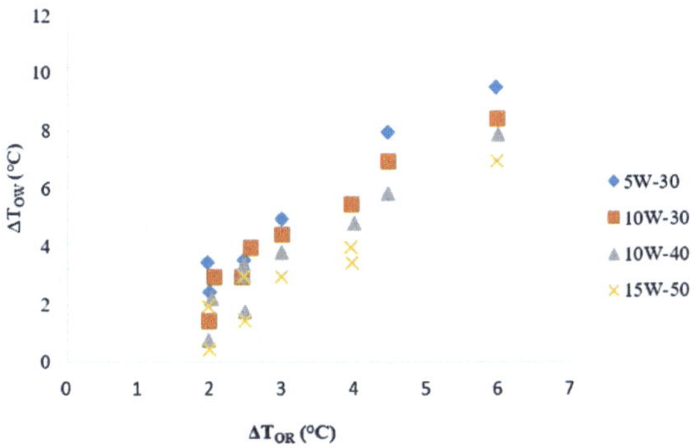

Fig. 2. Relation between ΔT_{ow} and ΔT_{or} at 1 N-m for nylon gear.

On the other hand, the temperature rises due to the higher sliding velocity, and this would lead to a reduction of the fluid viscosity, which further causes a decrease of μ according to Fig. 5. In addition to this, pressure dominating factors to control the friction directly influence the operating viscosity. As a result, it will increase the temperature and effect on friction which is also reported in previous investigation [11]. In the figure, the values of power loss were obtained at speed in between 500 to1000 RPM for various gear oil grades such as 5W30, 10W30, 10W40 and 15W50. In Fig. 6, it is visible that the higher the rotational speed, the higher power loss was recorded at different operating conditions. Figure 6 shows a comparison result of power dissipation between estimation and experiment.

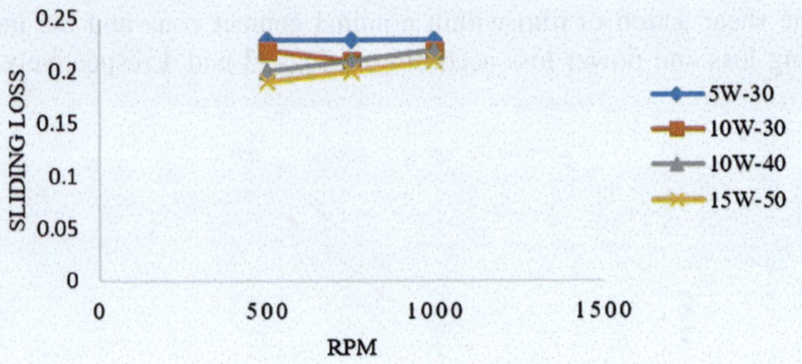

Fig. 3. Sliding loss variation with RPM for different oils (torque 2 N-m).

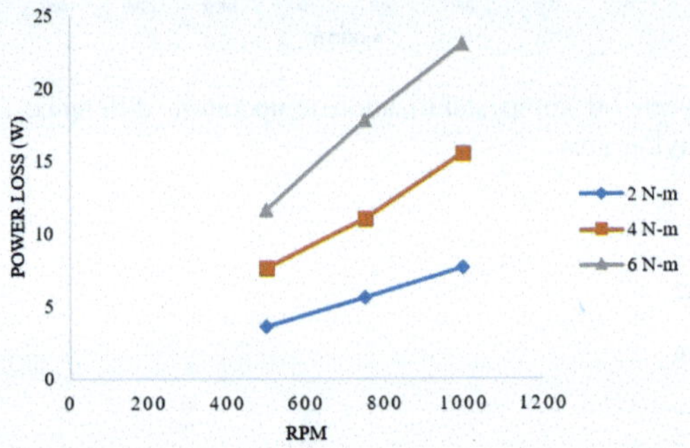

Fig. 4. Power loss variation with shaft speed (RPM) at different torques (oil 5W30).

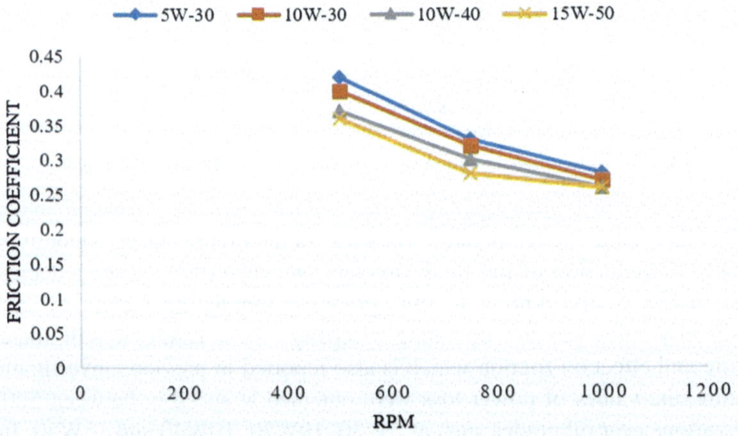

Fig. 5. Friction coefficient varies with shaft speed (RPM) for different types of gear oils.

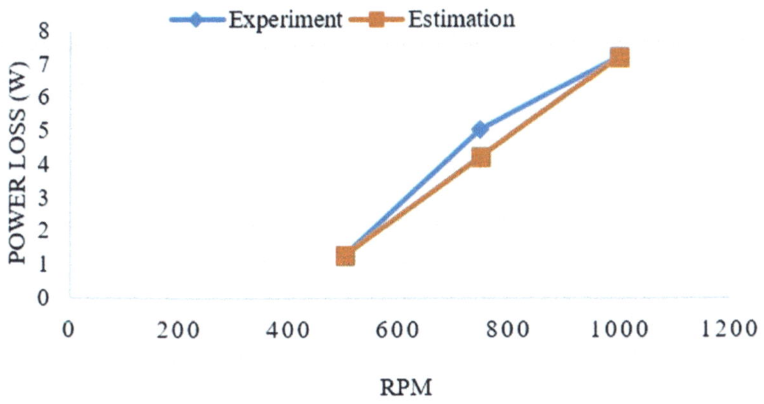

Fig. 6. Variation of power loss with shaft speed (RPM) (torque 2 N-m: oil 5W30).

4 Conclusions

In nylon gearing application, 15W50 gear oil exhibits lowest stabilization temperatures among all tested gear oils while 5W30 oil shows the maximum stabilization temperatures. On the other hand, gear oil 10W30 shows higher stabilization temperatures than gear oil 10W40. Experimental data also reveal that friction coefficient gradually decreases with increasing rotational speed for all tested oils. Furthermore, power loss of nylon gear increases with increasing sliding speed.

Acknowledgements. The authors are grateful to Dhaka University of Engineering and Technology (DUET), Gazipur for the financial support to complete this research work. The authors also acknowledge the support by the technical staff of department of mechanical engineering.

References

1. Michaelis, K., Höhn, B.-R., Hinterstoißer, M.: Influence factors on gearbox power loss. Ind. Lubric. Tribol. **63**, 46–55 (2011)
2. Höhn, B.-R., Michaelis, K., Otto, H.-P.: Flank load carrying capacity and power loss reduction by minimized lubrication. Gear Technol. 53–62 (2011)
3. Höhn, B.-R., Michaelis, K.: Influence of oil temperature on gear failures. Tribol. Int. **37**, 103–109 (2004)
4. Singh, A.K., Siddhartha, Singh, P.K.: An investigation on the thermal and wear behavior of polymer based spur gears. Tribol. Int. **118**, 264–272 (2018)
5. Singh, A.K., Siddhartha, Singh, P.K.: Polymer spur gears behaviors under different loading conditions: a review. Proc. Inst. Mech. Eng. Part J: J. Eng. Tribol. **232**, 210–228 (2018)
6. Mertens, A.J., Kumar, P., Senthilvelan, S.: The effect of the mating gear surface over the durability of injection-molded polypropylene spur gears. Proc. Inst. Mech. Eng. Part J: J. Eng. Tribol. **230**, 1401–1414 (2016)
7. Lu, Z., Liu, H., Zhu, C., Song, H., Yu, G.: Identification of failure modes of a PEEK-steel gear pair under lubrication. Int. J. Fatigue **125**, 342–348 (2019)

8. Ratanasumawong, C., Asawapichayachot, P., Phongsupasamit, S., Houjoh, H., Matsumura, S.: Estimation of sliding loss in a parallel-axis gear pair. J. Adv. Mech. Design Syst. Manuf. **6**, 88–103 (2012)
9. Drozdov, Y.N., Gavrikov, Y.A.: Friction and scoring under the conditions of simultaneous rolling and sliding of bodies. Wear **11**, 291–302 (1968)
10. O'Donoghue, J.P., Cameron, A.: Friction and temperature in rolling sliding contacts. ASLE Trans. **9**, 186–194 (1966)
11. Liu, H., Zhu, C., Sun, Z., Zhang, Y., Song, C.: Coefficient of friction of a starved lubricated spur gear pair. J. Mech. Sci. Technol. **30**, 2171–2177 (2016)

Manufacturing Automation

Effect of Grouser Angle of Attack on Performance of Adjustable Robot Wheel Assistive Grouser

Siti Suhaila Sabarudin[1], Ahmad Najmuddin Ibrahim[1(✉)], and Yasuhiro Fukuoka[2]

[1] Faculty of Mechanical and Manufacturing Engineering, University Malaysia Pahang, Pekan Campus, 26600 Pekan, Pahang Darul Makmur, Malaysia
anajmuddin@ump.edu.my

[2] Department of Mechanical Systems Engineering, Ibaraki University College of Engineering, Hitachi 3168511, Japan

Abstract. Wheeled rover mobile robot was designed to help human with the task that is out of human capability. Usually, it was used for driving over rough terrain for example on unconsolidated sandy dune incline. Normally rover was equipped with fixed grousers that were attached on its wheel but this type of wheel has a problem which is it tends to slip and sink into the sand. This problem happens when the wheel rotates and the grouser moves the sand from below of the wheel to the back of the wheel. This situation caused the sand accumulated behind the wheel. Previous researcher has designed an "assistive" grouser with adjustable angle to minimize the sand movement and subsequent sinkage to prevent this problem. The interaction between the rotation motion of an assistive grouser and the sand movement cannot be seen clearly during the experiment. The purpose of this study is to investigate the interaction between the rotation motion of single grouser and the sand movement by using computer simulation. Discrete Element Method (DEM) is used for the simulation process. From this simulation, the effect of grouser movement towards generated resistance force by the sand particles was observed. In high slip condition where the grouser rotates in a static position, when there is higher number of particles move upward toward the surface as the grouser rotated, it will cause the wheel to dig the sand surface. It has high tendency for the wheel to getting stuck in real experiment.

Keywords: Assistive grouser · Discrete Element Method · Wheel rover

1 Introduction

1.1 Overview

In the middle of this twentieth century, the development of mobile robot has been rise in order to help human with the task that is out of human capability. An example of mobile robot are wheeled robot, tracked robot, legged robot and etc. Usually wheeled robot used for driving on unconsolidated surface terrain. For example NASA "Spirit"

M. N. Osman Zahid et al. (Eds.): iMEC-APCOMS 2019, LNME, pp. 605–610, 2020.
https://doi.org/10.1007/978-981-15-0950-6_92

Mars rover in 2009. But one of the challenges when using wheeled rover is it tends to get stuck in soft sandy inclines which caused mobility failure. It was happened to NASA "Spirit" Mars where it was embedded into the sand and unable to move forward when trying to move over a sandy slope [1].

This kind of problem has been lead to terramechanics modelling studies. Terramechanics is a study of soil properties, specifically the interaction of wheel or tracked vehicles on various surfaces. This field of terramechanics began to emerge as a result of the interest in land locomotion mechanics generated by the pioneering work of Dr. Bekker [2]. There are three common terramechanics methods which is Bekker Method, dynamic Bekker Method and Discrete Element Method (DEM). The comparison between this three methods has been discuss by William Smith in his study [3].

1.2 Study Purpose

An effort to improve the mobility performance on sandy surface has been made by the previous researcher. A new prototype design of wheeled rover by attaching "assistive" grouser to conventional wheel rover [4, 5] has been built. The performance of wheel rover that attached with assistive grousers and wheel rover that attached with fixed grousers has been compared. From the experiment result, it shows that the rover with assistive grouser managed to minimize the sand movement under the rotating wheel. This type of grouser can be used to minimize the tendency of wheel rover getting stuck into the sand.

However, during the experimental test, the interaction between the grouser and the sand movement cannot be seen clearly. So the purpose of this study is to investigate the interaction between the rotation motion of single grouser and the sand movement by using computer simulation. Discrete Element Method (DEM) is used for the simulation process. From the simulation, we can observe the parameters that affect the sand movement.

2 Methodology

In this simulation study, conventional wheel rover attached with fixed grouser and modified wheel rover attached with assistive grouser was designed. Both wheel rovers were simulated at a level incline (flat surface) where the force of gravity component parallel to the incline. The simulation was done with the wheel rotate at a static position and constant 1 rpm speed (100% slip). The focus is to observe the effect between the rotation motion of grousers and the sand movement. Table 1 shows the parameter of the wheel rovers. The grouser length that has been chosen for the simulation is based on the previous experimental test [5]. Figure 1 shows the simulation of conventional and modified wheel rover design at flat surface (0-degree slope). The blue and orange arrow shows the direction of rotation. The observation starts when the grouser enters the sand surface until it exits the sand surface. To assist the analysis of the result, each recorded data was divided into three parts which is grouser angle range θ_1, θ_2 and θ_3. The values of the grouser angle range as in Table 2. The values are different because it depends on the position of the grouser (refer Fig. 1).

Table 1. Wheel rover parameter [5].

Wheel type/parameter	Conventional wheel rover	Modified wheel rover
Wheel diameter	408 mm	408 mm
Wheel width	90 mm	90 mm
Grouser type	Fixed grouser	Assistive grouser
Number of grouser	1	1
Grouser length	20 mm and 80 mm	50 mm and 90 mm
Grouser width	90 mm	90 mm
Grouser thickness	2 mm	2 mm

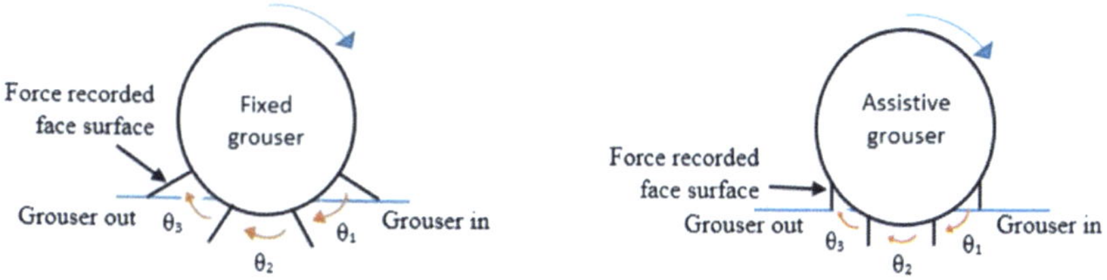

(a) Conventional wheel rover (b) Modified wheel rover

Fig. 1. Simulation of wheel rotation at 0-degree inclination slope (flat surface) for (a) conventional wheel rover and (b) modified wheel rover.

Table 2. Values of grouser angle range.

	Conventional wheel rover (degrees)	Modified wheel rover (degrees)
θ_1	0–49	0–47
θ_2	49–98	47–94
θ_3	98–147	94–141

3 Result and Discussion

3.1 Average Force Generated

In this section, a discussion regarding the force that was generated when the grouser and the sand particles interact will be explained. To assist the simulation analysis result, the average force data was recorded for X-axis and Y-axis. The force recorded is the resistance force experienced by the grouser plate surface. For this simulation, only one surface was recorded which is the face surface that experience the highest resistance force as shown in Fig. 1. Based on Fig. 1, positive force at X-axis means that the force is directed to the back of the wheel (to the left of Fig. 1). Positive force at Y-axis means the force is directed to upward toward the surface. The data was divided into three parts of grouser angle range as in Table 2.

A wheel or vehicle will experience different traction force and resistance force depends on the type of surface. Basically traction force is a force used to generate motion between a body and a tangential surface. If the traction force is higher, it may help the body to move forward. While resistance force is a force that resisting the motion of the body when it move on a surface. If the force generated by sand-grouser interaction at X-axis direction is higher toward the back of the wheel, it can be assumed that the force helps the rover to move forward. If the force generated by sand-grouser interaction at Y-axis direction is higher to the downward, it can be assumed that the sand particles are mostly moving upward toward the surface. This kind of situation may lead the wheel to start digging the sand surface.

At 0-degree inclination slope (flat surface), Fig. 2 below shows the average force generated when the grouser starts to enter the sand surface and exit the sand surface for X-axis direction and Y-axis direction. CWR refer to conventional wheel rover and MWR refer to modified wheel rover.

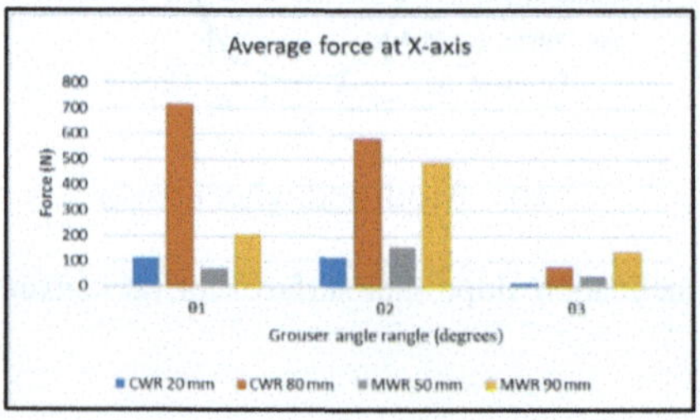

(a) X-axis direction.

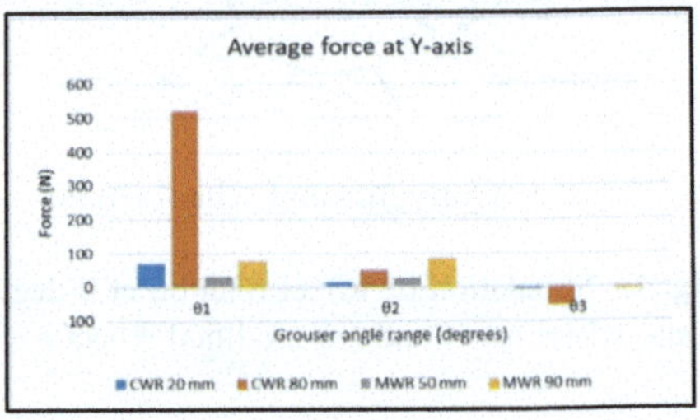

(b) Y-axis direction.

Fig. 2. Average force generated at 0-degree inclination slope for all grouser type and length for (a) X-axis direction and (b) Y-axis direction.

From Fig. 2(a), CWR 80 mm has a higher average force compared to CWR 20 mm at all grouser angle range. The highest average force is at grouser angle range θ_1 when the grouser starts to enter the sand surface. The average force from the sand-grouser interaction becomes smaller as the grouser continues to rotate (from θ_2 to θ_3) because most of the sand has been push toward the back of the wheel at the beginning of the rotation. For MWR, 90 mm length has higher average force compared to MWR 50 mm for all grouser angle range. The highest average force is at θ_2. For MWR the grouser moves at constant angle which is 0-degree to the vertical surface (refer Fig. 1-assistive grouser). At θ_1 and θ_3 the average force generated from sand-grouser interaction is smaller compared to at θ_2 because at θ_1 and θ_3, the grouser moves translationally without rotation, mainly in the Y-axis direction, while at θ_2 mainly in the X-axis direction.

For CWR, 80 mm length has the highest positive value at X-axis. It can be assumed that it has the highest traction to help the wheel to move forward compared to CWR 20 mm length. While MWR 90 mm has the second highest positive value at X-axis but smaller compared to CWR 80 mm. In practical use however, an 80 mm long grouser is too long for use on a wheel, while a long MWR can be folded upwards when on a hard surface terrain.

From Fig. 2(b), CWR 80 mm has the highest positive average force at Y-axis compared to CWR 20 mm at all grouser angle range. It can be assumed that most of the sand particles are moving downward as the grouser start to rotate. At θ_3, CWR 80 mm has higher negative force. This higher negative force shows that a large number of the sand particles are moving upward toward the surface. This condition may lead the wheel to start digging into the sand. For MWR, 90 mm length has higher average force compared to 50 mm length at all grouser angle range. At θ_3, MWR 90 mm has smaller negative force. It can be assumed that the number of sand particles that move upwards smaller.

From the Y-axis graph at grouser angle range θ_3, CWR 80 mm shows the highest negative average force compare to MWR 90 mm. The higher the negative value, the larger the number of sand particles moving upward toward the surface. In this kind of situation where the wheel is rotate at constant speed and static position (high slip), it may lead the wheel to start digging into the sand and getting stuck in the real experiment. It can also be assumed that, maybe CWR 80 mm is not suitable to be used in this kind of situation.

4 Conclusion

The main purpose of this simulation study is to understand the interaction between the moving grouser and the sand. By doing the experimental test alone, the parameter that affects the wheel cannot be seen clearly. In this simulation, conventional wheel rover (CWR) with fixed grouser and modified wheel rover (MWR) with assistive grouser was design. The length chosen is based on the experimental test.

From the simulation result at 0-degree slope, at X-axis direction, CWR 80 mm has the highest average force compared to the other length. It can be assumed that it has the highest traction force to move forward compared to the other length. Beside, MWR 90 mm has the second highest average. But at Y-axis, CWR 80 mm shows smaller positive average force at the middle of the rotation (θ_2) and higher negative values toward the end of the rotation (θ_3). This shows a large number of sand particles are moving upward toward the surface starting from the middle of the rotation toward the end. In real experiment, this could cause the wheel to start dig the sand surface.

For MWR 90 mm at Y-axis, the value of average force increase from θ_1 to θ_2 but slightly smaller at θ_3. When the wheel starts to rotate until the middle of the rotation, it shows positive and higher values of average force which means that the sand particles are moving downward. At θ_3 it shows a slightly negative value which means a smaller number of sand particles are moving upward. This is because MWR has constant grouser angle which is 0-degree to the vertical so by maintaining the angle during the rotation it mainly generated force horizontally. In real experiment, it could prevent the wheel from digging into the sand.

In high slip condition where the grouser rotates in a static position, when there is a higher number of particles move upward toward the surface when the grouser is rotated, it will cause the wheel to dig the sand surface. It has high tendency for the wheel to getting stuck in real experiment.

Acknowledgement. This research is fully supported by FRGS grant (FRGS/1/2018/TK03/UMP/02/6) and UMP internal grant (RDU180384, PGRS190373). The authors fully acknowledge the Malaysian Ministry of Education (MOE) and Universiti Malaysia Pahang for the approved fund which makes this important research viable and effective. Discrete Element Method (DEM) simulations and analysis were conducted using EDEM® Version 3.0 bulk material simulation software provided by DEM.

References

1. Arvidson, R.E., Bell, J., Bellutta, P., et al.: Spirit mars rover mission: overview and selected results from the northern home plate winter haven to the side of scamander crater. J. Geophys. Res. E: Planets **115**(9) (2010). [E00F03]
2. Wong, J.Y.: Terramechanics and Off-Road Vehicle Engineering: Terrain Behaviour, Off-Road Vehicle Performance and Design, 2nd edn. Butterworth-Heinemann, Oxford (2009)
3. Smith, W., Melanz, D., Senatore, C., Iagnemma, K., Peng, H.: Comparison of DEM and traditional modeling methods for simulating steady-state wheel-terrain interaction for small vehicles. In: 7th Americas Conference of the ISTVS, Tampa, FL, USA, pp. 4–7 (2013)
4. Ibrahim, A.N., Fukuoka, Y.: Effect of assistive grouser mechanism on lightweight rover power consumption pattern on steep soft sand inclines. In Intelligent Manufacturing and Mechatronics, pp. 343–351. Springer, Singapore (2018)
5. Ibrahim, A.N., Aoshima, S., Shiroma, N., Fukuoka, Y.: The effect of assistive anchor-like grousers on wheeled rover performance over unconsolidated sandy dune inclines. Sensors **16** (9), 1507 (2016)

Parametric Study of CNG-DI Engine Operational Parameters by Using Analytical Vehicle Model

Mohd Fadzil Abdul Rahim(✉), Abdul Aziz Jaafar, Rizalman Mamat, and Zahari Taha

Faculty of Mechanical and Manufacturing, Universiti Malaysia Pahang, 26600 Pekan, Pahang, Malaysia
mfadzil@ump.edu.my

Abstract. A parametric study of engine's influential parameters on compressed natural gas direct injection (CNG-DI) engine's torque was performed by using a dynamic, analytical model of the CNG-DI vehicle. The objective of the study is to analyze the effect of the selected six parameters on the engine's total brake torque. The simulations were carried out to mimic the speed-sweep test procedure of the vehicle on a chassis dynamometer. Based on the actual test setup, a ramp input is provided on the power pedal and the vehicle is allowed to accelerate under the influence of dynamometer inertia load only. Based on the results, the most influential parameter on the maximum engine torque is the injection pressure. The maximum predicted engine mean brake torque at 60 bar injection pressure is about 120 Nm. The second influential parameter is the injection duration, where the maximum predicted engine brake torque is about 100 Nm. Both parameters are related to the controlling amount of the fuel-injected, which affected the amount of energy released into the cylinder. The third influential parameter is the ignition timing, where the maximum pressure predicted is closed to 70 bar. It can be concluded that the magnitude of brake torque is sensitive to the amount of fuel supplied for combustion. This comprehensive model is suitable for parametric analysis regardless of the expensive computing time.

Keywords: Dynamic modelling · Transient simulation · Gas Direct-Injection

1 Introduction

A new configuration of compressed natural gas direct injection (CNG-DI) engine is developed. A model is required to predict the theoretical engine torque and to identify the influential parameters on engine's brake torque. Engine parameters such as the ignition timing [1], the injection duration [2] and the injection timing [3] are highly influence parameters which affect the engine torque. The gas injection pressure also affected engine performance since injector is operated in a choke condition [4]. Previously, most studies utilized moderate injection pressure at a value of 20 bar [5, 6].

Nowadays the experimentation cost for parametric investigations is expensive hence encouraged the used of model-based analysis. In the field of IC engine

M. N. Osman Zahid et al. (Eds.): iMEC-APCOMS 2019, LNME, pp. 611–616, 2020.
https://doi.org/10.1007/978-981-15-0950-6_93

modelling, the mean value model and experimental model are too simplified, thus hindered the evaluation of parameters interaction [7]. Analytical engine model based on thermodynamic single-zone modelling is preferable to analyze detail interaction between parameters. Its consisted the solutions of in-cylinder thermodynamic processes such as combustion, heat transfer, intake and exhaust flow dynamics [8], throttle dynamics [7], and work transfer from the piston to the crankshaft [9]. Recently, the model is also expected to be able to simulate the transient condition of engine operation [10, 11] to gain a more realistic insight into actual driving condition.

Based on the discussed findings, this study is carried out with the purpose to identify the influential engine parameters affecting the engine's brake torque by using a comprehensive analytical model of CNG-DI prototype. There are six parameters have been selected which are the ignition timing, injection duration, injection timing, throttle positioning rate, gas injection pressure, initial vehicle speed, and vehicle mass.

2 Modelling Method

The developed analytical engine model in the current study is made of two types of model (1) Simulink model (2) Simscape model. Simulink blocks symbolize basic mathematical operations. And linked Simulink blocks are equivalent to a mathematical model or representation of a system under study. Whereas, Simscape blocks dedicated to the modelling of physical components in their actual physical properties.

2.1 The Simulink Model

The Simulink model consisted of the following; the throttle dynamics model [7], the intake manifold model [7], the intake and exhaust valve dynamics [8], the combustion model [12], the heat transfer model [8], and the empirical model of convective heat transfer coefficient [8]. The variation of cylinder volume and cylinder surface area are solved by the solution of kinematics and dynamics of the crank slider mechanism which consider the piston-pin and crankshaft offset [9]. All the sub-models are incorporated within the framework of the modified first law of thermodynamics to solve the in-cylinder pressure and temperature. The crankshaft dynamics are modelled based on the approach taken by Zweiri [9].

2.2 The Simscape Model

The SimScape model consists of the clutch model, the simple manual transmission model, the final drive model and finally attached to a vehicle model. The input to these models is the acceleration and torque from the crankshaft. Figure 1 presents the overall layout of the model developed in Matlab-Simulink environment.

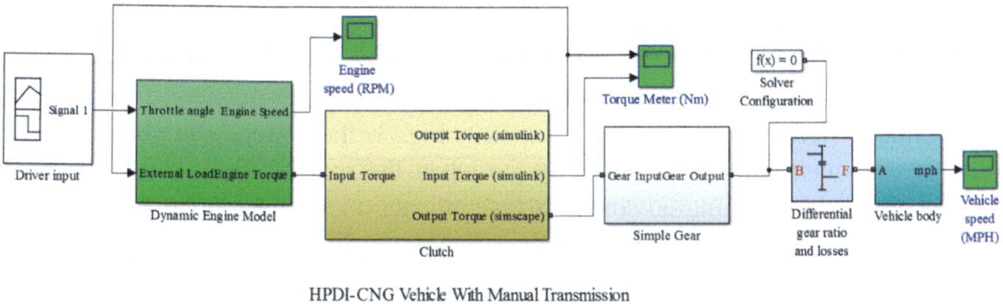

Fig. 1. The overall model layout in Matlab-Simulink

3 Simulation Procedure

Six control parameters were manipulated and studied. The selected parameters are presented in Table 1.

Table 1. Cases description for the parametric study

Case	Description	Manipulated parameter ranges
1	Effect of throttle opening rate	2 s to 18 s period
2	Effect of ignition timing advance	−5 to −20 deg BTDC
3	Effect of injection pressure variation	The mass flow rate of 20 bar to 60 bar setup
4	Effect of increased injection duration	0.02 s to 0.1 s
5	Effect of different vehicle mass	1050 kg to 1250 kg
6	Effect of different initial speed	0 mph to 20 mph

The measured mass flow rates are corresponding to the gas injection pressure setup has been recorded from independent injector testing on the bench [4]. The vehicle model had been simulated in an acceleration mode, mimicking the speed sweep test procedure on a chassis dynamometer. The vehicle was given ramp input. The recorded simulation time was set to 30 s (Table 2).

Table 2. Specification of the baseline case for vehicle simulation analysis

Description	Magnitude
Throttle opening rate	0.6885 rad/s
Ignition timing advance	Default
Injection pressure	40 bar (mass flow rate = 0.7 g/s)
Injection duration	Default
Vehicle mass	1015
Initial speed	0 mph

4 Results and Discussion

Based on Fig. 2, the increased throttle rate opening affect the maximum mean brake torque timing except its magnitude. Based on Fig. 3, there were slightly noticeable increments of maximum mean brake torque when the spark timings were advanced. The effect of ignition timing advanced is insignificant due to the combustion speed of CNG is slower compared to the gasoline. Based on Fig. 4, the injection pressure significantly affects the maximum mean brake torque. The highest mean brake torque was obtained by the highest injection pressure of 60 bar at a value about 120 Nm. The maximum brake torque value obtained is 19% lower than the maximum value obtained by the port injection gasoline engine which was rated at a value of 148 Nm. Based on Fig. 5, the longest injection duration produced the highest maximum mean brake torque at a value of about 100 Nm. The effect of injection duration and injection pressure increment are constructive because both parameters control the amount of fuel supply for combustion. A larger amount of fuel supplied for combustion, more heat is released by combustion and consequently increased the amount of mean brake torque. Based on Fig. 6, it is obvious that the effect of the vehicle initial speed is modest on the resultant net mean brake torque produced insignificance difference. The increment of vehicle mass decreased the maximum net mean brake torque but the difference is insignificant as presented by Fig. 7.

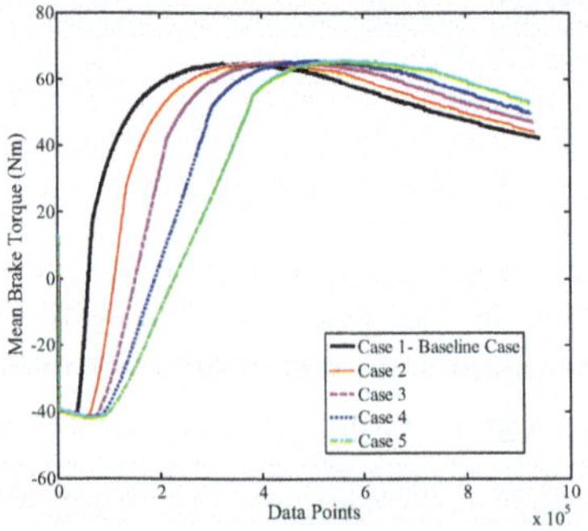

Fig. 2. Effect of the throttle positioning rate on the engine mean brake torque

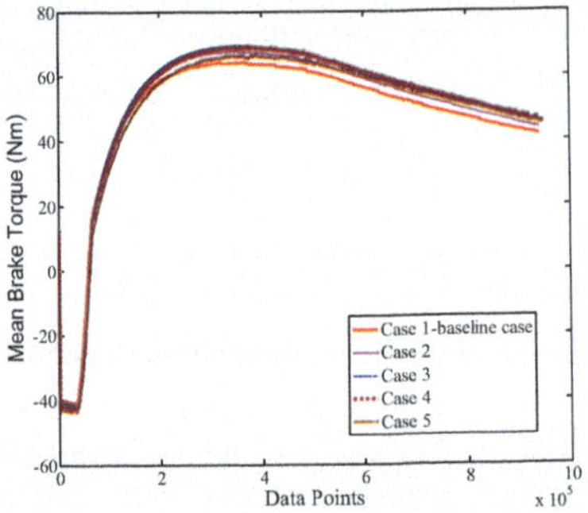

Fig. 3. Effect of ignition timing on engine mean brake torque

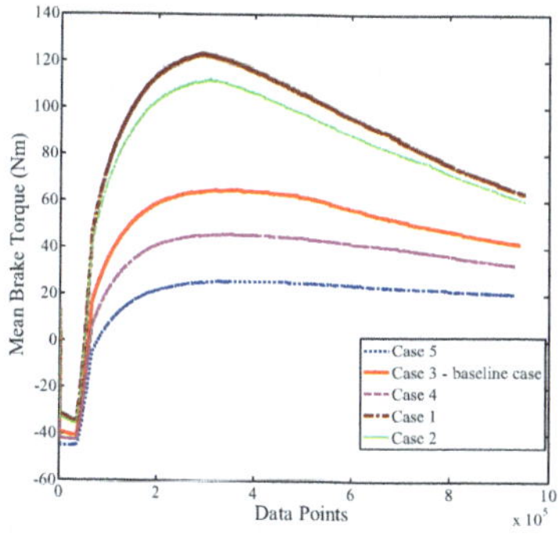

Fig. 4. Effect of gas injection pressure on engine mean brake torque

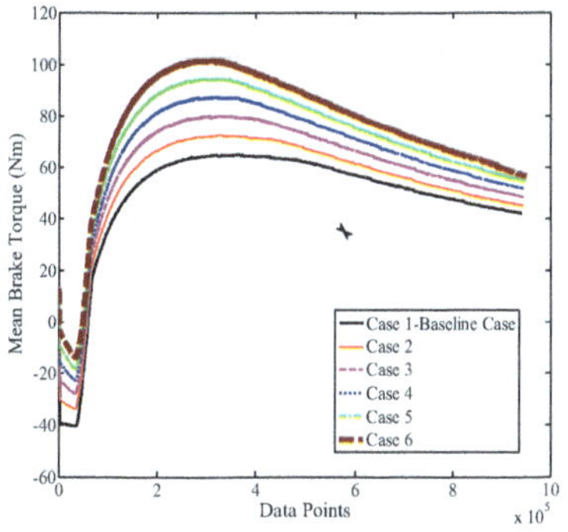

Fig. 5. Effect of injection duration on engine mean brake torque

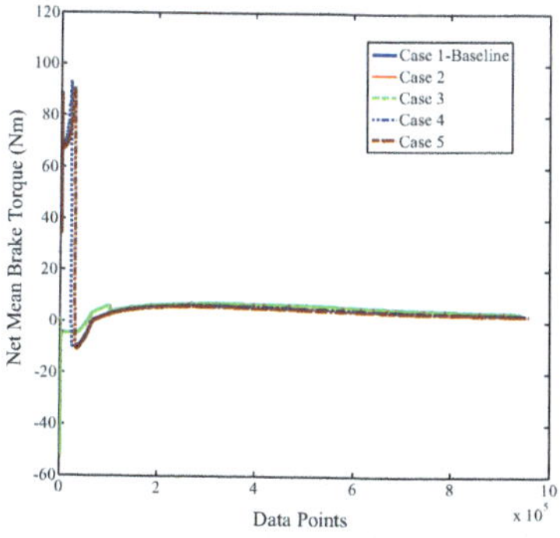

Fig. 6. Effect of initial vehicle speed on the engine net mean brake torque

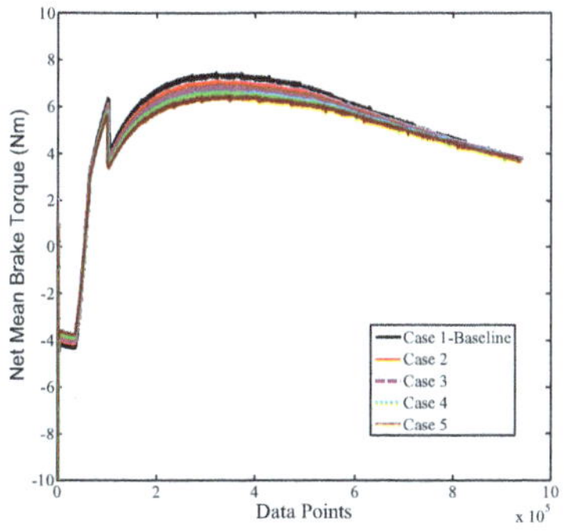

Fig. 7. Effect of vehicle mass on the engine net mean brake torque

5 Conclusion

The most influential parameter on the maximum engine torque is the injection pressure. Higher injection pressure resulted in higher engine mean brake torque. The maximum predicted engine mean brake torque at 60 bar is about 120 Nm. The second influential parameter is the injection duration where the maximum predicted engine brake torque is about 100 Nm. Both parameters are related to the controlling amount of fuel injected into the cylinder which affects the air to fuel ratio of the cylinder mixture. The third influential parameter is the ignition timing where the maximum pressure predicted is closed to 70 bar. Contrary to the characteristics of a gasoline engine where ignition timing is highly affected engine torque and speed. This is mostly due to the longer combustion period model for CNG fuel combustion which resulted with 'less sensitive' performance of engine output torque. The model is not suited for the control oriented analysis since it is computationally demanding. However, it details and comprehensiveness is a major contribution for the development of a complete vehicle model.

Acknowledgement. We would like to express our sincere gratitude to The Ministry of Education and Universiti Malaysia Pahang who has funded the study under research grant numbered FRGS/1/2017/TK03/UMP/03/2 and RDU1703145, members of Automotive Laboratory, and Innovative Manufacturing, Mechatronics and Sports laboratory which facilitate the study from the beginning.

References

1. Maji, S., Ranjan, R., Sharma, P.B.: Comparison of emissions and fuel consumption from CNG and gasoline fueled vehicles - effect of ignition timing. SAE Technical Paper Series (2000
2. Beccari, S., Pipitone, E., Cammalleri, M., Genchi, G.: Model-based optimization of injection strategies for SI engine gas injectors. J. Mech. Sci. Technol. **28**, 3311–3323 (2014)
3. Liu, Y.F., Liu, B., Liu, L., Zeng, K., Huang, Z.H.: Combustion characteristics and particulate emission in a natural-gas direct-injection engine: effects of the injection timing and the spark timing. Proc. Inst. Mech. Eng. Part D J. Automob. Eng. **224**, 1071–1080 (2010)
4. Taha, Z., Rahim, M.F.A., Mamat, R.: Injection characteristics study of high-pressure direct injector for compressed natural gas (CNG) using experimental and analytical method. In: IOP Conference Series: Materials Science and Engineering, vol. 257, p. 12057 (2017)
5. Kalam, M.A., Masjuki, H.H.: An experimental investigation of high performance natural gas engine with direct injection. Energy **36**, 3563–3571 (2011)
6. Hassan, M.K., Aris, I., Zainuddin, K.N., Alina, N.A.: Torque and power of CNGDI engine with two different piston crown shapes. J. Appl. Sci. Res. **5**, 949–954 (2009)
7. Pezouvanis, A.: Engine modelling for virtual mapping (2010). http://hdl.handle.net/10454/4419
8. Sitthiracha, S.: An analytical model of spark ignition engine for performance prediction. In: The 20th Conference of Mechanical Engineering Network of Thailand, pp. 1–60 (2006)
9. Zweiri, Y.H., Whidborne, J.F., Seneviratne, L.D.: Detailed analytical model of a single-cylinder diesel engine in the crank-angle domain. Proc. IMechE, Part D J. Automob. Eng. **215**, 1197–1216 (2001)
10. Grahn, M., Olsson, J.O., McKelvey, T.: A diesel engine model for dynamic drive cycle simulations. IFAC Proc. **18**, 11833–11838 (2011)
11. Grahn, M., Johansson, K., McKelvey, T.: Data-driven emission model structures for diesel engine management system development. Int. J. Engine Res. **15**, 906–917 (2014)
12. Rousseau, S., Lemoult, B., Tazerout, M.: Combustion characterization of natural gas in a lean burn spark-ignition engine. Proc. Inst. Mech. Eng. Part D J. Automob. Eng. **213**, 481–489 (1999)

Author Index

M. N. Osman Zahid et al. (Eds.): iMEC-APCOMS 2019, LNME, pp. 617–619, 2020.
https://doi.org/10.1007/978-981-15-0950-6

The manufacturer's authorised representative in the EU is Springer Nature Customer Service Centre GmbH, Europaplatz 3, 69115 Heidelberg, Germany. If you have any concerns regarding our products, please contact ProductSafety@springernature.com

Printed and bound by CPI Group (UK) Ltd, Croydon, CR0 4YY
30/09/2026
02225446-0018